Inside Rhinoceros®

Ron K. C. Cheng

THOMSON
DELMAR LEARNING™

Australia Canada Mexico Singapore Spain United Kingdom United States

Inside Rhinoceros®

Ron K. C. Cheng

Vice President, Technology and Trades SBU:
Alar Elken

Editorial Director:
Sandy Clark

Senior Acquisitions Editor:
James DeVoe

Senior Development Editor:
John Fisher

Marketing Director:
Cynthia Eichelman

Channel Manager:
Fair Huntoon

Marketing Coordinator:
Sarena Douglass

Production Director:
Mary Ellen Black

Production Manager:
Larry Main

Production Editor:
Thomas Stover

Technology Project Specialist:
Kevin Smith

Editorial Assistant:
Mary Ellen Martino

Freelance Editorial:
Carol Leyba, Daril Bentley

Cover Design:
Cammi Noah

Printed in Canada
1 2 3 4 5 XXX 07 06 05 04 03

For more information, contact:
Delmar Learning
Executive Woods
5 Maxwell Drive
Clifton Park, NY 12065-8007

Or find us on the World Wide Web at
www.delmarlearning.com

Library of Congress Cataloging-in-Publication Data
Cheng, Ron.
Inside Rhinoceros 3 / Ron K.C. Cheng.
p. cm.
ISBN 1-4018-5063-4
1. Computer graphics. 2. Three-dimensional display systems. 3. Rhino (Computer file) I. Title.
T385.C4745 2003 006.6--dc21
2003009779

NOTICE TO THE READER

Publisher does not warrant or guarantee any of the products described herein or perform any independent analysis in connection with any of the product information contained herein. Publisher does not assume, and expressly disclaims, any obligation to obtain and include information other than that provided to it by the manufacturer.

The reader is expressly warned to consider and adopt all safety precautions that might be indicated by the activities herein and to avoid all potential hazards. By following the instructions contained herein, the reader willingly assumes all risks in connection with such instructions.

The publisher makes no representation or warranties of any kind, including but not limited to, the warranties of fitness for particular purpose or merchantability, nor are any such representations implied with respect to the material set forth herein, and the publisher takes no responsibility with respect to such material. The publisher shall not be liable for any special, consequential, or exemplary damages resulting, in whole or part, from the reader's use of, or reliance upon, this material.

About the Author

Ron K. C. Cheng leads the Product Design Unit of the Industrial Center of The Hong Kong Polytechnic University, where he is involved in various types of projects (including industrial projects, engineering projects, integrated learning projects, and manufacturing projects), computer-aided design, and development of computer-based learning materials. Cheng has also written quite a number of books for Autodesk Press, including works on Mechanical Desktop, AutoCAD, and Inventor.

Acknowledgments

Grateful acknowledgment is made to Jerry Hambley, Bill Adamoski, John Novak, and Bruce Weirich for their reviews of content and helpful suggestions. I would also like to thank the Delmar team for their efforts in association with this project, in particular, acquisitions editor James DeVoe, developmental editor John Fisher, editor Daril Bentley, and design-and-production specialist Carol Leyba.

Contents

Introduction

Rhinoceros (or Rhino, as it is popularly known) is a Windows-based 3D surface modeling program that continues to gain in popularity. Rhino is based on the popular NURBS (nonuniform rational B-spline) mathematics. NURBS is a computer modeling paradigm that makes it easy to build curved surfaces and organic, free-form shapes, which are the rule rather than the exception in today's design world.

Using Rhino tools, you construct, edit, transform, and analyze curves, surfaces, and solids. Because curves alone have limited application in computerized design and manufacturing, you use Rhino curve tools mainly for building frameworks for subsequent construction of surfaces and solids. It is these underlying frameworks that allow models constructed with Rhino to have maximum utility in the real world of manufactured designs.

In addition to its use as a modeler, Rhino is used to construct polygon meshes capable of coping with various applications of rendered solids and surfaces, as well as to output rendered images and 2D engineering drawings of 3D modeled objects. Used in conjunction with the Flamingo rendering tool, Rhino also provides you with the ability to produce photorealistic images as final output. All of these capabilities are covered in this book.

Audience and Prerequisites

This book is designed for students, designers, and engineers who wish to gain a general understanding of computer modeling concepts and to learn how to construct free-form organic shapes that address both upstream and downstream design/manufacturing considerations. These objectives are achieved in the context of learning to use Rhino and its associated Flamingo rendering tool as a computer modeling system for industrial and engineering design.

Computer modeling experience and a basic knowledge of design concepts are always helpful in learning any new software program. However, there really are no prerequisites in either case in regard to using this book, which explains basic design concepts, techniques, and considerations as it teaches you from the beginning of the learning curve how to use Rhino.

Philosophy and Approach

To bridge the theoretical and software-oriented approaches to computer modeling, this book provides a balanced presentation combining theory, concepts, and tutorials. Theory is relatively useless without hands-on experience, and vice versa. This book begins with an overview of the three basic types of computer models and the Rhino interface. The book then addresses the limitations of wireframe models, discusses various types of curves and underscores why learning how to construct curves is important, and shows you how to construct curves using Rhino, including the use of digitizers.

Subsequent chapters continue to pursue this logical progression from the conceptual to the practical, and from the basic to the more complex. Following the delineation of wireframe models and curves, the book continues with an in-depth examination of surface modeling via discussion and hands-on tutorials that show you step by step how to construct surfaces using Rhino.

Following surface modeling, the book provides detailed discussion of the various types of solid modeling methods, the forms in which Rhino represents solids, and the techniques and tools used to construct solids in Rhino. To consolidate what you have learned to this point, a latter chapter reviews curves, surfaces, and solids together. Finally, the book addresses rendering, drawing output, and outputting to various file formats.

Content

This book is written to Release 3 of Rhinoceros, although the concepts, techniques, and most of the exercises covered in this book do not require the use of Release 3. The chapters of this book deal with both concepts and methodology. In addition, each chapter includes an overview of the chapter's topic and content, a chapter summary, and review questions. Chapter 1 introduces computer modeling methods, as well as an examination of the Rhino user interface, including its key functional components.

Chapters 2 through 4 examine the limitations of wireframe modeling, introduce the use of curves in surface and solid modeling, and explore general methods of working with curves. After learning wireframe modeling and curves, in chapters 5 and 6 you learn more detailed surface modeling concepts, as well as how to construct, edit, transform, and analyze surfaces using Rhino.

Chapter 7 covers various solid modeling methods, as well as the manner in which Rhino represents solids in the computer. Chapter 8 addresses two main presentation methods: using shaded and rendered images, and producing 2D engineering drawings. Finally, Chapter 9 consolidates what you have learned by guiding you through a series of digital modeling projects.

Features and Conventions

This edition includes a companion CD-ROM at the back of the book (see "About the Companion CD-ROM" at the end of this introduction). Exercises in the book are supported by content found on the companion CD-ROM.

Italic font in regular text is used to distinguish certain command names, code elements, file names, directory and path names, user input, and similar items. Italic is also used to highlight terms and for emphasis.

The following is an example of the monospaced font used for code examples (i.e., command statements) and computer/operating system responses, as well as passages of programming script.

```
var myimage = InternetExplorer ? parent.
cell : parent.document.embeds[0];
```

The following are the design conventions used for various "working parts" of the text. In addition to these, you will find that the text incorporates many exercises and examples.

> ***NOTE:*** *Information on features and tasks that requires emphasis or that is not immediately obvious appears in notes.*

> ***TIP:*** *Tips on command usage, shortcuts, and other information aimed at saving you time and work appear like this.*

About the Companion CD-ROM

The companion CD-ROM found at the back of this book contains all Rhino files used in conjunction with exercises, as well as a trial version of Rhinoceros Release 3.

CHAPTER 1

Introduction to Digital Modeling and Rhinoceros

Introduction

This chapter introduces concepts involved in 3D digital modeling, outlines the key functions of Rhinoceros (commonly know as Rhino), and serves to familiarize you with the Rhino user interface.

Objectives

After studying this chapter, you should be able to:

- ❐ Explain the principles of digital modeling
- ❐ Describe the key functions of Rhino
- ❐ Use the Rhino user interface

Overview

To facilitate the process of designing a product or system, you use various media to capture your design ideas. Initially, you use sketches to represent the subject of the design. To elaborate on this design, you use models. Models can be physical or digital. Making a physical model can often be time consuming, or simply not feasible. To construct a digital model, you use computer technology and computer-aided design applications. This chapter provides an overview of digital modeling and introduces you to the Rhino interface and Rhino's basic functions and methods of operation.

Digital Modeling

The purpose of constructing a digital model of an object is to represent it in the computer in digital form in order to facilitate design, analysis, and downstream computerized operations. Using computer-aided design and rendering applications, you represent the object's geometry, texture, and color. To make better use of the computer and computer-aided design applications, you need to know the various ways models are represented in the computer, the types of modeling tools available, and the techniques for using these tools. There are three ways to represent a 3D object in the computer: as a wireframe model, as a surface model, or as a solid model. These model types are explored in the sections that follow.

Wireframe Model

In the history of digital modeling, the 3D wireframe model is the earliest type of 3D model. It is the most primitive type of 3D object. In essence, a wireframe model is a set of unassociated curves assembled in 3D space. The curves serve only to give the pattern of a 3D object. There is no relationship between the curves. Therefore, the model does not have any surface information or volume information. It has only data that describe the edges of the 3D object. Because of the limited information provided by the model, the use of wireframe models is very confined. Figure 1-1 shows a wireframe model of a joypad, and the same model in a cutaway view.

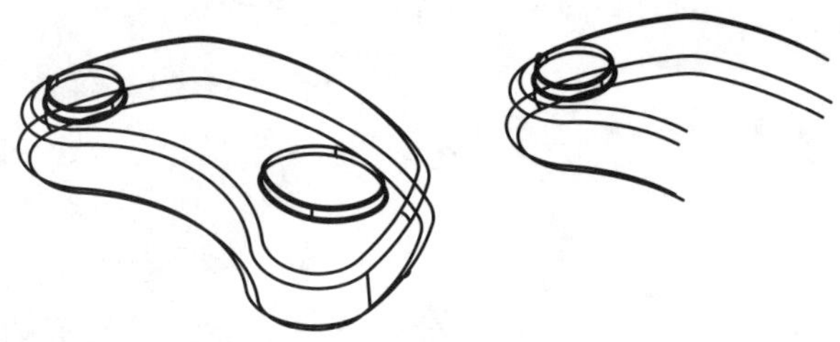

Fig. 1-1. Wireframe model and its cutaway view.

Surface Model

A surface is a mathematical construct represented as a thin sheet without thickness. A surface model is a set of surfaces assembled in 3D space to represent a 3D object. When compared to a 3D wireframe model, a surface model has in addition to edge data information on the contour and silhouette of the 3D object. Surface models are typically used in computerized manufacturing systems and in the generation of photorealistic rendering or animation. Figure 1-2 shows a surface model of a car body, a cross section of the model, and a rendered image of the sectioned model.

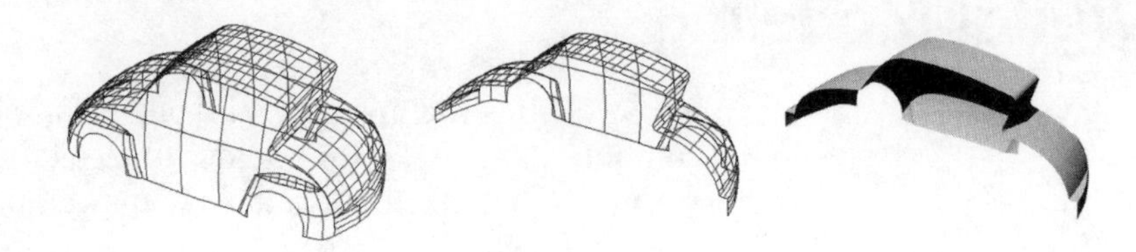

Fig. 1-2. Surface model, cross section of the model, and a rendered image of the sectioned model.

Solid Model

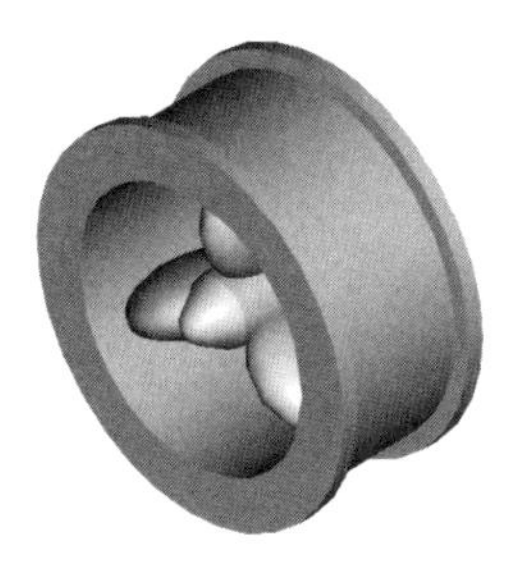

Fig. 1-3. Solid model and cross section of the model.

In regard to information, a 3D solid model is superior to the other two models because a solid model in a computer is a complete representation of the object. It integrates mathematical data that includes surface and edge data, as well as data on the volume of the object the model describes. In addition to visualization and manufacturing, solid modeling data is used in design calculation. Figure 1-3 shows a solid model of a wheel and a cross section of the model.

Assembly Model

To facilitate visualization of how various component parts of a product or system can or should be put together, you assemble individual components in an assembly model. Figure 1-4 shows the assembly of a toggle clamp.

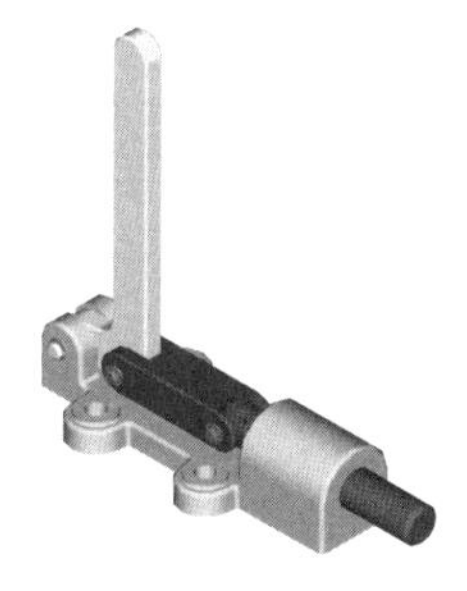

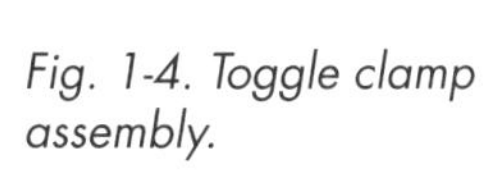

Fig. 1-4. Toggle clamp assembly.

Engineering Drawing

To represent a 3D object in a 2D drawing sheet, you use an orthographic engineering drawing. If you already have a 3D digital model, you use the computer to generate orthographic views of the model. Figure 1-5 shows a 2D engineering drawing generated from the assembly of the toggle clamp.

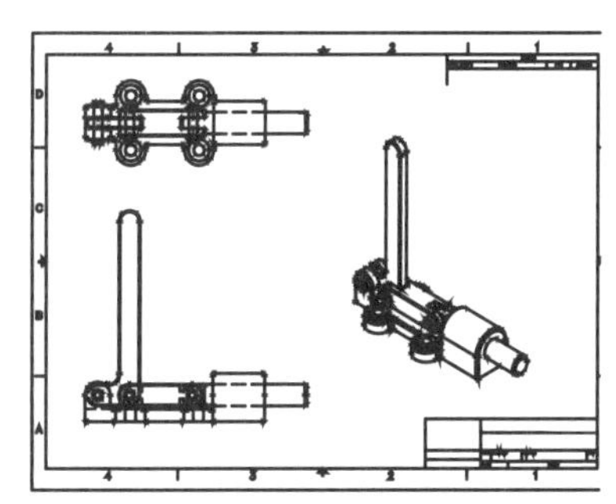

Fig. 1-5. 2D drawing generated from the toggle clamp assembly.

Downstream and Upstream Operations

Constructing 3D models and producing 2D drawings from 3D models is not necessarily the end of the digital modeling process. In a computerized manufacturing system, for example, the same digital model can (and often should be) designed to be used in all downstream operations, such as finite element analysis, rapid prototyping, CNC (computer numeric control) machining, and computerized assembly.

To enhance illustration of the 3D object, you construct renderings and animations. Because these operations may be done using different types of computer applications, and because each application may use a unique type of data format, the digital modeling system must enable the conversion of the 3D digital model into various file formats.

On the other hand, the digital modeling system must enable the opening of various types of file formats so that digital models constructed in other systems can be used for further elaboration of the design. Figure 1-6 shows a rapid prototyping machine making a 3D object, and figure 1-7 shows the CNC machining of a free-form 3D object.

Fig. 1-6. Rapid prototyping machine making a rapid prototype from a 3D model.

Fig. 1-7. CNC machining of a free-form 3D object.

Rhino Functions

Rhino is a 3D digital modeling application that enables you to construct six types of objects: points, curves, NURBS (nonuniform rational basis spline) surfaces, polysurfaces, solids, and polygon meshes. To facilitate downstream computerized operations and reuse of existing digital models constructed using some other computer application, you can export

Rhino models to various file formats and import various file formats into Rhino. You will learn more about this later in the book.

Curves and Points

Wireframe models by themselves have limited utility in design and manufacture because they are simply a set of unassociated curves. However, curves and points are required in many surface and solid construction operations. Therefore, you need to learn how to construct curves and points for the purpose of making surfaces and solids. Using Rhino, you construct points and various types of 3D curves. To construct free-form surfaces, NURBS curves are required. You will learn about NURBS curves in chapters 2 through 4.

3D Surfaces

There are two basic ways to represent a surface in the computer: using a NURBS surface to exactly represent the surface or using a polygon mesh to approximate the surface. By joining a set of contiguous NURBS surfaces, you create a polysurface. You will learn about NURBS surfaces and polysurfaces in chapters 5 and 6, and about polygonal meshes in Chapter 6. Figure 1-8 shows a NURBS surface, and figure 1-9 shows a polygonal mesh.

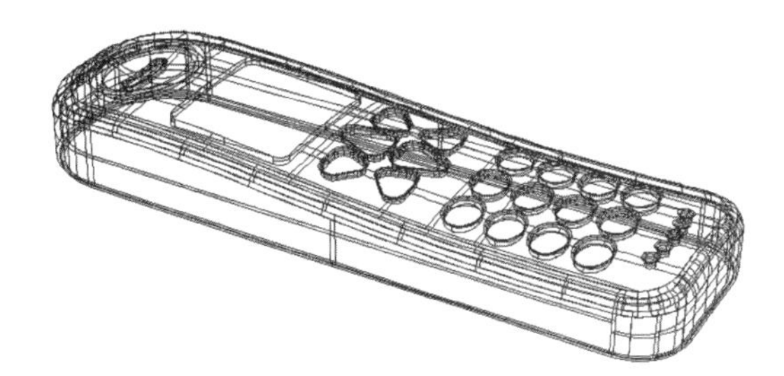

Fig. 1-8. NURBS surface model of a mobile phone casing.

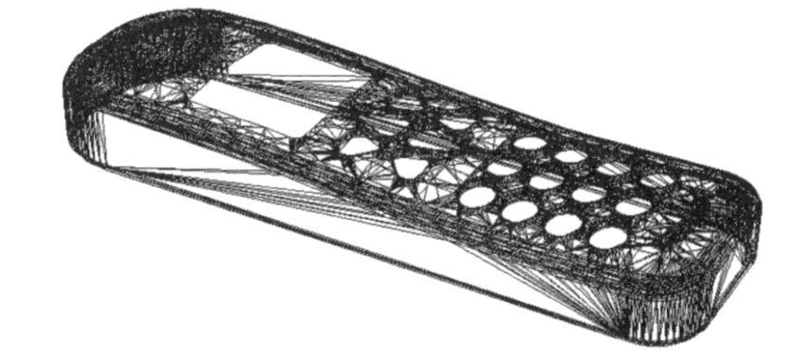

Fig. 1-9. Polygonal mesh model of the same mobile phone casing.

3D Solids

A Rhino solid is a closed-loop polysurface (i.e., a set of joined surfaces with no gap or opening). You construct Rhino solids in two basic ways: directly, using the solid modeling tools, or by converting a set of contiguous NURBS surfaces into a solid volume by joining them. You will learn more about solid modeling in Chapter 7. Figure 1-10 shows a Rhino solid.

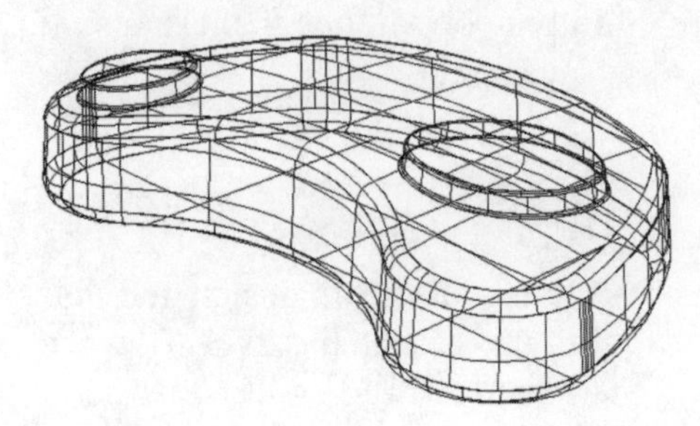

Fig. 1-10. Rhino solid.

A closed-loop polysurface is two or more surfaces joined to create an object that has a volume. In Rhino, this could also be a single surface with edges that are collapsed to a single point (such as a sphere). A torus is another example of a single surface that is also a solid.

Rendering

You output photorealistic renderings from surfaces, polysurfaces, and polygon meshes. You will learn about rendering in Chapter 8. Figure 1-11 shows the rendering of a mobile phone casing.

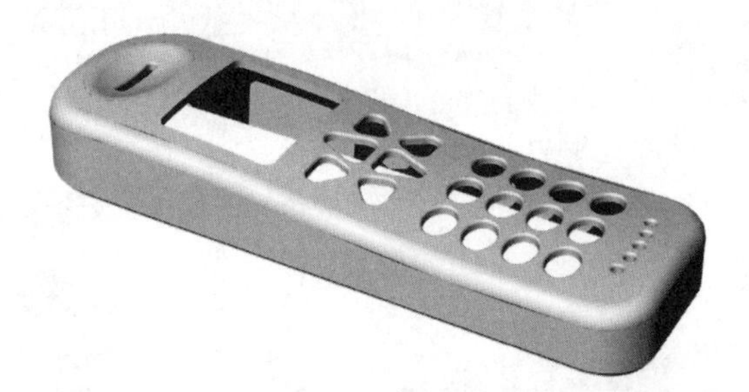

Fig. 1-11. Rendering of a mobile phone casing.

2D Drawings

To illustrate a 3D object on a 2D sheet, you typically generate a 2D drawing from the digital model of the 3D object and add appropriate dimensions and annotations to the drawing. You will learn about generation of 2D drawings from digital models in Chapter 8.

Starting Rhino

Now that you are familiar with some rudiments of the Rhino environment, let's try a bit of hands-on examination. Start Rhino by selecting the Rhinoceros 3.0 icon from your desktop. In the Application window, shown in figure 1-12, you will find five major areas: standard Windows title bar, main pull-down menu, command line interface, graphics area, and status bar. In addition, there are a number of toolbars. These components are described in the sections that follow.

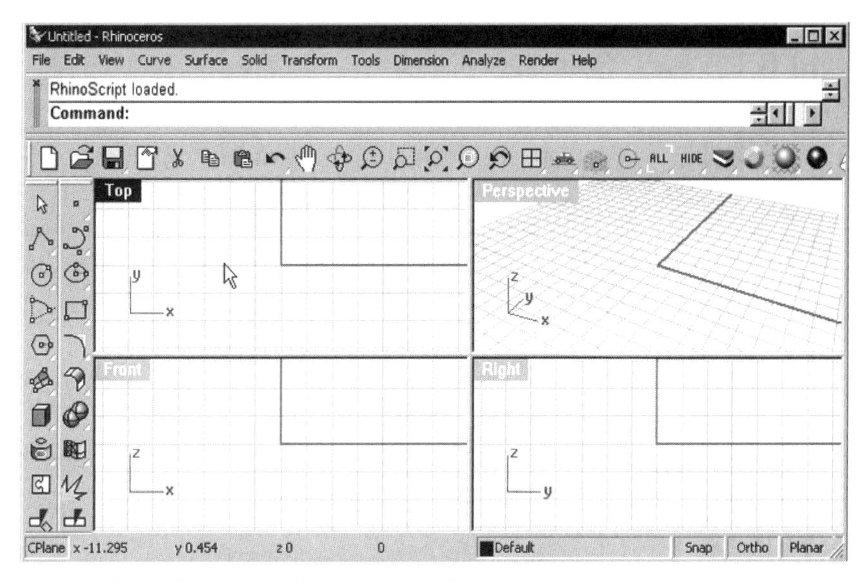

Fig. 1-12. Rhino Application window.

Standard Windows Title Bar

At the top of the Application window there is the standard Windows title bar. This title bar functions no differently and contains nothing different than the basic Windows title bar.

Main Pull-down Menu

Below the standard Windows title bar is the main pull-down menu, which contains twelve options: File, Edit, View, Curve, Surface, Solid, Transform, Tools, Dimension, Analyze, Render, and Help. The functions of these options are outlined in Table 1-1.

Table 1-1: Pull-down Menu Options and Their Functions

Pull-down Menu Option	Function
File	For working on files and templates
Edit	For editing points, curves, surfaces, and solids
View	For manipulating display settings and establishing construction planes
Curve	For constructing points and curves
Surface	For constructing NURBS surfaces
Solid	For constructing solids
Transform	For transforming objects you have constructed
Tools	Provides various types of useful tools

Pull-down Menu Option	Function
Dimension	For constructing 2D drawings and adding annotations
Analyze	Helps you analyze objects you have constructed
Render	For shading and rendering
Help	Provides useful help information

To perform commands and operations, you select options from pull-down menus and cascading menus. The following are examples of command sequences for performing specific operations.

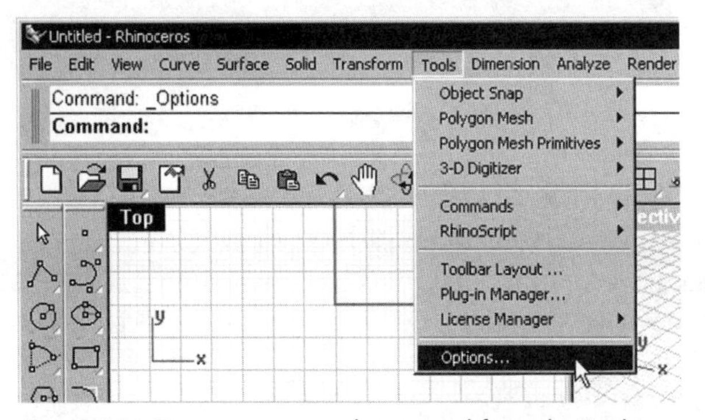

Fig. 1-13. Options command activated from the Tools pull-down menu.

1. Select Tools > Options. The Options command from the Tools pull-down menu is activated, as shown in figure 1-13.
2. Select Solids > Sphere > 3 Points. The 3 Points command from the Sphere cascading menu of the Solid pull-down menu is used to construct a three-point sphere. The Sphere menu shown in figure 1-14 is an example of a cascading menu.

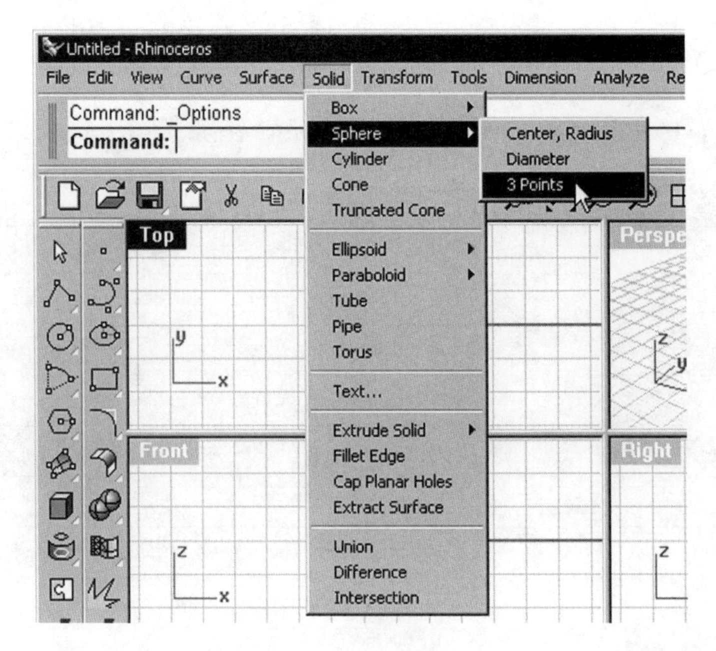

Fig. 1-14. Sphere cascading menu for constructing a three-point sphere.

Command Line Interface

Below the main pull-down menu is the command line interface, which provides a place for textual interaction. Here, you run a command by typing the command name or alias of the command and then pressing the Enter key or the space bar. After a command is run, further prompts or instructions will appear in this area or in any associated pop-up dialog boxes. Command names are not case sensitive, and therefore you can use any combination of small and capital letters to specify a command name. An "auto-complete" function is incorporated here at the command line. That is, if you type a command here you do not have to type the full name. For example, after you type *ci*, the system will automatically complete the typing to display the word *circle*. Figure 1-15 shows the command line interface.

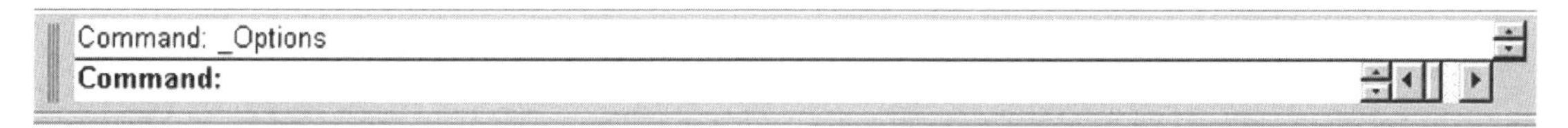

Fig. 1-15. Command line interface.

Graphics Area

The graphics area is where you construct your model. After you perform a command, you select a location in the graphics area for constructing any number of types of geometric objects. This area can be divided into a number of viewports, which can be docked or floating. Figure 1-16 shows a four-viewport configuration (i.e., Top, Front, Right, and Perspective viewports).

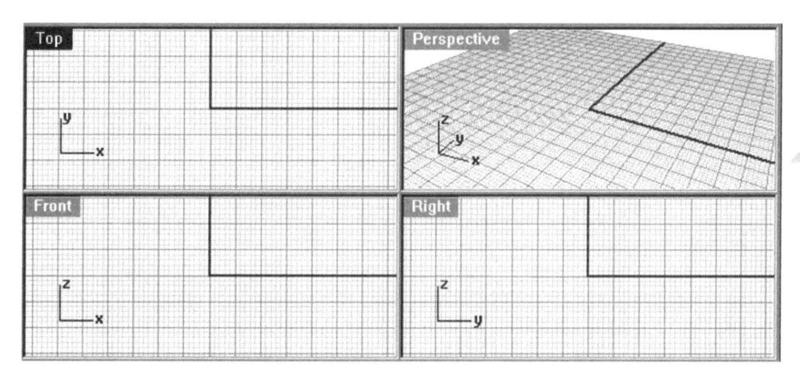

Fig. 1-16. Graphics area showing a four-viewport configuration.

Initially, the number of viewports and their orientation are determined by the viewport setting of the template file you use to start a new file. To use a template, you start a new file by selecting New from the File pull-down menu. This accesses the Open Template File dialog box (shown in figure 1-17), in which you select a template.

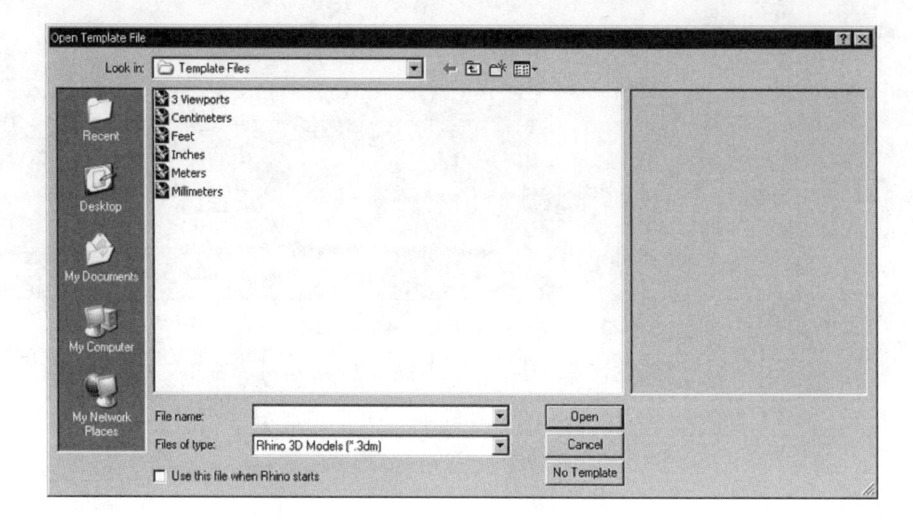

Fig. 1-17. Open Template File dialog box.

To set the viewport configuration while working on a file, as indicated in figure 1-18, you use the Viewport Layout cascading menu of the View pull-down menu or the Viewport Layout toolbar.

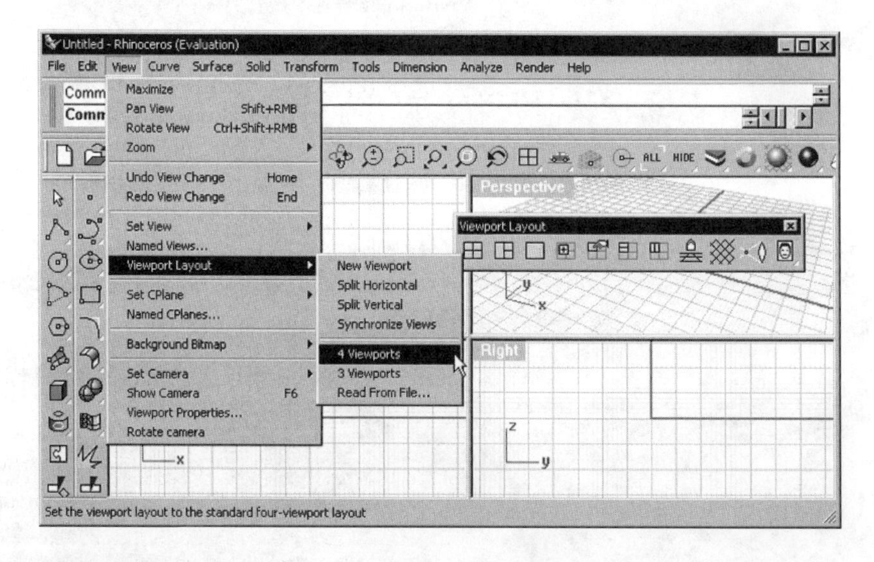

Fig. 1-18. Setting the viewport configuration.

Status Bar

At the bottom of the Application window is the status bar. It shows the location of the cursor marker, the current layer, and the state of the drawing aids. The status bar contains eight panes, the functions of which are outlined in Table 1-2.

Table 1-2: Status Bar Panes and Their Functions

Pane	Function
World/CPlane	Toggles the coordinate display to show either World coordinates or construction plane coordinates
Coordinate Display	Shows the coordinates of the mouse pointer
Distance	Displays the distance from the mouse pointer to the last picked point
Layer	Shows the current layer
Snap	Constrains the cursor movement to the specified snap intervals
Ortho	Constrains the cursor movement to be orthogonal or some other preset direction
Planar	Constrains the cursor movement to be parallel to the current construction plane from the last point
Osnap	Toggles the display of the Osnap dialog box (Osnap stands for "object snap")

Toolbars

Buttons on toolbars represent commands in a graphical way. To run a command, you select a button on a toolbar. Because there are many commands and toolbars, displaying all of them would take up the entire screen display. Therefore, only the Standard (see figure 1-19), Main1, and Main2 (see figure 1-20) toolbars are displayed by default.

Fig. 1-19. Standard toolbar.

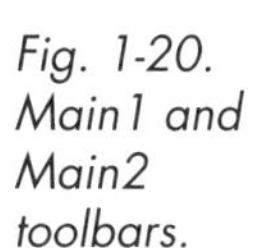

Fig. 1-20. Main1 and Main2 toolbars.

To find out which toolbars are available, and to display them on the screen, you use the Toolbar command by typing the command name at the command line interface or by selecting Tools > Toolbar Layout > Edit from the main pull-down menu. In the Toolbars dialog box, shown in figure 1-21, check the box next to a toolbar to display that toolbar.

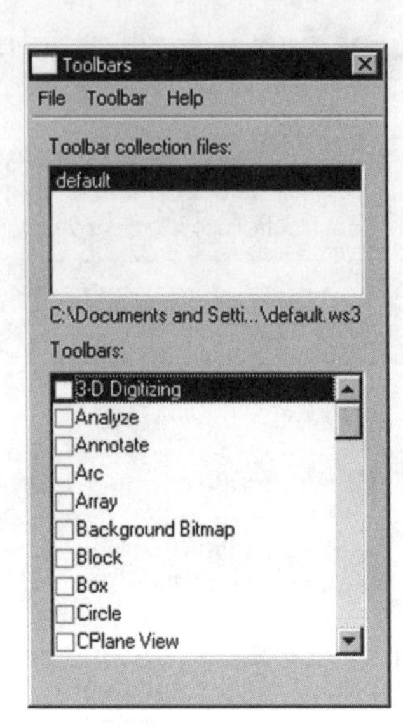

Fig. 1-21. Toolbars dialog box.

Command Interaction

To summarize, there are three ways to run a Rhino command.

- Make selections from the main pull-down menu
- Type a command at the command line interface
- Select a button on the toolbar

Left-click and Right-click

Normally, your pointing device (mouse) has two buttons (left and right). You use the left button to select an item from the main pull-down menu, a button on a toolbar, or a check box in a dialog box, as well as to specify a location on the graphics area.

Depending on where you place your cursor, right-clicking has different effects. Some toolbars have two commands sharing a single button. For example, the commands Zoom Extents and Zoom Extents All Views, shown in figure 1-22, share a single button on the Standard toolbar. Left-clicking activates the Zoom Extents command, and right-clicking starts the Zoom Extents All Viewports command.

When you place the cursor over the command line interface and right-click, it opens a pop-up menu listing a "history" of recently performed commands (the most recent command at the top), an example of which is shown in figure 1-23. You can repeat any previous command by selecting the command you want to repeat from the pop-up menu and left-clicking. Initially, the command history pop-up menu is blank.

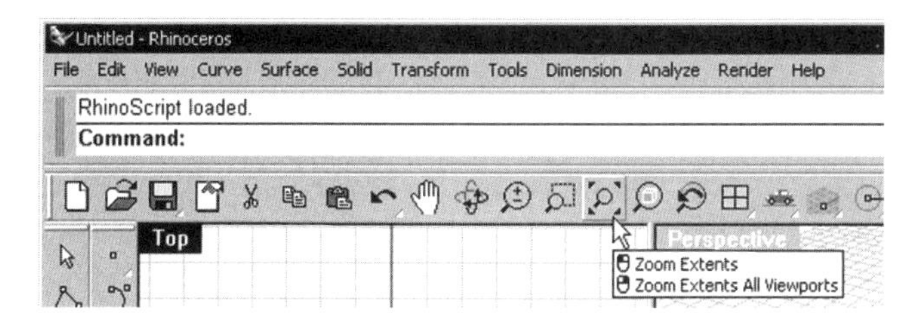

Fig. 1-22. Zoom Extents and Zoom Extents All Viewports commands.

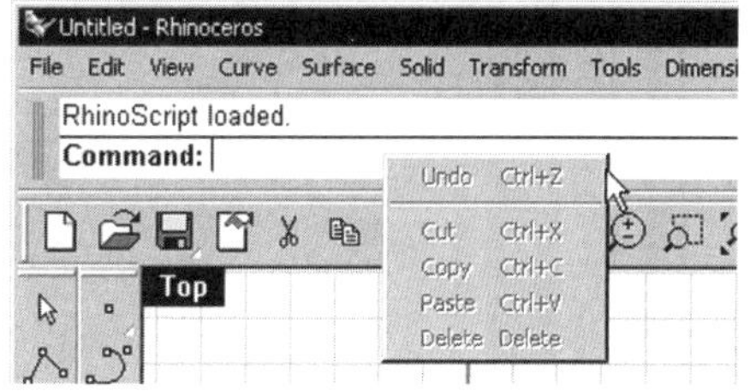

Fig. 1-23. Command history pop-up menu.

Several commands are accessed by placing the cursor over the graphics area and right-clicking in one of three ways. Simply right-clicking repeats the last command. Right-clicking and holding down the mouse button for a while (or holding down the Shift key) accesses the Pan command. Right-clicking and holding down the Ctrl key accesses the Zoom command.

Middle Mouse Button

If your mouse has the third (middle) button, pressing it will access a menu or toolbar based on the settings in the Mouse tab (shown in figure 1-24) of the Options dialog box (accessible by selecting Tools > Options). If the third button can be slid forward and backward, doing so will zoom in and out on the display.

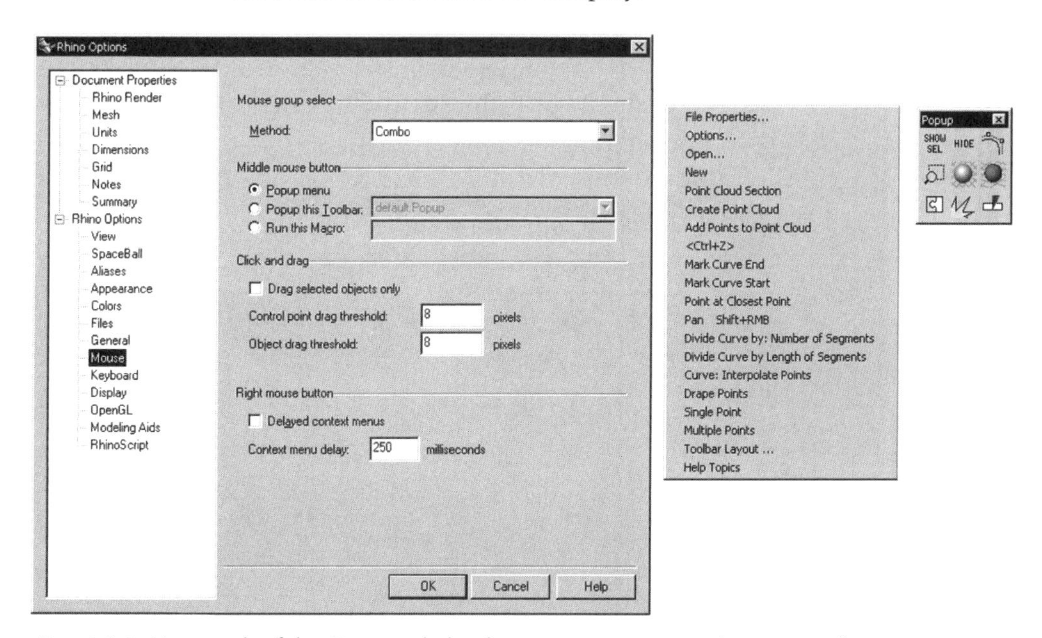

Fig. 1-24. Mouse tab of the Options dialog box, pop-up menu, and pop-up toolbar.

Construction Plane Concept

To construct geometric objects, you select options from the pull-down menu, toolbar, or command line interface. You must also select a location for the geometric object to be constructed. To specify a location using the pointing device, you pick a point in one of the viewports. To precisely specify a point, you key in a set of coordinates at the command line interface. (You will learn the use of coordinate systems in the next section.)

When you pick a point in one of the viewports, you have selected a point on an imaginary construction plane corresponding to the selected viewport. If you pick a point in the Top viewport, you have selected a point on a construction plane parallel to the Top viewport and passing through the origin. (The Perspective viewport has the same construction plane.) Similarly, if you pick a point in the Front viewport, you have selected a point on a construction plane parallel to the Front viewport and passing through the origin. You can construct objects on more than one construction plane. For example, to construct two circles on two different construction planes, you would perform the following steps.

1. From the main pull-down menu, select Curve > Circle > CENTER, Radius. Alternatively, select the *Circle: CENTER, Radius* option from the Circle toolbar, or input the following command at the command line interface.

   Command: *Circle*

2. Move the cursor over the Top viewport.

Indicating that the viewport is active, the label of the viewport (top left-hand corner) should be highlighted and the labels in the other viewports should be grayed out, as shown in figure 1-25.

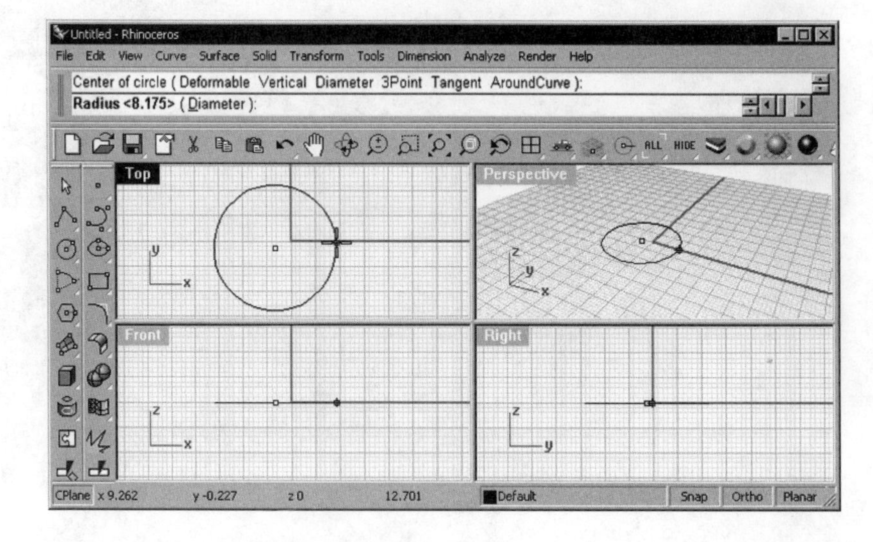

Fig. 1-25. Circle being constructed on the construction plane corresponding to the Top viewport.

3. Pick a point to specify the center location.
4. Pick to specify a point on the circumference. A circle is constructed on a construction plane parallel to the Top viewport.
5. Press the Enter key to repeat the Circle command.
6. Move the cursor over the Front viewport.

Now the label of the Front viewport is highlighted. You are working on a construction plane parallel to the Front viewport.

7. Pick a point specifying the center location.
8. Pick to specify a point on the circumference.

A circle is constructed on a construction plane parallel to the Front viewport, as shown in figure 1-26.

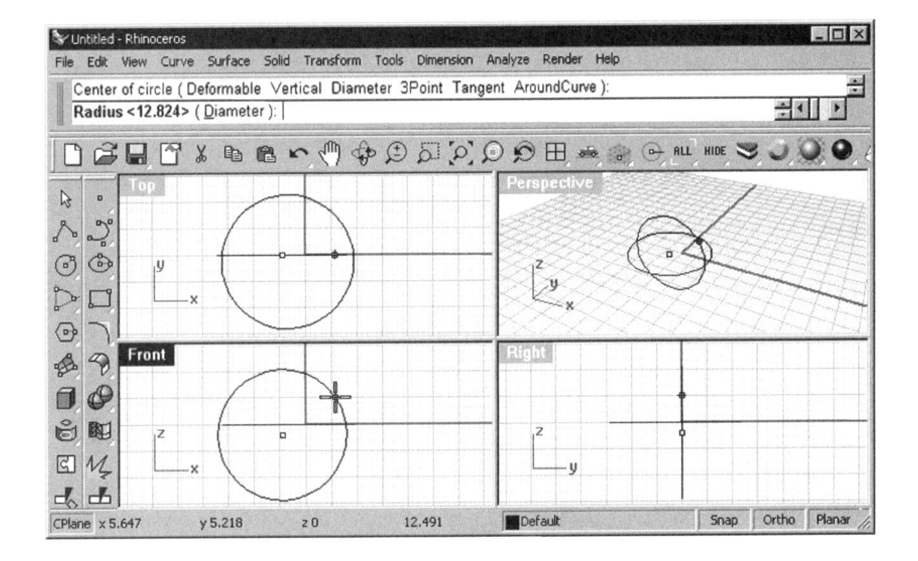

Fig. 1-26. Circle being constructed on a construction plane (the front construction plane) parallel to the Front viewport.

To summarize, the orientation of the objects you construct depends on the active construction plane, and the active construction plane depends on which viewport you select. In a four-viewport display, there are three construction planes. With the exception of the Perspective viewport, which has the same construction plane as the Top viewport, each viewport has a construction plane parallel to itself. In addition to the default construction planes, you can construct new construction planes using one of the options in the Set CPlane cascading menu of the View pull-down menu, as shown in figure 1-27. You will learn how to set up construction planes in Chapter 4.

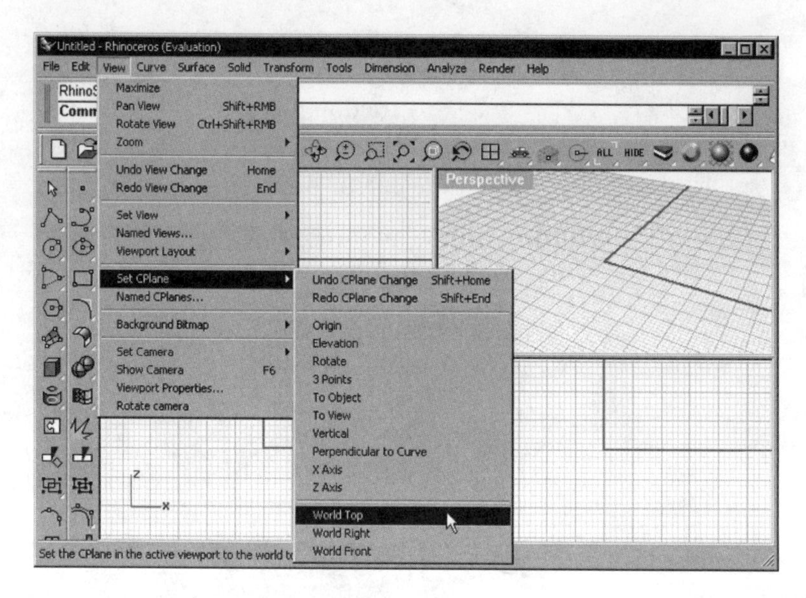

Fig. 1-27. Commands for manipulating the construction plane.

Coordinate Systems

There are two coordinate systems: the construction plane coordinate system (corresponding to the active viewport) and World coordinate system (independent of the active viewport). The construction plane coordinate system corresponds to the construction plane. In each viewport, a red line and a green line depict, respectively, the X and Y axes of the construction plane in the viewport. (The color of these lines, and other color settings, are configurable via the Color tab of the Rhino Options dialog box, accessed by selecting Tools > Options from the main pull-down menu.) The Z direction is perpendicular to the construction plane.

In addition to the red and green lines, there is a World Axes icon in the lower left-hand corner of the respective viewport. The World Axes icon is in the shape of a tripod. The lines on the tripod depict the absolute X, Y, and Z axes of the World coordinate system.

To specify a location using the command line interface, you indicate the nature of the Cartesian coordinate system or polar coordinate system you want to use via the nature of the coordinate input. For the Cartesian coordinate system options, you specify X and Y values or X, Y, and Z values (separated by a comma or commas, as shown in Table 1-3). If you specify only the X and Y values, the Z value is assumed to be zero. For the polar coordinate system options, you specify a distance value and an angular value, separated by a less than (<) sign. For either coordinate system, if you want to specify a point relative to the last selected point, you prefix the coordinate with the letter *r*. If you want to use a World coordinate, you prefix the coordinate with the letter *w*. Coordinate systems that operate under Rhino, and what each specifies, are summarized in Table 1-3.

Table 1–3: Coordinate Systems and Their Functions

Coordinate System	Example	Specifies
Construction plane Cartesian system	2,3	A point 2 units in the X direction and 3 units in the Y direction from the origin (Z is zero)
—	2,3,4	A point 2 units in the X direction, 3 units in the Y direction, and 4 units in the Z direction from the origin
Construction plane polar system	2<45	A point 2 units at an angle of 45 degrees from the origin
Relative construction plane Cartesian system	r2,3	A point 2 units in the X direction and 3 units in the Y direction from the last reference point
Relative construction plane polar system	r4<60	A point 4 units at an angle of 60 degrees from the last reference point
World Cartesian system	w3,5	A point 3 units in the absolute X direction and 5 units in the absolute Y direction from the absolute origin, regardless of the location of the current construction plane
World relative Cartesian system	wr3,6	A point 3 units in the absolute X direction and 6 units in the absolute Y direction from the last reference point, regardless of the location of the current construction plane
World polar system	w4<30	A point 4 units at an angle of 30 degrees on the absolute XY plane from the absolute origin, regardless of the location of the current construction plane
World relative polar system	wr5<45	A point 5 units at an angle of 45 degrees on the absolute XY plane from a reference point, regardless of the location of the current construction plane

The Concept of Layers

The term *layer* originates from manual drafting. It refers to overlay of clear transparent sheets. It is a grouping mechanism in which groups of objects are drawn on different transparent sheets. By removing or overlaying the sheets, you control which set or sets of objects are shown on a drawing.

In computer-aided design, layers are not physical sheets. You might say they are conceptual layers. You set up layers in a file and place objects on different layers. By turning layers on or off, you control the display of the objects on the screen. In addition to on and off, you can lock a layer so that objects placed on the layer can be seen and snapped to but cannot be selected or manipulated (such as moving or erasing). In a multi-layer Rhino file setup, you can move objects from one layer to another. To try editing a layer, perform the following steps.

1 From the main pull-down menu, select Edit > Layers > Edit Layers, or select Edit Layers from the Standard toolbar (as shown in figure 1-28). Alternatively, right-click on the Layer (pane) option on the status bar.

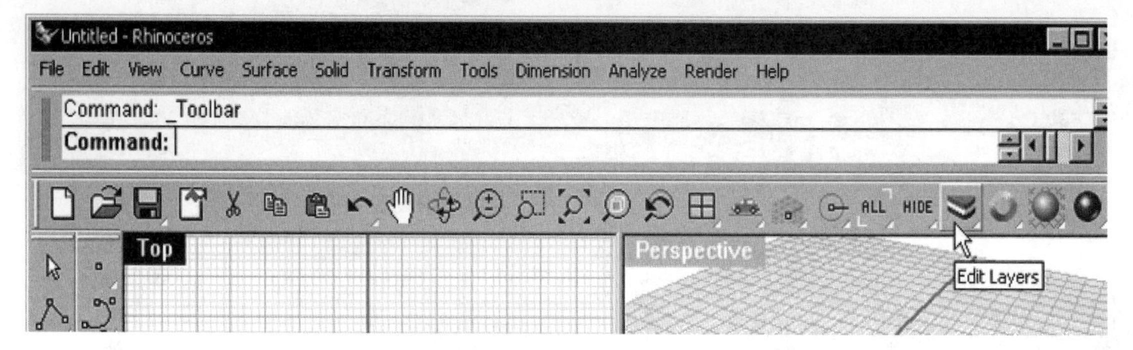

Fig. 1-28. Selecting Edit Layers from the Standard toolbar.

2 In the Layers dialog box, shown in figure 1-29, add new layers, delete existing layers, set visibility of layers, lock objects on layers, and define the color and material properties of objects placed on a layer as desired.

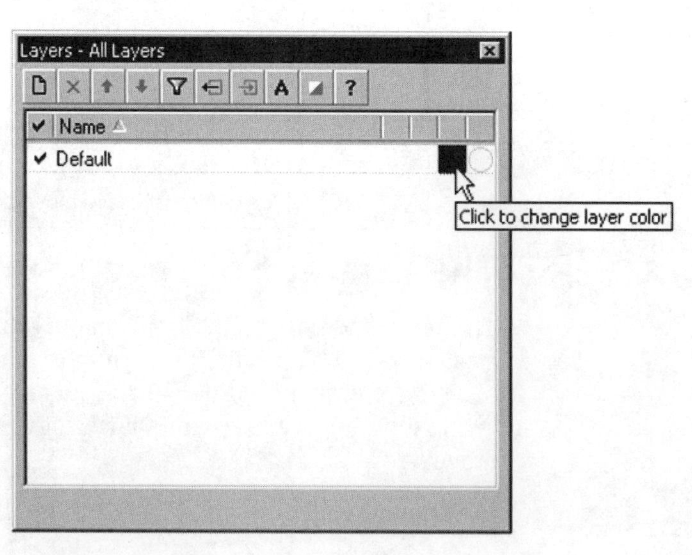

Fig. 1-29. Layers dialog box.

3 If you wish to set the color of objects on a layer, select a layer and then click in the Color column of the Select Color dialog box, shown in figure 1-30.

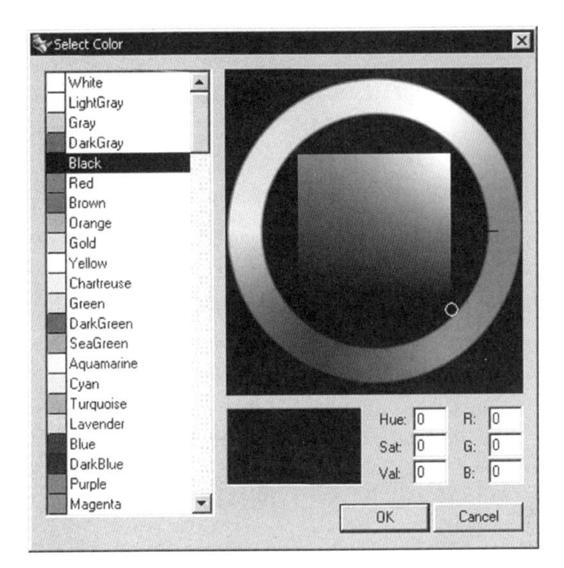

Fig. 1-30. Select Color dialog box.

4 In the Select Color dialog box, specify a color by selecting a color from the color swatch or by inputting hue, saturation, and value (RGB, or red, green, and blue) values in these respective fields.

5 Specify the material (visual appearance of as rendered) of the object by selecting the layer and then the Material column of the Edit Layer dialog box. This accesses the Material Properties dialog box, shown in figure 1-31. In the Material Properties dialog box, establish the settings as desired. You will learn more about color and material in Chapter 8.

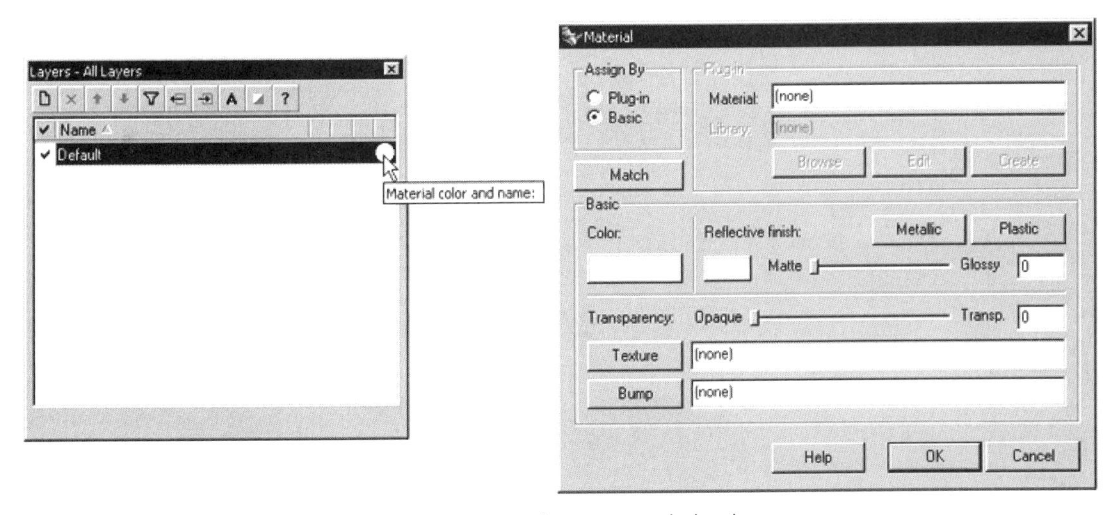

Fig. 1-31. Material Properties dialog box.

Visibility

Other than turning off a layer to make objects residing on that layer invisible, you can hide selected objects by selecting Edit > Visibility > Hide and then selecting the objects. You can unhide a hidden object to make it visible again. You can also lock an object, whereby it is visible but cannot be modified.

Group and Block

Objects can be grouped so that you can select them collectively. To group objects, select Edit > Groups > Group. To combine a number of objects so that they are treated as a single object, select Edit > Blocks > Create Block Definition.

Drawing Aids

Rhino offers several drawings aids: grid meshes, snapping to grids, planar mode, elevator mode, ortho mode, and object snap. These aids are described in the sections that follow.

Grid Mesh

A grid mesh of known spacing on the screen gives you a sense of the actual size of the current viewport. You control the spacing of the grid mesh via the Document Properties option, accessed by selecting File > Properties and then selecting the Grid tab in the Document Properties dialog box, shown in figure 1-32.

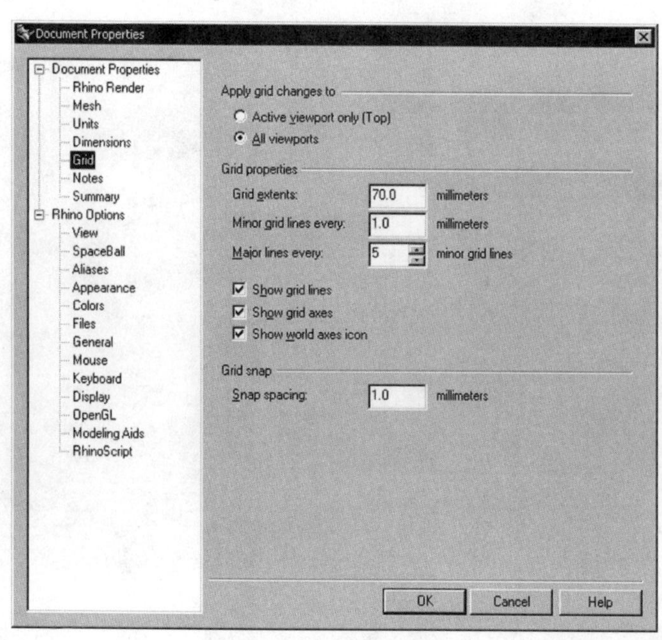

Fig. 1-32. Grid tab of the Document Properties dialog box.

In the Grid tab of the Document Properties dialog box, you control the display of grid lines and set their spacing. In addition, you control the display of the World Axes icon. To quickly turn on or off the grid mesh in a viewport, press the F7 key.

Snap to Grid

The grid display shown in the viewport is for visual reference only. Without the use of further aids, it is virtually impossible to select these points precisely using the pointing device. To restrict the movement of the cursor so that it will stop only at the grid intervals, you use the Snap option, or select or deselect the Snap button on the status bar.

Planar Mode

Because the active construction plane depends on the current working viewport, you may get an unexpected outcome if you change the active viewport in the middle of an active command. To restrict the current construction plane to that of the last selected point, you use planar mode (established via the Planar option), or by selecting or deselecting the Planar button on the status bar.

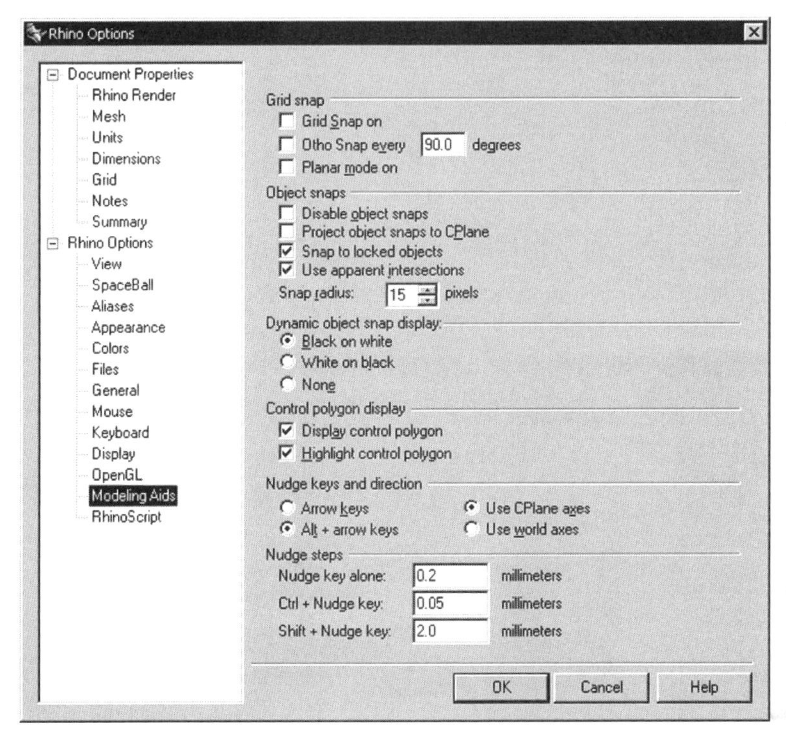

Fig. 1-33. Modeling Aids tab of the Options dialog box.

Elevator Mode

To make the cursor move perpendicularly, to the current construction plane, hold down the Ctrl key and pick, and then release the Ctrl key and drag the mouse.

Ortho Mode

To restrict cursor movement in a specified angular direction, you select or deselect Ortho on the status bar. To set the angular intervals, you use the Modeling Aids tab (shown in figure 1-33) of the Options dialog box, accessible by selecting Tools > Options.

Object Snap

To help locate the cursor in relationship to selected features of existing geometric objects, you use object snap. Feature aspects you snap to are end, near, point, midpoint, center, intersection, perp, tan, quad, and knot.

There are two ways to set object snap. You can temporarily set object snap mode (i.e., for each snap to a feature) by specifying the snap mode before you select an object. You set persistent object snap by selecting Osnap on the status bar to use the Snap option from the Osnap dialog box, shown in figure 1-34. Temporary object snap mode is activated by selecting Tools > Object Snaps and then setting this mode.

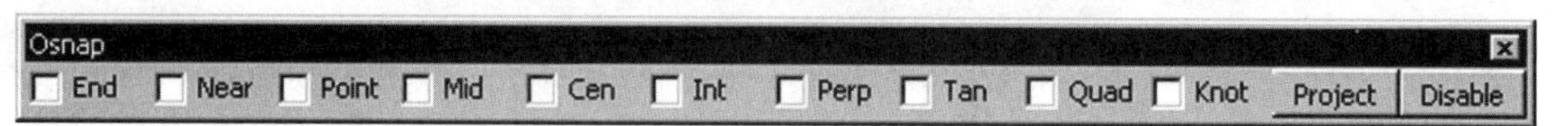

Fig. 1-34. Osnap dialog box.

Help System

The help system contains several categories: Help Topics, Frequently Asked Questions, Learn Rhino, Help on the Web, Command List, Feature Overview, and Technical Support. The sections that follow describe Help Topics and Frequently Asked Questions.

Help Topics

By selecting Help Topics from either the standard toolbar or Help pull-down menu, you gain access to the comprehensive Help dialog box, shown in figure 1-35. Here you will find all the information you need.

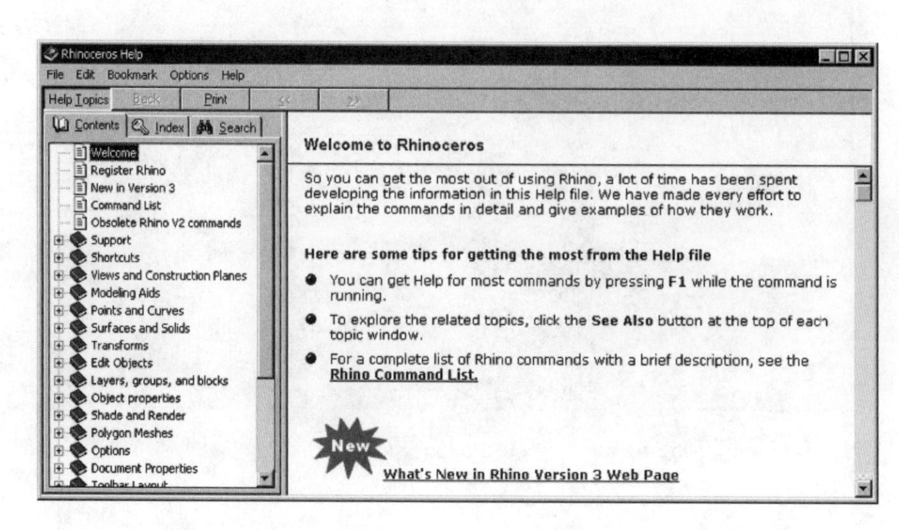

Fig. 1-35. Help dialog box.

Frequently Asked Questions

Selecting Frequently Asked Questions from the Help pull-down menu brings you to the web site *www.rhino3d.com/support/faq/*. This site includes the Frequently Asked Questions section, shown in figure 1-36.

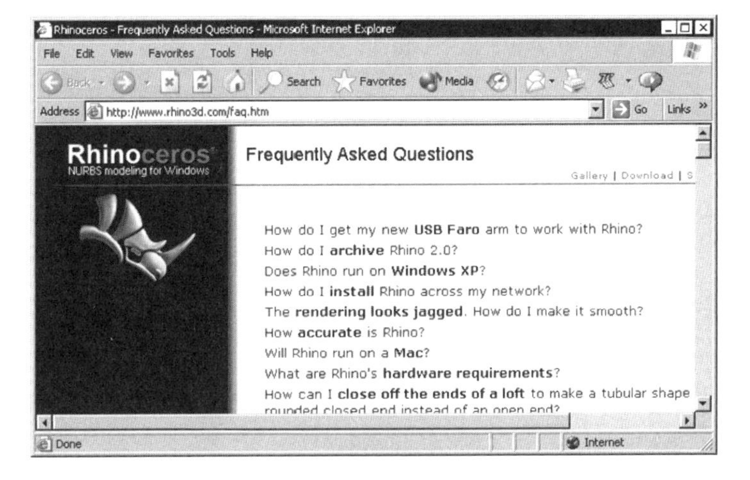

Fig. 1-36. Frequently asked questions.

Rhino Options and Document Properties

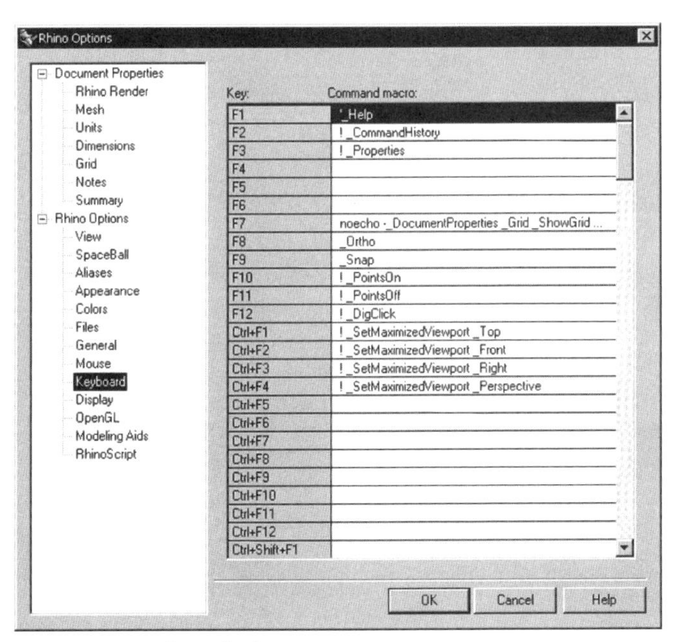

Fig. 1-37. Keyboard tab.

By now you should have a general understanding of Rhino's user interface and various drawing aids. Before you proceed to the other chapters to learn how to construct various types of objects, spend some time learning the meaning of various settings in the Rhino Options and Document Properties dialog boxes, which share the same dialog box. To access this dialog box, select File > Properties (or Tools) > Options. Some of these options are discussed in the sections that follow.

Shortcut Keys

Using the Keyboard tab, shown in figure 1-37, you can assign shortcut keys to the commands you use most frequently.

Command Aliases

The Aliases tab, shown in figure 1-38, enables you to set an alias for sets of command strings.

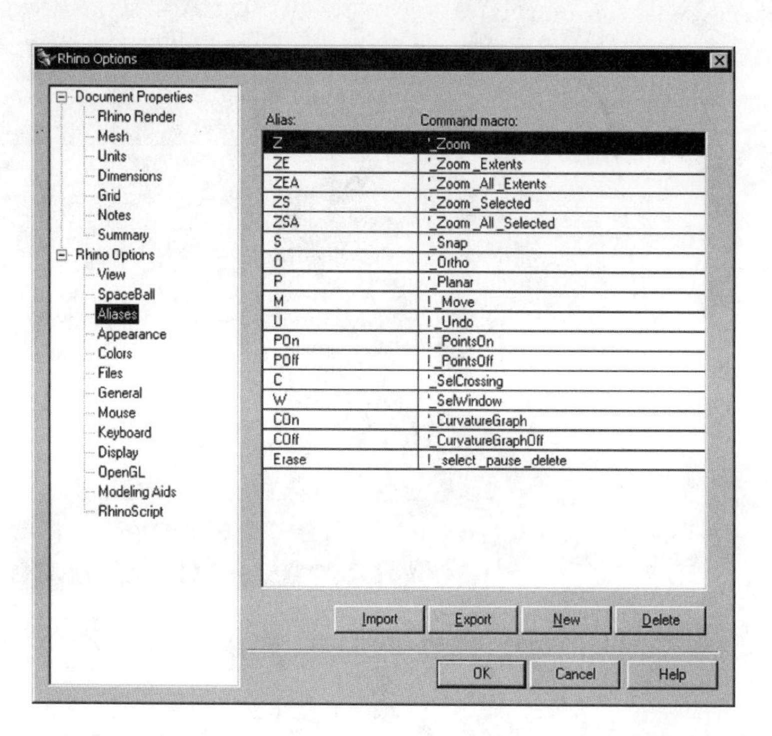

Fig. 1-38. Aliases tab.

Files Location

To set the template file and auto-save file location, you use the Files tab.

Appearance

In the Appearance tab, you set the color and appearance of the user interface.

Units

Before you start constructing a model, you should check the units of measurement so that the model you construct is compatible with any upstream and downstream operations. In the Units tab, you set the units of measurement and the tolerance of the model. In addition, you set the display tolerance.

File Handling

If your free-form models are constructed to facilitate downstream computerized operations, you can output them to various file formats for this purpose. On the other hand, you can reuse data constructed in upstream computer applications.

Many file types can be imported into and opened (a single operation) in Rhino to make possible the reuse of existing engineering designs constructed via other computer design applications. To facilitate engineering data exchange in which digital models you have constructed using Rhino can be used in other applications you can save the entire Rhino files or export selected objects as various file formats.

Basically, data you import or export can be categorized as four major types: curves and points, NURBS surfaces and polysurfaces, solids, and polygon meshes. The types of files you export depend on the types of objects you have constructed in Rhino. For example, you can export a solid to a surface or to a solid file, but you cannot export an open polysurface or polygon mesh to a valid solid file. When you input a solid, you get a polysurface. However, you will not get a NURBS surface if the source file is a polygon mesh. The types of files you can open and/or save in Rhino are outlined in Table 1-4.

Table 1–4: File Formats Rhino Can Open and/or Save

Open File Format	Save As File Format
Rhino 3D model (*.3dm)	Rhino 3 3D model (*.3dm)
—	Rhino 2 3D model (*.3dm)
3D Studio (*.3ds)	3D Studio (*.3ds)
LightWave (*.lwo)	LightWave (*.lwo)
Step (*.stp, *.step)	STEP (*.stp, *.step)
IGES (*.igs, *.iges)	IGES (*.igs, *.iges)
Raw Triangles (*.raw)	Raw Triangles (*.raw)
Stereolithography (*.stl)	Stereolithography (*.stl)
VDA (*.vda)	VDA (*.vda)
AutoCAD DWG (*.dwg)	AutoCAD drawing file (*.dwg)
AutoCAD DXF (*.dxf)	AutoCAD drawing exchange file (*.dxf)
Adobe Illustrator (*.ai)	Adobe Illustrator (*.ai)
Points File (*.asc, *.csv, *.txt, *.xyz)	Points File (*.txt)

Open File Format	Save As File Format
Points File (*.asc, *.csv, *.txt, *.xyz)	ACIS (*.sat)
—	Windows Metafile (*.wmf)
—	VRML (*.wrl, *.vrml)
—	Wavefront (*.obj)
—	XGL (*.xgl)
—	POV-Ray Mesh (*.pov)
—	GHS Export File (*.gif)
—	RenderMan (*.rib)
—	SLC (*.slc)
—	Moray UDO (*.udo)
—	Viewpoint (*.mts, *.mtx)
—	Parasolid (*.x_t)
—	Object Properties (*.csv)

Open, Import, and Insert

To reuse digital data from other applications, you can open a file, import a file, or insert a file. Naturally, opening a file converts the file to Rhino format, whereupon you can continue with the design. If you want to incorporate data into an existing Rhino file, you import or insert the file. Importing a file converts the imported data to individual Rhino curves or surfaces. Inserting a file also converts the data to Rhino format, but in block form. To decompose a block into individual elements, you explode the block.

Save and Export

To export an entire Rhino file, you save it in various file formats. To export selected objects of a Rhino file, you export them. While exporting, you can use the World coordinate system or specify an origin.

Save Methods and Worksession Manager

A Rhino file can be saved via five operations: Save, Save Small, Incremental Save, Save As, and Save as Template. By selecting File > Save, you save your file together with a preview and associated rendering meshes. By selecting File > Save Small, you save your file without the preview and rendering meshes. Rendering meshes are polymeshes used in rendering. The first time you render your model, a set of polymeshes is constructed. If you save these polymeshes, subsequent rendering time is shorter because the polymeshes are already there. By selecting File > Incremental Save, you save your model in a set of versions so that you can experiment with changes and retrieve previous versions. By selecting File > Save As, you save your model with a different file name. By selecting File > Save as Template, the model is saved as a template.

The Worksession manager enables a group of designers to work collaboratively on a project by dividing a large project into a number of smaller projects in terms of files of smaller size. Each designer works on and controls a file but can view other portions of the project. To use the Worksession manager, shown in figure 1-39, select File > Worksession Manager.

You can attach external files of various formats (formats that Rhino supports). Geometry in the attached file cannot be modified but can be used as input for constructing geometry.

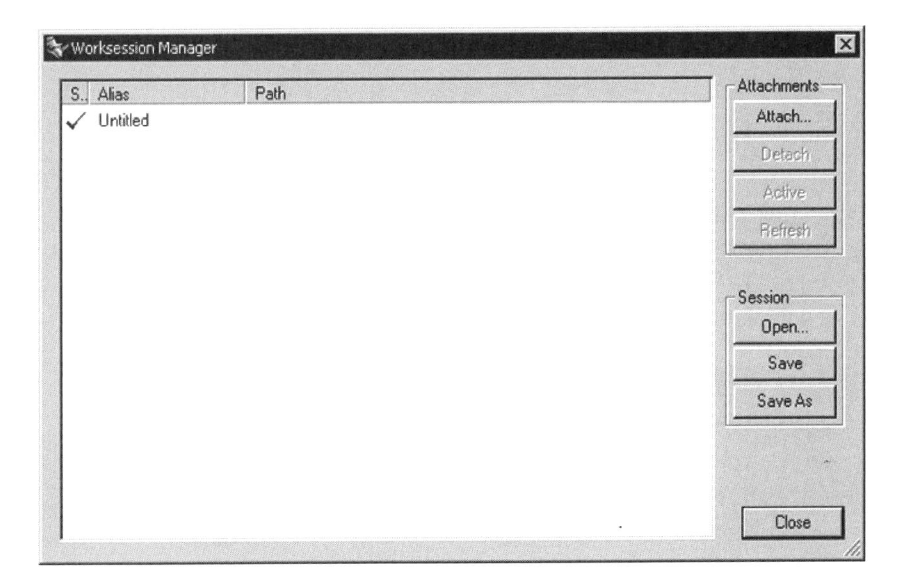

Fig. 1-39. Worksession manager.

Summary

Digital modeling is a representation of an object in the computer. There are three types of digital models: wireframe, surface, and solid. A wireframe model is a set of unassociated curves in 3D space that does not have any surface or volume information. A surface model is a set of assembled surfaces and is best suited to the representation of free-form objects. A solid model is the most comprehensive of the three types, containing information about the vertices, edges, surfaces, and volume of the object it represents.

Rhino is a 3D digital modeling tool. You use it to construct points, curves, surfaces, polysurfaces, solids, and polygon meshes. In addition, you output photorealistic renderings, 2D drawings, and file formats of various types for downstream computerized operations. You also reuse upstream digital models by opening various file formats.

The Rhino user interface contains five major areas: standard Windows title bar, main pull-down menu, command line interface, graphics area, and status bar. In addition, there are a number of toolbars. You perform commands via the main pull-down menu, command line interface, or toolbar. The graphics area is where you construct digital models.

When you use the mouse to select a point in one of the viewports in the graphics area, you select a point on a construction plane corresponding to the selected viewport. To input a point at the command line interface, you use either the construction plane coordinate system or the World coordinate system. To help you construct objects in the graphics area, you use various drawing aids. Rhino objects are organized into layers. Objects you construct are placed on the current layer.

Review Questions

1. Using simple sketches, illustrate the three types of digital models.
2. What types of objects can you construct using Rhino?
3. Give a brief account of the file formats supported by Rhino.
4. Describe the Rhino user interface.
5. Explain the concepts of the construction plane and the two types of coordinate systems.
6. Outline the content of the Rhino Help system.

CHAPTER 2

Wireframe Modeling and Curves: Part 1

Introduction

This chapter introduces key concepts involved in wireframe modeling, by way of explaining the importance of 3D curves in solid and surface modeling and illustrating various methods of constructing points, basic curves, and derived curves.

Objectives

After studying this chapter you should be able to:

- ❒ State the characteristics of a 3D wireframe model
- ❒ Appreciate the importance of 3D curves in making 3D solids and surfaces
- ❒ Use Rhino to construct point objects, basic curves, and derived curves

Overview

As you learned in Chapter 1, there are three basic ways of representing an object in the computer: wireframe modeling, surface modeling, and solid modeling. This chapter and the next two chapters are intended to provide you with a deep understanding of wireframe modeling. In this chapter you will examine the characteristics of wireframe modeling, explore the significance of curves in surface and solid modeling, and learn how to use Rhino as a tool for constructing point objects and various types of basic curves and derived curves. In the next two chapters you will learn more about Rhino curve tools. After equipping yourself with basic skills in 3D curve manipulation, in chapters 5 through 7 you will learn how to perform various methods of surface and solid construction.

Wireframe Modeling Concepts

A 3D wireframe model is a set of unassociated curves that represents an object by defining the object's edges. The rectangular block in figure 2-1 is shown as a wireframe model. In essence, this model has eight separate, unrelated curves. Between the curves, there is no information.

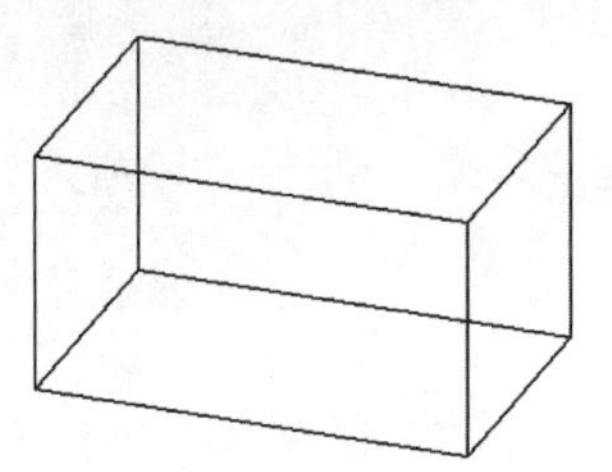

Fig. 2-1. Wireframe model representing a rectangular block.

Representation of 3D Objects by Wireframe Models

To illustrate curved surfaces in a wireframe model, such as a cylinder or a sphere, you construct curves depicting the contour of the surface. Figure 2-2 shows the wireframe model of a cylinder.

Because all curves in a 3D wireframe model are independent entities and there is no information between them, a wireframe model represents an object only implicitly. To perceive a 3D object by viewing a wireframe model, you combine your visual perception of the curves with added meaning from your imagination on how the curves relate to one another. Hence, the wireframe model shown in figure 2-1 has several meanings. It can be simply a set of curves; that is, just a framework. It can also be a box with an opening in one of its sides, as depicted in figure 2-3. Furthermore, it also implies a solid block, as depicted in figure 2-4.

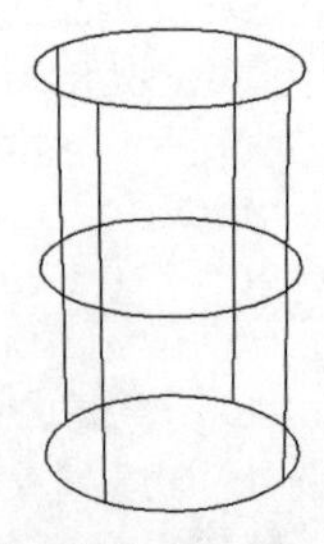

Fig. 2-2. Contour wireframe representing the curved surfaces of a cylinder.

Fig. 2-3. Box with an opening.

Fig. 2-4. Solid box.

Thinking About Wireframe Models

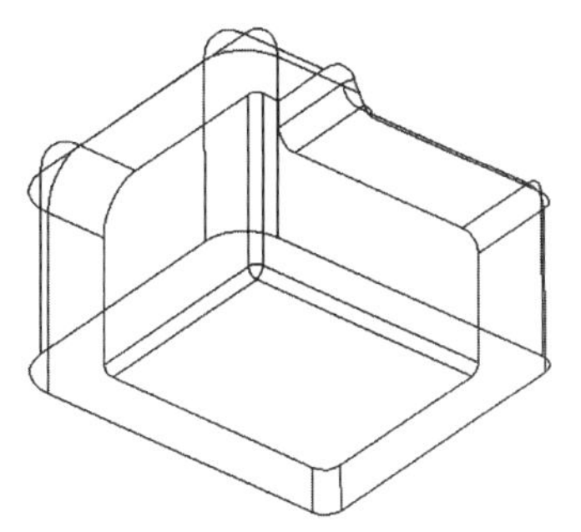

Fig. 2-5. Wireframe model of a complex object.

Prior to constructing the wireframe model of an object, you "deconstruct" the object into discrete curves by thinking about the outlines and silhouettes of the object and the edges of the object where two faces meet. To create the model, you construct the curves in accordance with your perception of the object's appearance. Because making wireframe models requires you to determine the locations of the vertices of the curves, the task of wireframe construction is tedious. Despite the laborious work required, objects are not completely represented by wireframe models. For example, construction of the fillet edges of the model shown in figure 2-5 is very time consuming.

Limitation of Wireframe Models

Because a wireframe model is simply a set of unrelated curves, there is no surface and volume information stored in the computer. As a result, a wireframe model has limited application in computerized downstream operations such as analysis, CNC machining, and rapid prototyping.

Curves for Surface and Solid Modeling

Although the use of curves alone in computer modeling is diminishing, you still need to learn how to construct 3D curves, because they are required in the creation of surfaces and solids; in particular, in free-form surface modeling.

Curves for Surface Modeling

There are many ways to construct surfaces in the computer. A fundamental method of creating free-form surfaces is to use a set of curves to define the silhouette and the contour of the surface and let the computer construct a surface patch on the curves. (You will learn surface modeling in chapters 5 and 6.) Figure 2-6 shows four 3D curves and a surface constructed from the curves.

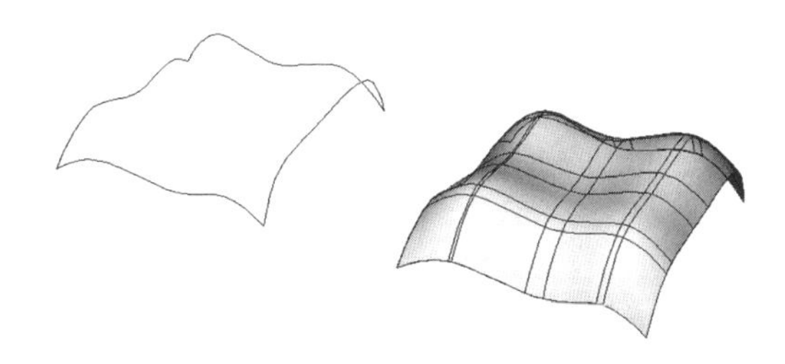

Fig. 2-6. Curves and surface constructed from them.

Curves for Solid Modeling

Among the many ways to represent a solid in the computer, one basic way is to construct a curve in a closed loop and sweep the curve in 3D space. (You will learn solid modeling in Chapter 7.) Figure 2-7 shows a closed-loop curve being extruded to form a solid.

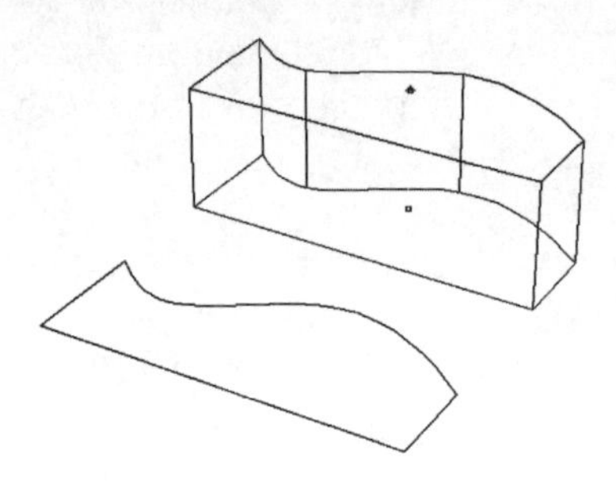

Fig. 2-7. Extruding a closed-loop curve to construct a solid.

Curves

To prepare yourself for making 3D solids and surfaces, you need to understand the characteristics of the curves and ways to construct them. Of the many types of curves you can use to construct 3D surfaces and solids, the spline is the most important because it enables you to construct complex free-form shapes. Spline curves are discussed in the section that follows.

Spline Curves

A spline is a way of defining a free-form curve by specifying two end points and two or more tangent vectors that control the profile of the curve. There are many mathematical methods of defining a spline. Some of the most popular methods are outlined in the sections that follow.

Polynomial Spline

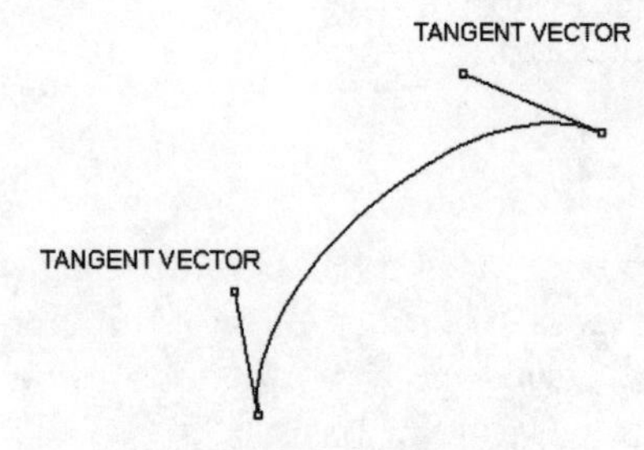

Fig. 2-8. Polynomial spline segment and its tangent vectors.

A polynomial spline is a set of spline segments. At the end points of each spline segment there is a tangent vector having a direction and magnitude, as shown in figure 2-8. The effect of the tangent defines the curvature of the segment. Because the tangent vector is described by a polynomial equation, the spline is called a polynomial spline. How the tangent vector affects the shape of the spline segment is determined by the degree of the polynomial equation. The overall shape of a polynomial spline is the combined effect of all of its segments.

B-spline

A B-spline is also a multi-segment spline curve. A connection point between two contiguous spline segments is called a knot. Being an extension of the polynomial spline, each spline segment is formed from the weighted sum of four local polynomial basis functions. Hence, the spline is called a basis spline (or B-spline). Four control points lying outside the spline control the shape of a B-spline segment, as shown in figure 2-9. Movement of a control point affects only four segments of the curve.

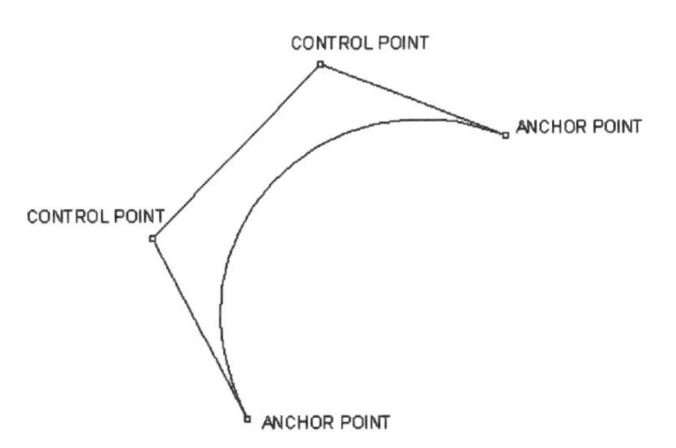

Fig. 2-9. B-spline segment and its control points.

Nonuniform Rational B-spline (NURBS)

A nonuniform rational B-spline curve is a derivative of the B-spline curve. It has two additional characteristics: it is nonuniform and rational. Unlike the uniform B-spline, in which a uniform parameter domain defines each spline segment, the parameter domain of a nonuniform B-spline need not be uniform.

Because of the nonuniform characteristic, different levels of continuity between the spline segments can be attained, unequal spacing between the knot points is allowed, and the spline can more accurately interpolate among a set of given points. With a rational form, the curve can better represent conic shapes. Using NURBS mathematics, you can trim a curve at any point and the curve will retain its original shape. NURBS curves and surfaces are used in most surface modeling tools.

Rhino Curves

Rhino curves are spline curves. The program uses NURBS mathematics to define curves and surfaces. To reiterate, a spline curve is a set of connected spline segments, and the joint between two contiguous spline segments is a knot. The degree of the polynomial basis equation, the control point location, and the weight of the control points determine the shape of each spline segment.

Polynomial Degree

The degree of a polynomial equation has a direct impact on the complexity of the shape of the curve. For example, a line is a degree 1 NURBS curve, a circle is a degree 2 NURBS curve, and free-form curve is a NURBS curve of degree 3 or above. You can raise or reduce the polynomial degree of a curve. Raising the degree of a polynomial spline curve does not change the curve's shape but does increase the number of control points.

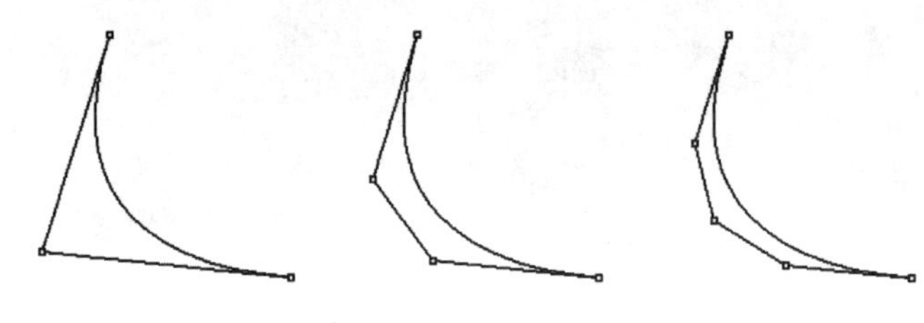

Fig. 2-10. Degree 2, 3, and 4 curves.

In turn, more control points enable you to modify the curve to create a more complex shape. On the other hand, reducing the degree decreases the number of control points and hence simplifies the shape of the curve. Figure 2-10 shows curves with increasing degrees of polynomial equation.

Control Point Location

The location of control points directly affects the contour of a curve. Moving control points changes the curve's shape, an example of which is shown in figure 2-11. Normally, control points lie outside the curve. In an open-loop curve, only the first and last control points coincide with the end points of the curve. In a closed-loop curve, all control points lie outside the curve, and example of which is shown in figure 2-12.

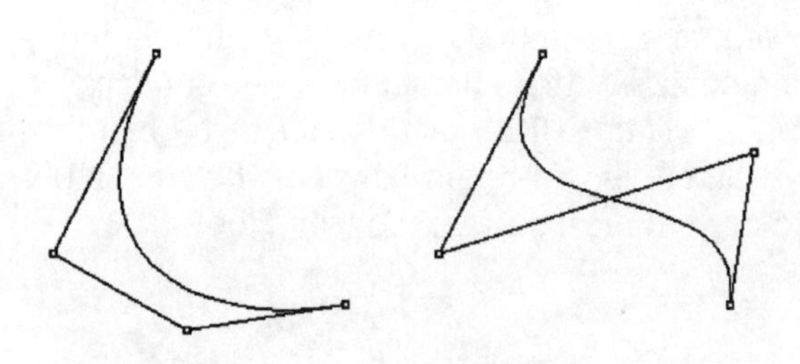

Fig. 2-11. Control point locations affecting the shape of the curve.

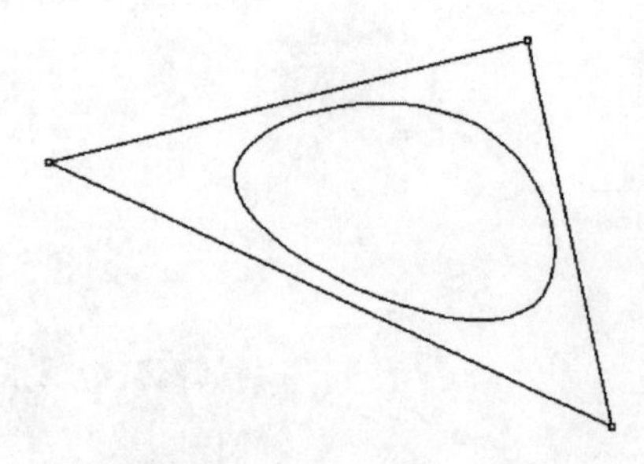

Fig. 2-12. Control points in a closed-loop curve.

Control Point Weight

The weight of control points also has a significant effect on the shape of a spline segment. You can regard the weight of a control point as a pulling force that pulls the spline curve toward the control point. The higher the weight, the closer the curve will be pulled to the control point, an example of which is shown in figure 2-13.

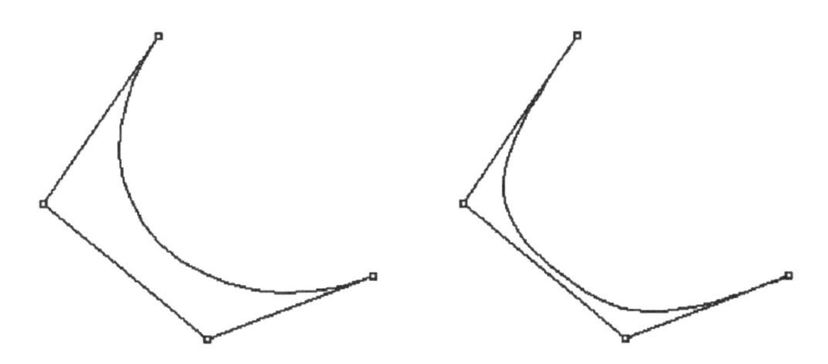

Fig. 2-13. Control points with different weights (from left to right, weight 1 and weight 7).

Knot

A spline curve is a set of spline segments, and a knot defines the junction between two contiguous spline segments. Adding knots to a curve increases the number of spline segments without changing the shape of the curve. However, having more segments means that the curve has more control points. Subsequently, you can modify the curve to create a more complex shape. Figure 2-14 shows a knot added to a curve.

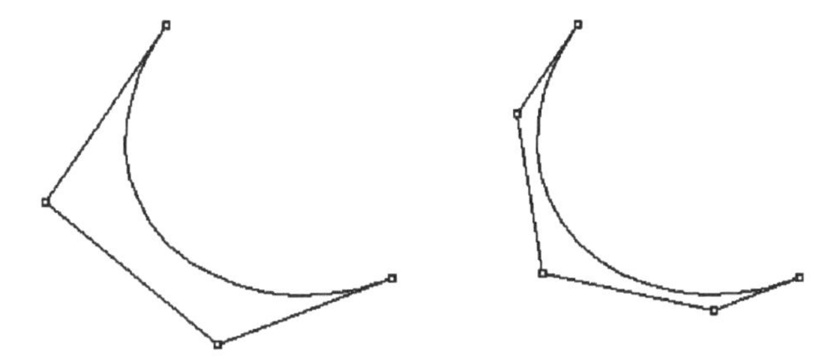

Fig. 2-14. More control points with additional knot (right).

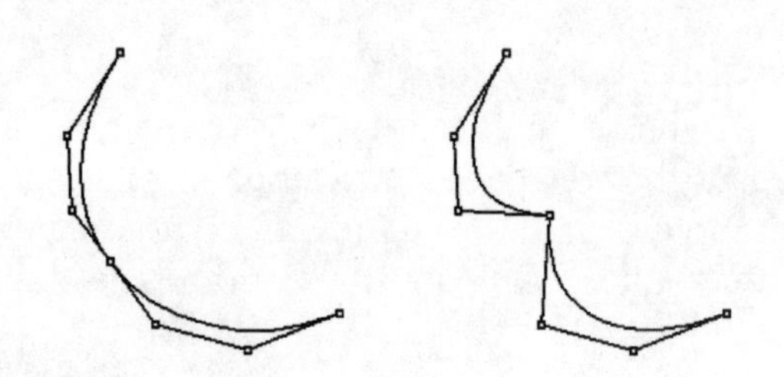

Fig. 2-15. Result of kink point moved.

Kink Point

A kink point is a special type of knot on a curve in which the tangent direction of the contiguous spline segments is not the same. A kink occurs when you join two curves with different tangent directions, or when you explicitly add a kink to a curve. Figure 2-15 shows the effect of moving a kink point of a curve.

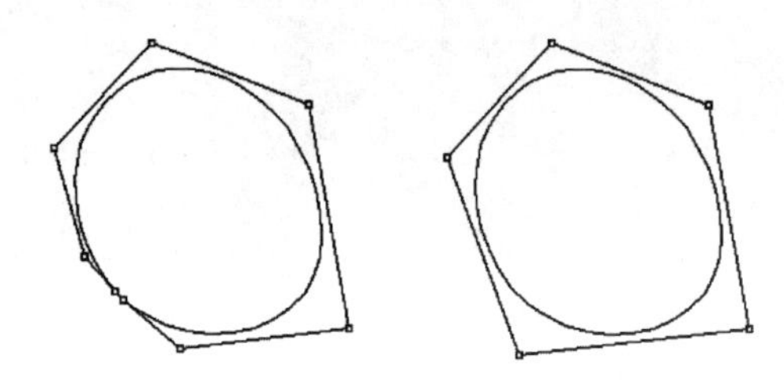

Fig. 2-16. Closed-loop curve with a kink and a periodic curve.

Periodic Curve

A closed curve with no kink point is called a periodic curve. Figure 2-16 shows a closed-loop curve with a kink and a periodic curve.

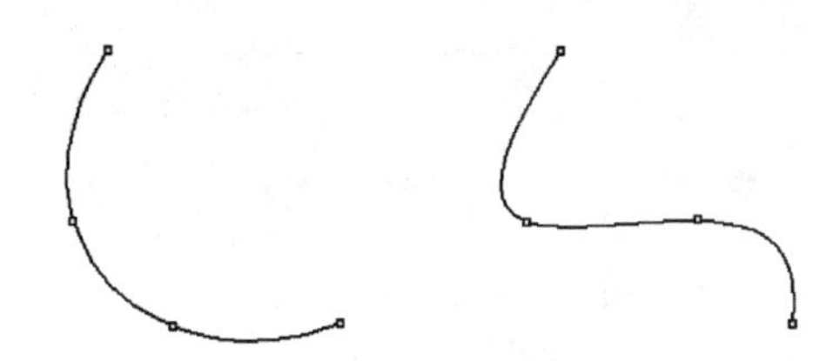

Fig. 2-17. Curve modified by moving edit points.

Edit Point

Because control points are normally not lying along the curve, you may find it difficult to make a curve pass through designated locations by manipulating the control point. To add flexibility in modifying a NURBS curve, Rhino enables you to use edit points along the curve. Edit points are independent of degree, control points, and knots of a curve. You move the edit points to change the shape of the curve, an example of which is shown in figure 2-17.

Handlebar Editor

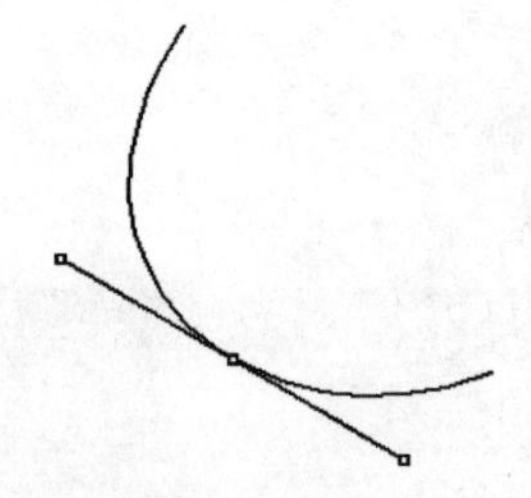

Fig. 2-18. Use of the handlebar.

Edit points enable you to modify a curve only in a limited scope. To change the tangent direction and the location of any point along a curve you use the handlebar, which is a special type of editing tool. It consists of a point on the curve and two tangent lines. Moving the central point of the handlebar changes the location of a point along the curve. Selecting and dragging the end points of the handlebar changes the weight and tangent direction of the curve at a selected location. An example of use of the handlebar is shown in figure 2-18.

Rhino Curve Construction Tools

Using Rhino, you construct point objects, basic curves, and derived curves, and edit and transform curves. To examine existing curves, you use analysis tools. These construction elements and tools are explored in the sections that follow.

Command Menu

As mentioned in Chapter 1, you use Rhino commands in three ways: by selecting an item from the main pull-down menu, by clicking on a button on a toolbar, or by typing a command at the command line interface. You will find the tools for creating point objects and NURBS curves in the Curve pull-down menu. To edit and transform curves, you use, respectively, the Edit and Transform pull-down menus. To examine and evaluate curves, you use the Analysis pull-down menu. These pull-down menu options are shown in figure 2-19.

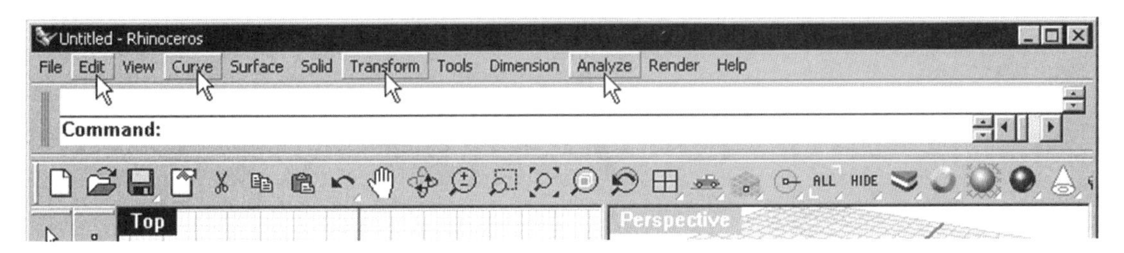

Fig. 2-19. Pull-down menu options.

Toolbars dedicated to curve construction, edit, transform, and analysis are integrated in the main toolbar, which is displayed by default. Clicking and holding on a button on the toolbar that has a small white triangle in its lower right-hand corner accesses a fly-out toolbar. By selecting and dragging the header bar of a fly-out toolbar, you place the fly-out toolbar on your desktop as a floating toolbar. Figure 2-20 shows the main toolbars and the Point fly-out toolbar.

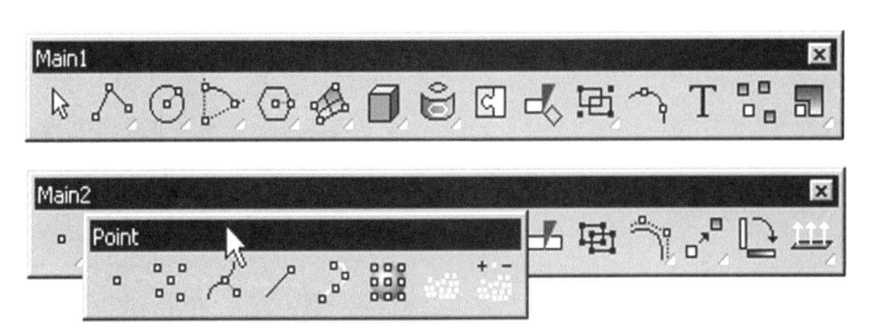

Fig. 2-20. Dragging the Point fly-out toolbar.

Input for Individual Point and Basic Curve Construction

To construct individual point objects and basic curves, you use a command and specify a location. To specify a location, you use your pointing device to select a point in one of the viewports, or input a set of coordinates at the command line interface.

As explained in Chapter 1, there is a construction plane in each of the viewports. Selecting a point in one of the viewports specifies a location on the corresponding construction plane. To key in a set of coordinates, you use either the construction plane coordinate system or the World coordinate system.

The X and Y directions of the construction plane coordinates correspond to the X (red in color) and Y (green in color) axes of the construction plane. Therefore, typing the same construction plane coordinates (other than the origin) at the command line interface results in different locations in 3D space, depending on where the cursor is placed; for example, on the Top viewport as opposed to the Front viewport. In the following exercises you will use the pointing device to select points in viewports.

Construction of Point Objects and Basic Curves

A point is a node. Basic curves follow basic geometric patterns. There are eleven types of basic curves: line, free-form, circle, arc, ellipse, parabola, polygon, rectangle, conic, helix, and spiral. These eleven types of curves are shown in figures 2-21 through 2-23.

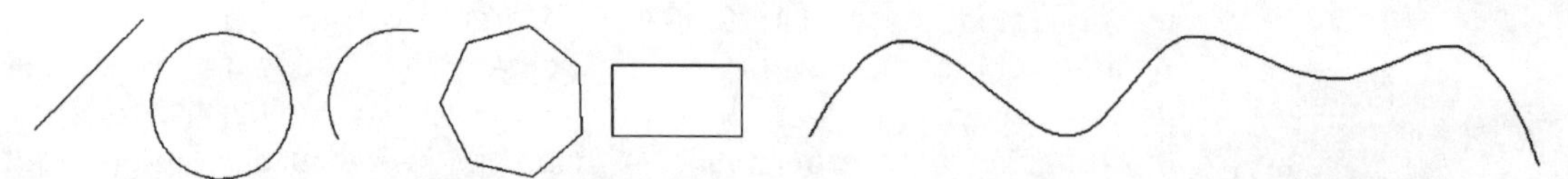

Fig. 2-21. Line, circle, arc, polygon, and rectangle (left to right).

Fig. 2-22. Free-form curve.

Fig. 2-23. Ellipse, parabola, conic, helix, and spiral (left to right).

Point Object

Figure 2-24 shows the Point toolbar (showing various point construction commands) and the V3 Repository toolbar (showing the command for constructing a point cloud section). Using point objects, you specify locations in 3D space, define vertices and definition points of curves, and portray locations on a surface. Point construction commands (including the point cloud section) are discussed in the sections that follow.

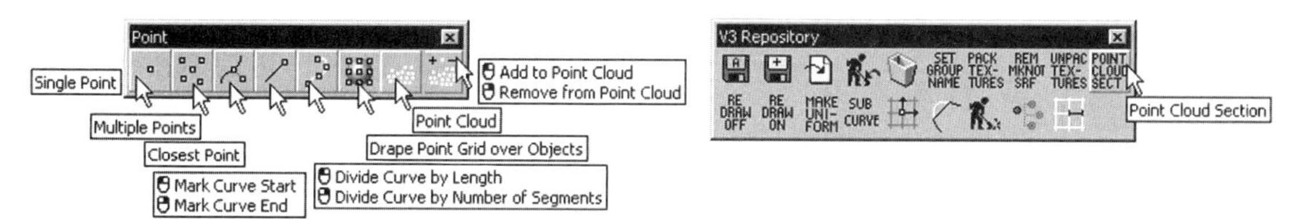

Fig. 2-24. Point toolbar and V3 Repository toolbar.

Table 2-1 outlines the functions of point and point cloud construction commands and their locations in the main pull-down menu and respective toolbars.

Table 2-1: Point Construction Commands and Their Functions

Toolbar Button	From Main Pull-down Menu	Function
Single Point	Curve > Point Object > Single Point	Constructing a single point object
Multiple Points	Curve > Point Object > Multiple Points	Constructing a series of point objects
Closest Point	Curve > Point Object > Closest Point	Constructing a point on a curve that is closest to a specified location
Drape Point Grid Over Objects	Curve > Point Object > Drape Point	Constructing a matrix of points on the surfaces described by a rectangle
Mark Curve Start	Curve > Point Object > Mark Curve Start	Constructing a point at the start point of a curve
Mark Curve End	Curve > Point Object > Mark Curve End	Constructing a point at the end point of a curve
Divide Curve by Length	Curve > Point Object > Divide Curve by > Length of Segments	Constructing a series of points along a curve by specifying the distances between consecutive points
Divide Curve by Number of Segments	Curve > Point Object > Divide Curve by > Number of Segments	Constructing a series of points along a curve by specifying the number of segments

Toolbar Button	From Main Pull-down Menu	Function
Point Cloud	Curve > Point Cloud > Create Point Cloud	Grouping a set of selected points to form a single point cloud object
Add to Point Cloud	Curve > Point Cloud > Add Points	Adding selected points to a point cloud object
Remove from Point Cloud	Curve > Point Cloud > Remove Points	Removing selected points from a point cloud object
Point Cloud Section	Curve > Point Cloud > Point Cloud Section	Generating curves from a point cloud object

Multiple Points

To construct a set of point objects for making curves, perform the following steps.

1. Start a new file. Use the metric (Millimeters) template file.
2. Double click on the Top viewport title to maximize the viewport.
3. Select Curve > Point Object > Multiple Points, or click on the Multiple Points button on the Point toolbar.
4. Select locations A, B, C, and D in accordance with figure 2-25. (The exact location of points in this exercise is unimportant.)
5. Press the Enter key to terminate the command.

Four point objects are constructed.

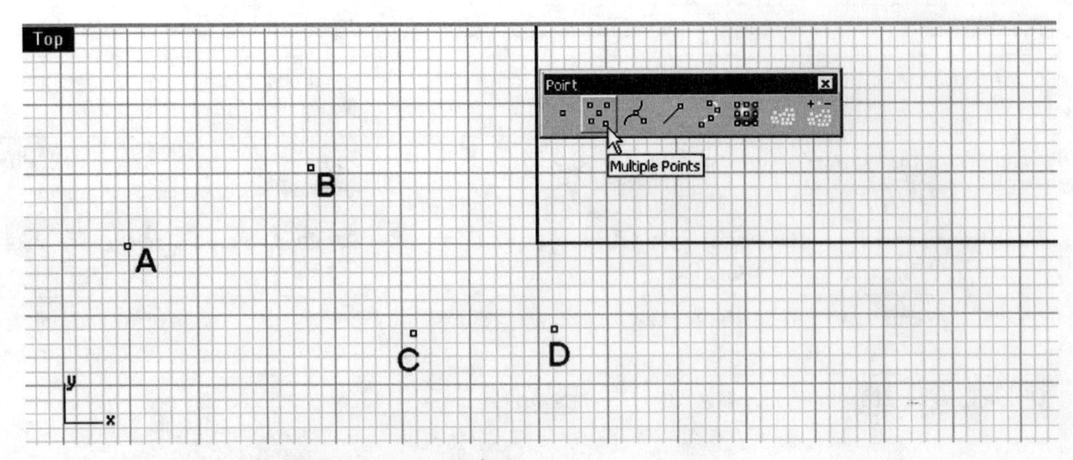

Fig. 2-25. Four point objects.

NOTE: *If you are using the Evaluation version, the number of times you can perform a save is limited to 25. To be able to see the result of* *your work on this surface model in later stage of this book, do not save your files until you work on the projects in Chapter 9. Opening a new file, starting a new file, or exiting Rhino automatically closes your current file.*

6 If you want to save your file, save it as *Point1.3dm.*

Points Near a Curve and Along a Curve

To construct a point object near a curve and along a curve, perform the following steps.

1 Start a new file. Use the metric (Millimeters) template file.
2 Double click on the Top viewport title to maximize the viewport.
3 Select Curve > Free-Form > Interpolate Points, or click on the *Curve: Interpolate Points* button on the Curve toolbar.
4 Select locations A, B, C, D, and E, indicated in figure 2-26, and then press the Enter key. (The exact location of points in this exercise is unimportant.)

A free-form curve is constructed.

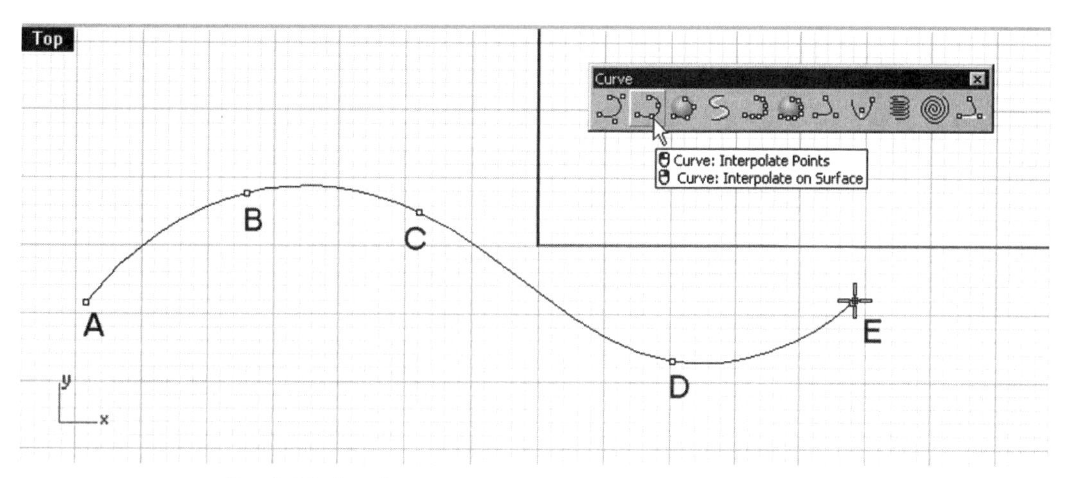

Fig. 2-26. Free-form curve.

5 Select Curve > Point Object > Closest Point, or click on the Closest Point button on the Point toolbar.
6 Select curve A, indicated in figure 2-27, and press the Enter key.
7 Select location B, indicated in figure 2-27.

A point closest to the selected location is constructed along the curve.

8 Select Edit > Undo to undo the last command.

Construct a point at the start point and a point at the end point of the curve, as follows.

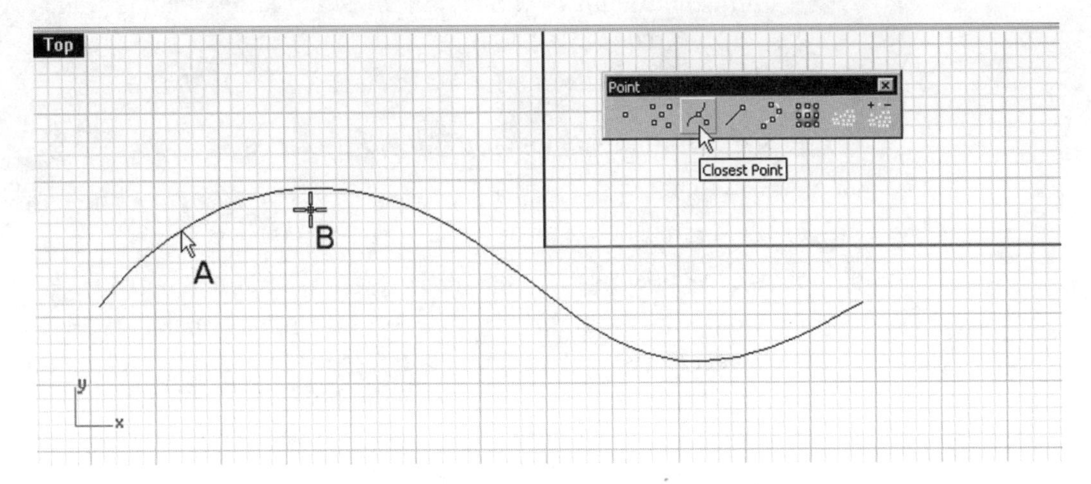

Fig. 2-27. Point closest to a curve being constructed.

9 Select Curve > Point Object > Mark Curve Start, or left-click on the *Mark Curve Start/Mark Curve End* button on the Point toolbar.
10 Select curve A, indicated in figure 2-28, and press the Enter key.
11 Select Curve > Point Object > Mark Curve End, or right-click on the *Mark Curve Start/Mark Curve End* button on the Point toolbar.
12 Select curve A, indicated in figure 2-28, and press the Enter key.

Points are constructed at the start point and end point of the curve.

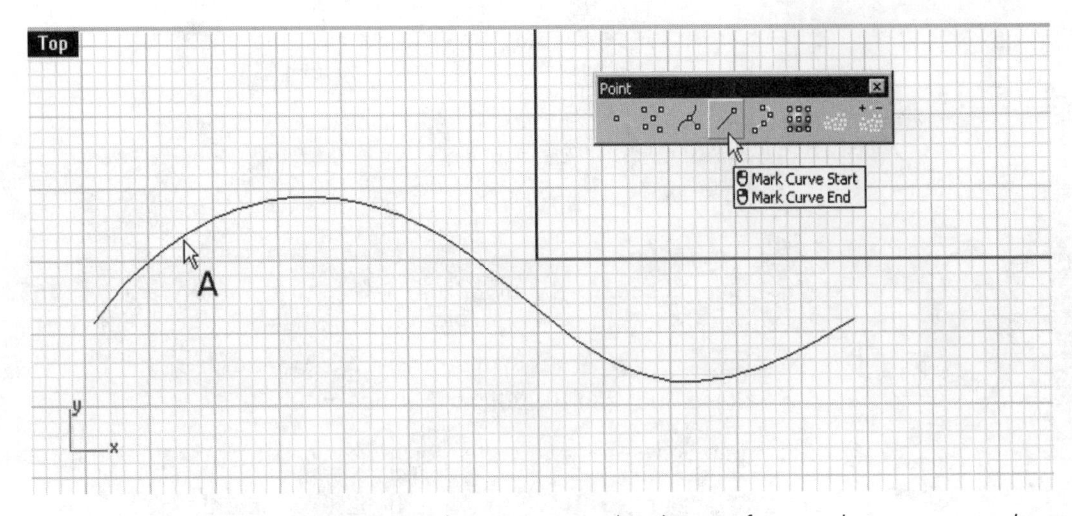

Fig. 2-28. Points at the start point and end point of a curve being constructed.

13 Select Edit > Undo twice to undo the last two points. (Do not over-undo.)

14 Construct a number of points along the curve, as follows.

15 Select Curve > Divide Curve by > Length of Segments, or left-click on the *Divide Curve by Length/Divide Curve by Number of Segments* button on the Point toolbar.

16 Select curve A, indicated in figure 2-29, and press the Enter key.

17 Type *10* at the command line interface to specify a length.

A set of points at 10-unit intervals are constructed along the curve.

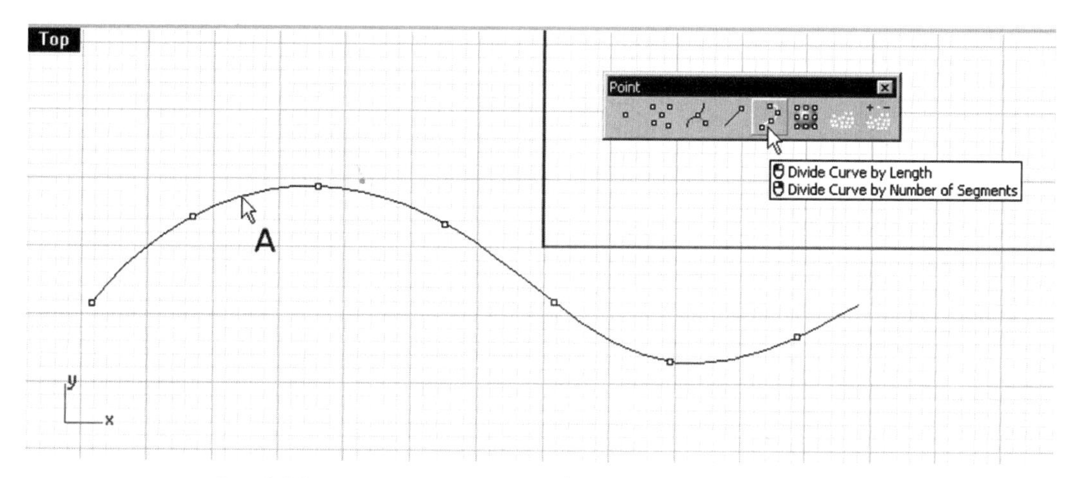

Fig. 2-29. Points at 10-unit intervals.

18 Select Edit > Undo to undo the last command.

19 Select Curve > Divide Curve by > Number of Segments, or right-click on the *Divide Curve by Length/Divide Curve by Number of Segments* button on the Point toolbar.

20 Select curve A, indicated in figure 2-30, and press the Enter key.

21 Type *6* at the command line interface to specify 6 segments.

A total of 7 points are constructed, as shown in figure 2-30.

22 If you want to save your file, save it as *Point2.3dm*.

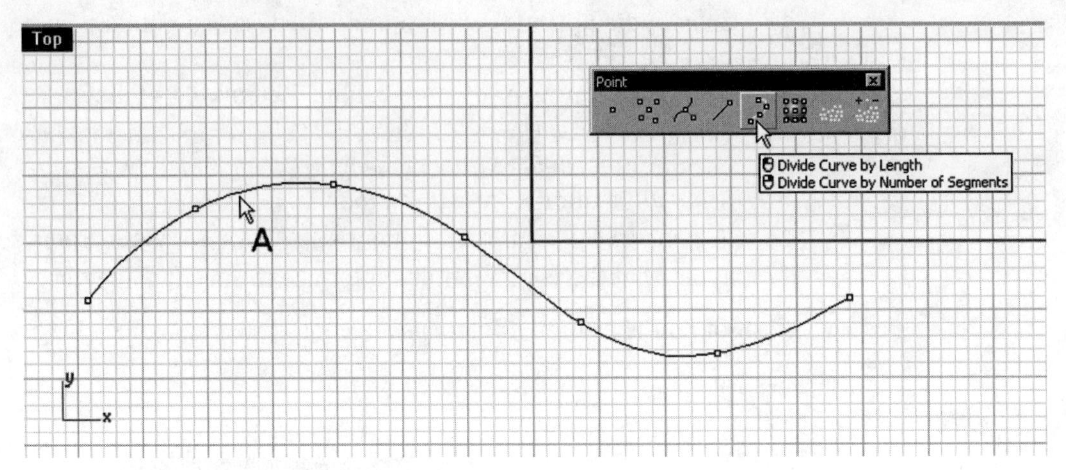

Fig. 2-30. Points equally spaced along the curve.

Drape Points and Point Cloud Manipulation

To construct drape points and manipulate a point cloud, perform the following steps.

1. Select File > Open.
2. In the *Files of type* field of the Open dialog box, select IGES (*.igs*, *.iges*), and select the file *PointDrape.igs* from the *Chapter 2* folder on the companion CD-ROM.
3. Double click on the Top viewport title to maximize the viewport.
4. Click on Zoom Extents on the Standard toolbar.
5. Select Curve > Point Object > Drape Points, or click on the Drape Point Grid Over Objects button on the Point toolbar.
6. Select points A and B, indicated in figure 2-31.

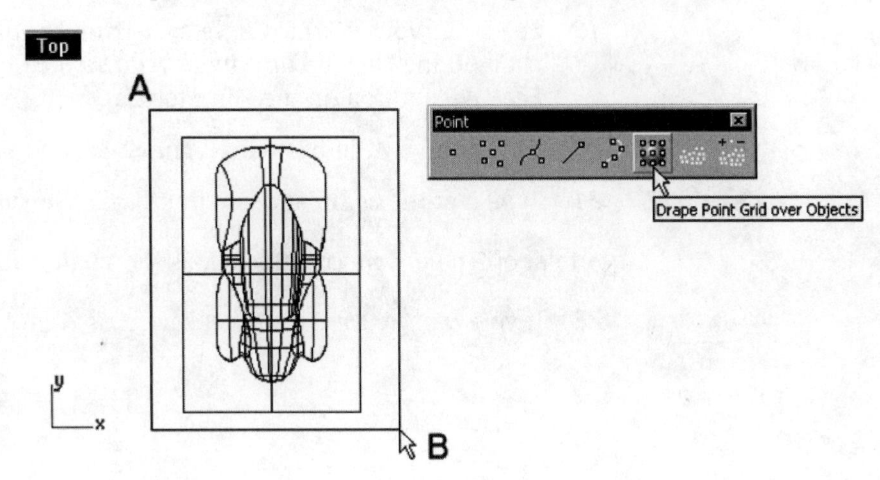

Fig. 2-31. Drape points being constructed.

7. Select Edit > Layers > Edit Layers, or click on the Edit Layers button on the Standard toolbar.
8. In the Layers dialog box, check the *Hidden layer* option along the *IGES level 0* layer to turn off the layer, as shown in figure 2-32.
9. Click on the Checkmark icon on the dialog box to close it.

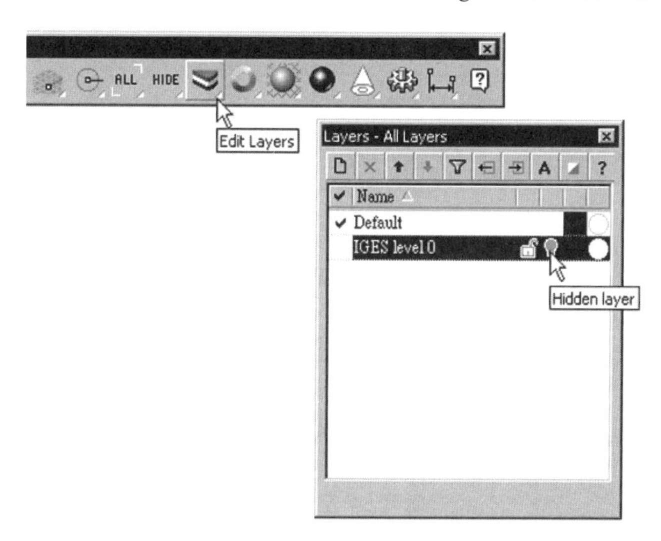

Fig. 2-32. A layer being turned off.

A drape point grid is constructed and a layer is turned off. Construct a point cloud set, as follows.

10. Select Curve > Point Cloud > Create Point Cloud, or click on the Point Cloud button on the Point toolbar.
11. Select locations A and B, indicated in figure 2-33, and press the Enter key.

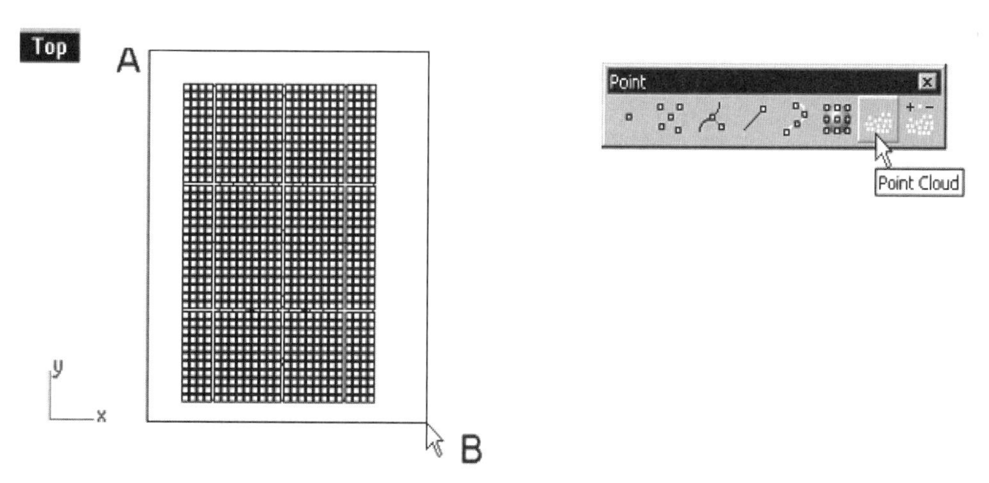

Fig. 2-33. Point cloud being constructed from a set of selected points.

12 Construct curves from the point cloud, as follows.

13 Select Curve > Point Cloud > Point Cloud Section, or click on the Pont Cloud Section button on the V3 Repository toolbar.

14 Select the point cloud and press the Enter key.

15 Click on the OK button on the Point Cloud Section Options dialog box.

16 Select locations A and B, indicated in figure 2-34, to describe a section line across the point cloud. (If the section line is too far away from the points, a prompt "Too few points within minimum distance of requested section plane" will be displayed at the command line interface. If you see this message, try again.)

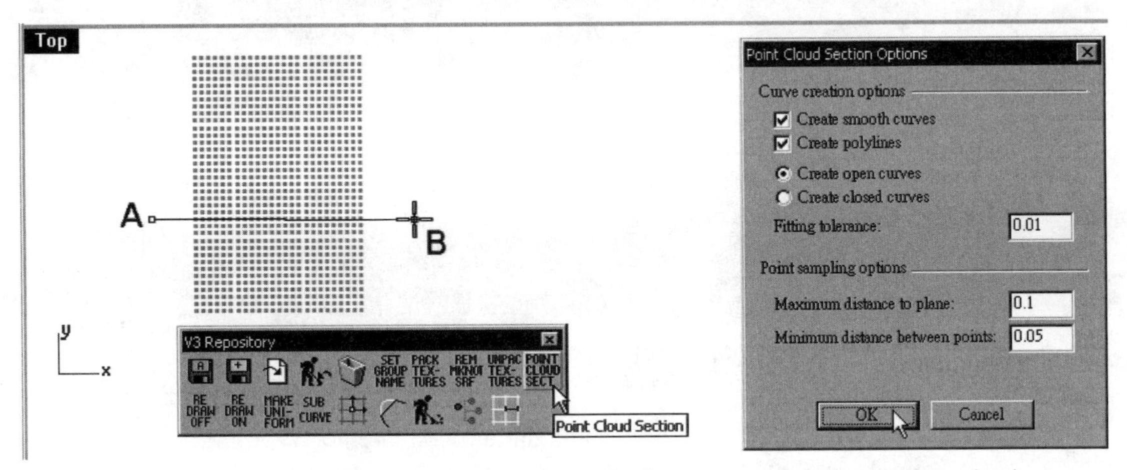

Fig. 2-34. Point cloud section curves being constructed from a point cloud.

17 Continue to select pairs of start and end points.

18 Press the Enter key when finished.

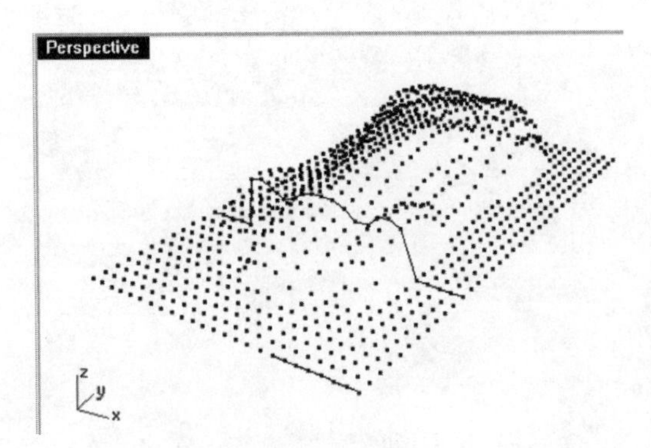

Fig. 2-35. Point cloud curves.

19 Double click on the Top viewport title to return to a four-viewport configuration, and then double click on the Perspective viewport title to maximize the viewport.

A number of curves are constructed, as shown in figure 2-35.

20 If you want to save your file, save it as *PointDrape.3dm*.

In the previous steps, opening an IGES file to obtain a set of surfaces, constructing a point cloud from the surfaces, and constructing point cloud curves serve to demonstrate how drape points, point clouds, and point cloud curves are constructed.

In reality, point clouds are usually obtained by digitizing a real physical model for the purpose of reconstructing the surfaces of the model in the computer. You will learn the use of digitizing in Chapter 4.

Line, Polyline, and Arc

Both line segments and polylines are degree 1 curves. A line or a set of contiguous line segments are separate objects, but a polyline is a single object consisting of a set of joined contiguous line segments. An arc is a degree 2 curve. Figure 2-36 shows the Lines toolbar and the Arc toolbar. You use the tools found on these toolbars to construct lines, polylines, and arcs.

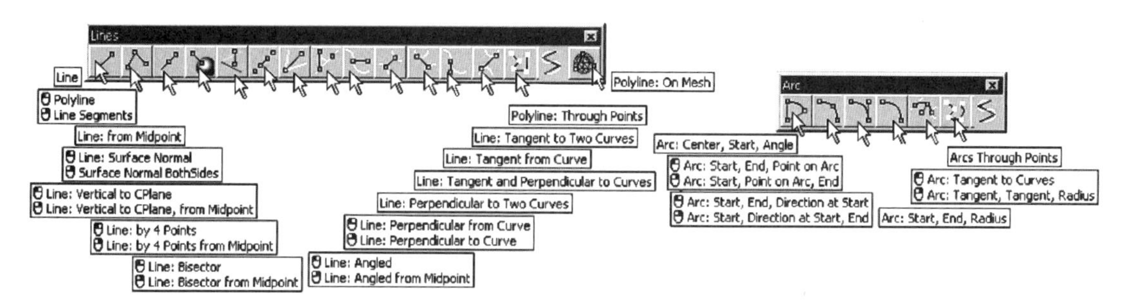

Fig. 2-36. Lines toolbar and Arc toolbar.

Table 2-2 outlines the functions of line, polyline, and arc construction commands and their locations in the main pull-down menu and respective toolbars.

Table 2-2: Line, Polyline, and Arc Construction Commands and Their Functions

Toolbar Button	From Main Pull-down Menu	Function
Line	Curve > Line > Single Line	Constructing a single line segment by specifying two end points or a midpoint and an end point
Polyline	Curve > Polyline > Polyline	Constructing a string of connected line segments
Line Segments	Curve > Line > Line Segments	Constructing a string of line segments

Toolbar Button	From Main Pull-down Menu	Function
Line: from Midpoint	—	Constructing a line by specifying its midpoint and one of the end points
Line: Surface Normal	Curve > Line > Normal to Surface	Constructing a line normal to a selected surface
Surface Normal Both Sides	—	Constructing a line normal to a selected surface by specifying the midpoint and one of the end points
Line: Vertical to CPlane	Curve > Line > Vertical to CPlane	Constructing a line normal to the current construction plane
Line: Vertical to CPlane, from Midpoint	—	Constructing a line normal to the current construction plane by specifying the midpoint and one of the end points
Line: by 4 Points	Curve > Line > From 4 Points	Constructing a line by using two reference points to specify a direction and selecting two end points on the direction line
Line: by 4 Points from Midpoint	—	Constructing a line by using two reference points to specify a direction and selecting two end points on the direction line to indicate the start point and one of the end points of the line
Line: Bisector	Curve > Line > Bisector	Constructing a line bisecting two selected reference lines
Line: Bisector, from Midpoint	—	Constructing a line bisecting two selected reference lines by specifying the midpoint and one of the end points
Line: Angled	Curve > Line > Angled	Constructing a line at a specified angle to a reference line
Line: Angled from Midpoint	—	Constructing a line at a specified angle to a reference line by specifying its midpoint and one of the end points
Line: Perpendicular from Curve	Curve > Line > Perpendicular from Curve	Constructing a line perpendicular to a selected curve
Line: Perpendicular to Two Curves	Curve > Line > Perpendicular to 2 Curves	Constructing a line perpendicular to two selected curves
Line: Tangent from Curve	Curve > Line > Tangent from Curve	Constructing a line tangent to a selected curve
Line: Tangent to Two Curves	Curve > Line > Tangent to 2 Curves	Constructing a line tangent to two selected curves

Toolbar Button	From Main Pull-down Menu	Function
Polyline: Through Points	Curve > Polyline > Through Points	Constructing a curve through a set of selected points (polyline if degree of polynomial is set to 1)
Line: Tangent and Perpendicular to Curves	Curve > Line > Tangent, Perpendicular	Constructing a line tangent to a curve and perpendicular to another curve
Polyline: On Mesh	Curve > Polyline > Polyline on Mesh	Constructing a polyline on the face of a polygon mesh
Arc: Center, Start, Angle	Curve > Arc > CENTER, Start, Angle	Constructing an arc by specifying the center, the start point, and the arc's angle
Arc: Start, End, Point on Arc	Curve > Arc > Start, End, Point	Constructing an arc by specifying the start point, the end point, and a point along the arc
Arc: Start, End, Direction at Start	Curve > Arc > Start, End, Direction	Constructing an arc by specifying the start point, the end point, and the start point's tangent direction
Arc: Start, End, Radius	Curve > Arc > Start, End, Radius	Constructing an arc by specifying the start point, end point, radius, and orientation of the arc
Arc: Tangent, Tangent, Radius	Curve > Arc > Tangent, Tangent, Radius	Constructing an arc with a specified radius and tangent to two curves
Arc: Start, Point on Arc, End	Curve > Arc > Start, Point, End	Constructing an arc by specifying the start point, a point along the arc, and the end point

Line Segments and Polylines

To construct a single line segment by specifying two points, perform the following steps.

1 Start a new file. Use the metric (Millimeters) template.

2 Select Curve > Line > Single Line, or click on the Line button on the Lines toolbar.

3 Select point A in the Top viewport and point B in the Front viewport, as shown in figure 2-37.

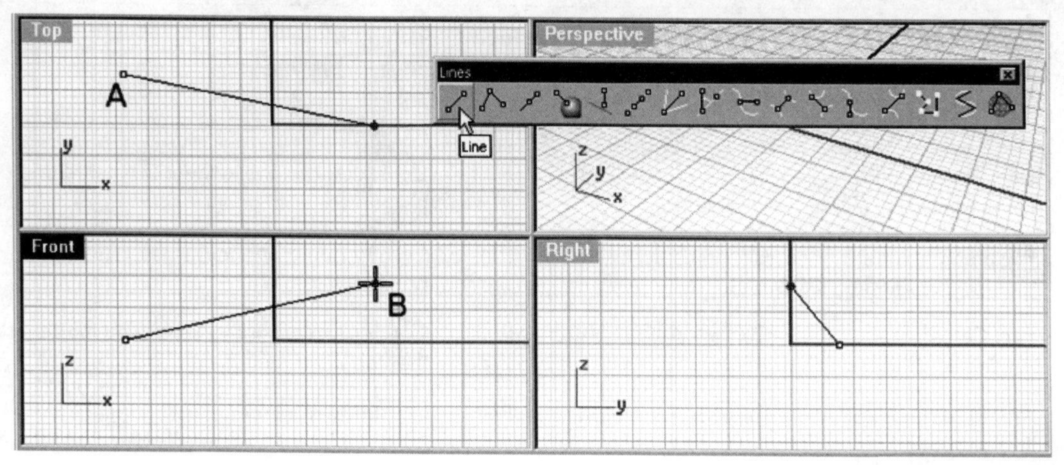

Fig. 2-37. Single line segment.

A single line segment with its end points defined by selected points on two construction planes is constructed. To construct a line perpendicular to the construction plane of the Top viewport, continue with the following steps.

4 Select Curve > Line > Vertical to CPlane, or *left*-click on the *Line: Vertical to CPlane/Line: Vertical to CPlane, from Midpoint* button on the Lines toolbar.

5 Select point A in the Top viewport to specify a location, and select point B in the Front viewport to specify the height. This creates a line perpendicular to the construction plane, as shown in figure 2-38. (Height is defined by the distance between the first selected point and the second selected point.)

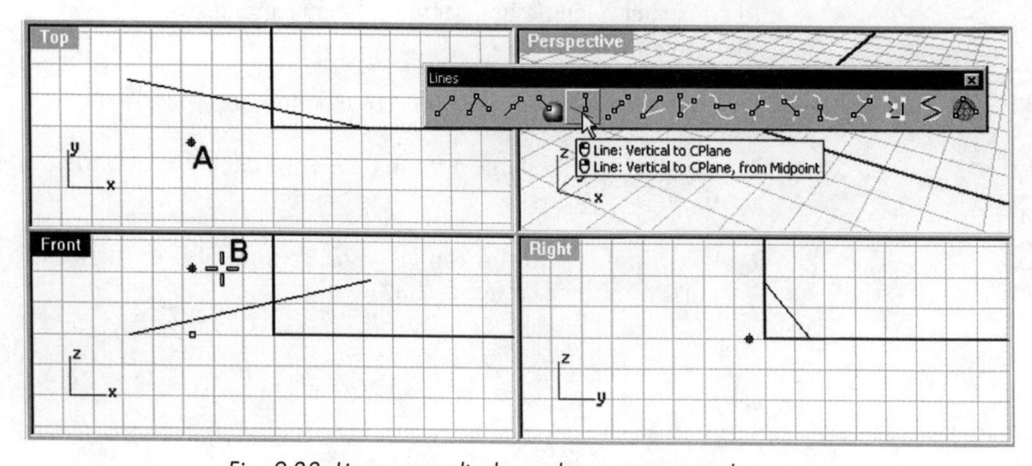

Fig. 2-38. Line perpendicular to the construction plane.

6 Hold down the Ctrl key and select the two line segments you just constructed.
7 Select Edit > Delete.

To construct a set of connected line segments, continue with the following steps.

8 Maximize the Top viewport.
9 Select Curve > Line > Line Segments, or right-click on the *Polyline/Line Segments* button on the Lines toolbar.
10 Select points A, B, C, D, and E, indicated in figure 2-39, and then press the Enter key.

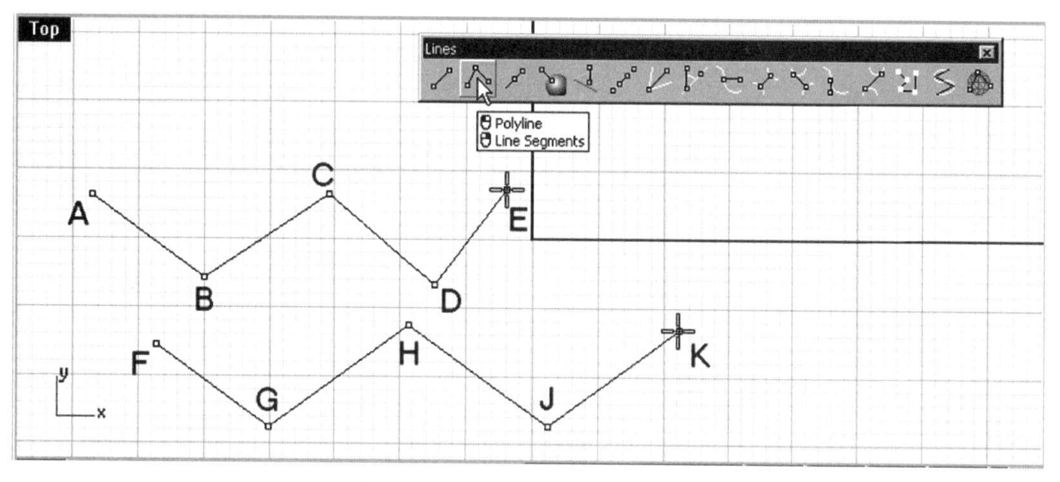

Fig. 2-39. Set of connected line segments and a polyline.

To construct a polyline, continue with the following steps.

11 Select Curve > Polyline > Polyline, or left-click on the *Polyline/Line Segments* button on the Lines toolbar.
12 Select points F, G, H, J, and K, indicated in figure 2-39, and then press the Enter key.

Connected Line Segments and Polylines

At a glance, the connected line segments and the polyline seem to be similar. In fact, the polyline is a single object and the connected line segments are separate line segments with their end points coincident with each other. To separate the polyline into individual line segments, you explode it. To convert a set of connected line segments to a single polyline, you join the line segments, as follows.

1 Select Edit > Join, or click on the Join button on the Main1 toolbar.
2 Select the connected line segments AB, BC, CD, and DE, shown in figure 2-39, and then press the Enter key.

The line segments are joined to become a polyline.

3 Select Edit > Explode, or left-click on the *Explode/Extract Surfaces* button on the main toolbar.
4 Select the polyline FGHJK and press the Enter key.

The polyline is exploded to show individual line segments.

5 If you want to save your file, save it as *Lines.3dm*.

Four-point Line

To construct a line in a direction defined by two points with the end points nearest to two selected points (i.e., a four-point line), perform the following steps.

1 Start a new file. Use the metric (Millimeters) template.
2 With reference to figure 2-40, construct four point objects A, B, C, and D.
3 Check the Osnap pane on the status bar to display the Osnap dialog box.
4 In the Osnap dialog box, check the Point box to establish object snap mode. This way, the cursor will snap to point objects.

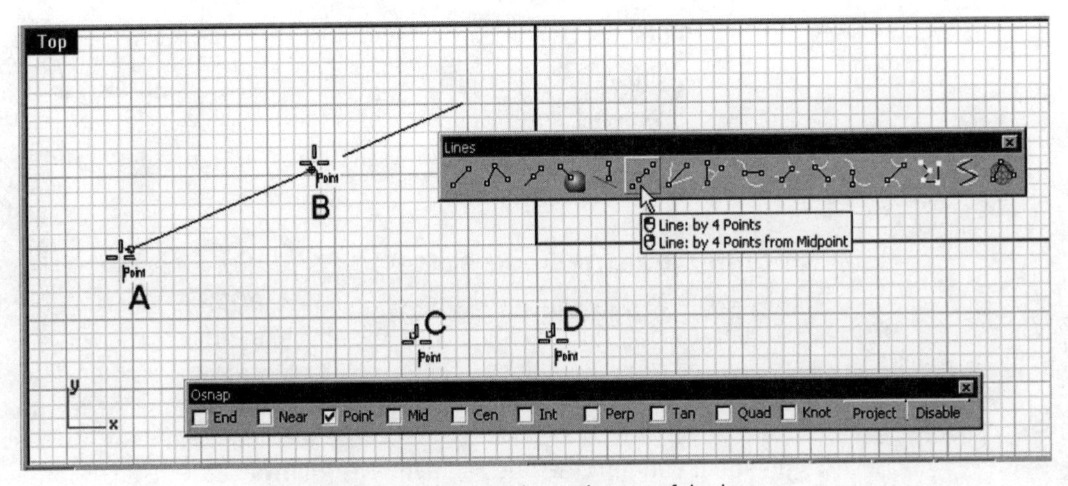

Fig. 2-40. Point indicating the end point of the line.

5 Select Curve > Line > From 4 Points, or left-click on the *Line: by 4 Points/Line: by 4 Points from Midpoint* button on the Lines toolbar.

6 Select points A and B, indicated in figure 2-40, to indicate a direction.

7 Select points B and C, indicated in figure 2-40, to specify the start point and end point of the line.

A line is constructed.

8 If you want to save your file, save it as *Lines4pt.3dm.*

Line at an Angle

To construct a line at an angle to a reference line, perform the following steps.

1 Start a new file. Use the metric (Millimeters) template.
2 Maximize the Top viewport.
3 Check the Osnap pane in the status bar.
4 In the Osnap dialog box, check the End box and clear the other boxes.
5 Select Curve > Line > Single Line, or click on the Line button on the Lines toolbar.
6 Select points A and B, indicated in figure 2-41, to construct a line segment.

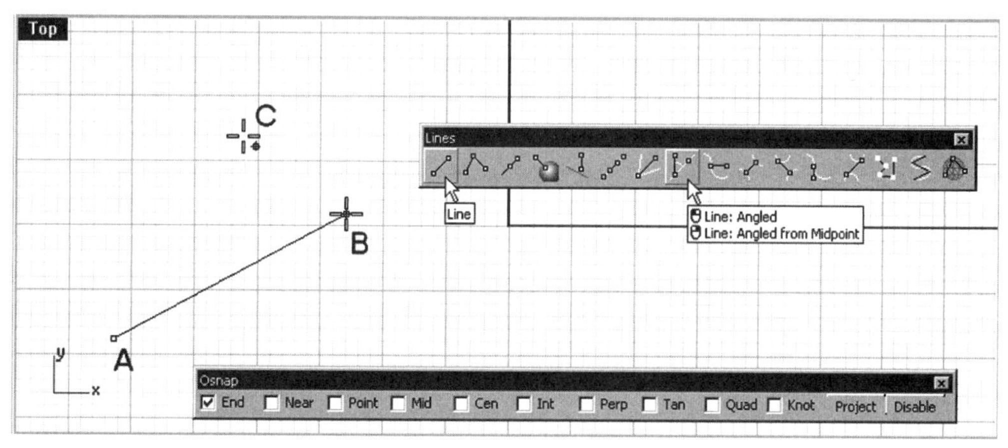

Fig. 2-41. Line segment.

7 Select Curve > Line > Angled, or left-click on the *Line: Angled Line/Line: Angled from Midpoint* button on the Lines toolbar.
8 Select end points A and B, indicated in figure 2-41.

9 Type *25* to specify an angle.
10 Select point C, indicated in figure 2-41.

A line at 25 degrees on the selected line is constructed.

Line Bisector

To construct a line bisector, perform the following steps.

1 Select Curve > Line > Bisector, or left-click on the *Line: Bisector/ Line: Bisector from Midpoint* button on the Lines toolbar.
2 Select end points A and B, indicated in figure 2-42, to specify the first reference line (start of the bisector line and start of the angle to be bisected).
3 Select end points A and C, indicated in figure 2-42, to indicate the end of the angle to be bisected.
4 Select point D, indicated in figure 2-42.

A line bisector is constructed.

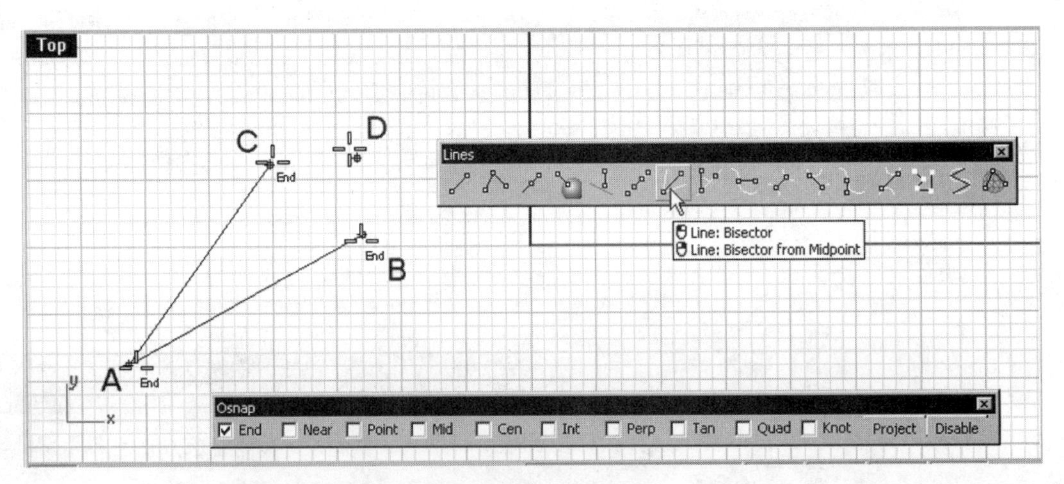

Fig. 2-42. Line bisector being constructed.

Polyline Through a Set of Points

To construct a polyline passing through a set of points, perform the following steps.

1 Select Curve > Point Objects > Multiple Points, or click on the Multiple Points button on the Point toolbar.

2 Select locations A, B, C, D, and E, indicated in figure 2-43, and then press the Enter key.
3 Select Curve > Polyline > Polyline Through Points, or click on the Polyline Through Points button on the Lines toolbar.
4 Select points A, B, C, D, and E, indicated in figure 2-43, and then press the Enter key.

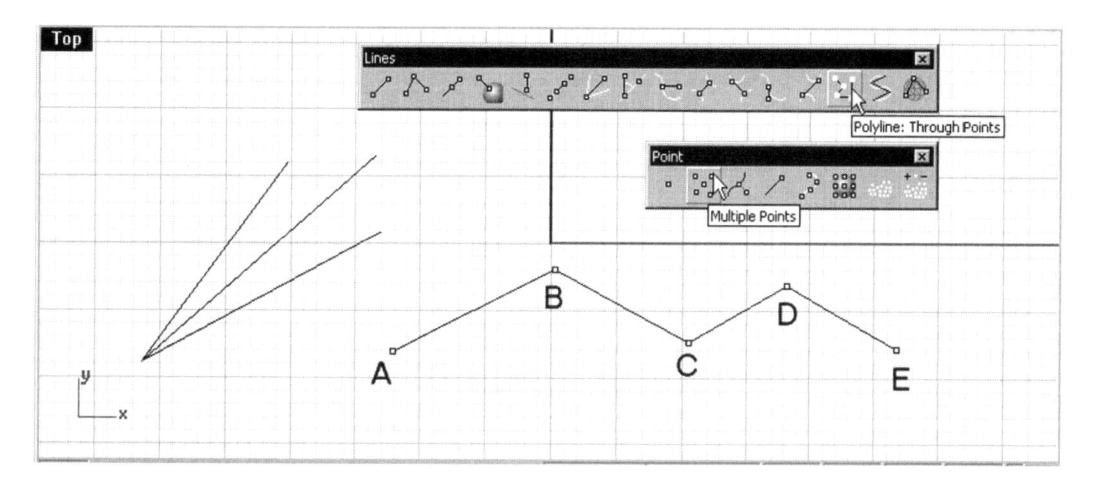

Fig. 2-43. Points.

A polyline is constructed. Note that this command is the same command as Curve > Free-Form > Fit To Points, with the exception that the polyline command has the curve's degree of polynomial set to 1 and the free-form curve command set to 3.

5 If you want to save your file, save it as *LinePt.3dm.*

Line Perpendicular to a Curve

To construct a line perpendicular to a curve, perform the following steps.

1 Start a new file. Use the metric (Millimeters) template.
2 Maximize the Top viewport.
3 Select Curve > Free-Form > Interpolate Points, or left-click on the *Curve: Interpolate Points/Curve: Interpolate on Surface* button on the Curve toolbar.
4 Select locations A, B, C, D, and E, indicated in figure 2-44, and press the Enter key to construct a U-shape free-form curve.
5 Press the Enter key to repeat the command.
6 Select locations F, G, and H, indicated in figure 2-44, and press the Enter key to construct another free-form curve.

7 Select Curve > Point Object > Single Point, or click on the Single Point button on the Point toolbar.

8 Select location J, indicated in figure 2-44.

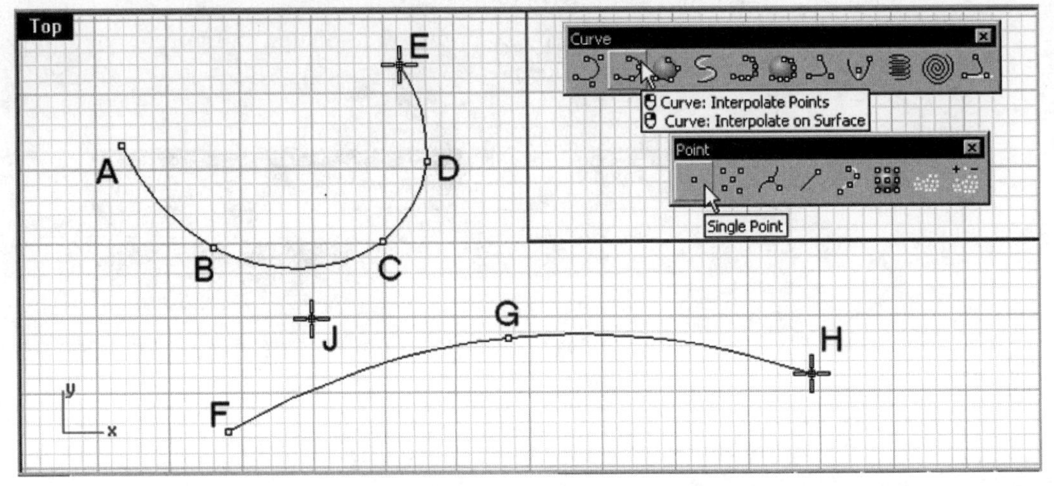

Fig. 2-44. Curves and point.

Construct a line from the point perpendicular to a curve, as follows.

9 Click on the Point button on the Osnap toolbar.

10 Select Curve > Line > Perpendicular from Curve, or right-click on the *Line: Perpendicular from Curve/Line: Perpendicular to Curve* button on the Lines toolbar.

11 Select point A and then curve B, indicated in figure 2-45.

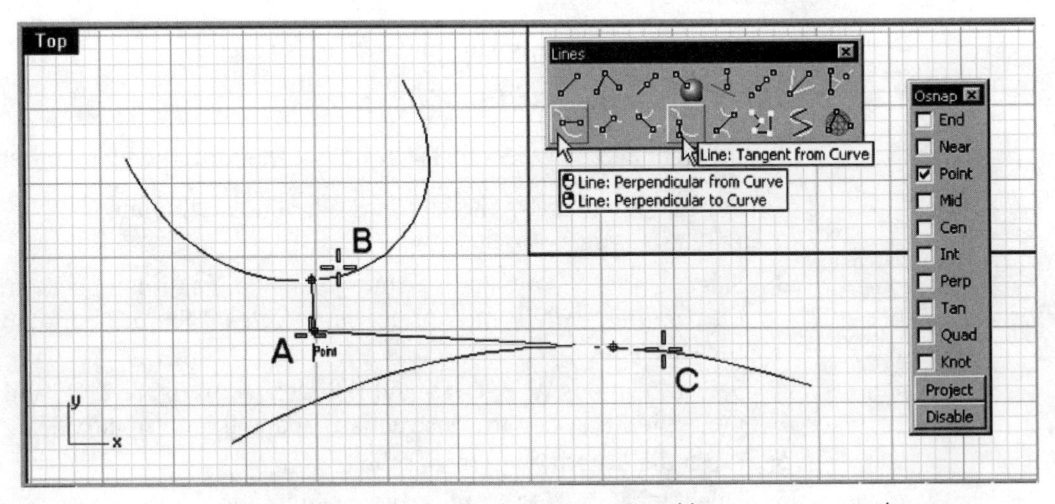

Fig. 2-45. Line perpendicular to a curve and line tangent to another curve.

Line Tangent to a Curve

To construct a line tangent to a curve, perform the following steps.

1. Select Curve > Line > Tangent from Curve, or click on the *Line: Tangent from Curve* button on the Lines toolbar.
2. Select point A and then curve C, indicated in figure 2-45.

Line Perpendicular to Two Curves

To construct a line perpendicular to two curves, perform the following steps.

1. Select Curve > Line > Perpendicular to 2 Curves, or click on the Line: Perpendicular to Two Curves button on the Lines toolbar.
2. Select curve A and then curve B, indicated in figure 2-46.

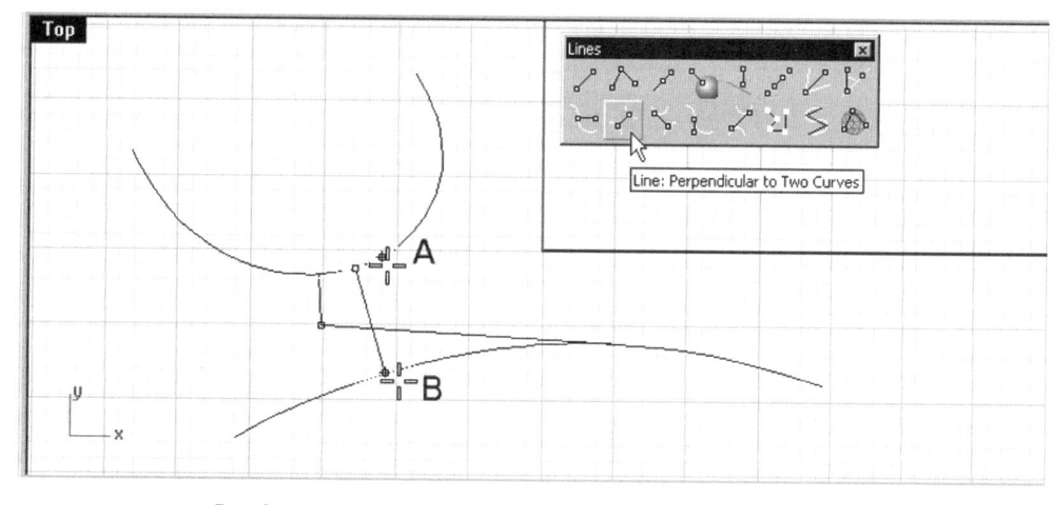

Fig. 2-46. Line perpendicular to two curves.

Line Tangent to Two Curves

To construct a line tangent to two curves, perform the following steps.

1. Select Curve > Line > Tangent to 2 Curves, or click on the *Line: Tangent to Two Curves* button on the Lines toolbar.
2. Select curves A and B, indicated in figure 2-47.

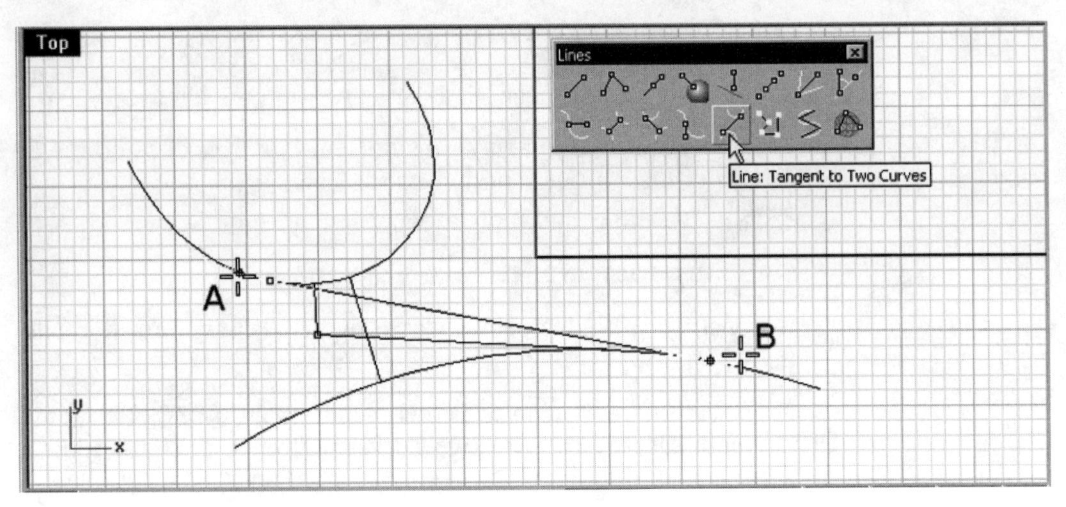

Fig. 2-47. Line tangent to two curves.

Line Tangent to a Curve and Perpendicular to Another Curve

To construct a line tangent to a curve and perpendicular to another curve, perform the following steps.

1. Select Curve > Line > Tangent, Perpendicular, or click on the *Line: Tangent and Perpendicular to Curves* button on the Lines toolbar.
2. Select curve A where the tangent point will be placed, as shown in figure 2-48.

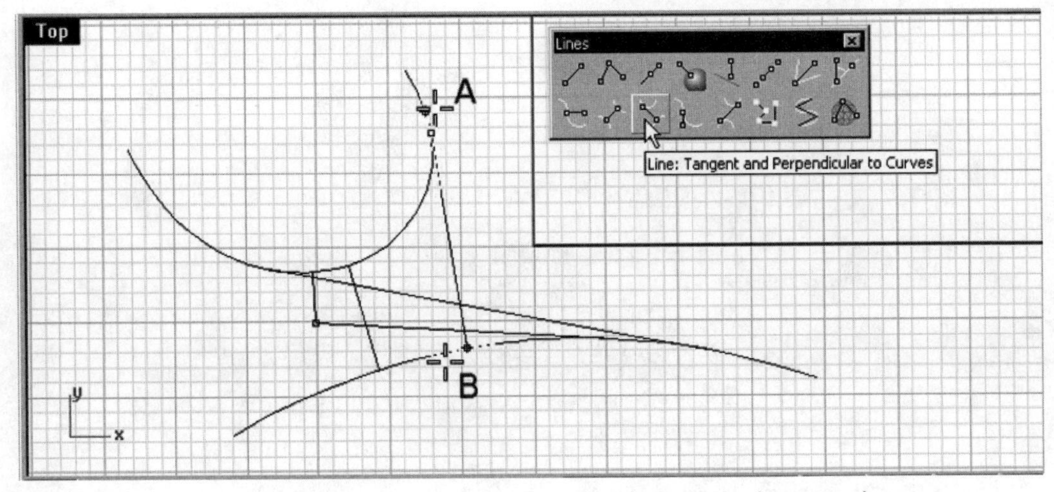

Fig. 2-48. Line tangent to a curve and perpendicular to another curve.

3 Select curve B where the perpendicular point will be placed, as shown in figure 2-48.

4 If you want to save your file, save it as *Lines2.3dm*.

Arc

As explained previously, there are many ways to construct an arc. In the following, you will construct arcs in two ways.

1 Start a new file. Use the metric (Millimeters) template.

2 Maximize the Top viewport.

3 Select Curve > Arc > Center, Start, Angle, or click on the *Arc: Center, Start, Angle* button on the Arc toolbar.

4 Select points A, B, and C, indicated in figure 2-49, to specify the center, the start point, and the angle of the arc.

5 Select Curve > Arc > Start, End, Direction, or click on the *Arc: Start, End, Direction at Start/Arc: Start, Direction at Start, End* button on the Arc toolbar.

6 Select points D, E, and F, indicated in figure 2-49, to specify the start point, the end point, and the direction of the arc at the start point.

Two arcs are constructed.

7 If you want to save your file, save it as *Arc.3dm*.

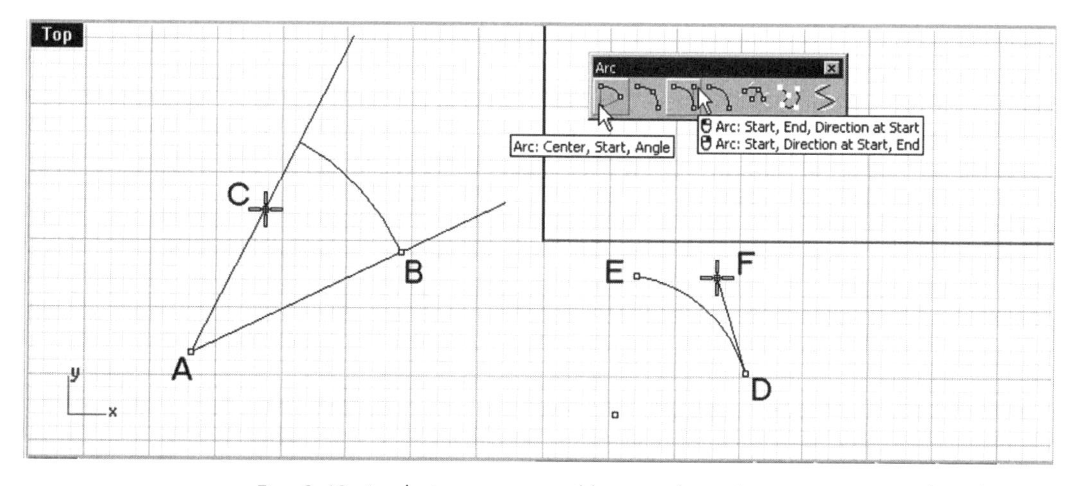

Fig. 2-49. Arc being constructed by specifying the center, start, and angle.

Free-form Curve

A flexible way to define a curve for making free-form surfaces is to use a free-form curve. A free-form curve is a very important tool for making surfaces. By default, a free-form curve is a degree 3 NURBS curve. In the next chapter, you will learn how to increase or decrease its polynomial degree.

There are eight ways to define a free-form curve. Figure 2-50 shows the Curve toolbar (which contains buttons for seven types of free-form curves, as well as conic, parabola, helix, and spiral), the Lines toolbar (and its option Convert Curve to Polyline), and the Arc toolbar (and its option Convert Curve to Arcs).

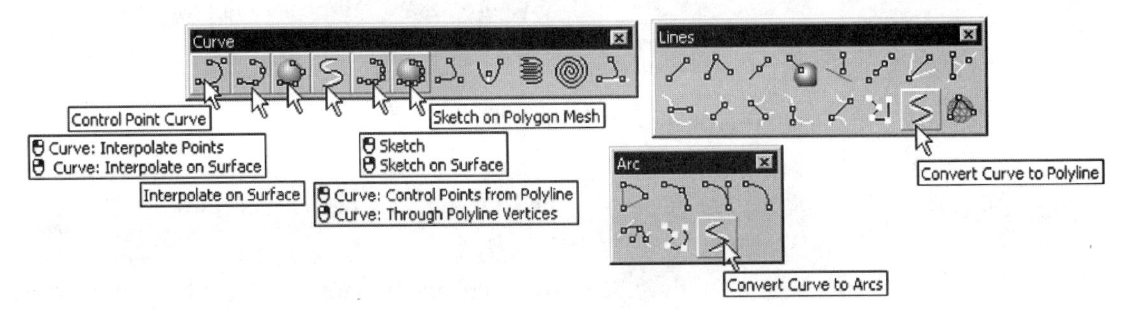

Fig. 2-50. Curve, Lines, and Arc toolbars.

Table 2-3 outlines the functions of free-form curve commands and their locations in the main pull-down menu and respective toolbars.

Table 2-3: Free-form Curve Commands and Their Functions

Toolbar Button	From Main Pull-down Menu	Function
Control Point Curve	Curve > Free-Form > Control Points	Constructing a spline curve by specifying control points
Curve: Interpolate Points	Curve > Free-Form > Interpolate Points	Constructing a spline curve passing through specified points
Sketch	Curve > Free-Form > Sketch	Using the pointing device (mouse) as a drawing pen to sketch a spline curve
Interpolate on Surface	Curve > Free-Form > Interpolate on Surface	Constructing a spline curve passing through specified points on a selected surface
Sketch on Surface	Curve > Free-Form > Sketch on Surface	Using the pointing device (mouse) as a drawing pen to sketch a spline curve on a selected surface

Toolbar Button	From Main Pull-down Menu	Function
Sketch on Polygon Mesh	Curve > Free-Form > Sketch on Polygon Mesh	Using the pointing device (mouse) as a drawing pen to sketch a spline curve on a polygon mesh
Curve: Through Polyline Vertices	Curve > Free-Form > Fit to Points	Constructing a spline curve passing through selected point objects
Curve: Control Points from Polyline	Curve > Free-Form > Fit to Polyline	Constructing a spline curve with its control point passing through the end points of a selected polyline
Convert Curve to Polyline	Curve > Convert > Curve to Lines	Constructing a set of joined lines from a curve
Convert Curve to Arcs	Curve > Convert > Curve to Arcs	Constructing a set of joined arcs from a curve

Control Point Curve and Interpolate Curve

In the following you will construct a series of point objects. You will then construct two free-form curves by specifying control points and points through which the curve has to interpolate. To create the series of point objects, perform the following steps.

1. Start a new file. Use the metric (Millimeters) template file.
2. Double click on the Top viewport to maximize the viewport.
3. Select Curve > Point Object > Multiple Points, or click on the Multiple Points button on the Point toolbar.
4. Select locations A, B, C, and D, indicated figure 2-51, and then press the Enter key.

Points are constructed. Copy the points by continuing with the following steps.

5. Select Transform > Copy, or click on the Copy button on the Transform toolbar.
6. Select locations A and B, indicated in figure 2-52, and then press the Enter key.
7. Select point C, indicated in figure 2-52, as the point to copy from. This is the base point of copy.

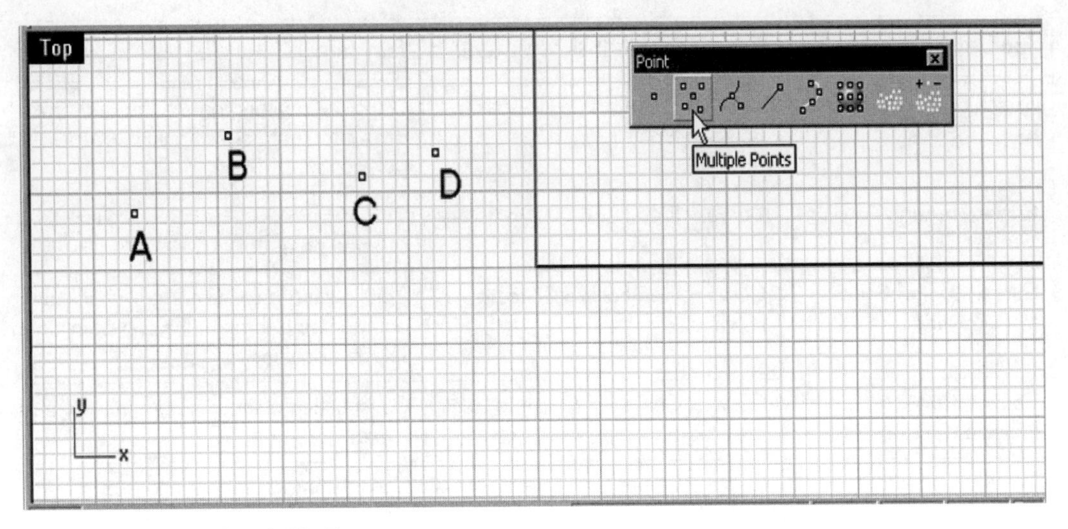

Fig. 2-51. Four points.

8 Select point D, indicated in figure 2-52, as the point to copy to. Distance and direction between this point and the base point indicate the location of the copied objects in relation to the original objects.

9 Press the Enter key to terminate the command.

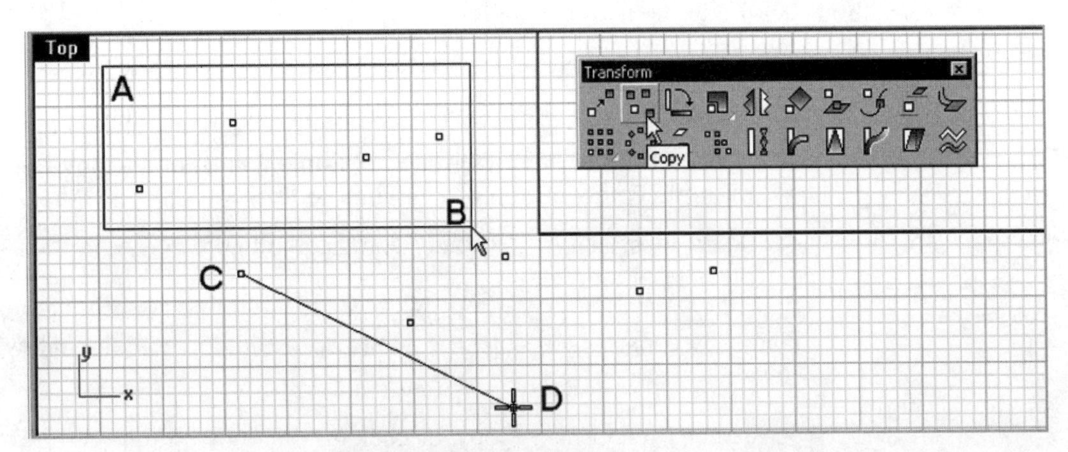

Fig. 2-52. Base point of copy selected.

The points are copied. Construct a free-form curve by specifying control points, as follows.

10 Click on the Osnap button on the status bar.

11 In the Osnap dialog box, check the Point box.

12 Select Curve > Free-Form > Control Points, or click on the Control Point Curve button on the Curve toolbar.

13 Select points A, B, C, and D, indicated in figure 2-53, and then press the Enter key to terminate the command.

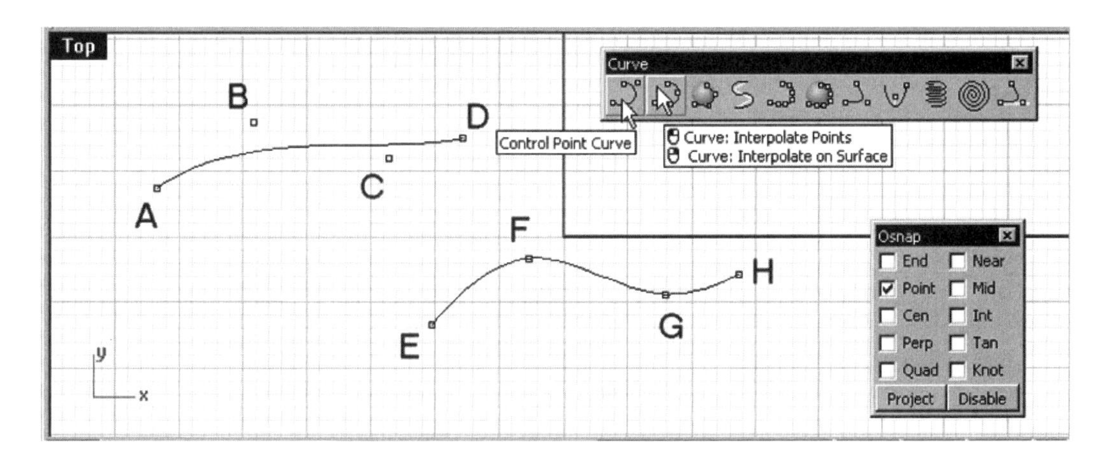

Fig. 2-53. Control point curve and interpolated curve.

A free-form curve is constructed. Except the end points, the curve does not pass through the control point locations. Construct a free-form curve to interpolate selected points, as follows.

14 Select Curve > Free-Form > Interpolate Points, or click on the *Curve: Interpolate Points* button on the Curve toolbar.

15 Select points E, F, G, and H, indicated in figure 2-53, and press the Enter key to terminate the command.

A free-form curve passing through the selected points is constructed. Now the point objects are not required. To keep these objects and hide them in the display, change the free-form curves to *Layer 01*, set the current layer to *Layer 01*, and turn off the default layer, as follows.

16 Select Edit > Layers > Edit Layers, or click on the Edit Layers button on the Layer toolbar. (The Layer toolbar is a fly-out from the Standard toolbar.)

17 In the Layers dialog box, shown in figure 2-54, click on the *Create new layer* button.

18 Fig. 2-54. Layers dialog box.

19 Check the box next to the *Layer 01* option to set it as the current layer.

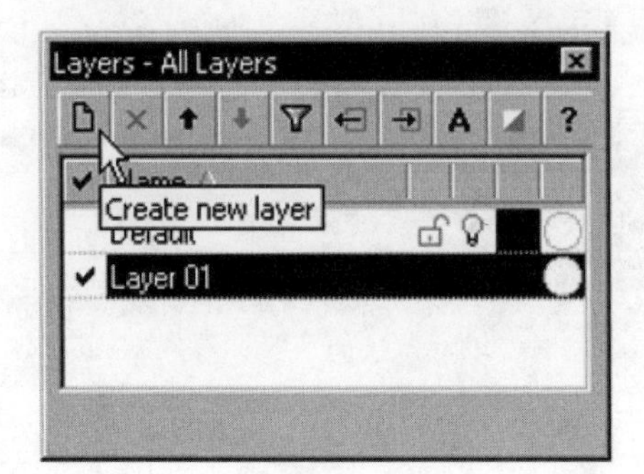

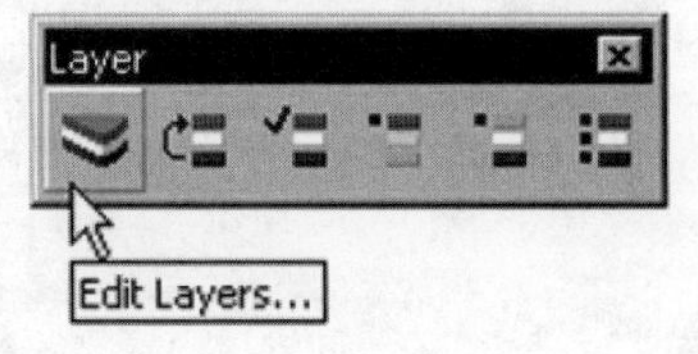

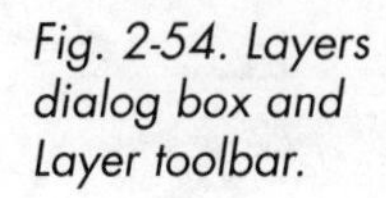
Fig. 2-54. Layers dialog box and Layer toolbar.

20 Click on the X icon on the dialog box to close it.

21 Select Edit > Layers > Change Object Layer, or click on the Change Layer button on the Layer toolbar.

22 Select the free-form curves and press the Enter key.

23 In the *Layer for objects* dialog box, select *Layer 01*, and then click on the OK button. (See figure 2-55.)

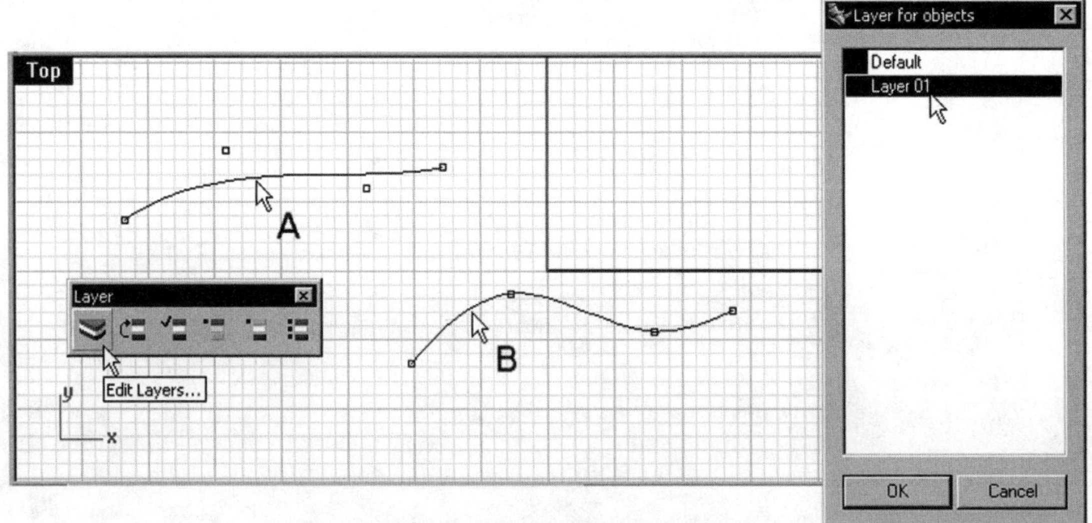

Fig. 2-55. Free-form curves layer being changed.

The free-form curves are moved to *Layer 01*. Set the current layer to *Layer 01* and turn off the default layer where the point objects reside, as follows.

24 Select Edit > Layers > Edit Layers, or click on the Edit Layers button on the Layer toolbar.

25 In the Layers dialog box, click on the *Hidden layer* button, indicated in figure 2-56, to turn off the default layer.

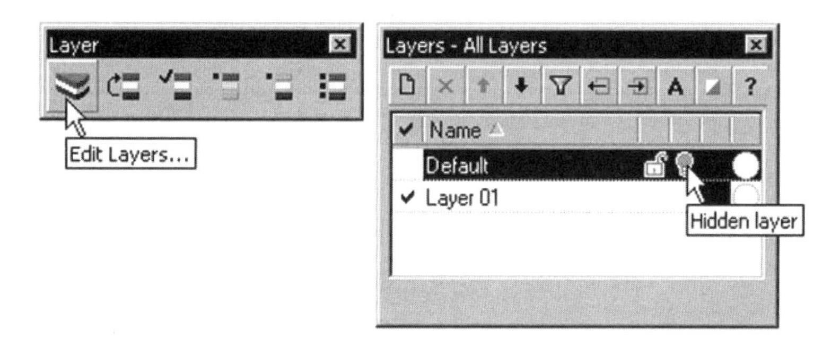

Fig. 2-56.Layer toolbar and Hidden layer *button of the Layers dialog.*

26 Close the dialog box.

To appreciate the difference between the two free-form curves, let's turn on their control points, as follows.

27 Select Edit > Control Points > Control Points On, or click on the *Control Points On/Points Off* button on the Point Editing toolbar.

28 Select the free-form curves and press the Enter key.

The control points are turned on, as shown in figure 2-57.

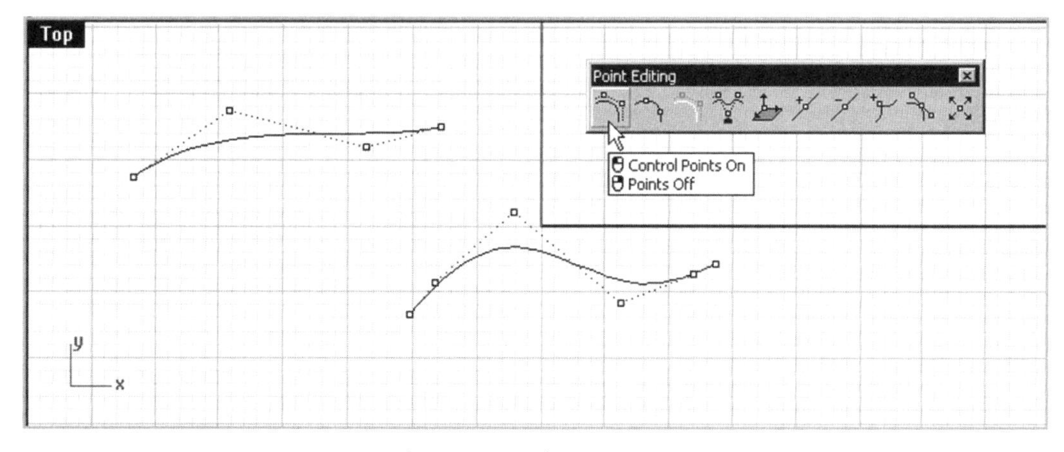

Fig. 2-57. Control points turned on.

The upper curve (curve constructed by specifying four control points) has the control point located at the specified location. The lower curve (curve constructed by specifying the interpolated points) has six control points, with the curve showing more segments. Turn off the control points, as follows.

29 Select Edit > Control Points > Control Points Off, or right-click on the *Control Points On/Points Off* button on the Point Editing toolbar.

30 If you want to save your file, save it as *FreeFormCurve.3dm.*

Constructing a Free-form Curve from a Polyline

A polyline is a series of joined line segments. Because line segments are degree 1 curves, a polyline is also a degree 1 curve. If you already have a polyline, you can construct a free-form curve to interpolate the end points of the polyline. On the other hand, if you have a free-form curve, you can convert it to a polyline. Such interconversion is basically required when you import data from a system that outputs polylines and export to a system that reads polylines.

1. Start a new file. Use the metric (Millimeters) template.
2. Maximize the Top viewport.
3. Select Curve > Polyline > Polyline, or click on the *Polyline/Line Segments* button on the Lines toolbar.
4. Select points A, B, C, D, and E, indicated in figure 2-58, and then press the Enter key.

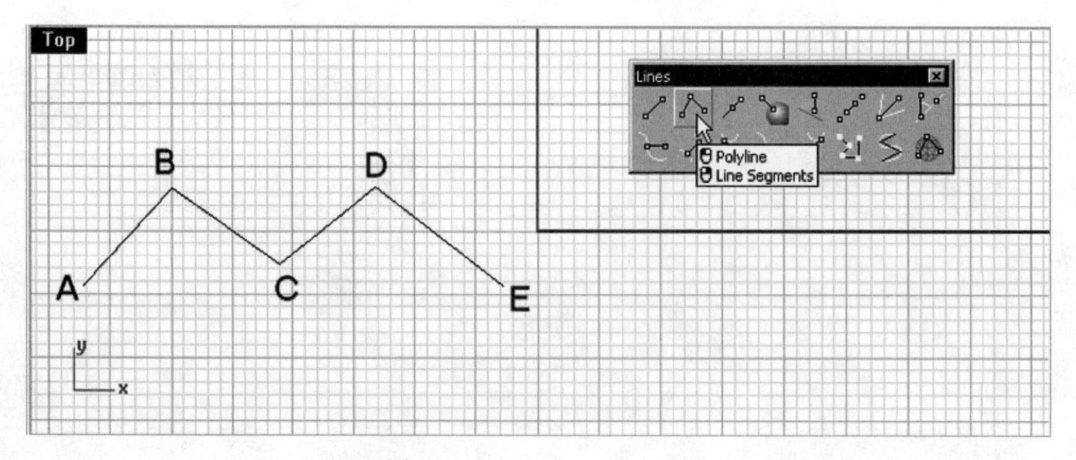

Fig. 2-58. Polyline.

5 Select Curve > Free-Form > Interpolate Polyline, or click on the *Curve: Control Points from Polyline/Curve: Through Polyline Vertices* button on the Curve toolbar.

6 Select the polyline and press the Enter key.

7 If you use the pull-down menu, type *C* at the command line interface to set the curve type, and then type *c*. Otherwise, skip this step.

8 Press the Enter key.

9 Select Curve > Free-Form > Interpolate Polyline, or right-click on the *Curve: Control Points from Polyline/Curve: Through Polyline Vertices* button on the Curve toolbar.

10 Select the polyline and press the Enter key.

11 If you use the pull-down menu, type *C* at the command line interface to set the curve type and then type *i*. Otherwise, skip this step.

12 Press the Enter key.

Two free-form curves are constructed: one using the vertices of the polyline as control points and the other one interpolated along the vertices of the polyline, as shown in figure 2-59.

13 If you want to save your file, save it as *Polyline2Curve.3dm*.

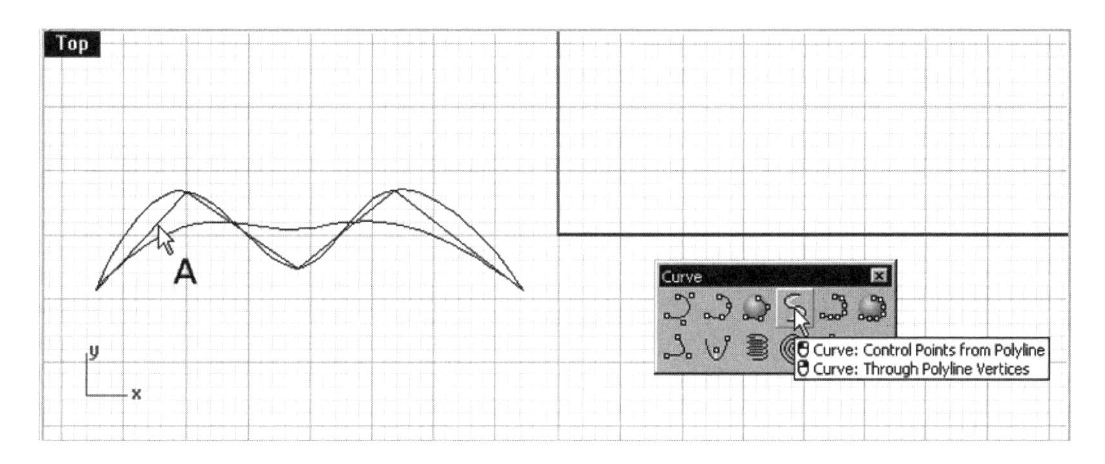

Fig. 2-59. Free-form curves constructed from a polyline in two ways.

Converting a Free-form Curve to Line Segments or Arc Segments

If you are going to export curves to another computerized system that does not support splines, you may consider converting the splines to lines and arcs. Perform the following to convert a curve to a polyline and a curve to a set of joined arcs.

1. Start a new file. Use the metric (Millimeters) template.
2. Maximize the Top viewport.
3. Select Curve > Free-Form > Interpolate Points, or click on the *Curve: Interpolate Points* button on the Curve toolbar.
4. Select locations A, B, C, D, and E, indicated in figure 2-60, and then press the Enter key.
5. Select Transform > Copy, or click on the Copy button on the Transform toolbar.
6. Select curve ABCDE, and then press the Enter key.
7. Select point F and then point G, indicated in figure 2-60, and press the Enter key.
8. Select Curve > Convert > Curve to Lines, or click on the Convert Curve to Polyline button on the Lines toolbar.
9. Select curve A, indicated in figure 2-61, and press the Enter key.
10. Type *A* at the command line interface to select the Angle Tolerance option.
11. Type *25* to specify the angular tolerance.
12. Type *T* at the command line interface to select the Tolerance option.
13. Type *5* to specify the tolerance.
14. Press the Enter key. The curve is converted to a polyline. The smaller the tolerance, the more will be the number of line segments in the polyline.
15. Select Curve > Convert > Curve to Arcs, or click on the Convert Curve to Arcs button on the Arc toolbar.
16. Select curve B, indicated in figure 2-61, and then press the Enter key.
17. Type *15* to specify the angular tolerance.
18. Type *T* at the command line interface to select the Tolerance option.
19. Type *5* to specify the tolerance.
20. Press the Enter key. The curve is converted to a set of connected arcs. The smaller the tolerance, the more will be the number of arc segments.

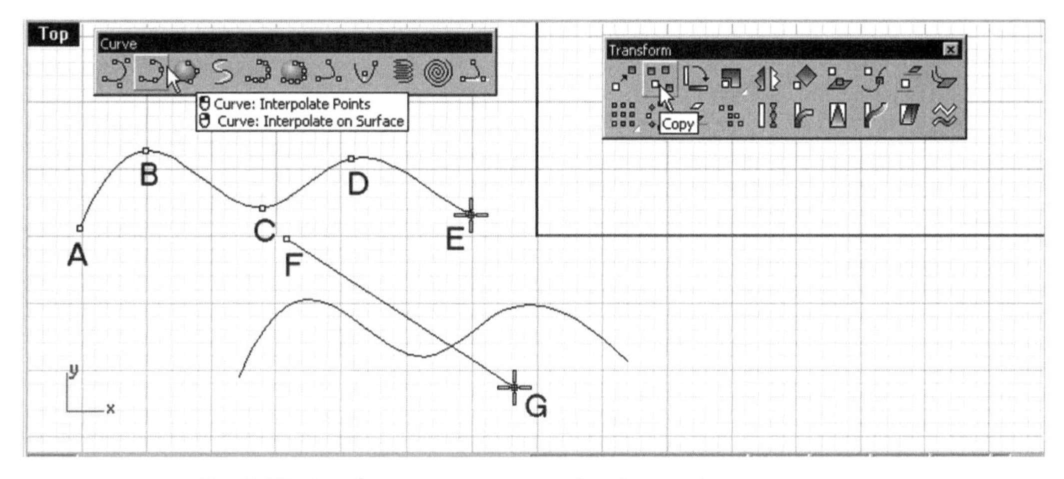

Fig. 2-60. Free-form curve constructed and copied.

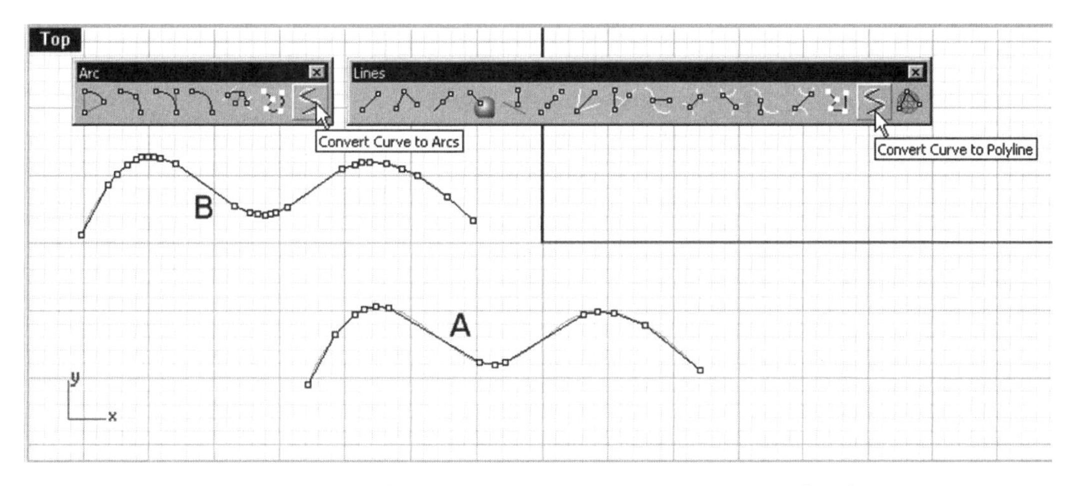

Fig. 2-61. A free-form curve converted to a polyline and a free-form curve converted to a set of connected arcs.

21 If you want to save your file, save it as *ConvertCurve.3dm*.

Sketch Curve and Free-form Curve

A sketch curve is also a type of free-form curve. Instead of inputting interpolate points or control points, you use the mouse as a drafting pen to construct a curve. Perform the following steps to construct a sketch curve, shown in figure 2-62.

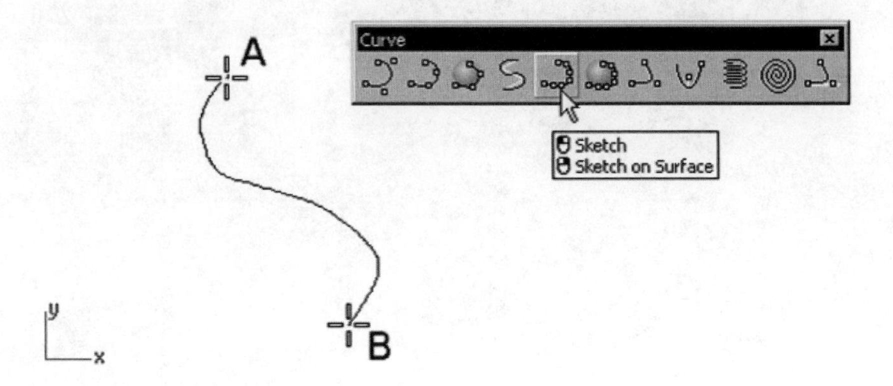

Fig. 2-62. Sketch curve being constructed.

1. Start a new file. Use the metric (Millimeters) template.
2. Maximize the Top viewport.
3. Select location A, shown in figure 2-62.
4. Hold down the left mouse button and drag the cursor to location B.
5. Release the mouse button.
6. Press the Enter key.

A free-form sketch curve is constructed. Unlike the other two methods (specifying the control points and specifying the edit points), this method of constructing a curve lets you use your mouse as an electronic drawing pen. It is particularly useful if you embed a background bitmap in the viewport as a reference. (You will learn how to embed a bitmap in Chapter 4.)

Constructing Lines and Curves on Surfaces and Polygon Meshes

Apart from the aforementioned methods of constructing line and free-form curves by picking points on the viewport, you can construct lines and curves on surfaces and polygon meshes. (You will learn how to construct surfaces and polygon meshes in later chapters.)

Line Normal to a Surface

To construct a line normal to a surface, perform the following steps.

1. Open the file *CurveOnSurface.3dm* from the *Chapter 2* folder of the companion CD-ROM.
2. Select Curve > Line > Normal to Surface, or click on the *Line: Surface Normal/Surface Normal Both Sides* button on the Lines toolbar.

3. Select surface A, indicated in figure 2-63.
4. Select location B, shown in figure 2-63, to indicate the start point of the line.
5. Select location C, shown in figure 2-63, to indicate the end point of the line.

A line normal to the surface is constructed.

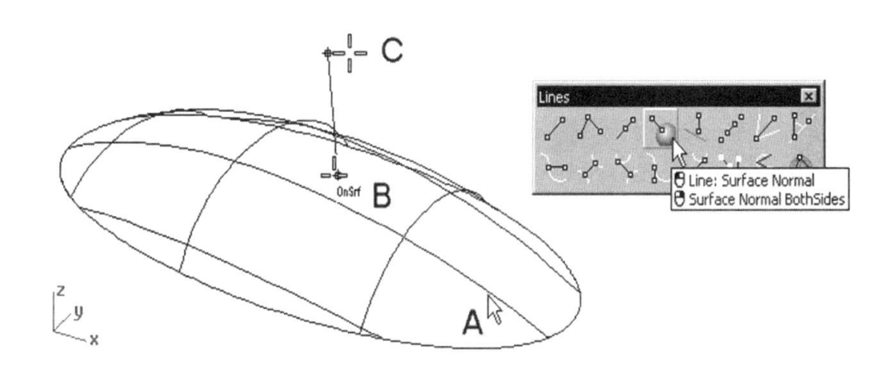

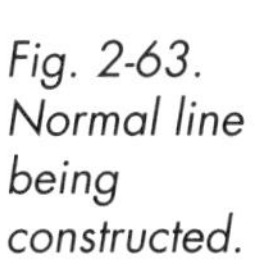
Fig. 2-63. Normal line being constructed.

Free-form Curve on a Surface

To construct a free-form curve on a surface, perform the following steps.

1. Select Curve > Free-Form > Interpolate on Surface, or click on the Interpolate on Surface button on the Curve toolbar.
2. Select surface A, indicated in figure 2-64.

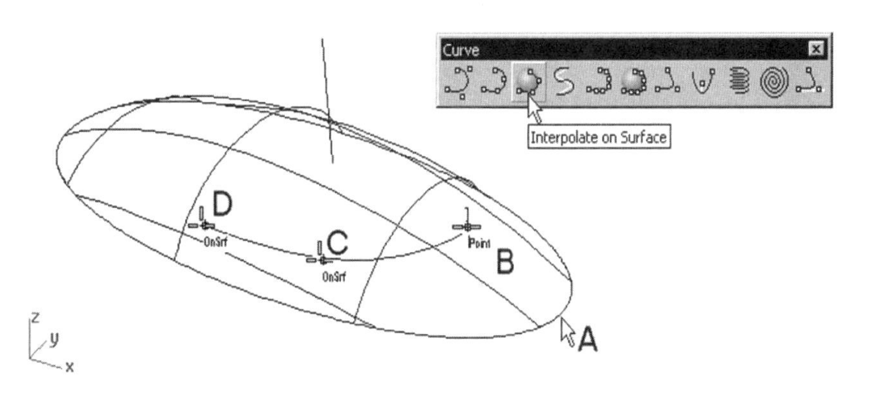

Fig. 2-64. Free-form curve being constructed on a surface.

3. Select locations B, C, and D, indicated in figure 2-64, and then press the Enter key.

A free-form curve is constructed on the surface.

Free-form Sketch Curve on a Surface

To construct a free-form sketch curve on a surface, perform the following steps.

1. Select Curve > Free-Form > Sketch on Surface, or right-click on the *Sketch/Sketch on Surface* button on the Curve toolbar.
2. Select location A (indicated in figure 2-65), hold down the mouse button, and drag to location B (indicated in figure 2-65).
3. Release the mouse button when finished.
4. Press the Enter key.

A free-form sketch curve is constructed on the surface.

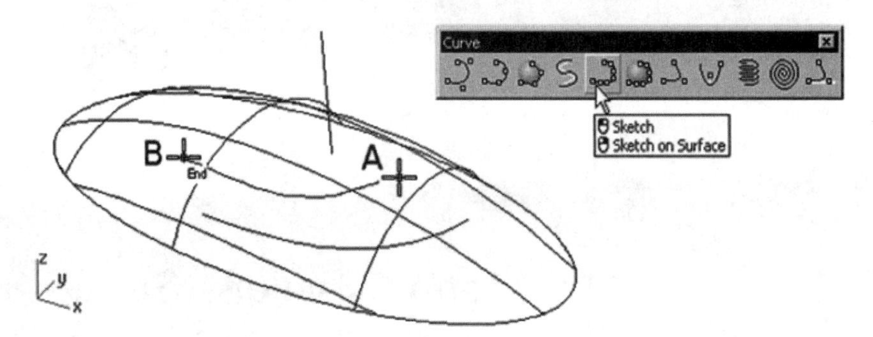

Fig. 2-65. Free-form sketch curve being constructed on a surface.

Free-form Sketch Curve on a Polygon Mesh

In addition to creating curves on NURBS surfaces, you can construct free-form sketch curves and polylines on a polygon mesh, as follows. A polygon mesh is an approximation of a smooth surface via a set of planar polygons.

1. Open the file *CurveOnMesh.3dm* from the *Chapter 2* folder of the companion CD-ROM.
2. Select Curve > Free-Form > Sketch on Polygon Mesh, or click on the Sketch on Polygon Mesh button on the Curve toolbar.
3. Select location A (indicated in figure 2-66) on the polygon mesh, hold down the mouse button, and drag to location B (indicated in figure 2-66).
4. Release the mouse button and press the Enter key.

A free-form sketch curve is constructed on a polygon mesh.

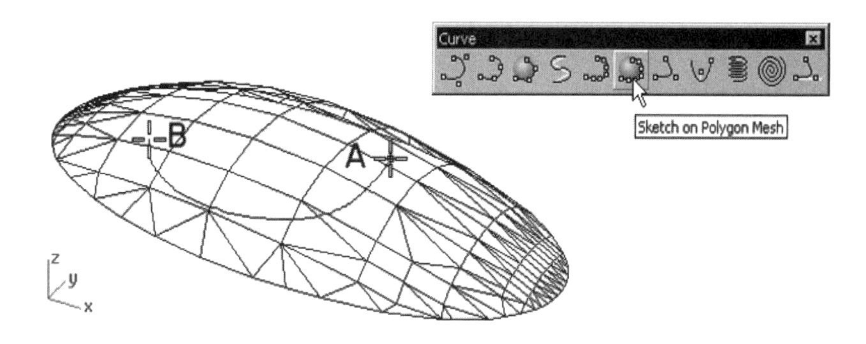

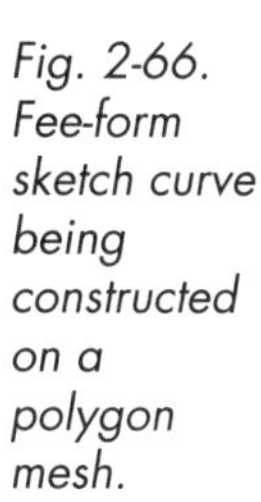
Fig. 2-66. Fee-form sketch curve being constructed on a polygon mesh.

Polyline on a Polygon Mesh

To construct a polyline on a polygon mesh, perform the following steps.

1. Select Curve > Polyline > On Mesh, or click on the *Polyline: On Mesh* button on the Lines toolbar.
2. Select polygon mesh A, indicated in figure 2-67.
3. Select locations B, C, and D, indicated in figure 2-67, and then press the Enter key.

A polyline is constructed on a polygon mesh.

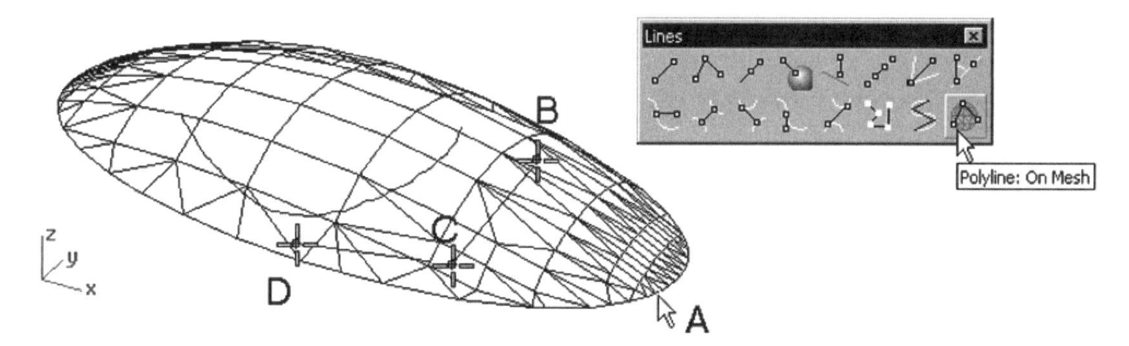

Fig. 2-67. Polyline being constructed on a polygon mesh.

Circle

Basically, a circle is a degree 2 curve. If you use the deformable circle, the degree of polynomial can be anywhere between 3 and 11, but such constructs are not true circles. There are a number of ways to construct a circle. A circle is constructed via the Circle toolbar, shown in figure 2-68.

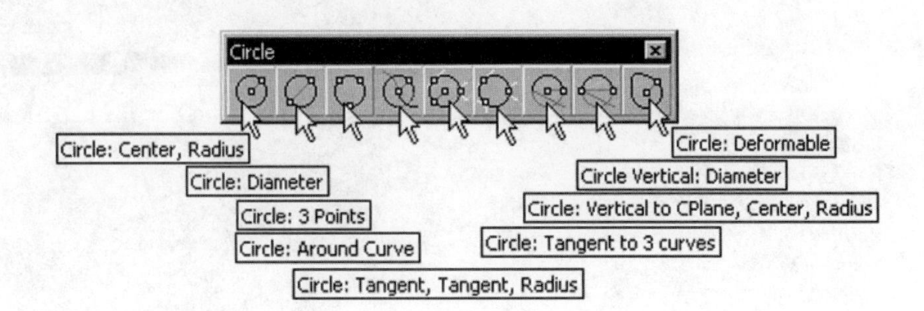

Fig. 2-68. Circle toolbar.

Table 2-4 outlines the functions of circle commands and their locations in the main pull-down menu and respective toolbars.

Table 2-4: Circle Commands and Their Functions

Toolbar Button	From Main Pull-down Menu	Function
Circle: Center, Radius	Curve > Circle > Center, Radius	Constructing a circle by specifying its center and radius
Circle: Diameter	Curve > Circle > Diameter	Constructing a circle by specifying two diametric points
Circle: 3 Points	Curve > Circle > 3 Points	Constructing a circle by specifying three points on the circumference
Circle: Around Curve	—	Constructing a circle around a selected curve by specifying a point on the curve and the radius
Circle: Tangent, Tangent, Radius	Curve > Circle > Tangent, Tangent, Radius	Constructing a circle by specifying two tangent curves and the radius
Circle: Tangent to 3 curves	Curve > Circle > Tangent to 3 Curves	Constructing a circle by specifying three tangent curves
Circle: Vertical to CPlane, Center, Radius	—	Constructing a circle with its plane perpendicular to the construction plane by specifying the center and radius
Circle Vertical: Diameter	—	Constructing a circle with its plane perpendicular to the construction plane by specifying two points on the diameter
Circle: Deformable	—	Constructing a circle of degree 3 by specifying the center and radius

Circle, Vertical Circle, and Circle Around Curve

To construct a circle along and perpendicular to a free-form curve, perform the following steps.

1. Start a new file. Use the metric (Millimeters) template.
2. Maximize the Top viewport.
3. Select Curve > Circle > Center, Radius, or click on the *Circle: Center, Radius* button on the Circle toolbar.
4. Select A and B, indicated in figure 2-69, to specify the center and a point on the circumference.

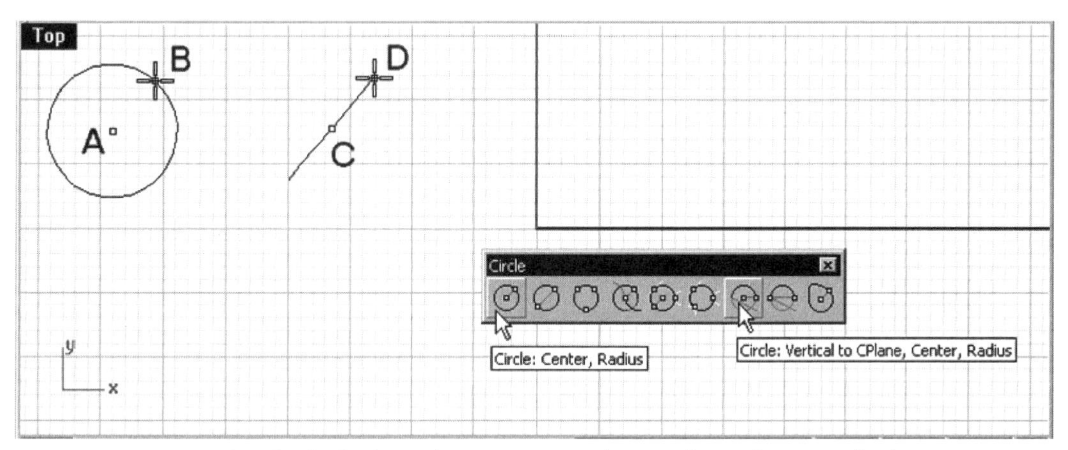

Fig. 2-69. Circle on the construction plane and a circle perpendicular to the construction plane.

5. Select Curve > Circle > Center, Radius and type *V* (V stands for vertical) at the command line interface, or click on the *Circle: Vertical to CPlane, Center, Radius* button on the Circle toolbar.
6. Select C and D, indicated in figure 2-69, to specify the center and the radius.
7. With reference to figure 2-70, construct a free-form curve interpolating along five points: A, B, C, D, and E.
8. Select Curve > Circle > Center, Radius and type *A* (A stands for AroundCurve) at the command line interface, or click on the *Circle: Around Curve* button on the Circle toolbar.
9. Select point F along a free-form curve to indicate the center location.
10. To specify a radius, you may type a value at the command line interface or select point G to indicate the radius.

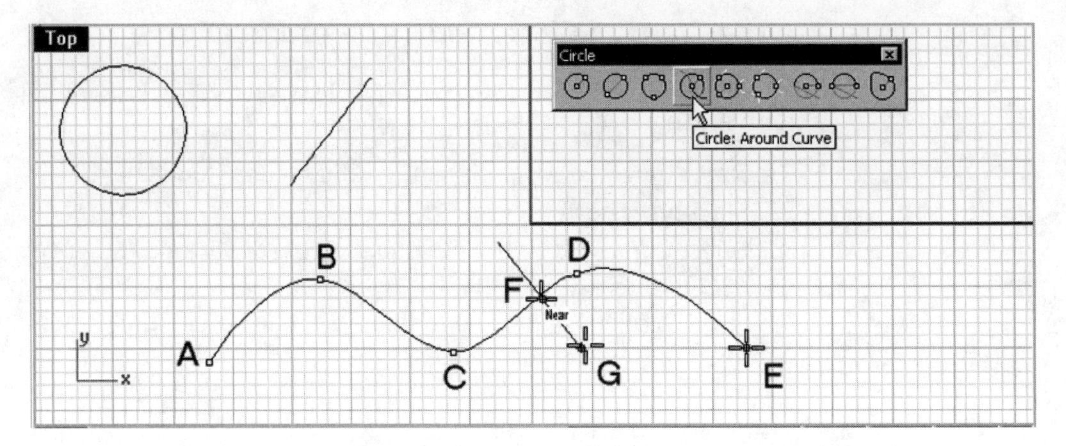

Fig. 2-70. Circle around a curve being constructed.

11 Double click on the Top viewport title to return to a four-viewport display, shown in figure 2-71.

12 If you want to save your file, save it as *Circle1.3dm*.

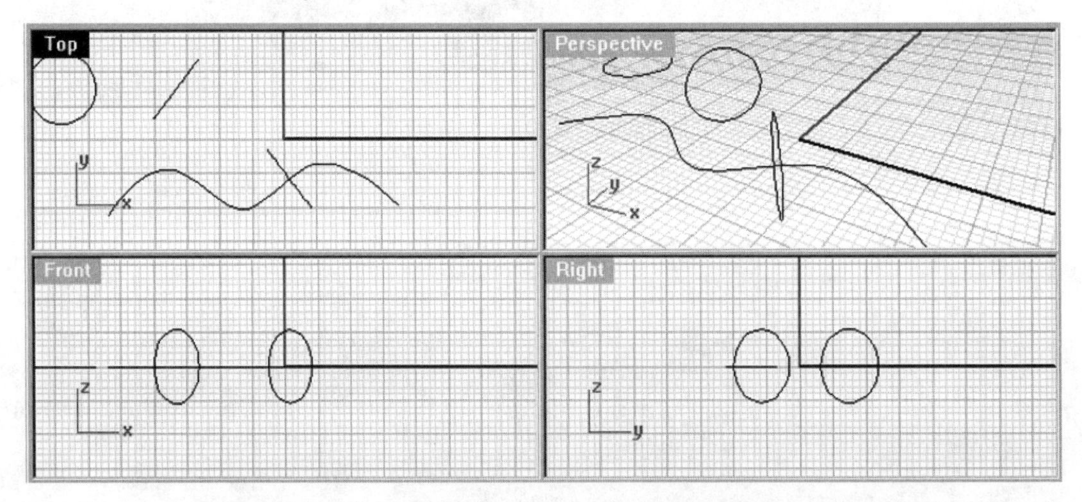

Fig. 2-71. Four-viewport display of circle constructed perpendicular to the selected curve.

Tangent Circles

To construct tangent circles, perform the following steps.

1 Start a new file. Use the metric (Millimeters) template.

2 Maximize the Top viewport.

3 With reference to figure 2-72, construct three free-form curves: A, B, and C.

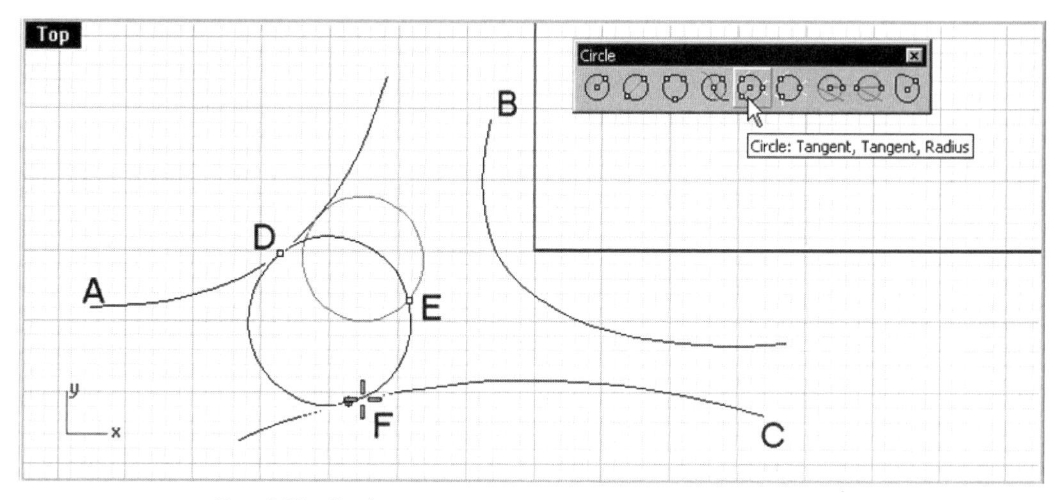

Fig. 2-72. Circle tangent to two curves.

4. Select Curve > Circle > Tangent, Tangent, Radius, or click on the *Circle: Tangent, Tangent, Radius* button on the Circle toolbar.
5. Select curve A at D, indicated in figure 2-72.
6. Select point E to specify the radius (distance between D and E).
7. Select curve C.
8. A circle tangent to two curves is constructed.
9. Select Curve > Circle > Tangent to 3 Curves, or click on the *Circle: Tangent to 3 Curves* button on the Circle toolbar.
10. Select curves A, B, and C, indicated in figure 2-73. A circle tangent to three curves is constructed.

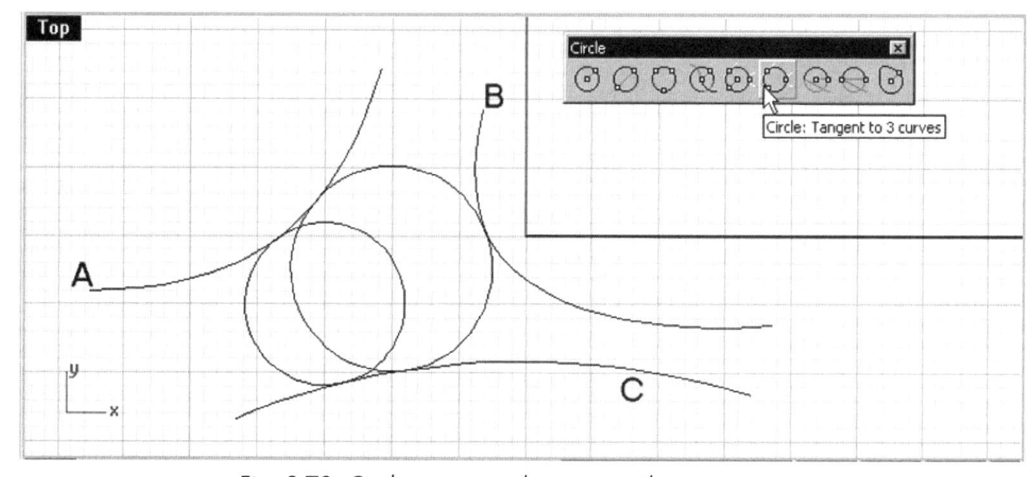

Fig. 2-73. Circle constructed tangent to three curves.

11. If you want to save your file, save it as *Circle2.3dm*.

Ellipse

An ellipse is also a degree 2 curve. There are five ways to construct an ellipse, which is done via the Ellipse toolbar, shown in figure 2-74.

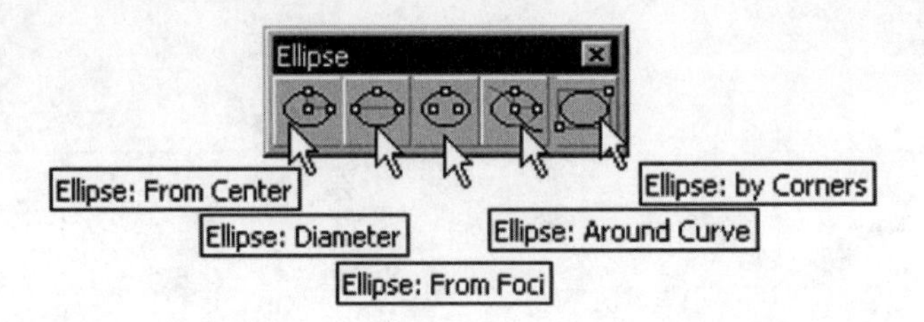

Fig. 2-74. Ellipse toolbar.

Ellipse Commands

Table 2-5 outlines the functions of ellipse commands and their locations in the main pull-down menu and respective toolbars.

Table 2-5: Ellipse Commands and Their Functions

Toolbar Button	From Main Pull-down Menu	Function
Ellipse: From Center	Curve > Ellipse > From Center	Constructing an ellipse by specifying the center, the first axis end point, and the second axis end point
Ellipse: Diameter	Curve > Ellipse > Diameter	Constructing an ellipse by specifying the end points of the first axis and the second axis end point
Ellipse: From Foci	Curve > Ellipse > From Foci	Constructing an ellipse by specifying the foci and a point on the ellipse
Ellipse: Around Curve	—	Constructing an ellipse around a selected curve by specifying a point on the curve as the center, the first axis end point, and the second axis end point
Ellipse: By Corners	—	Constructing an ellipse by specifying the diagonal points of an imaginary rectangle with edges tangent to the ellipse

Ellipse and Ellipse Around Curve

To construct an ellipse around the lower curve, perform the following steps.

1 Start a new file. Use the metric (Millimeters) template.

2 Maximize the Top viewport.
3 Select Curve > Ellipse > Diameter, or click on the *Ellipse: Diameter* button on the Ellipse toolbar.
4 Select points A and B to specify the diameter, and select point C to specify a point on the ellipse, as shown in figure 2-75.
5 With reference to figure 2-76, construct a free-form curve.

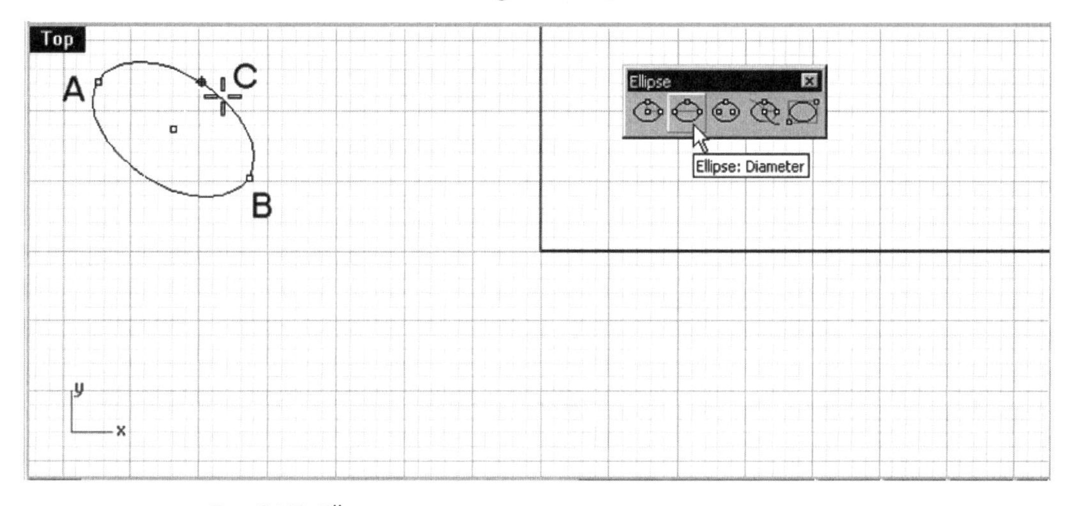

Fig. 2-75. Ellipse.

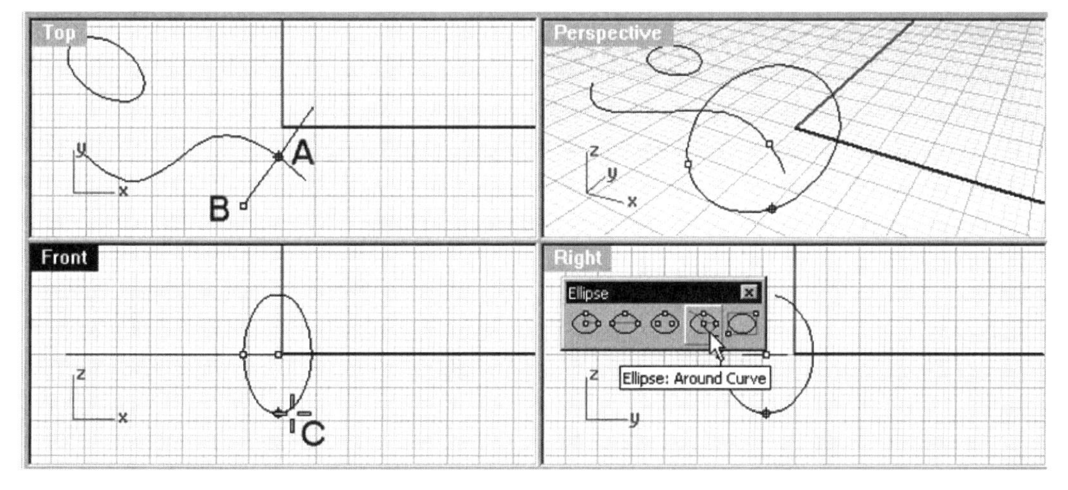

Fig. 2-76. Ellipse around curve.

6 Double click on the Top viewport title to return to a four-viewport display.

7 Select Curve > Ellipse > From Center and type *A* at the command line interface, or click on the *Ellipse: Around Curve* button on the Ellipse toolbar.

8 Select point A (indicated in figure 2-76) along the curve to indicate the center on the curve.

9 Select point B (indicated in figure 2-76) to indicate a radius (distance between A and B).

10 Move the cursor over the Front viewport and select point C (shown in figure 2-76) to indicate the other radius (distance from the center to the selected point).

An ellipse is constructed around a curve.

11 If you want to save your file, save it as *Ellipse.3dm*.

Parabola

A parabola is degree 2 curve. There are two ways to construct a parabola. A parabola is constructed via the Parabola button of the Curve toolbar, shown in figure 2-77.

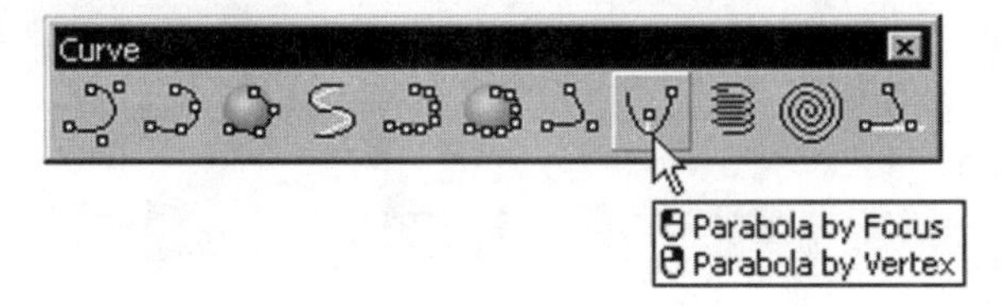

Fig. 2-77. Parabola button of the Curve toolbar.

Table 2-6 outlines the functions of parabola commands and their locations in the main pull-down menu and respective toolbars.

Table 2-6: Parabola Commands and Their Functions

Toolbar Button	From Main Pull-down Menu	Function
Parabola by Focus	Curve > Parabola > Focus, Direction	Constructing a parabola by specifying the focus point, direction, and a point on the parabola
Parabola by Vertex	Curve > Parabola > Vertex, Focus	Constructing a parabola by specifying the vertex point, focus point, and a point on the parabola

Perform the following steps to construct two parabola curves.

1 Start a new file. Use the metric (Millimeters) template.

2. Maximize the Top viewport.
3. Select Curve > Parabola > Focus, Direction, or click on the *Parabola: by Focus/Parabola by Vertex* button on the Curve toolbar.
4. Select points A and B, indicated in figure 2-78, to specify the focus point and the direction.
5. Select point C, indicated in figure 2-78, to specify one of the end points of the parabola.
6. Select Curve > Parabola > Vertex, Focus, or right-click on the *Parabola: by Focus/Parabola by Vertex* button on the Curve toolbar.
7. Select points D and E, indicated in figure 2-78, to specify the focus point and the direction.
8. Select point F, indicated in figure 2-78, to specify one of the end points of the parabola.

A parabola is constructed.

9. If you want to save your file, save it as *Parabola.3dm*.

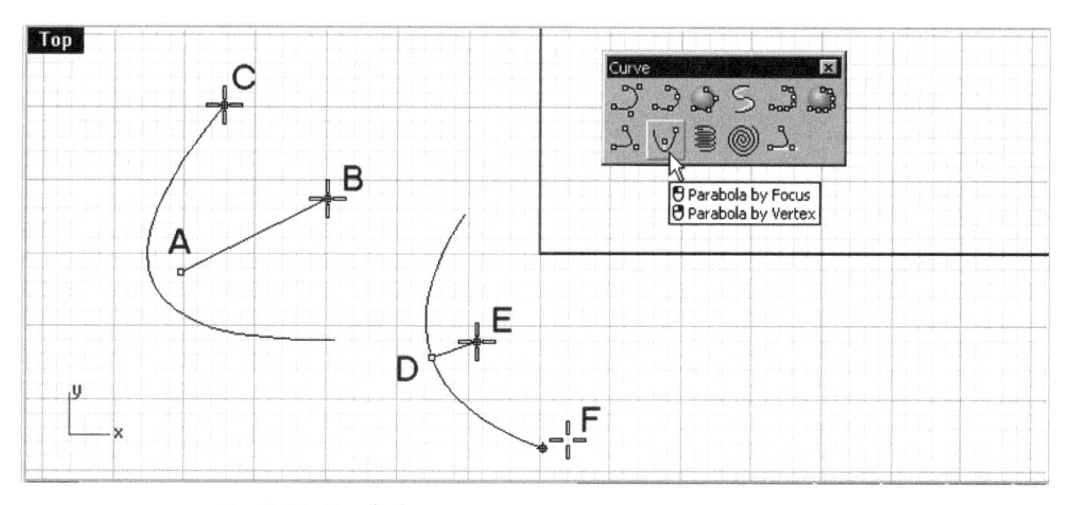

Fig. 2-78. Parabola curves.

Conic

A conic is also a degree 2 curve. You construct conics via the Conic button of the Curve toolbar, shown in figure 2-79.

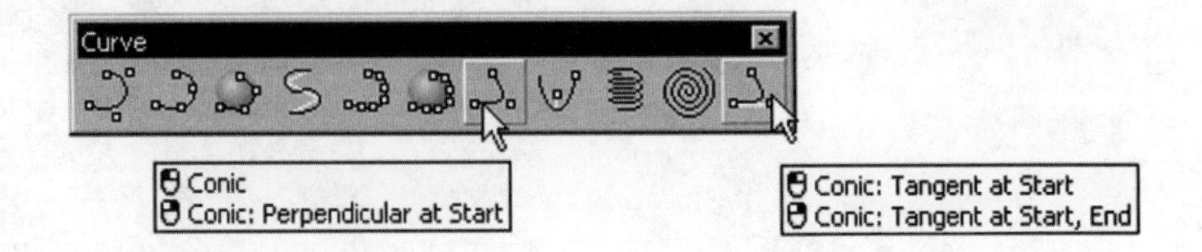

Fig. 2–79. Conic button of the Curve toolbar.

The Conic Command

Table 2-7 outlines the function of Conic commands and their locations in the main pull-down menu and respective toolbars.

Table 2-7: Conic Commands and Their Functions

Toolbar Button	From Main Pull-down Menu	Function
Conic	Curve > Conic	Constructing a conic by specifying the start point, end point, apex point, and a point on the conic
Conic: Perpendicular at Start	Curve > Conic	Constructing a conic with its start point perpendicular to a selected curve by selecting the end point of the curve as its first end point and specifying the other end point, apex point, and a point on the conic
Conic: Tangent at Start	Curve > Conic	Constructing a conic tangent to a selected curve by selecting the end point of the curve as its first end point and specifying the other end point, apex point, and a point on the conic
Conic: Tangent at Start, End	Curve > Conic	Constructing a conic tangent to two selected curves by selecting the end points of the curves as its end points and specifying the apex point and a point on the conic

Conic Curves

To construct conic curves in various ways, perform the following steps.

1. Start a new file. Use the metric (Millimeters) template.
2. Maximize the Top viewport.
3. With reference to figure 2-80, construct two free-form curves A and B and three point objects C, D, and E.
4. Check the End and Point boxes on the Osnap dialog box.

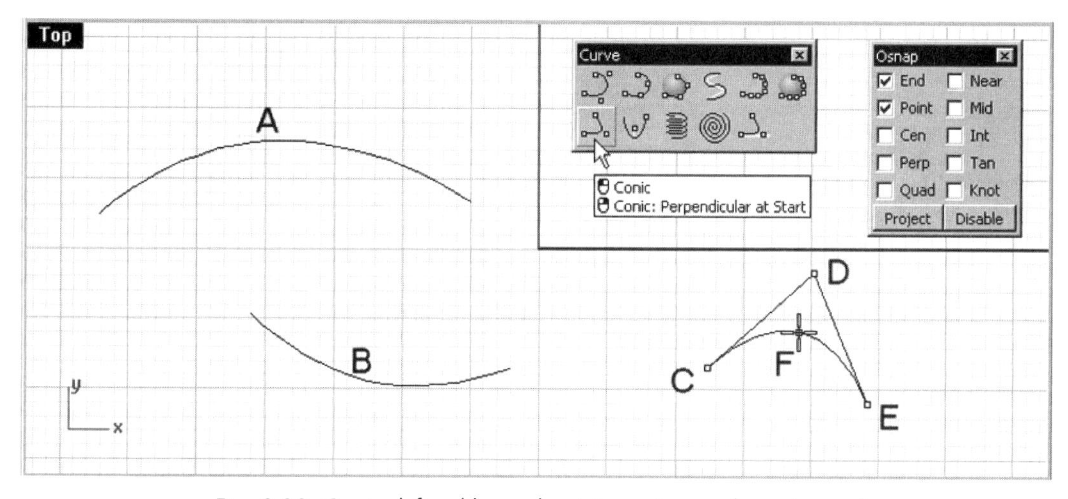

Fig. 2-80. Conic defined by end points, vertex, and a point on the curve.

5 Select Curve > Conic, or click on the *Conic/Conic: Perpendicular at Start* button on the Curve toolbar.

6 Select points C and D to indicate the end points of the conic.

7 Select point E to indicate the apex location.

8 Select location F to indicate a point on the conic curve.

Construct a conic that is perpendicular to a selected curve and a conic that is tangent to two curves, as follows.

9 Select Curve > Conic and type *P* at the command line interface, or right-click on the *Conic/Conic: Perpendicular at Start* button on the Curve toolbar.

10 Select end point A (indicated in figure 2-81) to specify the perpendicular end point.

11 Select location B (indicated in figure 2-81) to specify the other end point.

12 Select location C (indicated in figure 2-81) to specify the apex.

13 Select location D (indicated in figure 2-81) to specify a point on the conic curve.

14 Select Curve > Conic and type *T* at the command line interface, or right-click on the *Conic: Tangent at Start/Conic: Tangent at Start, End* button on the Curve toolbar.

15 Select end points E and F (indicated in figure 2-81) to specify the end points.

16 Select location G (indicated in figure 2-81) to specify a point on the conic curve.

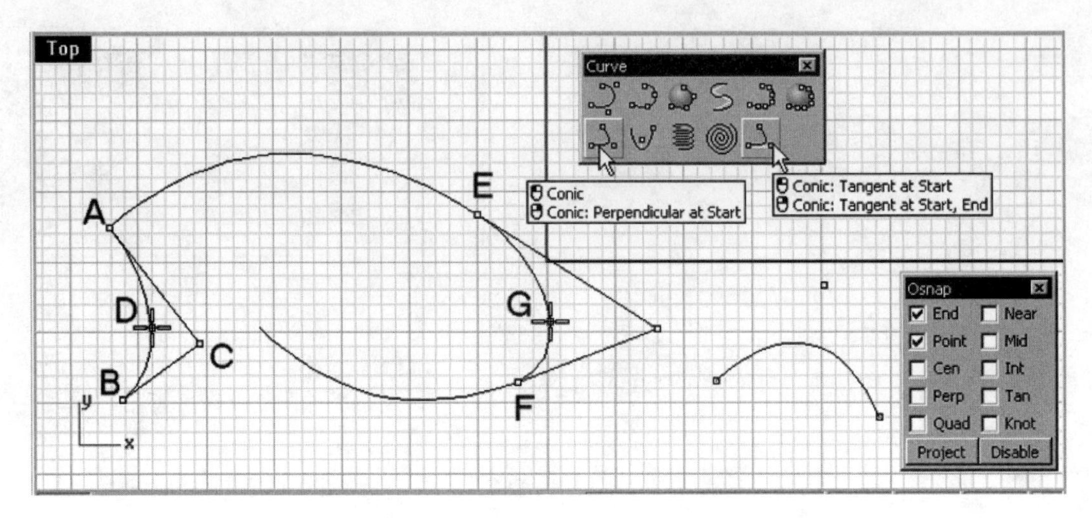

Fig. 2-81. Conic perpendicular to a curve and conic tangent to two curves.

17 Select Curve > Conic and type *T* at the command line interface, or click on the *Conic: Tangent at Start/Conic: Tangent at Start, End* button on the Curve toolbar.

18 Select end point A (indicated in figure 2-82) to specify the tangent end point.

19 Select location B (indicated in figure 2-82) to specify the other end point.

20 Select location C (indicated in figure 2-82) to specify the apex.

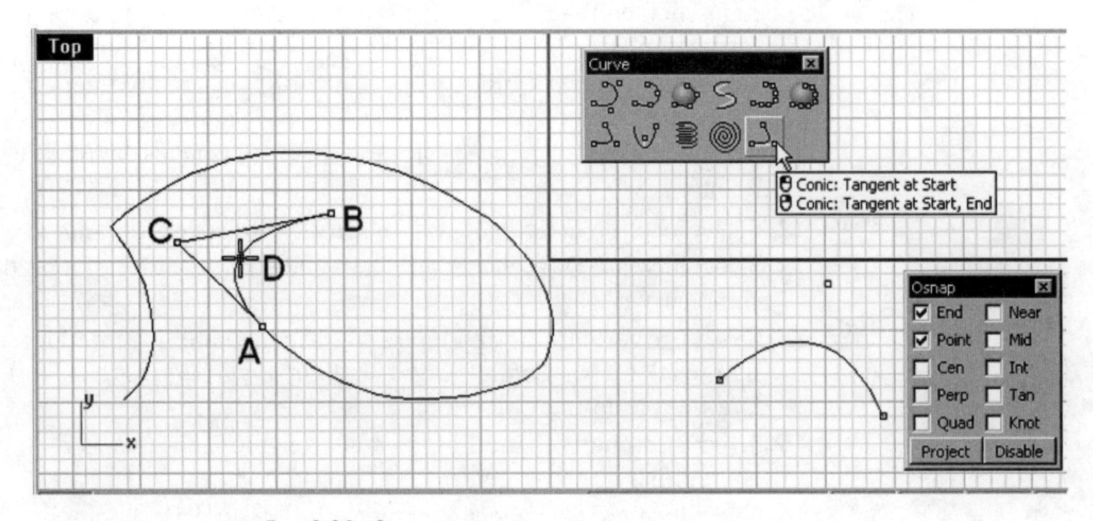

Fig. 2-82. Conic tangent to a curve.

21 Select location D (indicated in figure 2-82) to specify a point on the conic curve.

A conic curve is constructed.

22 If you want to save your file, save it as *Conic.3dm*.

Helix

A helix is a degree 3 curve. Using the Helix command, you can construct a helix with its axis lying on or perpendicular to the current construction plane. You can also construct a helix around a curve. The shape of a helix around a curve is the shape of the selected curve. Figure 2-83 shows the Helix button.

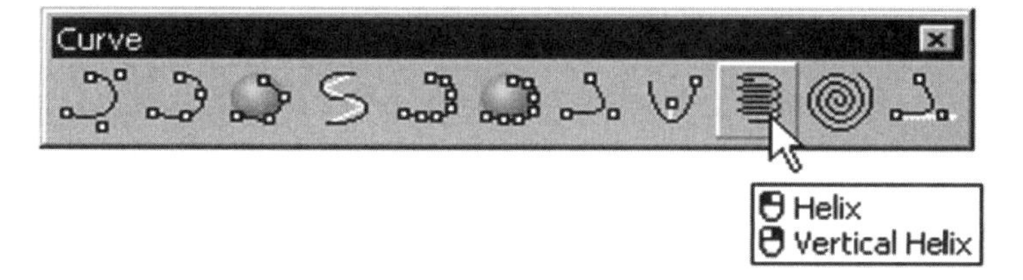

Fig. 2-83. Helix button of the Curve toolbar.

Helix Command

Table 2-8 outlines the functions of the helix commands and their locations in the main pull-down menu and respective toolbars.

Table 2-8: Helix Commands and Their Functions

Toolbar Button	From Main Pull-down Menu	Function
Helix	Curve > Helix	Constructing a helix with its axis on or perpendicular to the current construction plane by specifying the end points of the axis, the number of turns, and the height of the helix
Vertical Helix	Curve > Helix	Constructing a helix perpendicular to the current construction plane

Helix, Vertical Helix, and Helix Around Curve

In the following you will construct three helix curves.

1 Start a new file. Use the metric (Millimeters) template.

2 Select Curve > Helix, or click on the *Helix/Vertical Helix* button on the Curve toolbar.

3 Select locations A and B (indicated in figure 2-84) in the Top viewport to specify the axis end points.

4 Type *T* at the command line interface and type a number to specify the number of coils.

5 Select location C (indicated in figure 2-84) in the Top viewport to specify the radius of the helix.

6 Select Curve > Helix and type *V* at the command line interface, or right-click on the *Helix/Vertical Helix* button on the Curve toolbar.

7 Select location D (indicated in figure 2-84) in the Top viewport to specify an axis end point.

8 Select location E (indicated in figure 2-84) in the Front viewport to specify the other axis end point.

9 Type *T* at the command line interface and type a number to specify the number of coils.

10 Select location F (indicated in figure 2-84) in the Front viewport to specify the radius.

11 Construct a free-form curve A in the Top viewport, as shown in figure 2-85.

12 Select Curve > Helix, or click on the *Helix/Vertical Helix* button on the Curve toolbar.

13 Type *A* at the command line interface.

14 Select the free-form curve A (indicated in figure 2-85).

15 Type *T* at the command line interface and type a number to specify the number of coils.

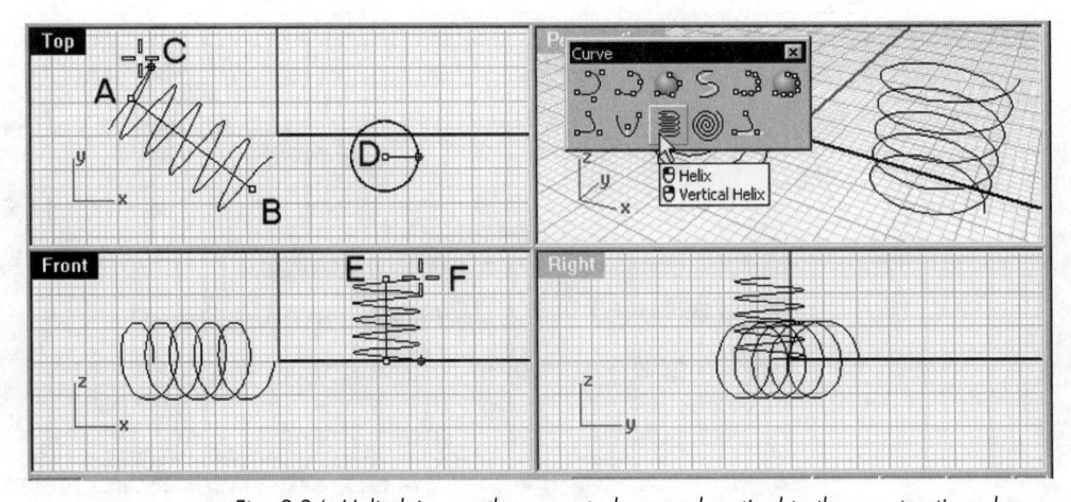

Fig. 2-84. Helix lying on the current plane and vertical to the construction plane.

16 Select location B (indicated in figure 2-85) to specify the radius of the helix.

17 If you want to save your file, save it as *Helix.3dm*.

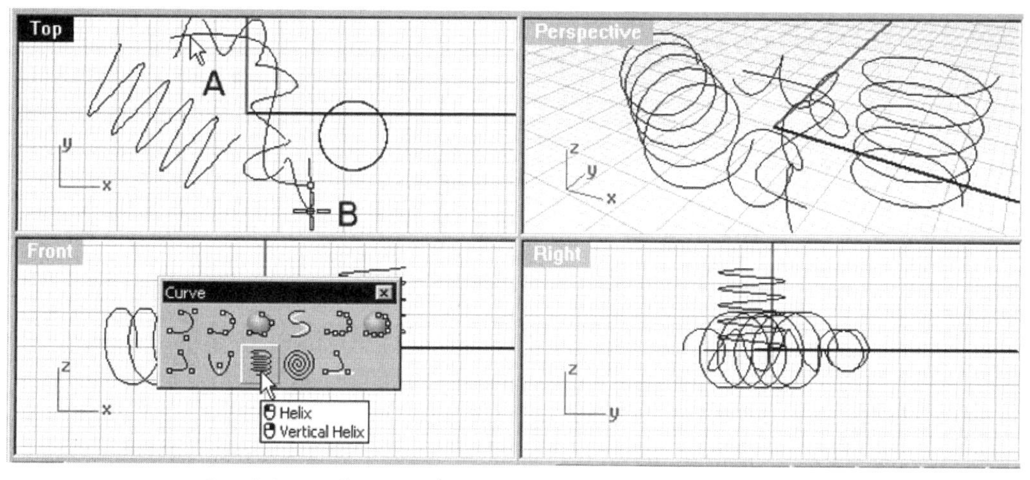

Fig. 2-85. Helix around curve.

Spiral

A spiral is a degree 3 curve. You can use the Spiral command to construct a 3D spiral or a flat spiral. You can construct a 3D spiral with a straight axis or around a curve. Figure 2-86 shows the Spiral button on the Curve toolbar.

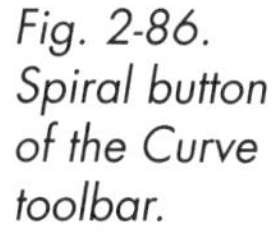

Fig. 2-86. Spiral button of the Curve toolbar.

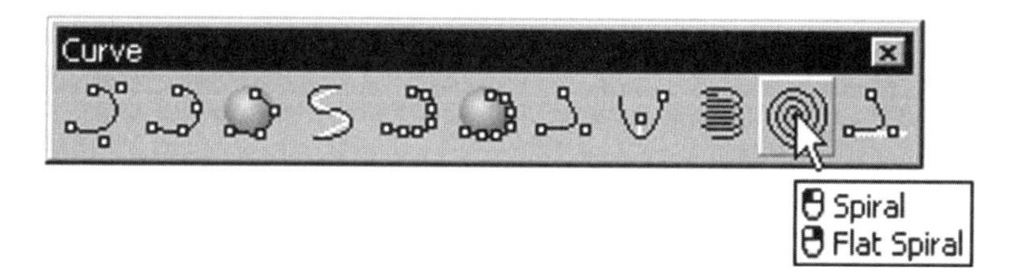

Spiral Commands

Table 2-9 outlines the function of the spiral commands and their locations in the main pull-down menu and respective toolbars.

Table 2-9: Spiral Commands and Their Functions

Toolbar Button	From Main Pull-down Menu	Function
Spiral	Curve > Spiral	Constructing a 3D spiral by specifying the axis, the radii at the axis end points, and the number of turns
Flat Spiral	Curve > Spiral	Constructing a flat spiral by specifying the axis, the radii at the axis end points, and the number of turns

Spiral, Flat Spiral, and Spiral Around Curve

In the following you will construct three spiral curves.

1. Start a new file. Use the metric (Millimeters) template.
2. Maximize the Top viewport.
3. Select Curve > Spiral, or click on the *Spiral/Flat Spiral* button on the Curve toolbar.
4. Select locations A and B (shown in figure 2-87) to indicate the end points of the axis.
5. Type *T* at the command line interface and type a number to specify the number of coils.
6. Select locations C and D (shown in figure 2-87) to indicate the first and second radii.
7. Select Curve > Spiral and type *F* at the command line interface, or right-click on the *Spiral/Flat Spiral* button on the Curve toolbar.

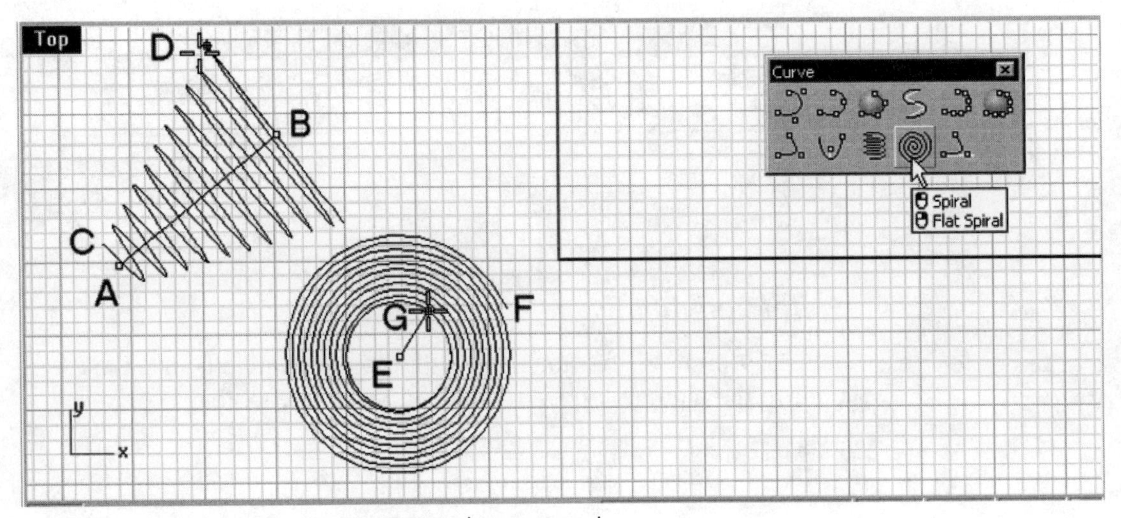

Fig. 2-87. Spirals constructed.

8 Select location E (shown in figure 2-87) to indicate the center of the spiral.
9 Select location F (shown in figure 2-87) to indicate the first radius.
10 Type *T* at the command line interface and type a number to specify the number of coils.
11 Select location G (shown in figure 2-87) to indicate the second radius.
12 With reference to figure 2-88, construct a free-form curve A.
13 Select Curve > Spiral, or click on the *Spiral/Flat Spiral* button on the Curve toolbar.
14 Type *A* at the command line interface.

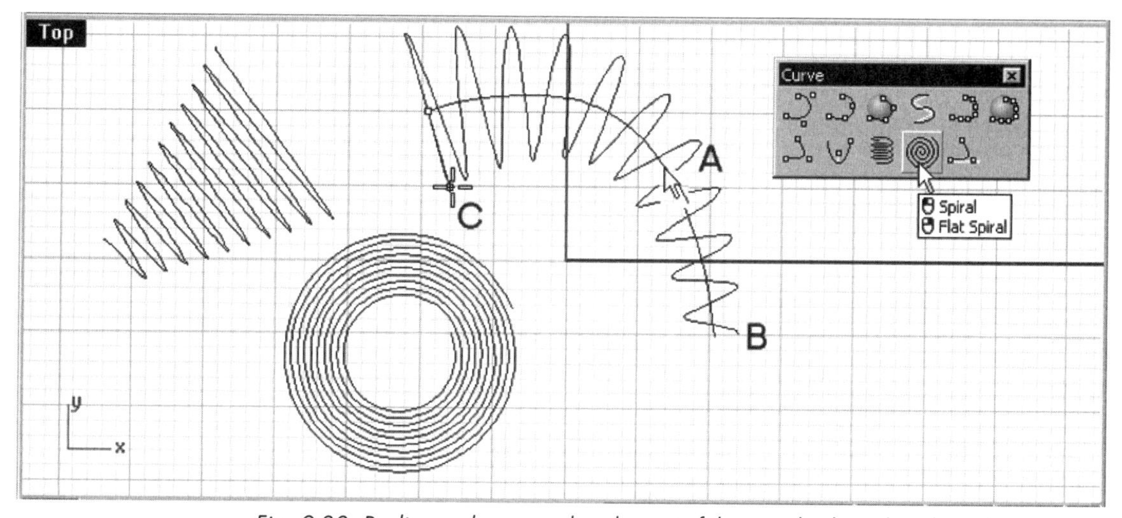

Fig. 2-88. Radius at the second end point of the spiral selected and number of turns specified.

15 Select curve A (indicated in figure 2-88).
16 Select location B (shown in figure 2-88) to indicate the first radius.
17 Type *T* at the command line interface and type a number to specify the number of coils.
18 Select location C (shown in figure 2-88) to indicate the second radius.

A spiral around the conic curve is constructed.

19 If you want to save your file, save it as *Spiral.3dm*.

Polygon

A polygon is a set of connected line segments. A polygon is a degree 1 curve. Using Rhino, you can construct two types of polygons: a regular polygon and a star-shaped polygon. You can construct a regular polygon in several ways: specifying the number of sides, the center, and a corner; specifying the center and a midpoint of an edge; or specifying two end points of an edge.

To construct a star, you specify the number of sides, corner of the star, and the second radius of the star. In addition to constructing a polygon or star on the current construction plane, you can construct them around a curve. To construct a regular polygon, you use the Polygon option from the Polygon toolbar, shown in figure 2-89.

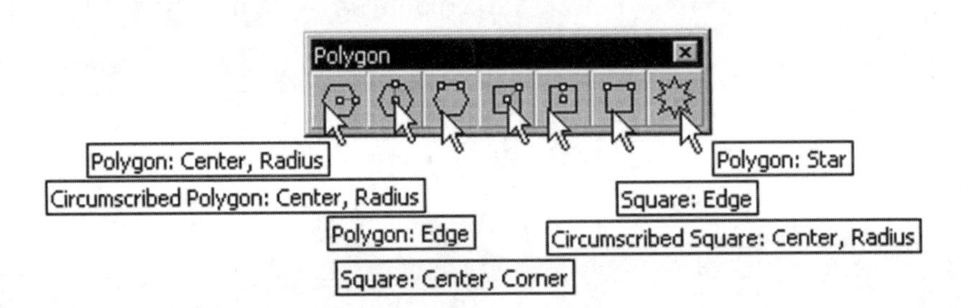

Fig. 2-89. Polygon toolbar.

Polygon Commands

Table 2-10 outlines the functions of polygon commands and their locations in the main pull-down menu and respective toolbars.

Table 2-10: Polygon Commands and Their Functions

Toolbar Button	From Main Pull-down Menu	Function
Polygon: Center, Radius	Curve > Polygon > Center, Radius	Constructing a polygon by specifying the number of sides, the center of the polygon, and the radius of the inscribed circle
Circumscribed Polygon: Center, Radius	—	Constructing a polygon by specifying the number of sides, the center of the polygon, and the radius of the circumscribed circle
Polygon: Edge	Curve > Polygon > By Edge	Constructing a polygon by specifying the number of sides and two end points of a side
Square: Center, Corner	Curve > Polygon > Center, Radius	Constructing a square by specifying the center of the square and the radius of the inscribed circle
Circumscribed Square: Center, Radius	—	Constructing a square by specifying the center of the square and the circumscribed circle

Toolbar Button	From Main Pull-down Menu	Function
Square: Edge	Curve > Polygon > By Edge	Constructing a square by specifying two end points of a side
Polygon: Star	Curve > Polygon > Star	Constructing a star by specifying the number of sides, the center of the star, the radius of the inscribed circle, and the second radius of the star

Polygon and Polygon Around Curve

To construct a regular polygon around a helix curve, perform the following steps.

1. Start a new file. Use the metric (Millimeters) template.
2. Maximize the Top viewport.
3. Select Curve > Polygon > Center, Radius, or click on the *Polygon: Center, Radius* button on the Polygon toolbar.
4. Select location A (indicated in figure 2-90) to specify the center.
5. Type *N* at the command line interface to set the number of sides.
6. Type *6* to set the number of sides to six.
7. Select location B (indicated in figure 2-90) to specify the radius.
8. Select Curve > Polygon > Star, or click on the *Polygon: Star* button on the Polygon toolbar.
9. Select location C (shown in figure 2-90) to indicate the center.
10. If you wish to change the number of sides, type *N* at the command line interface and then type a value.

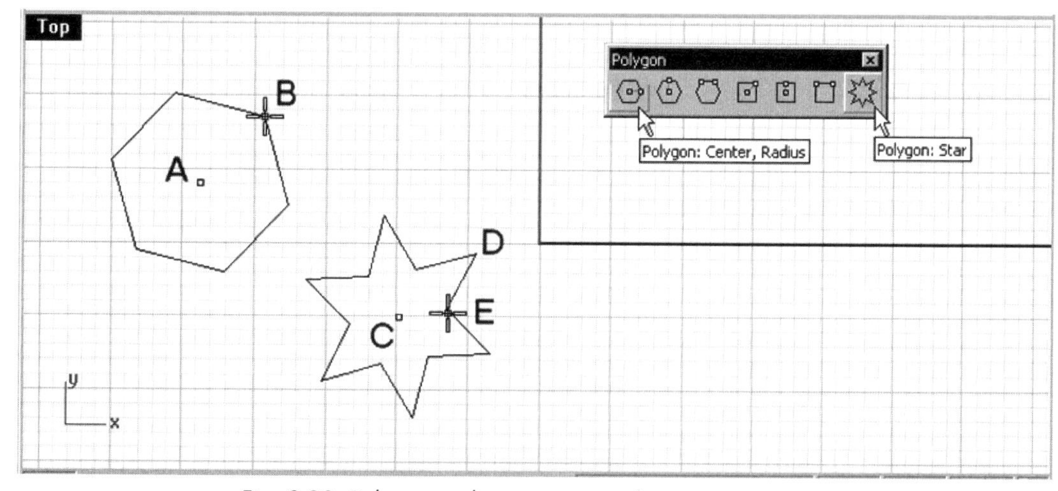

Fig. 2-90. Polygon and star constructed.

11 Select location D (shown in figure 2-90) to indicate the radius.

12 Select location E (shown in figure 2-90) to indicate the star radius.

13 Set the display to a four-viewport configuration by double clicking on the Top viewport label.

14 With reference to figure 2-91, construct a free-form curve in the Top viewport's label.

15 Select Curve > Polygon > Center, Radius, or click on the *Polygon: Center, Radius* button on the Polygon toolbar.

16 Type *A* at the command line interface.

17 Select curve A (indicated in figure 2-91) near its end point.

18 Select location B (shown in figure 2-91) to indicate the radius.

19 If you want to save your file, save it as *Polygon.3dm.*

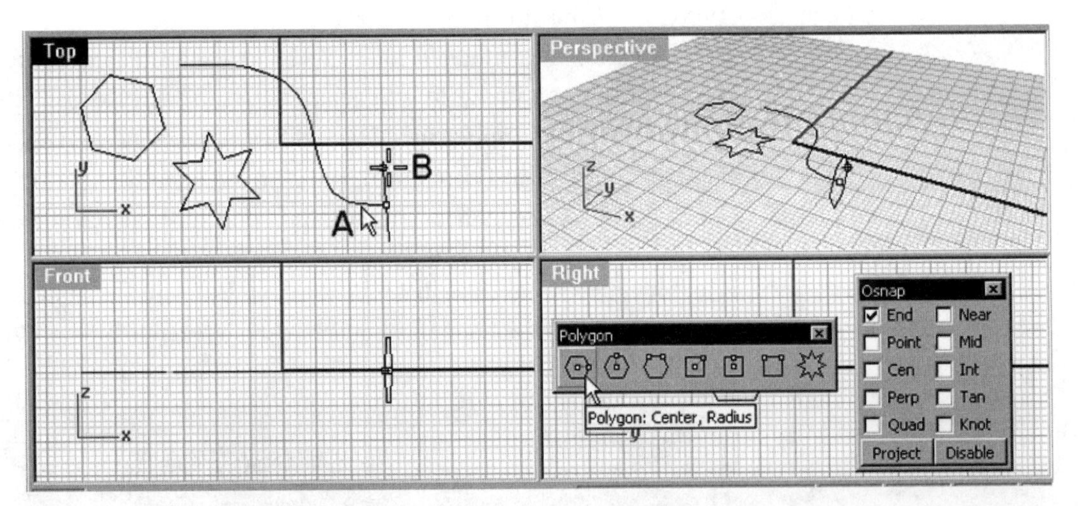

Fig. 2-91. Polygon around a curve.

Rectangle

Using Rhino, you can construct three types of rectangles. The first, consisting of four joined line segments, is a degree 1 curve. The second, consisting of four line segments and four arcs (filleted corners) joined, is a degree 2 curve. The third, consisting of four joined conic curves, is also a degree 2 curve. If you explode a rectangle, the first rectangle type will decompose into four separate line segments, the second will decompose into lines and arc segments, and the third will decompose into four conic curves. To construct a rectangle, you use various options of the

Rectangle command by accessing the pull-down menu or the Rectangle toolbar, shown in figure 2-92.

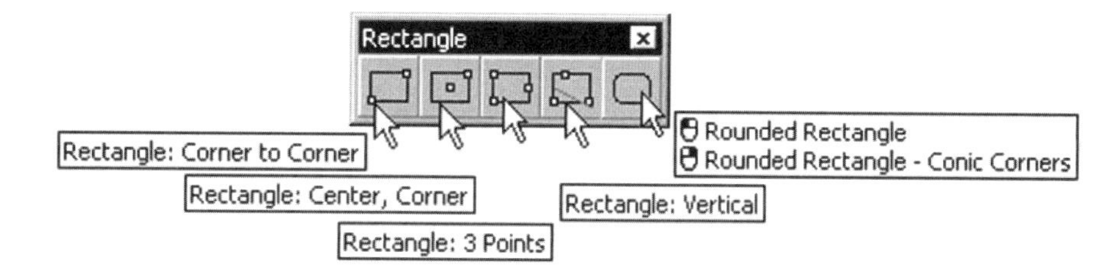

Fig. 2-92. Rectangle toolbar.

Rectangle Commands

Table 2-11 outlines the functions of rectangle commands and their locations in the main pull-down menu and respective toolbars.

Table 2-11: Rectangle Commands and Their Functions

Toolbar Button	From Main Pull-down Menu	Function
Rectangle: Corner to Corner	Curve > Rectangle > Corner to Corner	Constructing a rectangle by specifying its diagonal corners
Rectangle: Center, Corner	Curve > Rectangle > Center, Corner	Constructing a rectangle by specifying the center of the rectangular area and a corner of the rectangle
Rectangle: 3 Points	Curve > Rectangle > 3 Points	Constructing a rectangle by specifying two end points of an edge and the width
Rectangle: Vertical	Curve > Rectangle > Vertical	Constructing a rectangle perpendicular to the current construction plane by specifying two end points of an edge and the height of the rectangle
Rounded Rectangle/ Rounded	—	Constructing a rectangle that consists of four lines and four arc segments
Rectangle: Conic Corners	—	Constructing a rectangle that consists of four conic segments

Rectangle, Rounded Rectangle, and Conic Rectangle

To construct a rectangle, a rounded rectangle, and a conic rectangle, perform the following steps.

1. Start a new file. Use the metric (Millimeters) template.
2. Maximize the Top viewport.

3 Select Curve > Rectangle > Corner to Corner, or click on the *Rectangle: Corner to Corner* button on the Rectangle toolbar.
4 Select locations A and B (indicated in figure 2-93).
5 Select Curve > Rectangle > Corner to Corner and type *R* at the command line interface, or click on the *Rounded Rectangle/Rounded Rectangle: Conic Corners* button on the Rectangle toolbar.
6 Select locations C and D (indicated in figure 2-93).
7 Type a value at the command line interface to specify the radius, or select location E (indicated in figure 2-93).
8 Right-click on the *Rounded Rectangle/Rounded Rectangle: Conic Corners* button on the Rectangle toolbar.
9 Select locations A and B (indicated in figure 2-94).

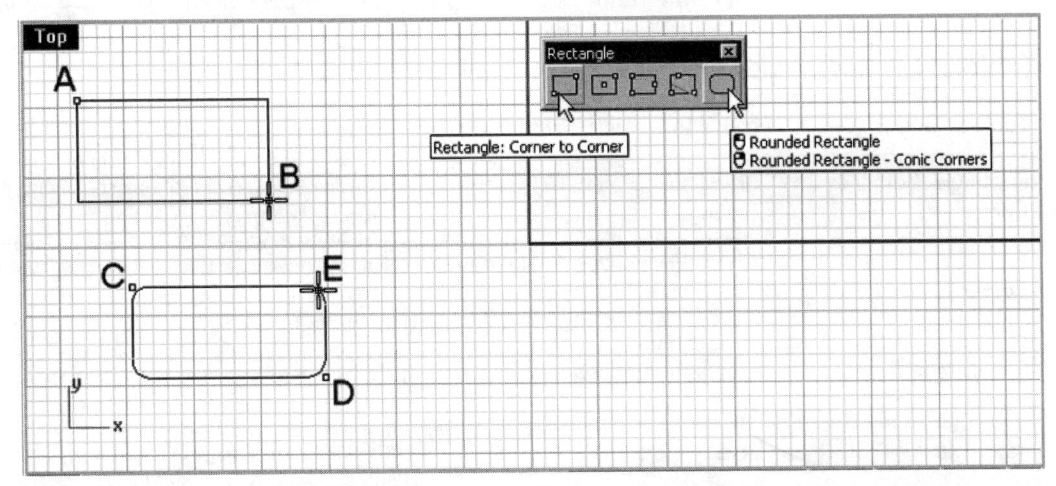

Fig. 2-93. Rectangles.

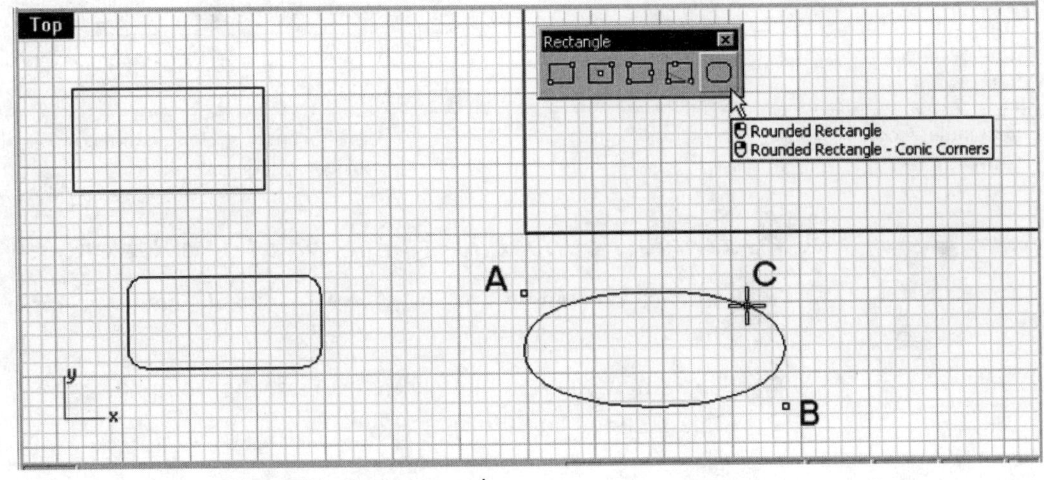

Fig. 2-94. Conic rectangle.

10 Select location C (indicated in figure 2-94).

11 If you want to save your file, save it as *Rectangle.3dm.*

Constructing Derived Curves

As well as constructing basic geometric curves, you can construct curves from existing objects, such as curves, surfaces, and solids. From existing curves, you can derive seven types of curves. You can extend a curve, construct a fillet or chamfer at the intersection of two curves, offset a curve, blend two curves, construct a curve from two planar curves residing on two viewports, or construct cross-section curves across a set of longitudinal profile curves. Figure 2-95 shows the Curve Tools toolbar and the Extend toolbar.

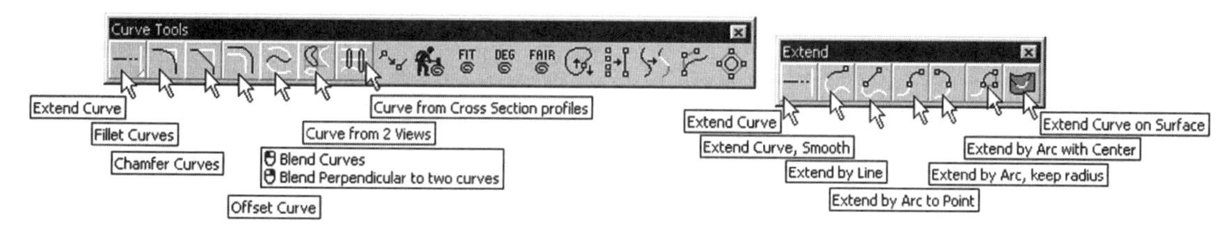

Fig. 2-95. Curve Tools and Extend toolbars.

Table 2-12 outlines the functions of curve tool extend commands and their locations in the main pull-down menu and respective toolbars.

Table 2-12: Curve Tool Extend Commands and Their Functions

Toolbar Button	From Main Pull-down Menu	Function
Extend Curve	Curve > Extend Curve > Extend Curve	Extending a curve to a boundary curve by attaching an arc to the curve
Extend Curve, Smooth	—	Lengthening a curve smoothly to a selected location
Extend by Line	Curve > Extend Curve > By Line	Extending a curve to a selected location by attaching a tangent line to the curve
Extend by Arc to Point	Curve > Extend Curve > By Arc to Point	Extending a curve to a selected location by attaching a tangent arc to the curve
Extend by Arc, Keep Radius	—	Extending a curve by attaching a tangent arc with a radius equal to the radius of curvature at the point of extension

Toolbar Button	From Main Pull-down Menu	Function
Extend by Arc with Center	Curve > Extend Curve > By Arc	Extending a curve by attaching a tangent arc that is specified by a center point and an end point
Extend Curve on Surface	Curve > Extend Curve > Curve on Surface	Extending a curve on a surface to the boundary edge of the surface
Fillet Curves	Curve > Fillet Curves	Constructing a tangent arc at the intersection of two curves
Chamfer Curves	Curve > Chamfer Curves	Constructing a beveled edge at the intersection of two curves
Offset Curves	Curve > Offset Curve	Constructing a curve that offsets at a distance from a selected curve
Blend Curves	Curve > Blend Curves	Constructing a tangent blend curve between two curves
Blend Perpendicular to Two Curves	Curve > Blend Curves	Constructing a perpendicular bend curve between two curves
Curve from 2 Views	Curve From 2 Views	Constructing a 3D curve from two planar curves residing on two different construction planes
Curve from Cross Section Profiles	Cross-Section Profiles	Constructing cross-section curves across a set of longitudinal profile curves

Extend

The Extend tool increases a curve's length. You extend a curve by extending it to a boundary curve, dragging an end of the curve to a new position, adding a straight line segment to the curve, adding an arc segment to the curve, or extending it to meet a surface. Start a new file. In the following you will construct two free-form curves. (By now, you should be able to use one of the methods delineated previously to construct a free-form curve.) Based on these curves, you will construct various types of derived curves.

1. Start a new file. Use the metric (Millimeters) template file.
2. Double click on the Top viewport title to maximize the viewport.
3. With reference to figure 2-96, construct two free-form curves A and B.
4. Select Curve > Extend Curve > Extend Curve, or click on the *Extend Curve, Smooth* button on the Extend toolbar.
5. Select curve A and press the Enter key.

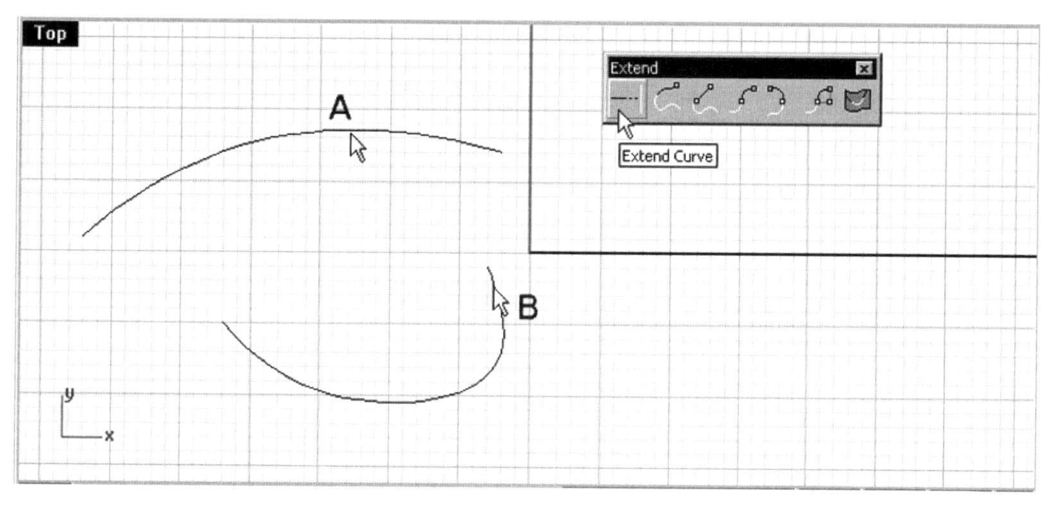

Fig. 2-96. Extended curve.

6 Select curve B to extend it to A.
7 Press the Enter key to terminate the command.

A curve is extended to a boundary curve.

8 Click on the *Extend Curve, Smooth* button on the Extend toolbar.
9 Select curve A (indicated in figure 2-97), and then select location B.

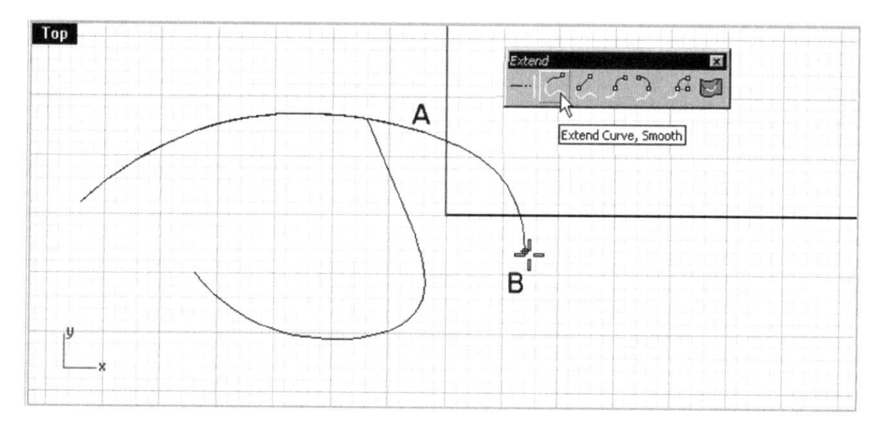

Fig. 2-97. Curve being extended by adding a spline segment.

The selected curve is extended by adding a spline segment at the end point.

10 Click on the Extend by Line button on the Extend toolbar.
11 Select curve A (indicated in figure 2-98), and then select location B.

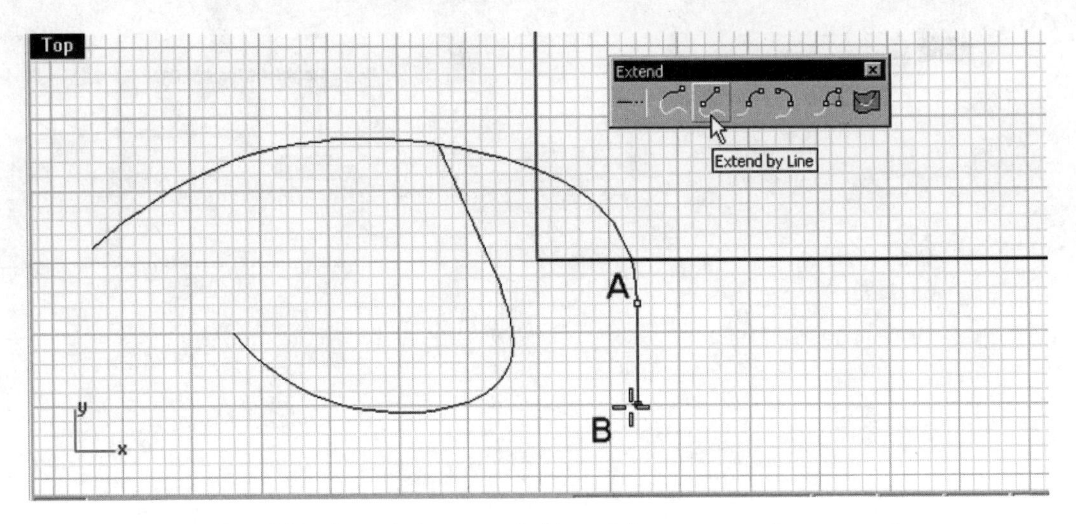

Fig. 2-98. Curve being extended by adding a tangent line segment.

The selected curve is extended by attaching to it a tangent line segment.

12 Click on the Extend by Arc to Point button on the Extend toolbar.

13 Select curve A (indicated in figure 2-99), and then select location B.

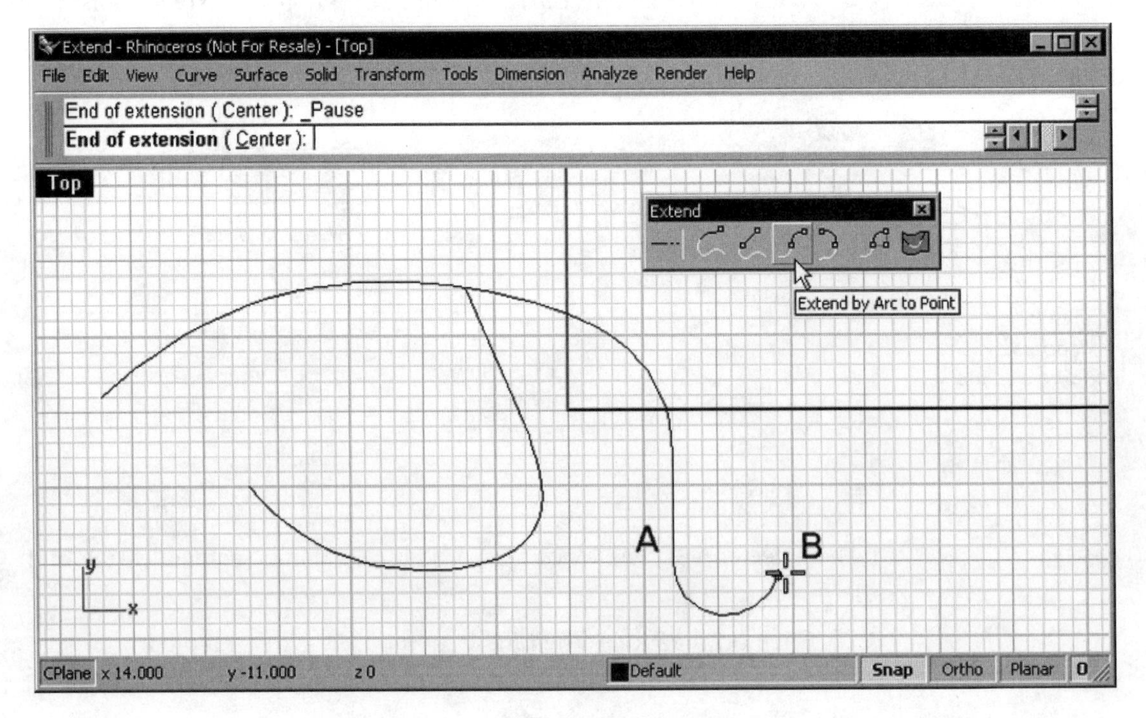

Fig. 2-99. Curve being extended by adding a tangent arc segment.

The selected curve is extended by attaching to it a tangent arc segment defined by the arc's end point.

14 Click on the *Extend by Arc, keep radius* button on the Extend toolbar.

15 Select curve A (indicated in figure 2-100), and then select location B.

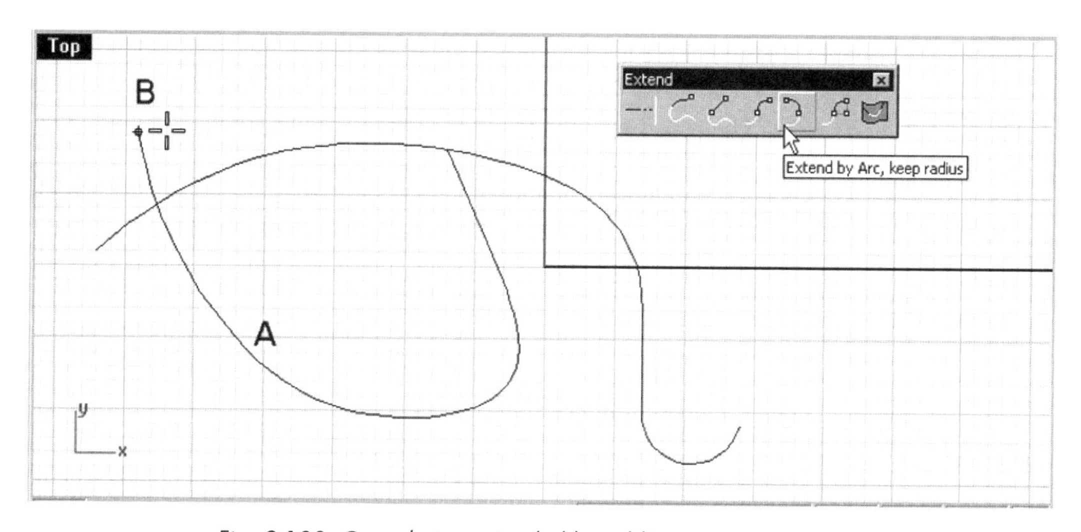

Fig. 2-100. Curve being extended by adding a tangent arc segment.

The selected curve is extended by attaching to it a tangent arc segment with a radius equal to the radius of curvature at the end point of the curve.

16 Click on the Extend by Arc with Center button on the Extend toolbar.

17 Select curve A (indicated in figure 2-101).

18 Select location B (indicated in figure 2-101) to specify the radius of the arc.

19 Select location C (indicated in figure 2-101) to specify the end point of the arc.

The selected curve is extended by attaching to it a tangent arc.

20 If you want to save your file, save it as *Extend1.3dm.*

To extend a curve on a surface to the surface's boundary edge, perform the following steps.

1 Open the file *Extend2.3dm* from the *Chapter 2* folder of the companion CD-ROM.

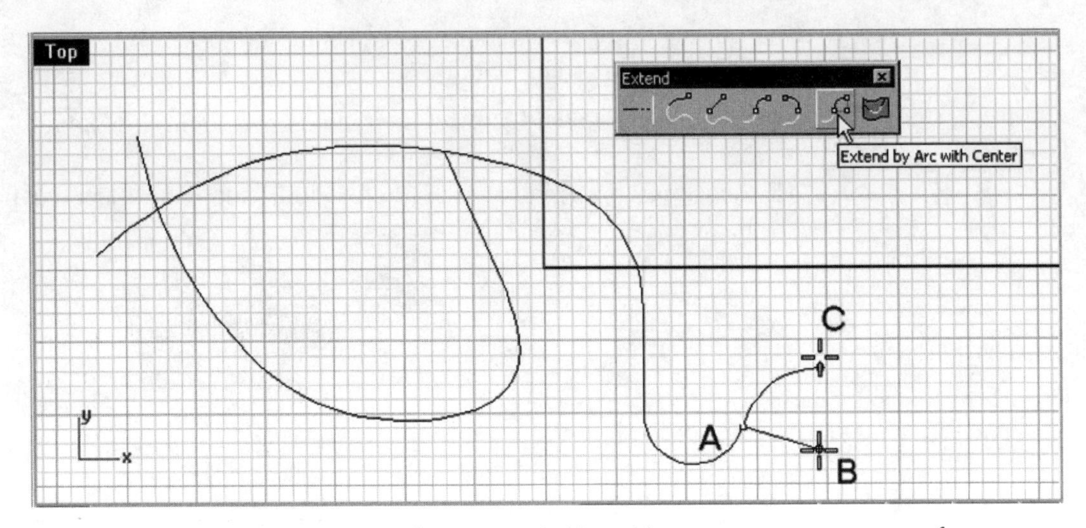

Fig. 2-101. Curve being extended by adding a tangent arc segment of designated radius.

2. Select Curve > Free-Form > Interpolate on Surface, or click on the Interpolate on Surface button on the Curve toolbar.
3. Select surface A (indicated in figure 2-102).
4. Select locations B, C, and D (indicated in figure 2-102) and press the Enter key.
5. Select Curve > Extend Curve > Curve on Surface, or click on the Extend Curve on Surface button.
6. Select curve A and surface B (indicated in figure 2-103).

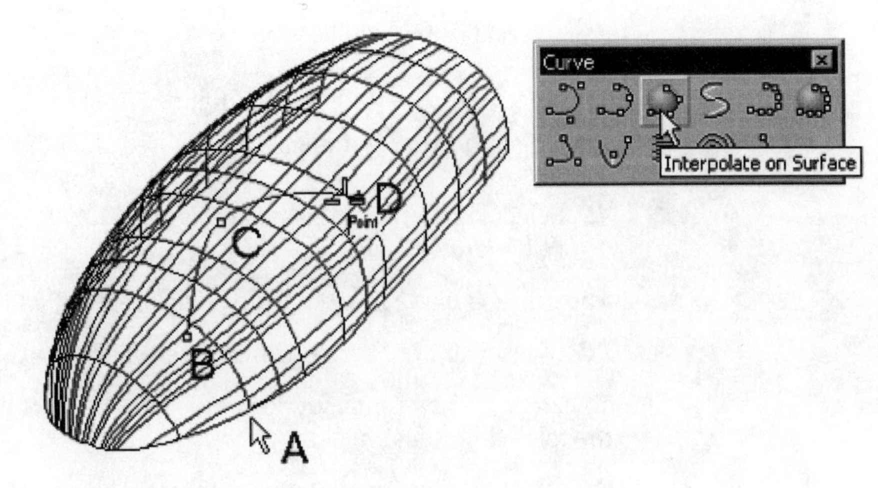

Fig. 2-102. Curve being constructed on a surface.

The curve on the surface is extended to the boundary edges of the surface.

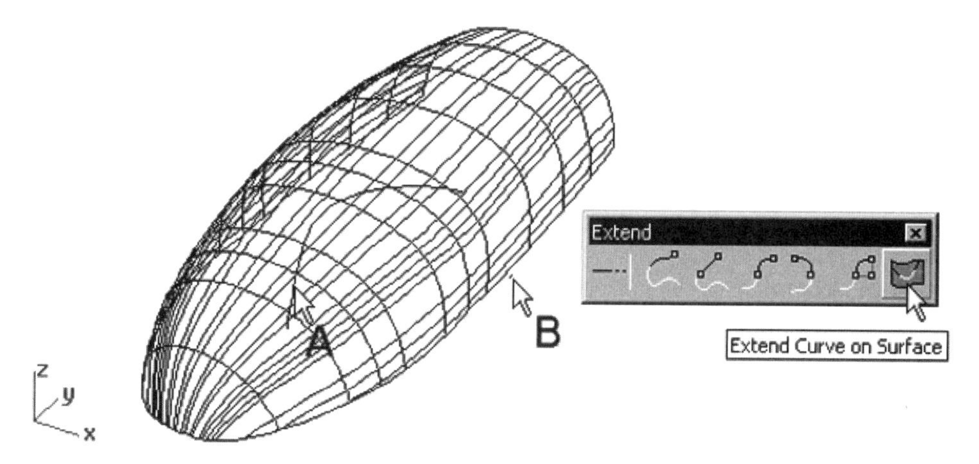

Fig. 2-103. Curve being extended.

Fillet

A fillet derives from two nonparallel curves. It is an arc segment tangent to two selected curves. To construct a fillet curve, perform the following steps.

1. Start a new file. Use the metric (Millimeters) template file.
2. Double click on the Top viewport title to maximize the viewport.
3. With reference to figure 2-104, construct two free-form curves.

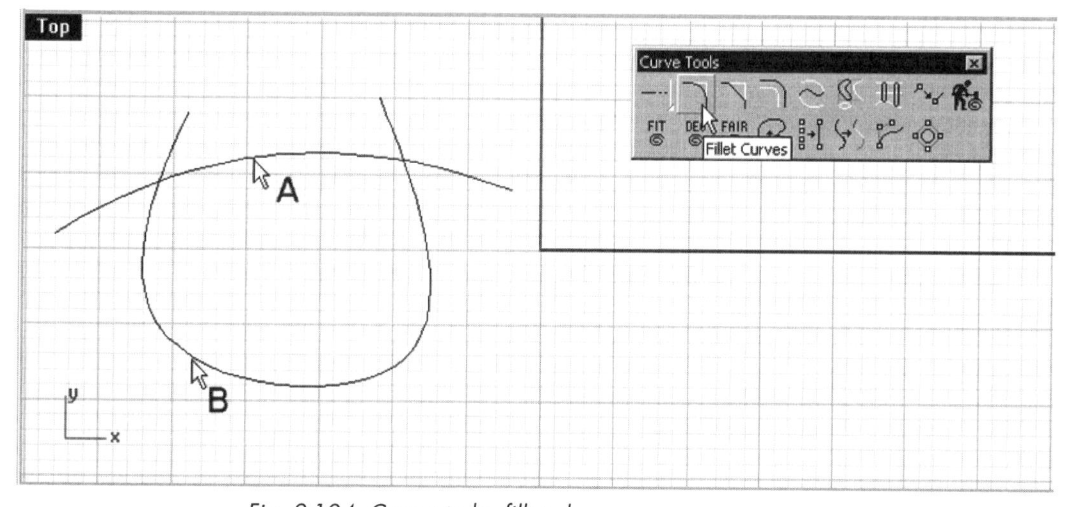

Fig. 2-104. Curves to be filleted.

4 Select Curve > Fillet Curves, or click on the Fillet Curves button on the Curve Tools toolbar.
5 Type *R* at the command line interface to use the Radius option.
6 Type *3* to specify a radius of 3 mm.
7 Select curves A and B, indicated in figure 2-104.

The selected curves are trimmed and a fillet curve is constructed.

Chamfer

A chamfer curve also derives from two nonparallel curves. It joins the selected curves with a beveled line. To construct a chamfer curve, perform the following steps.

1 Select Curve > Chamfer Curves, or click on the Chamfer Curves button on the Curve Tools toolbar.
2 Type *D* to use the Distance option.
3 Type *4* to set the first distance.
4 Type *3* to set the second distance.
5 Select curves A and B, indicated in figure 2-105.

A chamfer curve is constructed.

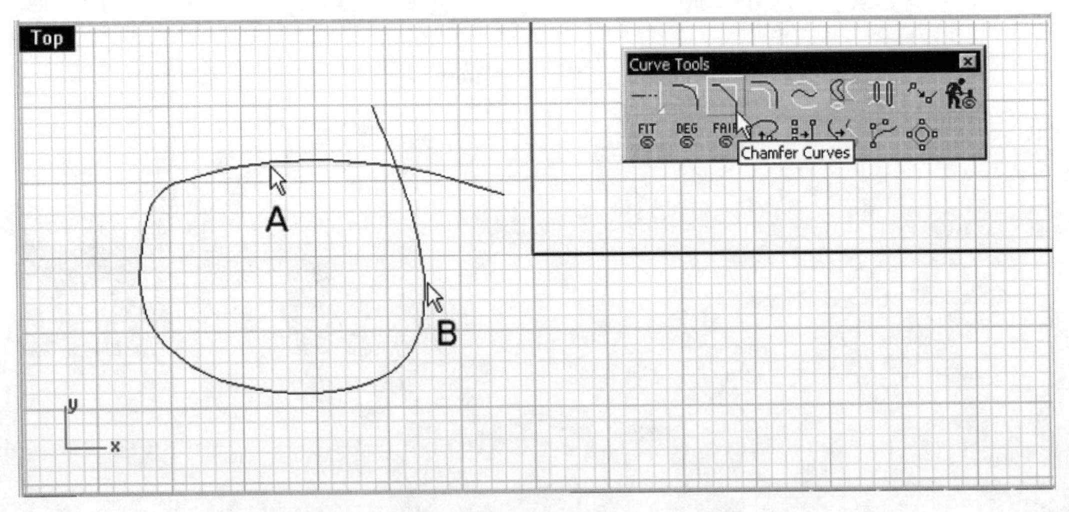

Fig. 2-105. Chamfer curve being constructed.

Offset

Offsetting derives a new curve at a specified distance from an existing curve. To construct offset curves, perform the following steps.

1. Select Curve > Offset Curve, or click on the Offset Curve button on the Curve Tools toolbar.
2. Type *D* to use the Distance option.
3. Type *4* to set the offset distance.
4. Select curve A and location B (indicated in figure 2-106).
5. Repeat steps 1 through 4 to offset curve A to location B (indicated in figure 2-107).

Two offset curves are constructed.

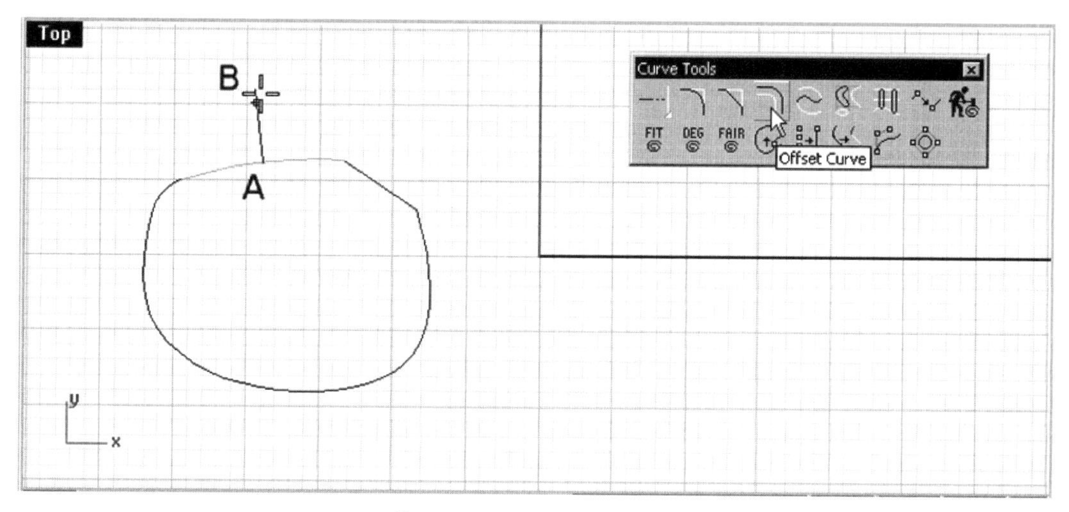

Fig. 2-106. Offset curve being constructed.

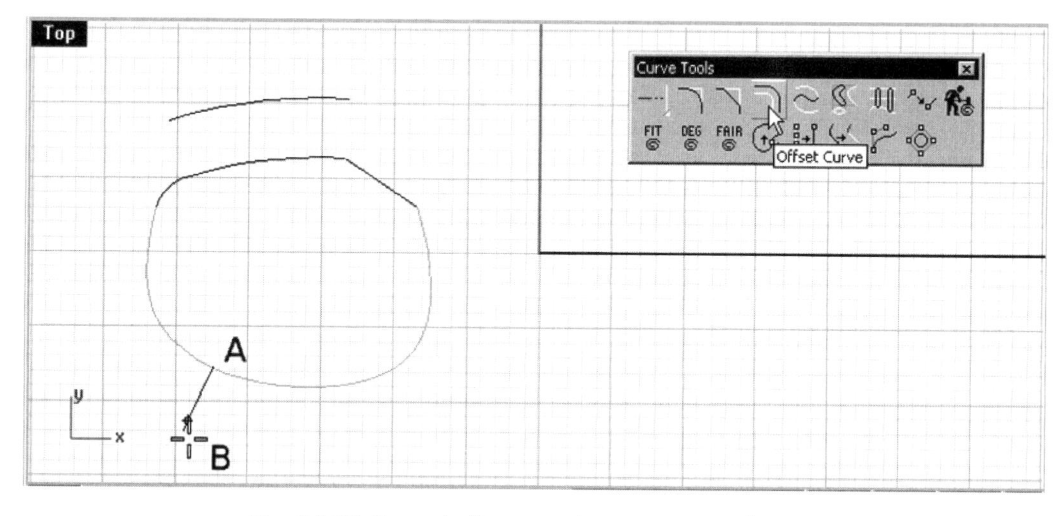

Fig. 2-107. Second offset curve being constructed.

Blend

Blending derives a curve that fits smoothly (curvature continuous) between two selected curves. To construct a blend curve, perform the following steps.

1. Click on the *Blend Curves/Blend Perpendicular to Two Curves* button on the Curve Tools toolbar.
2. Select curves A and B, indicated in figure 2-108

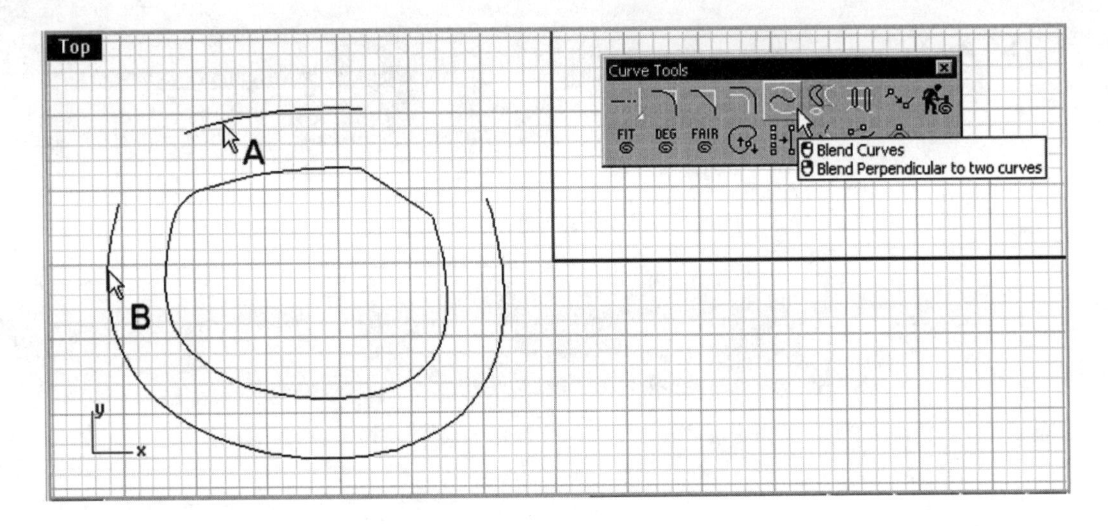

Fig. 2-108. Tangent blend curve being constructed.

A continuous curvature curve is constructed.

3. Check the End box on the Osnap dialog box.
4. Right-click on the *Blend Curves/Blend Perpendicular to Two Curves* button on the Curve Tools toolbar.
5. Select curve A and then end point B (indicated in figure 2-109).
6. Select curve C and then end point D (indicated in figure 2-109).

A perpendicular blend curve is constructed, as shown in figure 2-110.

7. If you want to save your file, save it as *DeriveCurve.3dm*.

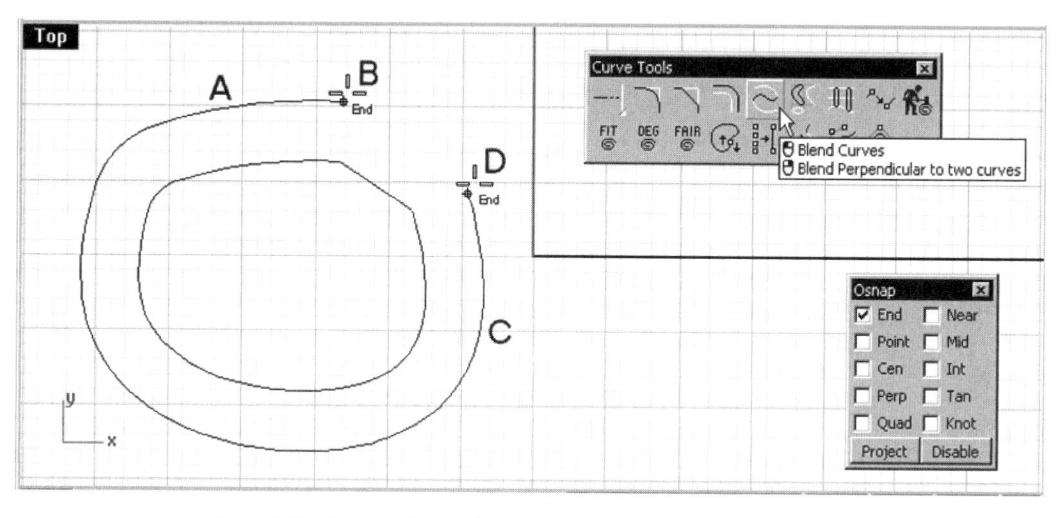

Fig. 2-109. Perpendicular blend curve being constructed.

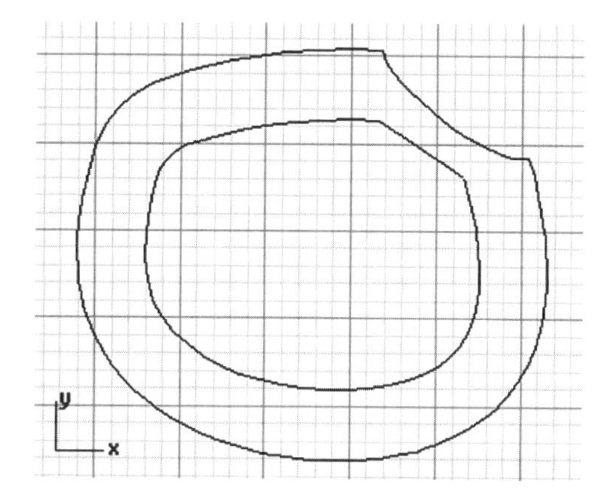

Fig. 2-110. Perpendicular blend curve.

3D Curve from Two Planar Curves

If you already have an idea of the shape of a 3D curve in two orthographic drawing views, you can first construct two planar curves residing on two adjacent viewports and derive the 3D curve from the planar curves. Perform the following steps to construct a 3D curve from two planar curves.

1. Start a new file. Use the metric (Millimeters) template file.
2. With reference to figure 2-111, construct a free-form curve on the Front viewport and another curve on the Right viewport. These curves represent the front and right views of a 3D curve.

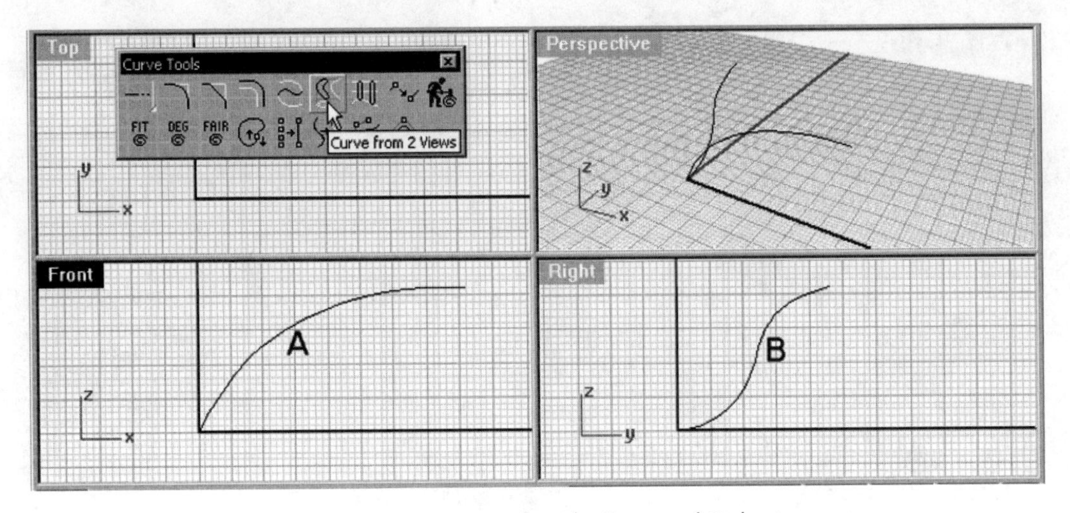

Fig. 2-111. Curves constructed on the Front and Right viewports.

3 Select Curve > Curve from 2 Views, or click on the Curve from 2 Views button on the Curve Tools toolbar.

4 Select curves A and B.

A 3D curve is derived, as shown in figure 2-112.

5 If you want to save your file, save it as *From2Curves.3dm.*

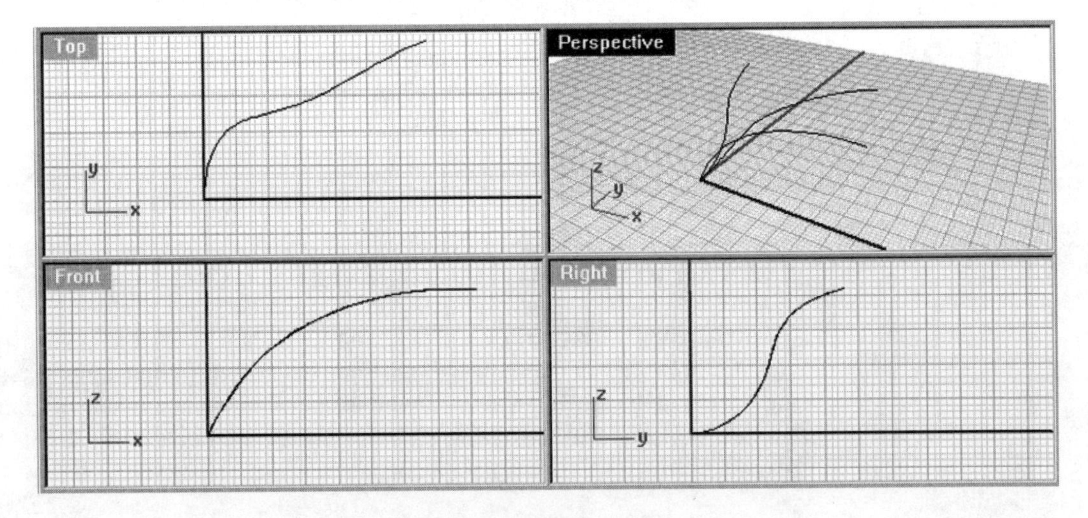

Fig. 2-112. 3D curve derived from two planar curves.

Cross-section Profiles

A cross-section profile is a closed-loop free-form curve interpolating the intersection points between a set of longitudinal curves and a specified section plane across the curves. Perform the following steps to construct cross-section profile curves.

1. Start a new file. Use the metric (Millimeters) template file.
2. With reference to figure 2-113, construct two free-form curves on the Top viewport and two more curves on the Front viewport.

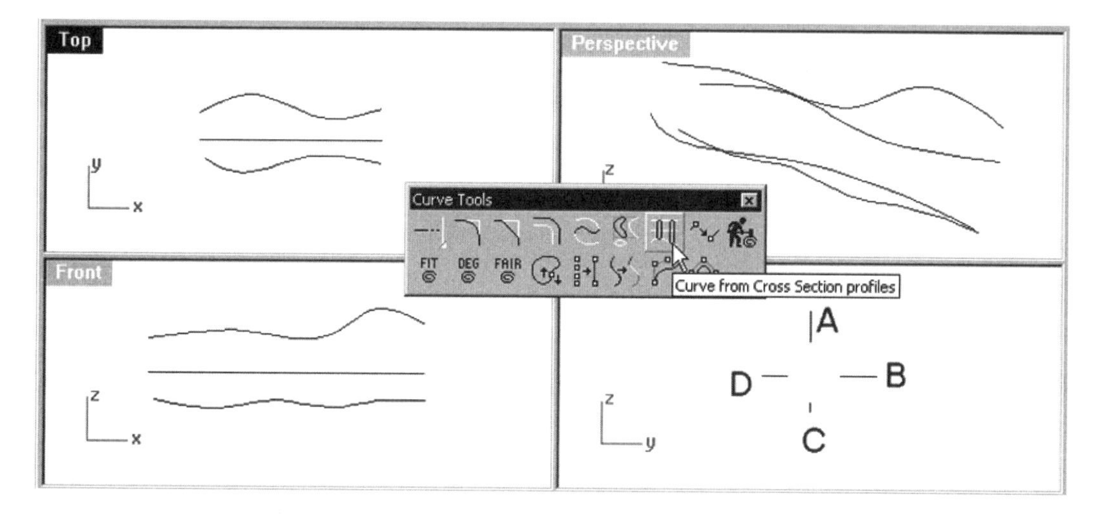

Fig. 2-113. Curves on the Top and Front viewports selected.

3. Select Curve > Cross Section Profiles, or click on the Curve from Cross Section Profiles button on the Curve Tools toolbar.
4. Select the profile curves indicated in figure 2-113 and press the Enter key.
5. Select locations A and B (indicated in figure 2-114).

A cross-section curve is constructed.

6. Select locations A and B (indicated in figure 2-115).

Another cross section curve is constructed.

7. Press the Enter key to terminate the command.

Two section profiles are constructed.

8. If you want to save your file, save it as *CrossSection.3dm*.

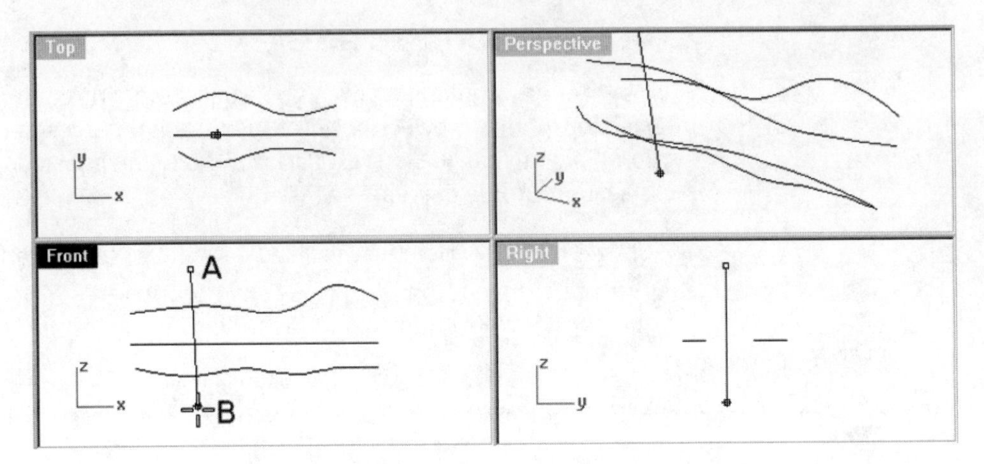

2-114. *First cross section being constructed.*

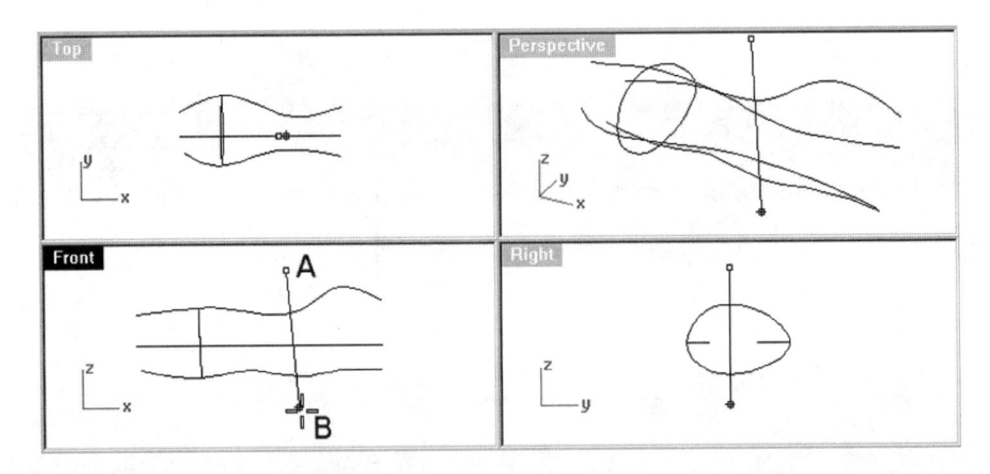

Fig. 2-115. Second cross section being constructed.

Summary

Among the three major types of computer models (wireframe, surface, and solid), the wireframe model has very limited application in design and downstream computerized manufacturing systems. Therefore, it is not sensible to exhaust our efforts in constructing complex 3D curves just to represent the edges and silhouettes of a 3D object. In this regard, curves are extremely useful and important in creating free-form surfaces and solids because you can use them as frameworks upon which surfaces and solids are constructed. Therefore, you need to have a good understanding of the characteristics of curves and the methods of constructing 3D curves.

To construct free-form surfaces and solids, NURBS (nonuniform rational basis spline) curves are commonly used in modeling applications. A NURBS curve is a set of polynomial spline segments. The shape of each spline segment in a curve is influenced by the degree of polynomial equation used to define the curve. As a designer, you may not need to know about the mathematics related to such equations, but you need to understand the relationship between degree of polynomial and the shape of a curve.

A line is degree 1, a circle is degree 2, and a free-form curve is degree 3 or above. The higher the degree, the more control points exist in a spline segment. For a curve of degree 3, each spline segment has four control points. Between two contiguous segments of a spline curve is a knot. Adding or removing knots increases or reduces the number of segments in a curve. A special type of knot on a curve is a kink, which is a junction between two contiguous segments where the tangent directions are not congruent. A closed curve without any kink is called a periodic curve. The overall shape of a NURBS curve is affected by the number of segments in the curve, the degree of polynomial equation, the location of control points, and the weight of the control points.

There are many ways to construct NURBS curves. Using Rhino, you can construct point objects, basic curves, and derived curves. Basic curves include lines, arcs, free-form curves, helixes, spirals, conics, ellipses, and parabolas. By deriving a curve from existing objects, you get a curve that has some form or shape correlation with existing objects. The objects from which you derive a curve can be curves, surfaces, or solids. Typically, you can derive curves from existing curves in seven ways: extending a curve, filleting two curves, chamfering two curves, offsetting a curve, blending two curves, deriving a 3D curve from two planar curves residing in two viewports, and deriving a set of cross-section curves across a set of longitudinal profile curves.

Review Questions

1. Outline the key concepts involved in wireframe modeling and explain the limitations of the wireframe model.
2. Explain the significance of curves in computer modeling.
3. With the aid of sketches, explain the characteristics and key features of a NURBS curve.
4. Give a brief account of the types of basic geometric point objects and curves you can construct using Rhino.
5. Depict the methods of deriving curves from existing curves.

CHAPTER 3

Wireframe Modeling and Curves: Part 2

Introduction

This chapter explores methods of constructing curve and point objects on existing objects. The chapter also shows you how to edit, transform, and analyze existing curves.

Objectives

After studying this chapter, you should be able to:

- ❒ Construct points and curves on existing objects
- ❒ Edit curves
- ❒ Transform curves
- ❒ Use various curve analysis tools

Overview

This chapter is a continuation of Chapter 2. Here you will learn how to construct points and curves on existing objects, and how to edit, transform, and analyze curves. Logically, you start making points and curves before you construct free-form NURBS surfaces. To detail your models and to be able to improvise during the design process, you derive points and curves from existing surfaces and construct more surfaces using the derived points and curves.

Point and Curve Construction and Analysis

Because the shape of the curves has a direct impact on the shape of the surfaces subsequently constructed, you may have to modify and transform the curves you construct. To examine the smoothness and other attributes of a curve, you use analysis tools.

Constructing Points and Curves from Objects

To develop your design from objects previously constructed, you derive points and curves. There are 13 ways to derive curves from existing surfaces. You construct such curves via the Curve From Object toolbar, shown in figure 3-1.

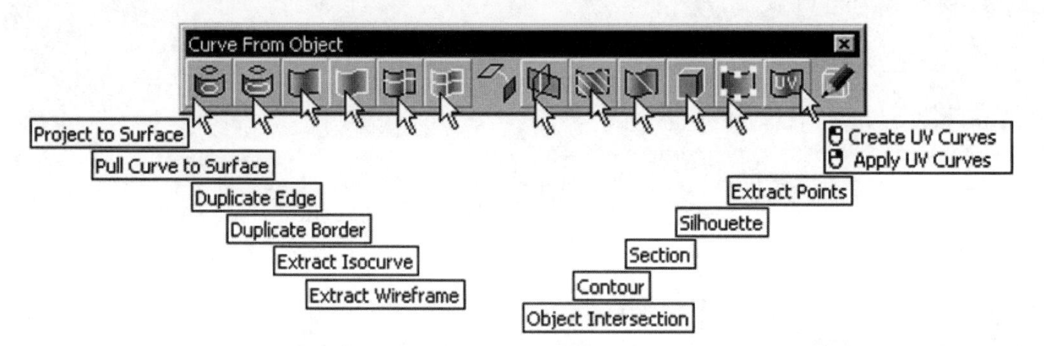

Fig. 3-1. Curve From Object toolbar.

Table 3-1 outlines the functions of the derived curve commands and their locations in the main pull-down menu and respective toolbars.

Table 3-1: Derived Curve Commands and Their Functions

Toolbar Button	From Main Pull-down Menu	Function
Project to Surface	Curve > Curve From Objects > Project	Constructing curves and points on surfaces by projecting selected curves and points in a direction perpendicular to the current construction plane
Pull Curve to Surface	Curve > Curve From Objects > Pullback	Constructing curves and points on surfaces by projecting selected curves and points normal to the selected surface
Duplicate Edge	Curve > Curve From Objects > Duplicate Edge	Constructing curves by duplicating selected edges of surfaces
Duplicate Border	Curve > Curve From Objects > Duplicate Border	Constructing curves by duplicating all edges of selected surfaces
Object Intersection	Curve > Curve From Objects > Intersection	Constructing curves or points at the intersection of selected surfaces and curves
Contour	Curve > Curve From Objects > Contour	Constructing a set of contour curves of selected surfaces

Toolbar Button	From Main Pull-down Menu	Function
Section	Curve > Curve From Objects > Section	Constructing section curves of selected surfaces
Silhouette	Curve > Curve From Objects > Silhouette	Constructing silhouette curves of selected surfaces
Extract Isocurve	Curve > Curve From Objects > Extract Isocurve	Constructing curves at specified locations along a selected surface that match surface isocurves
Extract Points	Curve > Curve From Objects > Extract Points	Constructing point objects at the control points of selected surfaces
Extract Wireframe	Curve > Curve From Objects > Extract Wireframe	Constructing a set of curves that match the edges and isocurves of selected surfaces
Create UV Curves	Curve > Curve From Objects > Create UV Curves	Mapping the edges and untrimmed edges of a selected surface on the current construction plane
Apply UV Curves	Curve > Curve From Objects > Apply UV Curves	Mapping selected planar curves on a selected surface

Projecting and Pulling a Curve to a Surface

Construction of 3D curves on a surface can be made easy by first constructing a 2D curve on one of the construction planes and then projecting it or pulling it toward the surface. The direction of projection is perpendicular to the construction plane, and the direction of pull is normal to the surface. To experiment with this functionality, perform the following steps.

1. Open the file *ProjectPull.3dm* from the *Chapter 3* folder on the companion CD-ROM.
2. Click on the Top viewport to set it as the current viewport.
3. Select Curve > Curve From Objects > Project, or click on the Project to Surface button on the Curve From Object toolbar.
4. Select curve A (indicated in figure 3-2) and press the Enter key.
5. Select surface B (indicated in figure 3-2) and press the Enter key.
6. Click on the Right viewport to set it as the current viewport.
7. Repeat steps 3 through 5 to project curve C (indicated in figure 3-2) onto surface D (indicated in figure 3-2).

Curves are projected onto the surfaces.

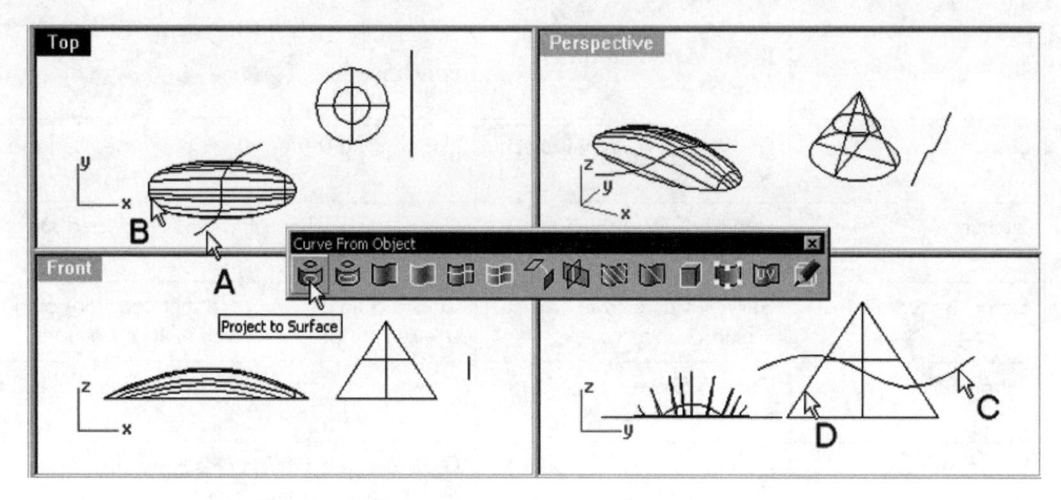

Fig. 3-2. Curves being projected onto surfaces.

8 Select Curve > Curve From Objects > Pullback, or click on the Pull Curve to Surface button on the Curve From Objects toolbar.
9 Select curve A (indicated in figure 3-3) and press the Enter key.
10 Select surface B (indicated in figure 3-3) and press the Enter key.
11 Repeat steps 8 through 10 to pull curve C onto surface D (indicated in figure 3-3).

Curves are pulled to the surface. Compare the result of pulling to that of projection.

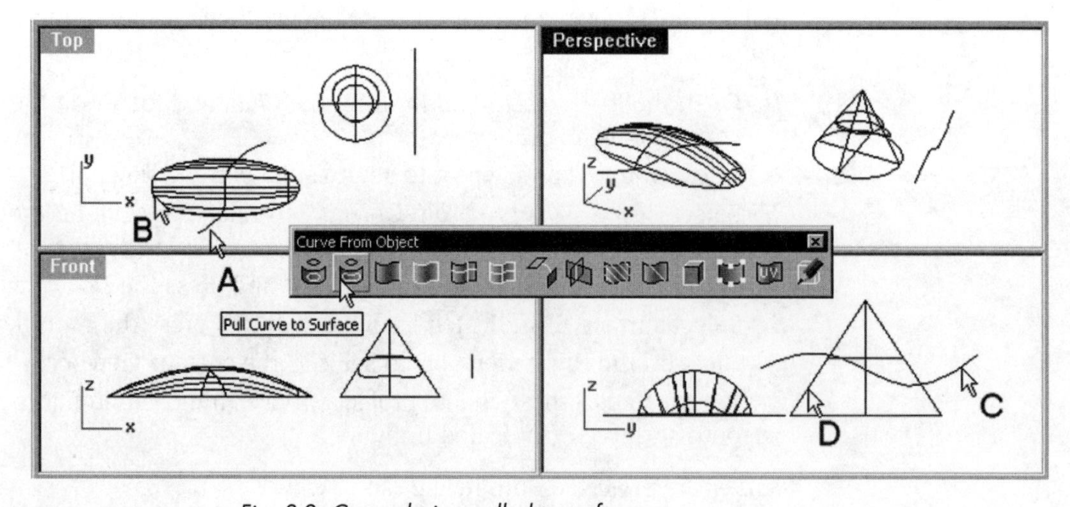

Fig. 3-3. Curves being pulled to surfaces.

Duplication of Surface Edge and Border

You can construct a 3D curve from an edge or the border of a surface. Superficially, an edge seems to be analogous to a border. However, an open surface can have multiple edges but only a single border. For example, a rectangular surface has four edges but just one continuous border. Perform the following steps to duplicate a surface's edge and border.

1. Open the file *EdgeBorder.3dm* from the *Chapter 3* folder on the companion CD-ROM.
2. Select Curve > Curve From Objects > Duplicate Edge, or click on the Duplicate Edge button on the Curve From Object toolbar.
3. Select edge A (indicated in figure 3-4) and press the Enter key.
4. Press the Esc key to ensure that nothing is selected.
5. Select Edit > Visibility > Hide, or click on the *Hide Objects/Show Objects* button on the Visibility toolbar.
6. Select surface B (indicated in figure 3-4) and press the Enter key.

An edge of the surface is duplicated and the surface is hidden. Now unhide the surface and duplicate its border, as follows.

7. Select Edit > Visibility > Show, or right-click on the *Hide Objects/Show Objects* button on the Visibility toolbar.
8. Select Curve > Curve From Objects > Duplicate Border, or click on the Duplicate Border button on the Curve From Object toolbar.
9. Select surface border A (indicated in figure 3-5) and press the Enter key.
10. Press the Esc key to ensure that nothing is selected.
11. Hide surface B (indicated in figure 3-5).

The border of the surface is duplicated and the surface is hidden.

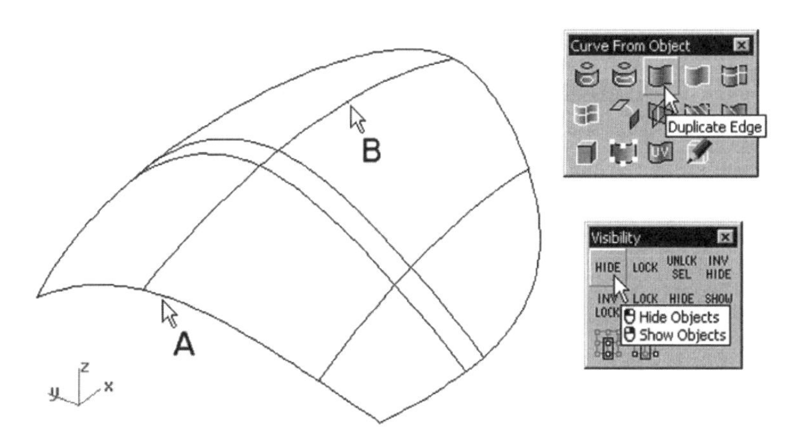

Fig. 3-4. Surface edge duplicated.

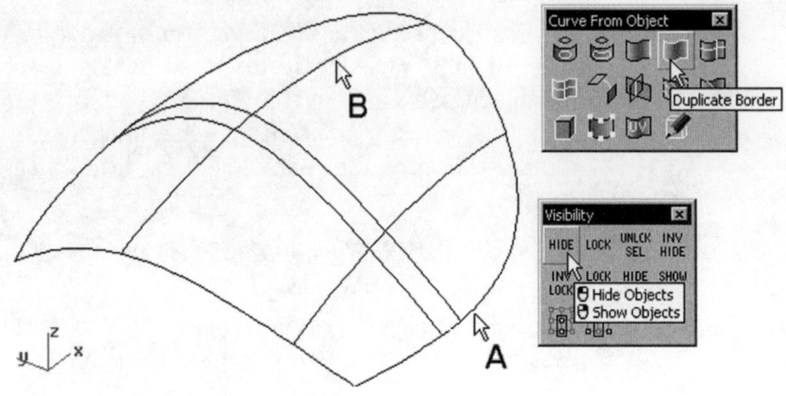

Fig. 3-5. Surface edge and border being duplicated.

Intersection of Surfaces/Curves

At the intersection of two surfaces or a surface and a curve you can construct an intersection curve or a point. Perform the following steps.

1. Open the file *Intersect.3dm* from the *Chapter 3* folder on the companion CD-ROM.
2. Select Curve > From Objects > Intersection, or click on the Object Intersection button on the Curve From Object toolbar.
3. Select surfaces A and B, and curve C (indicated in figure 3-6), and press the Enter key.

A curve is constructed at the intersection of surfaces A and B and two points are constructed at the intersection of surface A and curve C.

4. Hide the surfaces and curve to see the result, shown in figure 3-7.

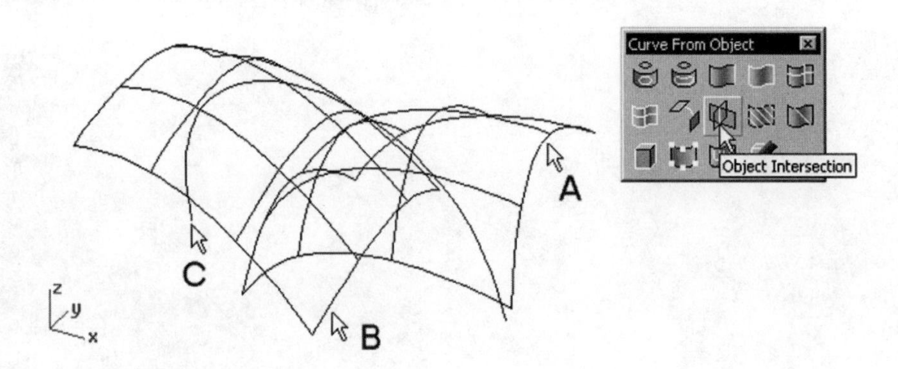

Fig. 3-6 Surface border being duplicated.

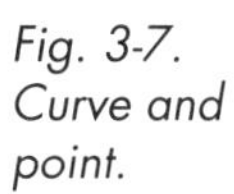
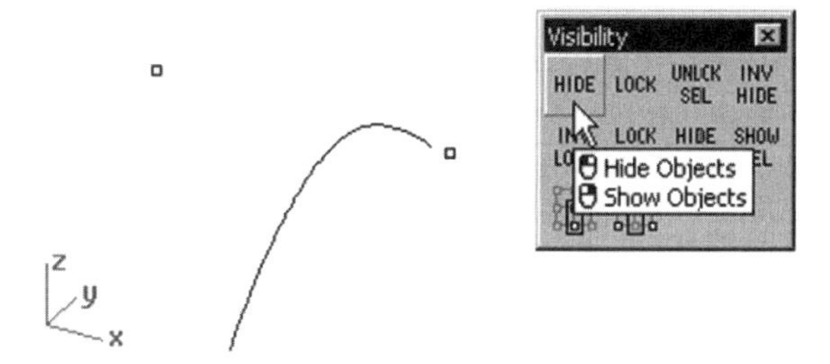

Fig. 3-7. Curve and point.

Contour Lines and Sections

Contour lines and sections are both curves derived on a surface. They are section lines cutting across a surface. Contour lines typically form a set of section lines spaced at a regular interval. To construct a set of contour lines across an ellipsoid surface, perform the following steps. Note that these commands also work on polygon meshes (you will learn about polygon meshes later in this chapter). You can construct a NURBS surface from a set of contour lines, and you can construct sections through a polygon mesh.

1. Open the file *ContourSection.3dm* from the *Chapter 3* folder on the companion CD-ROM.
2. Select Curve > Curve From Objects > Contour, or click on the Contour button on the Curve From Object toolbar.
3. Select surface A (indicated in figure 3-8) and press the Enter key.
4. Select points B and C (shown in figure 3-8) to indicate the contour plane base point and contour plane direction.
5. Select points D and E (shown in figure 3-8) to indicate the distance between contour lines.

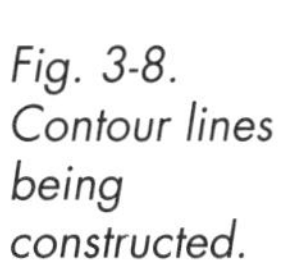
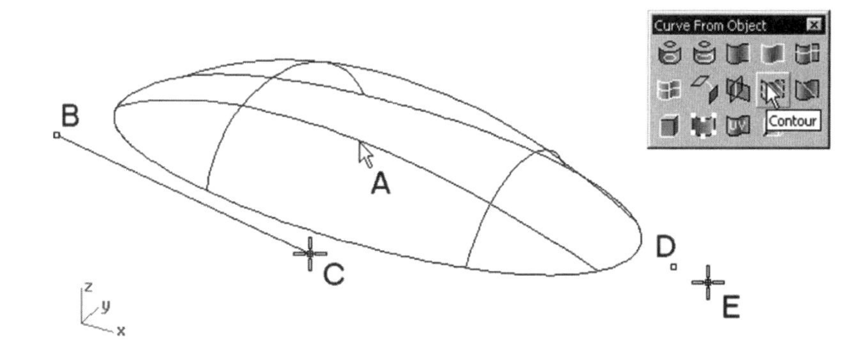

Fig. 3-8. Contour lines being constructed.

Contour lines are constructed along the ellipsoid surface. To construct a section across the surface, continue with the following steps.

6. Select Curve > Curve From Objects > Section, or click on the Section button on the Curve From Object toolbar.
7. Select surface A (indicated in figure 3-9) and press the Enter key.
8. Select points B and C (indicated in figure 3-9) to define a section plane.

A section curve is constructed.

9. Press the Enter key to terminate the command.

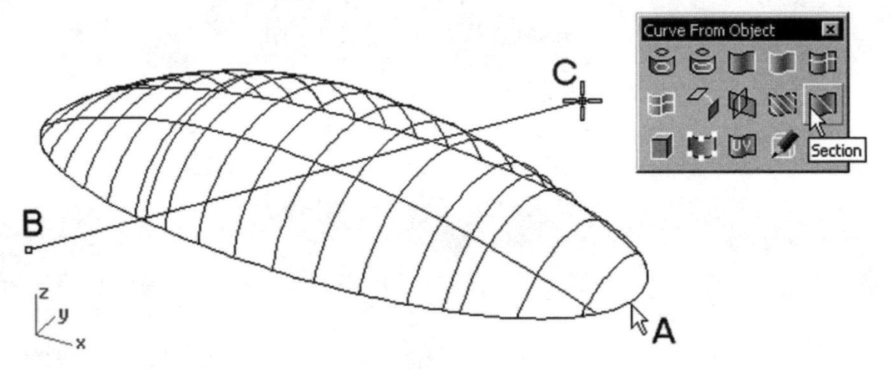

Fig. 3-9. Section line being constructed.

Silhouette of a Surface in a Viewport

The shape of the silhouette of a free-form surface resembles the edge of the shadow of the object projected in the viewing direction of the viewport. You see a different silhouette of the same object at different viewing angles. To construct a silhouette of an ellipsoid surface in a selected viewport, perform the following steps.

1. Open the file *Silhouette.3dm* from the *Chapter 3* folder on the companion CD-ROM.
2. Select Curve > Curve From Objects > Silhouette, or click on the Silhouette button on the Curve From Object toolbar.
3. Select surface A (indicated in figure 3-10) and press the Enter key.
4. Press the Esc key to ensure that nothing is selected.
5. Hide the surface to see the result, shown in figure 3-11.

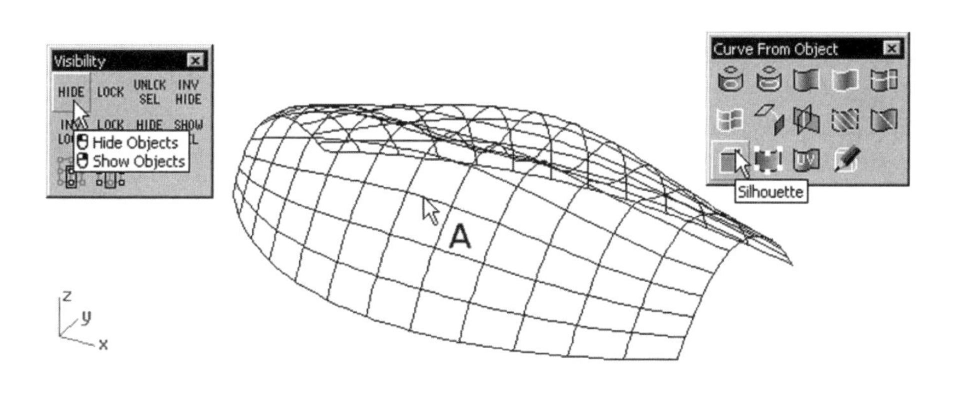

Fig. 3-10. Silhouette being constructed.

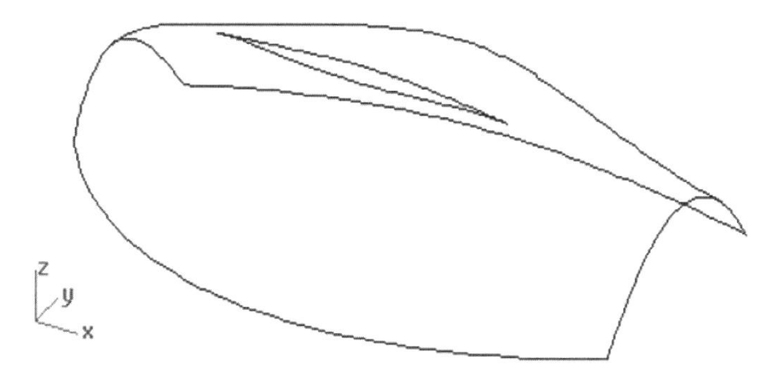

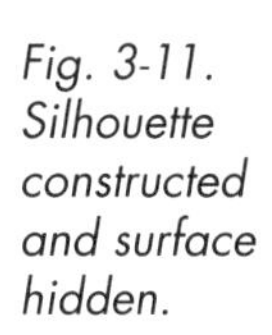

Fig. 3-11. Silhouette constructed and surface hidden.

Extracting Isocurves at Designated Locations of a Surface

Isocurve lines are directionally U and V lines displayed along a surface to help you visualize the profile and curvature of a surface. They are visual aids and not curves on the surface. To extract an isocurve at designated locations on the surface, perform the following steps.

1. Open the file *Extract.3dm* from the *Chapter 3* folder on the companion CD-ROM.
2. Select Curve > Curve From Objects > Extract Isocurve, or click on the Extract Isocurve button on Curve From Object toolbar.
3. Select edge A (indicated in figure 3-12) of the surface.
4. Select location B (indicated in figure 3-12) and press the Enter key.

An isocurve is constructed.

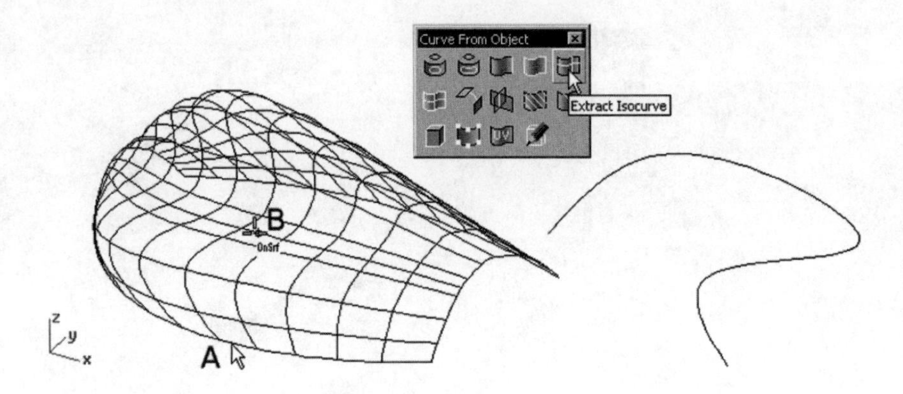

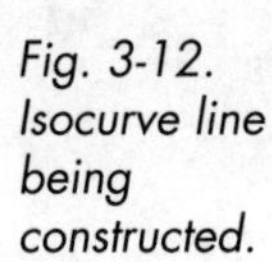

Fig. 3-12. Isocurve line being constructed.

Extracting a Wireframe from Isocurves of a Surface

To extract a wireframe from the isocurves of a surface for further development of your model, perform the following steps.

1. Select Curve > Curve From Objects > Extract Wireframe, or click on the Extract Wireframe button on the Curve From Object toolbar.
2. Select surface A (indicated in figure 3-13) and press the Enter key.

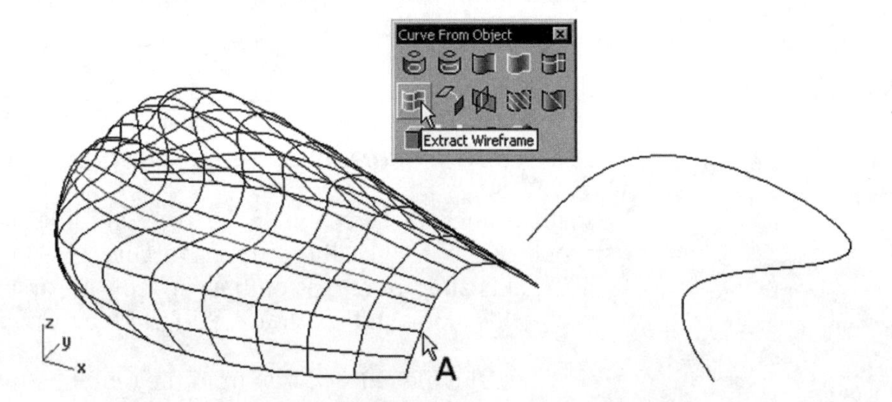

Fig. 3-13. Wireframe being extracted from the surface.

A wireframe is extracted from the surface. Because the extracted curves and the isocurve lines of the surface lie in the same location, you will not find any visual difference after you construct the curves. To see the difference, you can hide the surface and shade the viewport, as follows.

3. Press the Esc key to ensure that nothing is selected.
4. Shade the display by clicking on the *Shaded Viewport/Wireframe Viewport* button on the Standard toolbar.
5. Unhide the surface.

Extracting Control Points

Like curves, NURBS surfaces also have control points that govern their shape. To extract the control points of a surface and a curve, perform the following steps.

1. Select Curve > Curve From Objects > Extract Points, or click on the Extract Points button on the Curve From Object toolbar.
2. Select surface A and curve B (indicated in figure 3-14) and press the Enter key.

Points are constructed at the control point locations of the surface and curve.

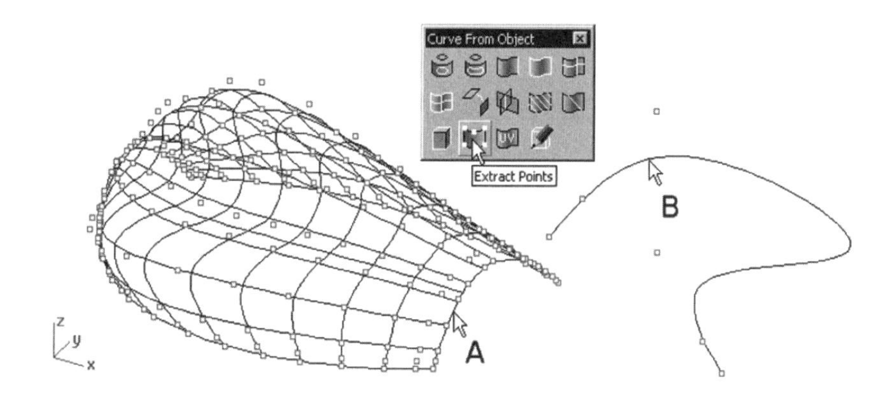

Fig. 3-14. Point objects being constructed.

Creating Reference U and V Curves on the X-Y Plane and Applying Planar Curves on Surfaces

To place a set of planar curves on a surface with reference to the U and V orientation of the surface, you first create a set of reference UV curves on the X-Y plane, construct a curve on the X-Y plane with reference to the reference UV curves, and apply the curve to the surface. Perform the following steps.

1. Open the file *ApplyUV.3dm* from the *Chapter 3* folder on the companion CD-ROM.
2. Select Curve > Curve From Objects > Create UV Curves, or click on the *Create UV Curves/Apply UV Curves* button on the Curve From Object toolbar.
3. Select surface A (indicated in figure 3-15).

4 The U and V curves of the selected surface are mapped onto the X-Y plane.
5 With reference to figure 3-15, construct curve B relative to the reference UV curves.
6 Select Curve > From Objects > Apply UV Curves, or right-click on the *Create UV Curves/Apply UV Curves* button on the Curve From Object toolbar.
7 Select curve B (indicated in figure 3-15) and press the Enter key.
8 Select surface A (indicated in figure 3-15).

Curve A is applied to the surface, becoming curve B, as shown in figure 3-16.

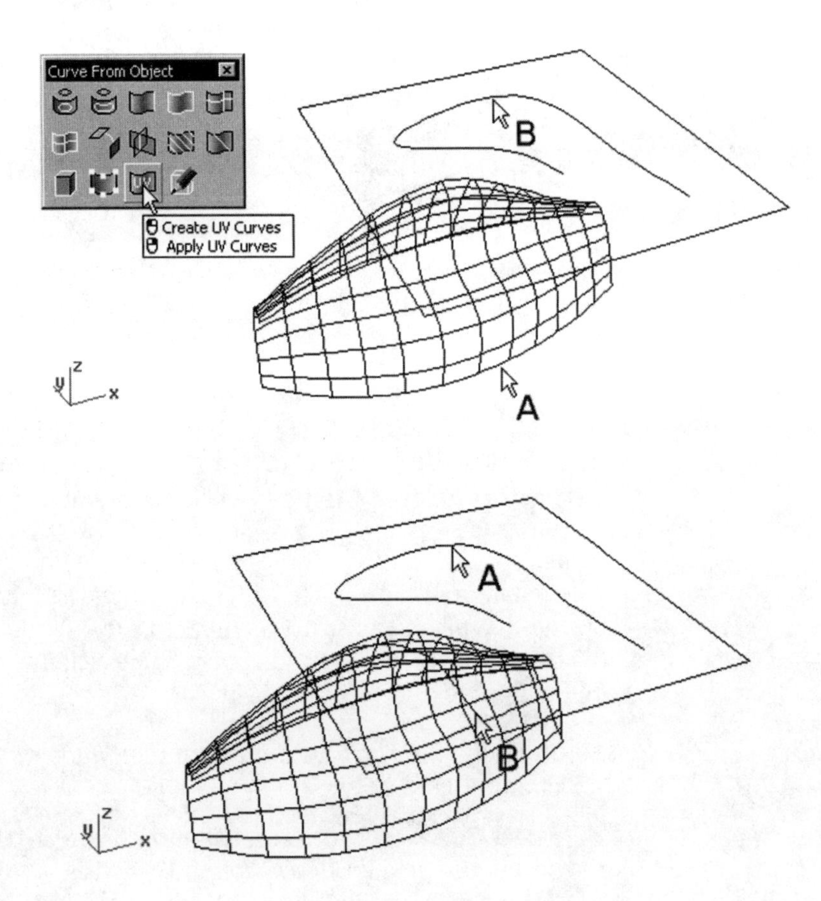

Fig. 3-15. Reference UV curves and a curve relative to the reference curves.

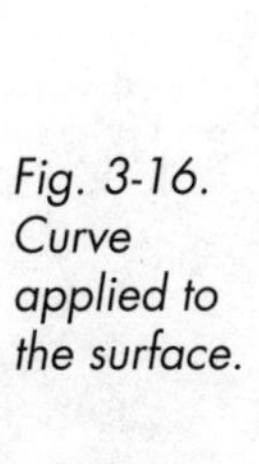

Fig. 3-16. Curve applied to the surface.

Editing Curves

In Chapter 2 and previously in this chapter you have learned various methods of constructing points and curves. This section explores methods of editing curves. These methods can be grouped into the following three main categories.

- Edit by joining, exploding, trimming, and splitting
- Edit by manipulating control points, edit points, knots, and points along curves
- More advanced editing concerning polynomial degree and fit tolerance

Joining, Exploding, Trimming, and Splitting

To meet various design needs, you may have to join two or more curves into one, explode a joined curve into individual curve components, trim a curve to remove unwanted portions, or split a curve into two curves. You perform these operations via the Join, Explode, Trim, and Split buttons of the Main1 and Main2 toolbars, shown in figure 3-17.

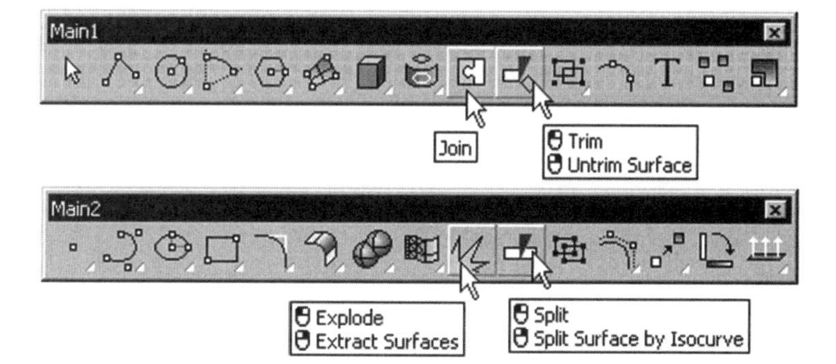

Fig. 3-17. Edit tools on Main1 and Main2 toolbars.

Table 3-2 outlines the Join, Explode, Trim, and Split commands and their locations in the main pull-down menu and respective toolbars.

Table 3-2: Join, Explode, Trim, and Split Commands and Their Functions

Toolbar Button	From Main Pull-down Menu	Function
Join	Edit > Join	Joining two or more curves into a single curve, or two or more surfaces into a polysurface
Trim	Edit > Trim	Trimming a curve or a surface

Toolbar Button	From Main Pull-down Menu	Function
Untrim Surface	—	Untrimming a trimmed surface
Explode	Edit > Explode	Exploding a joined curve into individual curve components or exploding a polysurface into individual surfaces
Extract Surfaces	—	Extracting individual surfaces from a polysurface without having to explode the polysurface
Split	Edit > Split	Splitting a curve into two curves
Split Surface by Isocurve	—	Splitting a surface along its isocurve

Joining Curves

To reflect our design intent, it is sometimes easiest to construct a number of contiguous curves and then join them to form a single curve. Perform the following steps.

1. Start a new file. Use the metric (Millimeters) template.
2. Maximize the Top viewport.
3. With reference to figure 3-18, construct three free-form curves A, B, and C.
4. Construct a fillet curve between curve A and B, a chamfer curve between B and C, and a blend curve between C and A, as shown in figure 3-19.

Join the six contiguous curves, as follows.

5. Select Edit > Join, or click on the Join button on the Main1 toolbar.
6. Select all curves. Because the selected curves form a closed loop, the command terminates automatically after you select the last curve. Otherwise, you have to press the Enter key to terminate the command.

The curves are joined.

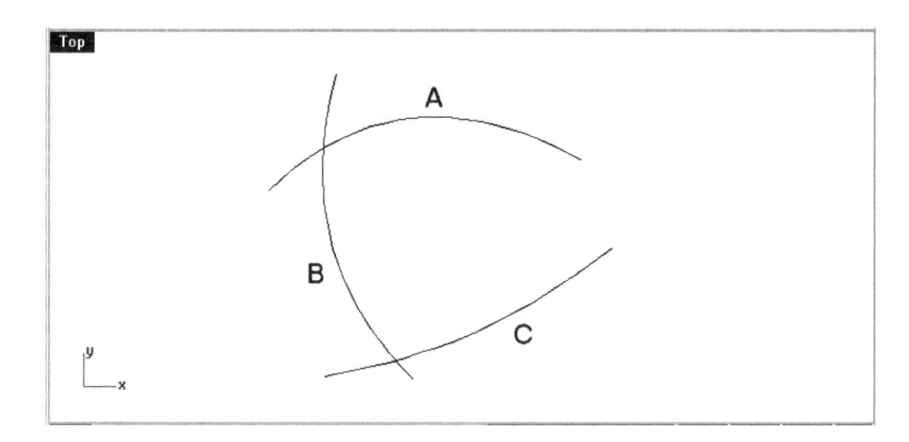

Fig. 3-18. Free-form curves.

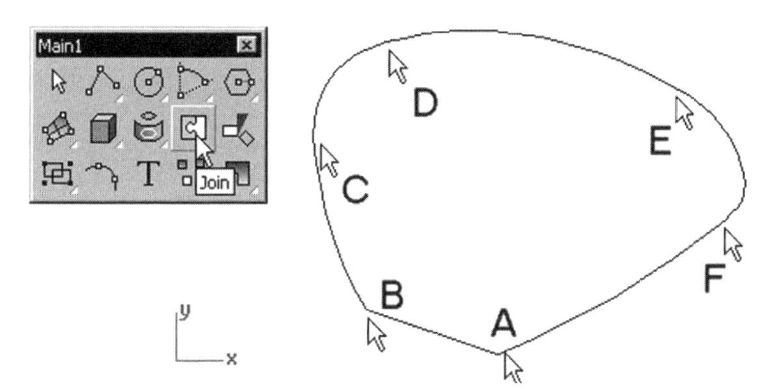

Fig. 3-19. Fillet, chamfer, and blend curves.

Continuity of Contiguous Curves

Continuity refers to the smoothness at the junction between two contiguous curves or surfaces. There are three types of continuity: positional (G0 continuity), tangent (G1 continuity), and curvature (G2 continuity).

In a G0 continuity joint, the control points at the end points of contiguous curves coincide. The chamfer joints at A and B indicated in figure 3-19 have G0 continuity.

In a G1 continuity joint, the tangent direction of the control points at the end points of the curves is the same. The fillet joints at C and D indicated in figure 3-19 have G1 continuity.

In a G2 continuity joint, the curvature and tangent direction of the control points at the end points of the curves are the same. The blend-curve joints at E and F indicated in figure 3-19 have G2 continuity.

The best type of curve for subsequent surface construction is the G2 curve. If you have a G0 or G1 curve, you may modify them by using the advanced editing tools, discussed later in this chapter.

Exploding a Joined Curve

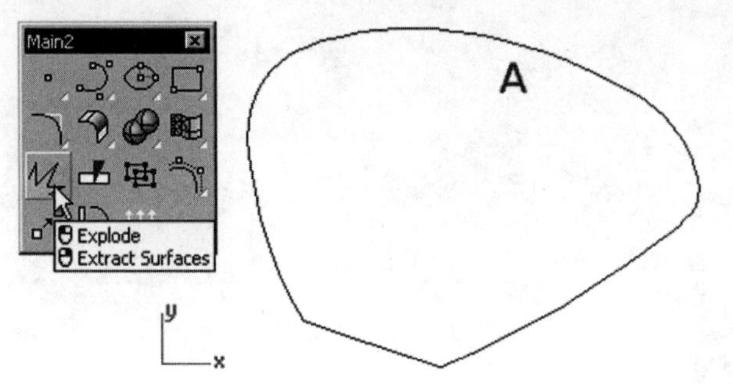

Fig. 3-20. Joined curve being exploded.

You can explode a joined curve into separate, individual curves, as follows.

1. Select the joined curve A (indicated in figure 3-20) and press the Enter key.
2. If you want to save your file, save it as *JoinExplode.3dm*.

Trimming Curves

Unwanted portions of a curve can be trimmed way. After trimming, the shape of the remaining portion of the curve remains unchanged. Perform the following steps to trim a curve.

1. Start a new file. Use the metric (Millimeters) template.
2. Maximize the Top viewport.
3. With reference to figure 3-21, construct two free-form curves A and B.

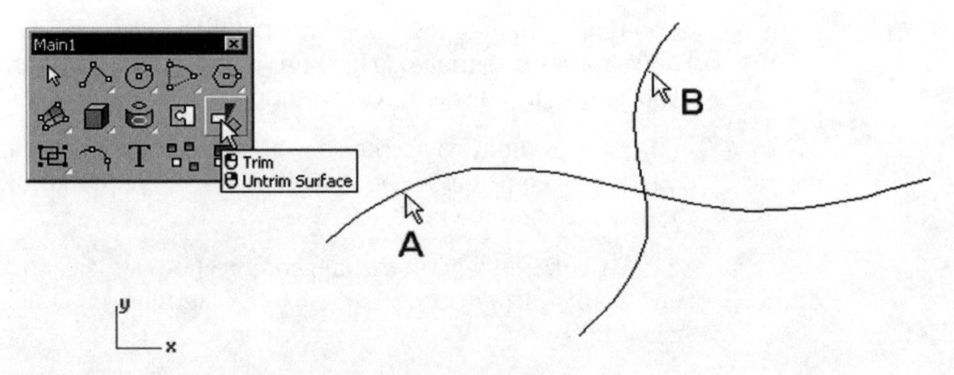

Fig. 3-21. Curve being trimmed.

4. Select Edit > Trim, or click on the Trim button on the Main2 toolbar.

5. Select curve A (indicated in figure 3-21) and press the Enter key. This is the cutting edge.
6. Select curve B (indicated in figure 3-21). This is the portion to be trimmed away.
7. Press the Enter key to terminate the command.

The curve is trimmed.

Splitting a Curve into Two Curves

To split a curve into two curves and keep the shape of the split curves unchanged, perform the following steps.

1. Select Edit > Split, or click on the *Split/Split Surface by Isocurve* button on the Main1 toolbar.
2. Select curve A (indicated in figure 3-22) and press the Enter key. This is the curve to be split into two curves.
3. Select curve B (indicted in figure 3-22). This is the cutting object.
4. Press the Enter key to terminate the command.

Curve A is split into two curves.

5. If you want to save your file, save it as *TrimSplit.3dm.*

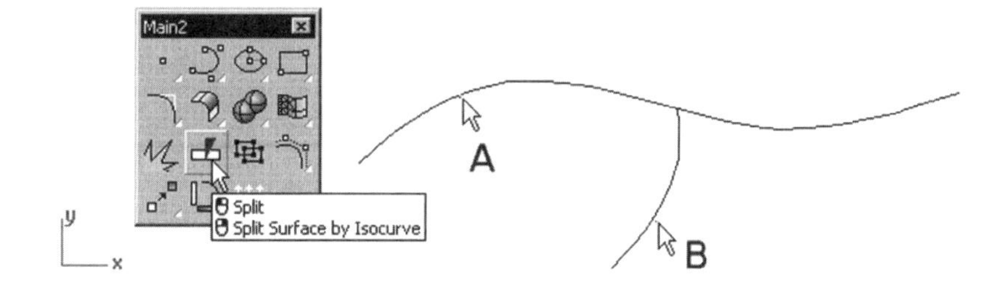

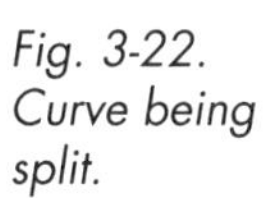

Fig. 3-22. Curve being split.

Point Editing

The shape of a curve is controlled by the location and weight of control points. Because the control points of a curve (except the two control points located at the end points of the curve, which we call anchor points) all lie outside the curve, it may be more predictable to manipulate the edit points along the curve.

Apart from manipulating the control points or the edit points, you can add or remove knots from a curve, add kink points to a curve, and use a handlebar to modify the tangency direction of selected locations along the curve. You perform these operations from the Point Editing toolbar, shown in figure 3-23.

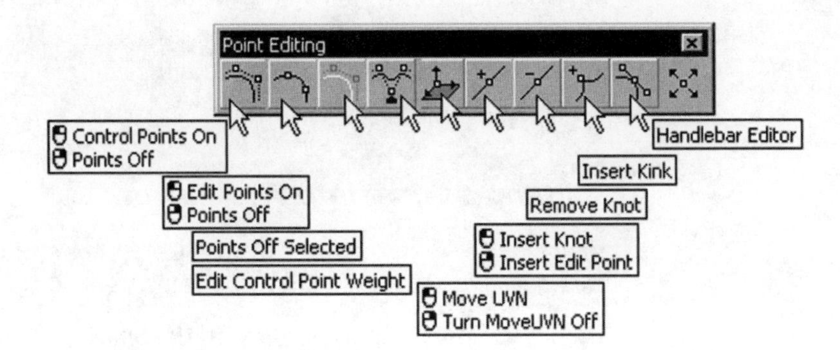

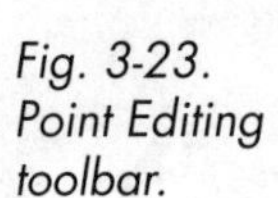
Fig. 3-23. Point Editing toolbar.

Table 3-3 outlines the point editing commands and their locations in the main pull-down menu and respective toolbars.

Table 3-3: Point Editing Commands and Their Functions

Toolbar Button	From Main Pull-down Menu	Function
Control Points On	Edit > Control Points > Control Points On	Turning on the control points of selected curves and surfaces
Points Off	Edit > Control Points > Control Points Off	Turning off the control points of selected curves and surfaces
Edit Points On	Edit > Control Points > Show Edit Points	Turning on the edit points of selected curves
Points Off	—	Turning off edit points
Points Off Selected	Edit > Control Points > Hide Points	Turning off the selected control points and edit points of curves, and selected control points of surfaces
Edit Control Point Weight	Edit > Control Points > Edit Weight	Adjusting the weight of selected control points
Move UVN/Turn Move UVN Off *	Transform > Move UVN	Moving selected control points of a surface in the U direction and V direction, as well as along the normal direction of a surface
Insert Knot	Edit > Control Points > Insert Knot	Inserting knot points to a curve or surface
Insert Edit Point	Edit > Control Points > Insert Edit Point	Inserting edit points in a curve

Toolbar Button	From Main Pull-down Menu	Function
Remove Knot	Edit > Control Points > Remove Knot	Removing knot points from a curve or surface
Insert Kink	Edit > Control Points > Insert Kink	Inserting kink points in a curve
Handlebar Editor	Edit > Control Points > Handlebar Editor	Using the handlebar to modify a curve or surface

* This command concerns surface editing. It is listed here because it shares the same toolbar with the other curve editing tools.

To change the shape of a curve by point editing, perform the following steps.

1. Start a new file. Use the metric (Millimeters) template.
2. Maximize the Top viewport.
3. With reference to figure 3-24, construct a free-form curve interpolating four points A, B, C, and D.

Fig. 3-24. Free-form curve.

Edit Point Manipulation

Edit points lie along a curve. To precisely change the shape of a curve, you display the edit points, select them, and drag them to new locations. To turn on the edit points of the curve and manipulate the points to modify the curve, perform the following steps.

1. Select Edit > Control Points > Show Edit Points, or click on the *Edit Points On/Points Off* button on the Point Editing toolbar.
2. Select curve A (indicated in figure 3-25) and press the Enter key.

The edit points are now turned on. Note that the number and locations of the edit points of your curve may not be the same as those shown in figure 3-25.

3 Select edit point A (indicated in figure 3-26), hold down the left mouse button, and drag the mouse to location B (indicated in figure 3-26).

The curve is modified.

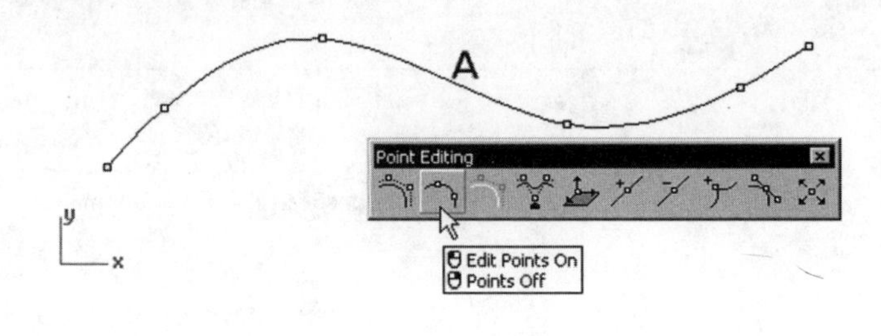

Fig. 3-25. Edit points turned on.

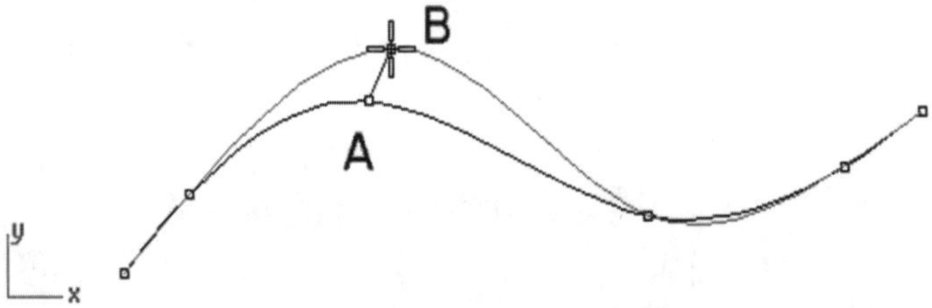

Fig. 3-26. Modified curve.

Control Point Manipulation

Control points let you control the shape of a curve in a more comprehensive way. You can change the location and the weight of the control points. In the following, you will turn off the edit points, turn on the control points, and modify the curve by manipulating the control points.

1 Right-click on the *Edit Points On/Points Off* button on the Point Editing toolbar.
2 Select Edit > Control Points > Control Points On, or click on the *Control Points On/Points Off* button on the Point Editing toolbar.
3 Select curve A (indicated in figure 3-27) and press the Enter key.

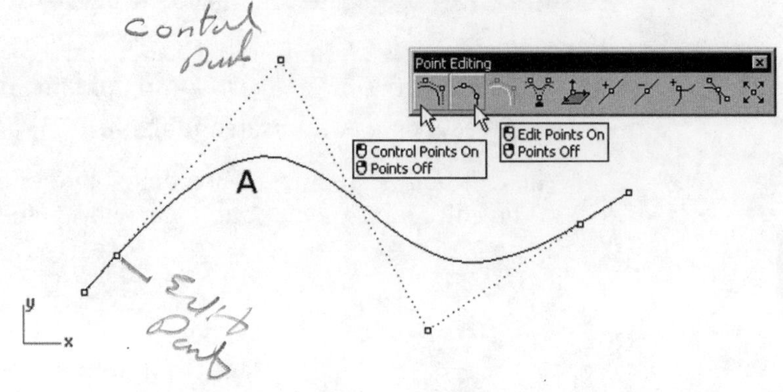

Fig. 3-27. Edit points turned off and control points turned on.

Control points are turned on. Note that with the exception of the first and last the control points lie outside the curve.

4. Select control point A, hold down the left mouse button, and drag the mouse to location B (indicated in figure 3-28).

The curve is modified. Compare the results of manipulating edit points and control points.

To edit the weight of a control point, continue with the following steps.

5. Select Edit > Control Points > Edit Weight, or click on the Edit Control Point Weight button on the Point Editing toolbar.
6. Select control point A (you can select multiple points) and press the Enter key.
7. In the Set Control Point Weight dialog box, shown in figure 3-29, set the weight of the selected point(s) to 7, and then click on the OK button.

Note that the curve is being pulled more closely to the selected control point.

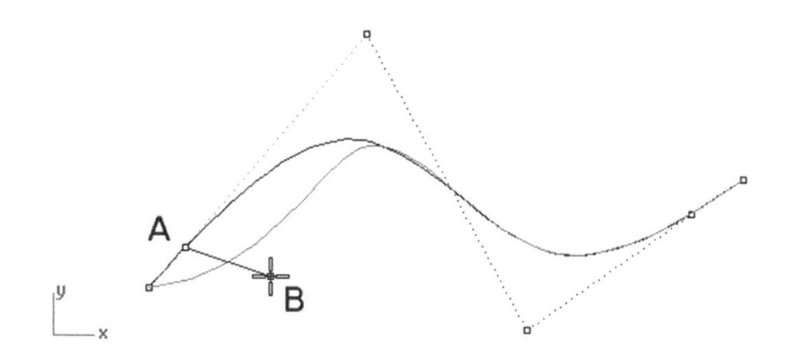

Fig. 3-28. Curve being modified by moving a control point.

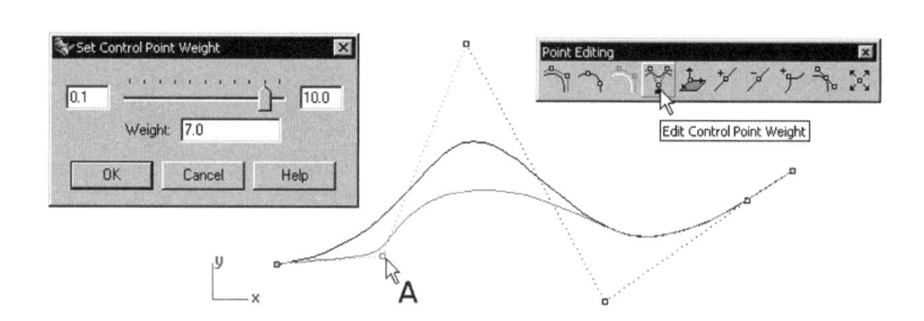

Fig. 3-29. Point weight being modified in the Set Control Point Weight dialog box.

Insertion of Knot Point

One way to increase the complexity of the shape of a curve is to change the number of spline segments by inserting knots along the curve. In the following, you will add a knot to the curve. Adding knots to a curve increases the number of spline segments. As a result, there will be more control points. However, the shape of the curve will not change until you manipulate the control points. Perform the following steps.

1. Select Edit > Control Points > Insert Knot, or click on the *Insert Knot/Insert Edit Point* button on the Point Editing toolbar.
2. Select curve A (indicated in figure 3-30).
3. Select point B (indicated in figure 3-30) along the curve where you want to insert a knot. Note that you can insert multiple control points.

A control point is inserted. Compare figure 3-31 to figure 3-29 in terms of the change in the number of control points.

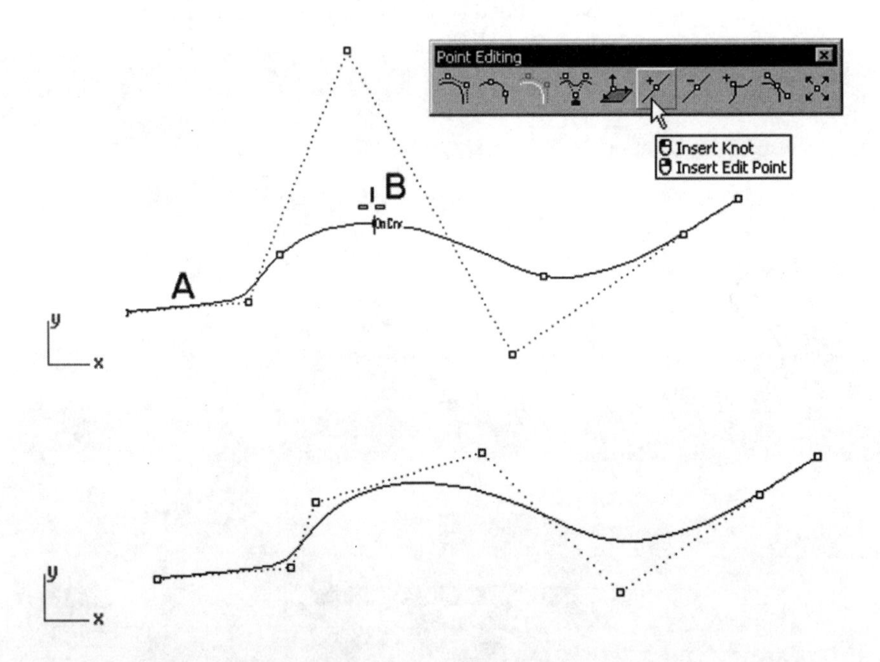

Fig. 3-30. Knot being inserted in the curve.

Fig. 3-31. Number of control points increased.

Insertion of Kink Point

A spline curve is analogous to bamboo: you can bend bamboo to a great degree and still retain a very smooth curve. If we continue to bend the bamboo further, it will fracture at a certain point. The fracture point is equivalent to a kink in a spline curve.

In the following, you will add a kink point to a curve. A kink point is a special type of knot on a curve in which the tangent directions of contiguous spline segments are not the same.

1 Select Edit > Point Editing > Insert Kink, or click on the Insert Kink button on the Point Editing toolbar.
2 Select curve A (indicated in figure 3-32).

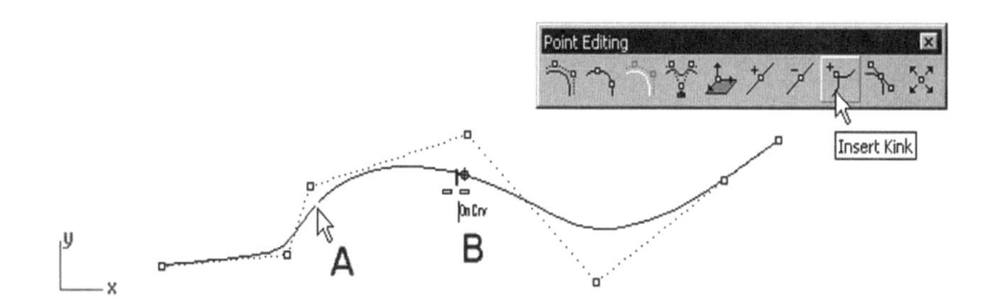

Fig. 3-32. Kink point being inserted.

3 Select point B (indicated in figure 3-32) along the curve where you want to insert a kink. Note that you can insert multiple kink points.
4 Press the Enter key to terminate the command.

A kink is inserted.

5 Select kink point A (indicated in figure 3-33), hold down the left mouse button, and drag the mouse to a new location to see the effect of a kink.

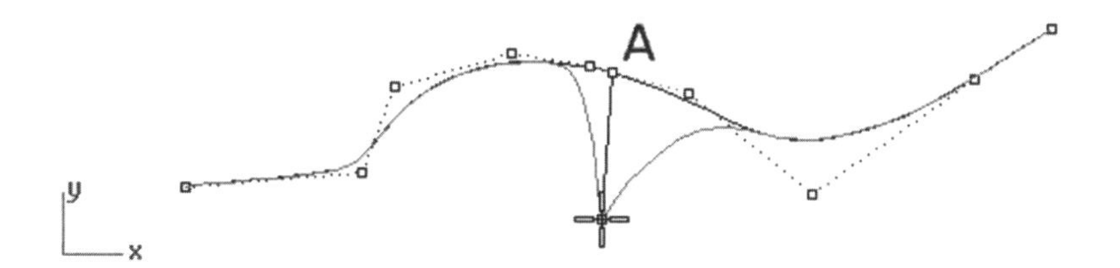

Fig. 3-33. Inserted kink point being moved.

Removal of Knot and Kink Points

To remove a kink point or to reduce the complexity of a curve, we remove the kink points or knots along the curve. Because removing knots on a curve reduces the complexity of the curve, the curve's shape will change. To remove a kink point or other type of knot from a curve, perform the following steps.

1. Select Edit > Control Points > Remove Knot, or click on the Remove Knot button on the Point Editing toolbar.
2. Select the curve.
3. Select knots A and B (indicated in figure 3-34) and press the Enter key.

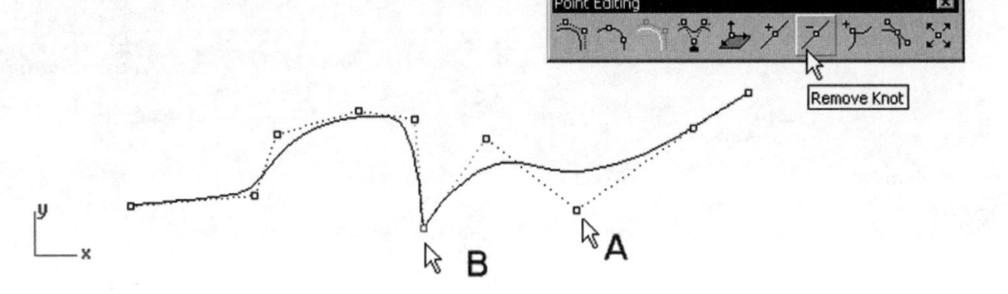

Fig. 3-34. Knots being removed.

A knot of the curve is removed, as shown in figure 3-35. After you remove a knot from a curve, the number of control points decreases. As a result, the shape of the curve is simplified and changed.

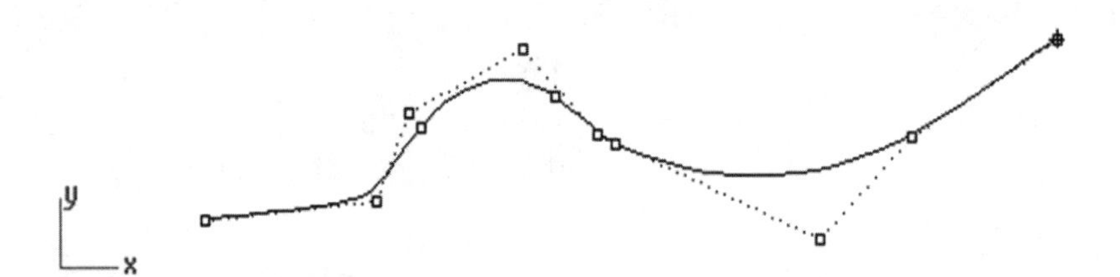

Fig. 3-35. Knots removed.

Handlebar Editor

Sometimes we want to modify the shape of a curve at a specific location along the curve. To make such a modification, we use the handlebar editor, as follows.

1. Select Edit > Point Editing > Handlebar Editor, or click on the Handlebar Editor button on the Point Editing toolbar.
2. Select the curve (indicated in figure 3-36).

The handlebar is displayed at the point where you selected the curve.

3. Click on an end point of the handlebar and drag, as indicated in figure 3-37.
4. Press the Enter key.

The curve is modified. Using a handlebar to edit the shape of a curve, the curve profile changes but the location of the selected point at the handlebar does not change.

5 If you want to save your file, save it as *PointEdit.3dm*.

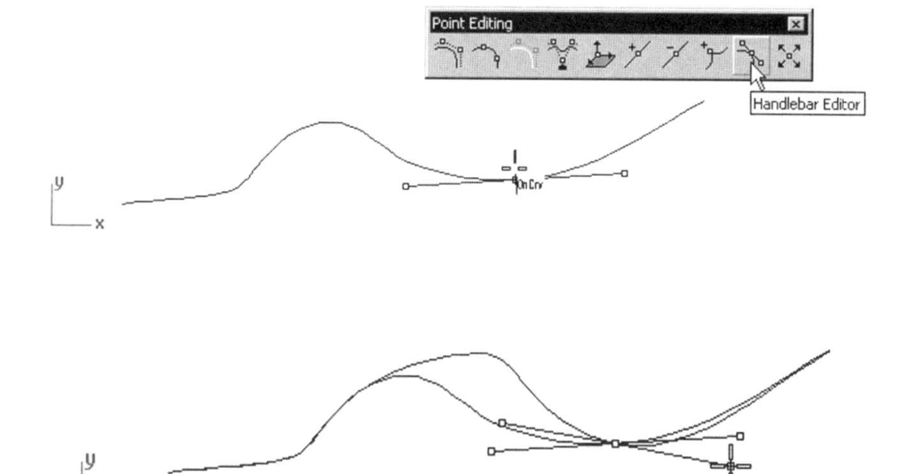

Fig. 3-36. Handlebar being displayed.

Fig. 3-37. Handlebar being manipulated.

Advanced Curve Editing

This section explores more advanced methods of modifying a curve. The advanced editing tools are located on the Curve Tools toolbar, shown in figure 3-38.

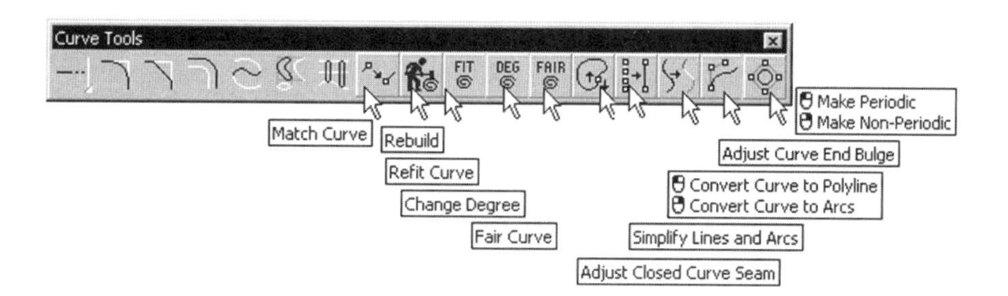

Fig. 3-38. Advanced editing tools on the Curve Tools toolbar.

Table 3-4 outlines the advanced curve editing commands and their locations in the main pull-down menu and respective toolbars.

Table 3-4: Advanced Curve Editing Commands and Their Functions

Toolbar Button	From Main Pull-down Menu*	Function
Match Curve	Curve > Curve Edit Tools > Match	Matching an end point of a curve with the end point of another curve by modifying the position, tangency, and/or curvature of the selected curves
Rebuild	Edit > Rebuild	Reconstructing the selected curve or surface with a new polynomial degree setting and new number of control points
Refit Curve	Curve > Curve Edit Tools > Refit to Tolerance	Constructing a curve from another curve with fewer or more control points by setting a new tolerance value for the curve
Change Degree	Edit > Change Degree	Modifying a curve or surface by changing its polynomial degree setting
Fair Curve	Curve > Curve Edit Tools > Fair	Removing large curvature variations of a curve by limiting the geometry change to a specified tolerance
Adjust Closed Curve Seam	Curve > Curve Edit Tools > Adjust Closed Curve Seam	Moving and flipping the direction of the seam point of a closed curve
Simplify Lines and Arcs	Curve > Curve Edit Tools > Simplify Lines and Arcs	Changing line and arc segments of a curve to NURBS line and arc segments
Convert Curve to Polyline	Curve > Convert > Curve to Lines	Converting a free-form curve to a polyline
Convert Curve to Arcs	Curve > Convert > Curve to Arcs	Converting a free-form curve to a set of arcs
Adjust Curve End Bulge	Edit > Adjust End Bulge	Adjusting the end bulge of a curve or a surface without changing the tangent direction and continuity of the curve
Make Periodic	Edit > Make Periodic	Constructing a closed-loop curve without a kink from a curve
Make Non-Periodic	—	Making a periodic curve or surface to become non-periodic

*Commands under the Curve menu are specific for curve editing, and commands under the Edit menu can be used for both curves and surfaces.

In the following, you will work with various curve editing tools.

1. Start a new file. Use the metric (Millimeters) template.
2. Maximize the Top viewport.
3. With reference to figure 3-39, construct two free-form curves.

Matching a Curve to Another Curve

If we have two separate curves, we can modify the shape of one of the curves to make it match the other curve, as follows.

1. Select Curve > Curve Edit Tools > Match, or click on the Match Curve button on the Curve Tools toolbar.
2. Select curve A (indicated in figure 3-39). This curve will be modified.

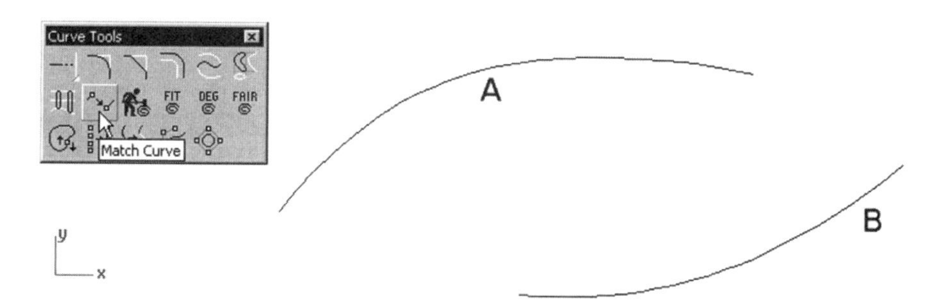

Fig. 3-39. Free-form curves.

3. Select curve B near its right end (indicated in figure 3-39).
4. Try out various options in the Match Curve dialog box, shown in figure 3-40.
5. Check the Tangency, *Average curves*, and *Preserve other end* options, and then click on the OK button.

The shape of the selected curve is modified to match the other curve.

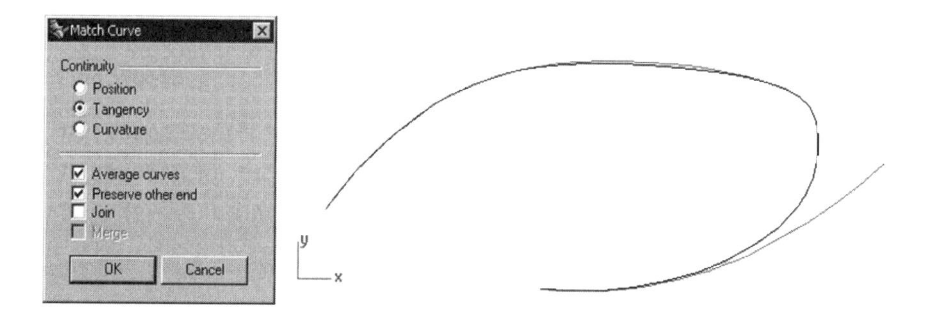

Fig. 3-40. Match Curve dialog box.

Rebuilding a Curve

After you have finalized the shape of a curve (via various editing methods, such as those you performed previously), you might need to smooth the curve by rebuilding it with a different number of control points and a different degree of polynomial. Rebuilding always removes kinks. It is a common technique used by designers to both simplify curves and remove potential problems with kinks. If you reduce the control point number, the curve simplifies and the shape changes. If you increase the control point number, the shape will not change until you manipulate the control points. Perform the following steps.

1. Select Edit > Join, or click on the Join button on the Main1 toolbar.
2. Select curves A and B (indicated in figure 3-41) and press the Enter key.

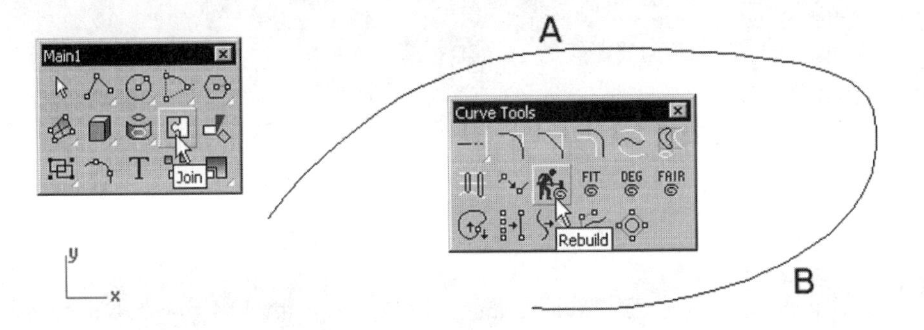

Fig. 3-41. Curves being joined and rebuilt.

3. Select Edit > Rebuild, or click on the Rebuild button on the Curve Tools toolbar.
4. Select the curve and press the Enter key.

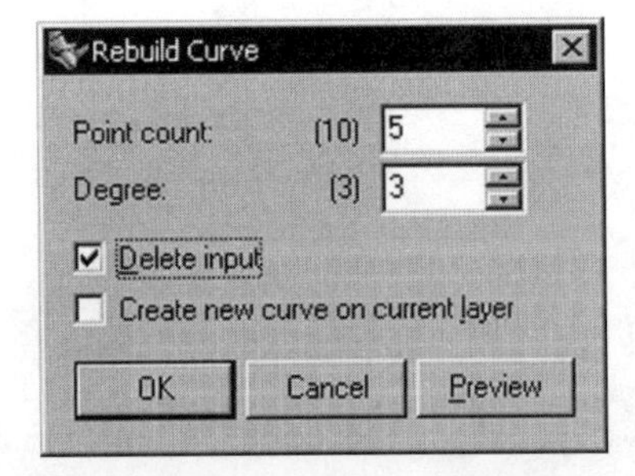

Fig. 3-42. Rebuild Curve dialog box.

5. In the Rebuild Curve dialog box, shown in figure 3-42, set the number of control points to 5. (The default value shown may not be the same as yours. This command always remembers the previous settings when last used.) Leave the degree unchanged.
6. Click on the OK button.

The selected curve is rebuilt with a curve consisting of a different number of control points.

Fitting a Curve from a Polyline

If we already have a polyline, we can "refit" a free-form curve to correspond to the polyline's tolerance setting. This is sometimes necessary when, for example, we input data from another system and in the process input polyline segments that do not correctly correspond with existing geometry. If we want to build a smooth surface, we need to refit the polyline to a free-form curve, as follows.

1 With reference to figure 3-43, construct a polyline line with four line segments.

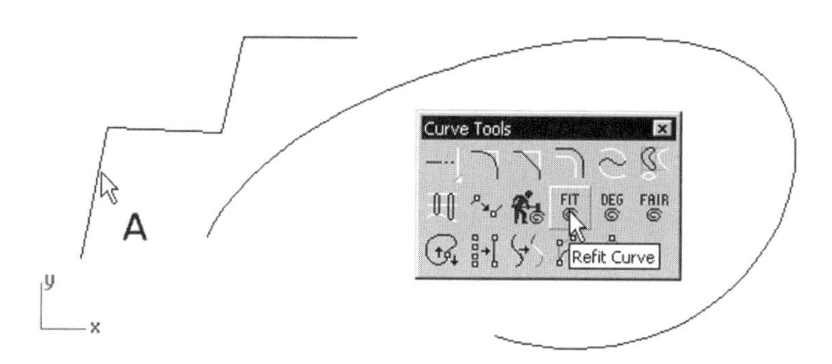

Fig. 3-43. Polyline constructed and being refitted.

2 Select Curve > Curve Edit Tools > Refit to Tolerance, or click on the Refit Curve button on the Curve Tools toolbar.

3 Select polyline A (indicated in figure 3-43) and press the Enter key.

You may keep the original curve, or delete it using the Delete Input option. Here, the input curve is deleted.

4 Accept the default degree of polynomial by pressing the Enter key.

A free-form curve is refit from the selected polyline, as shown in figure 3-44.

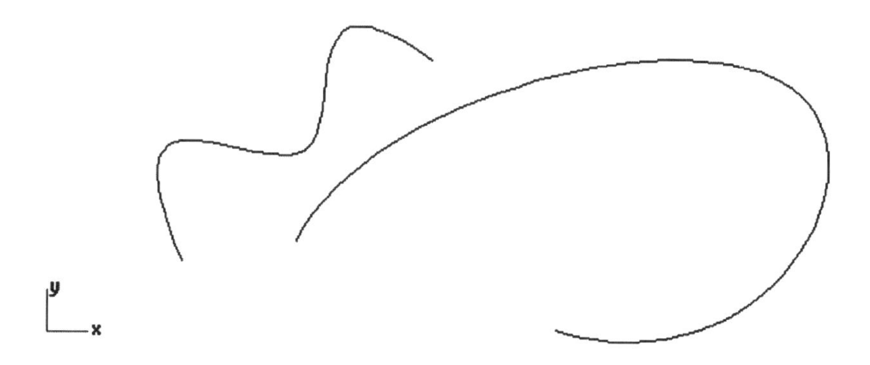

Fig. 3-44. Polyline refitted to a free-form curve.

Changing the Polynomial Degree of a Curve

As previously mentioned, a line is degree 1, an arc is degree 2, and the default polynomial degree of a free-form curve is 3. In the following, you will change the polynomial degree of a curve. To see the effect of the change, you will turn on the control points.

1. Select Edit > Control Points > Control Points On, or click on the *Control Points On/Points Off* button on the Point Editing toolbar.
2. Change the polynomial degree and watch the change in the number of control points in the curve take effect.
3. Select Edit > Change Degree, or click on the Change Degree button on the Curve Tools toolbar.
4. Select curve A (indicated in figure 3-45) and press the Enter key.

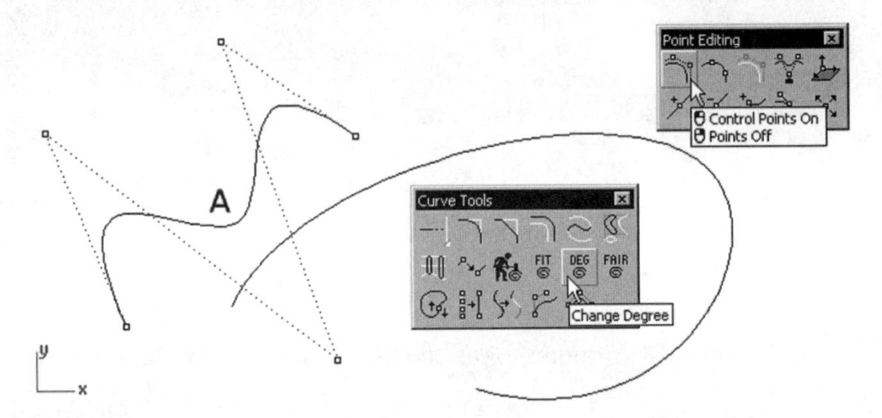

Fig. 3-45. Degree of polynomial of the selected curve being changed.

5. Type *4* at the command line area to specify a degree 4 curve.

The degree of polynomial of the selected curve is changed, as shown in figure 3-46. To reiterate, you do not need to turn on the control points to change the polynomial degree.

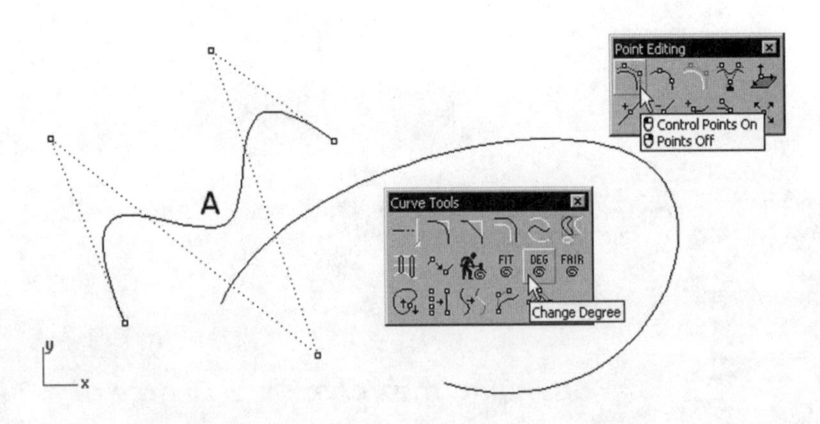

Fig. 3-46. Degree of polynomial changed.

Fairing a Curve

Fairing removes large curvature variation in a curve, as follows

1. Select Curve > Curve Edit Tools > Fair, or click on the Fair button on the Curve Tools toolbar.

2 Select curve A (indicated in figure 3-47) and press the Enter key.
3 Type *4* at the command line area to specify a tolerance of 4 mm.

The curve is faired, as shown in figure 3-48

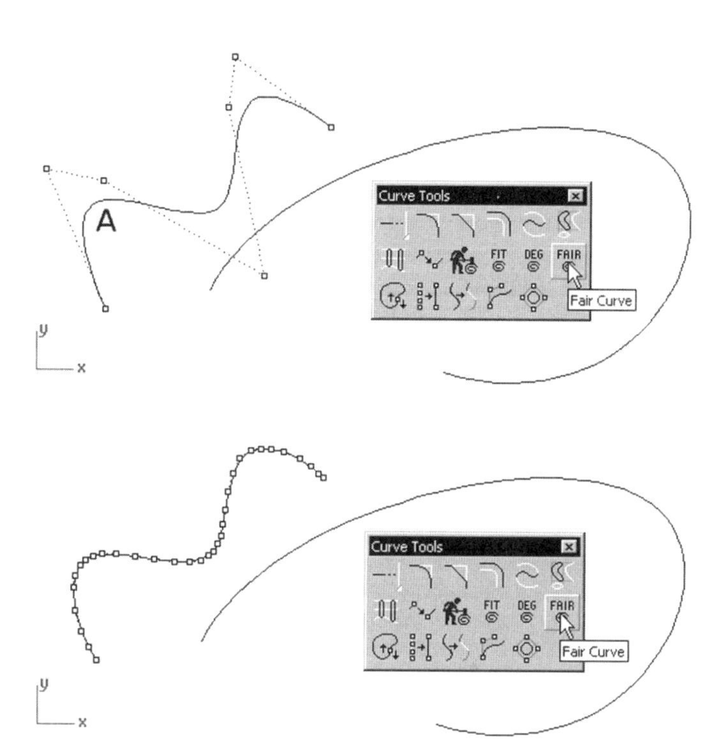

Fig. 3-47. Curve being faired.

Fig. 3-48. Faired curve.

Simplifying a Curve

If you extend a curve by adding a line segment or an arc segment to it, the degree of polynomial of the added segment will not be congruent with the curve itself. To replace such line or arc segments in a curve with a true NURBS curve, you simplify the curve. In the following, you will add an arc segment to a curve and then simplify it.

1 Select Curve > Extend Curve > By Arc, or click on the Extend by Arc with Center button on the Extend toolbar.
2 Select curve A (indicated in figure 3-49).
3 Select location B (shown in figure 3-49) to indicate the radius of the arc.
4 Select location C (shown in figure 3-49) to indicate the end point of the arc.

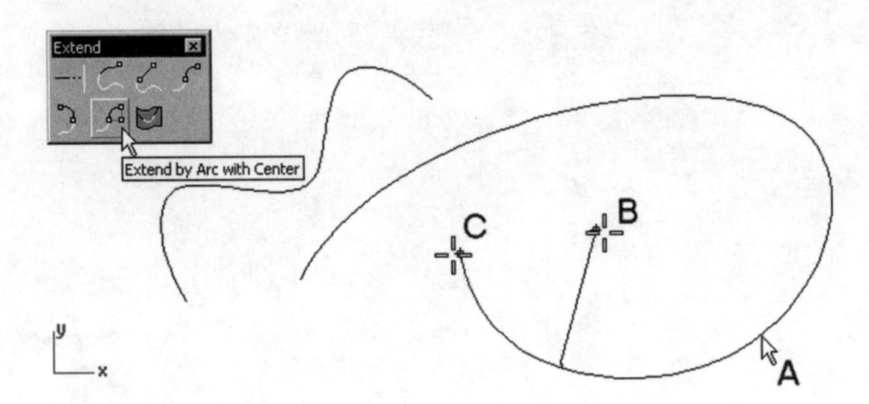

Fig. 3-49. Arc being added.

5 Select Curve > Curve Edit Tools > Simply Lines and Arcs, or click on the Simplify Lines and Arcs button on the Curve Tools toolbar.
6 Select curve A (indicated in figure 3-50) and press the Enter key.

The curve is extended with an arc segment, and the arc segment of the curve is replaced with a true NURBS segment. You will not find much change on the shape of the curve because the curve takes on the shape of the arc.

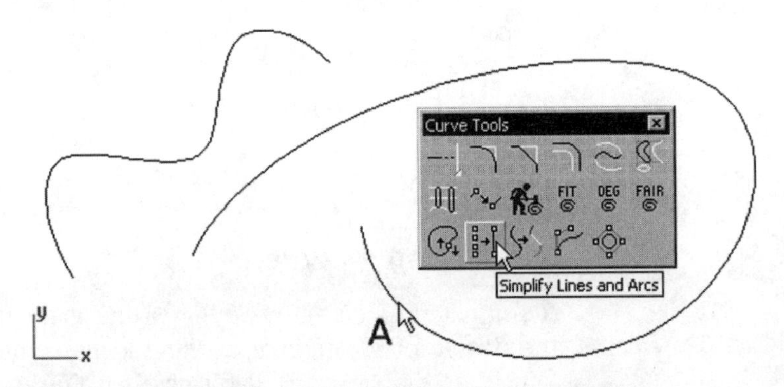

Fig. 3-50. Curve being simplified.

Editing End Bulge

To modify the shape of a curve without changing the tangent direction at the end points, you edit the end bulge, as follows. As a result, the continuity at the junction of the curves remains unchanged.

1 Select Edit > Adjust End Bulge, or click on the Adjust End Bulge button on the Curve Tools toolbar.
2 Select curve A (indicated in figure 3-51).
3 Select and drag the control points.
4 Press the Enter key when finished.

The end bulge is modified.

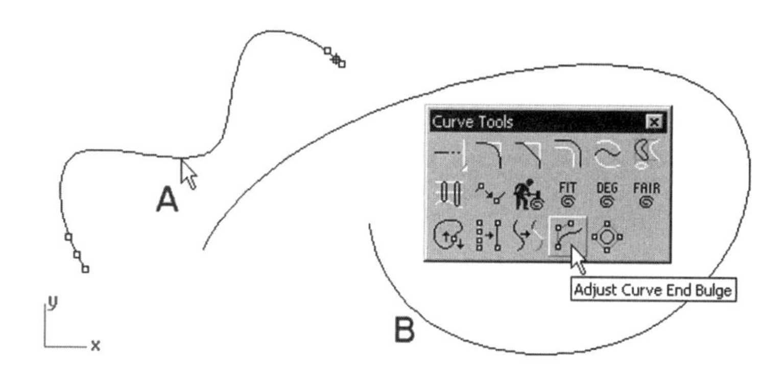

Fig. 3-51. End bulge being modified.

Making a Curve Periodic

A periodic curve is a closed curve with no kink point. If you make an open-loop curve periodic, the curve becomes a closed-loop curve. To make a curve periodic, perform the following steps.

1. Select Edit > Make Periodic, or click on the *Make Periodic/Make Non-Periodic* button on the Curve Tools toolbar.
2. Select curve B (indicated in figure 3-52) and press the Enter key.

The curve becomes a periodic curve.

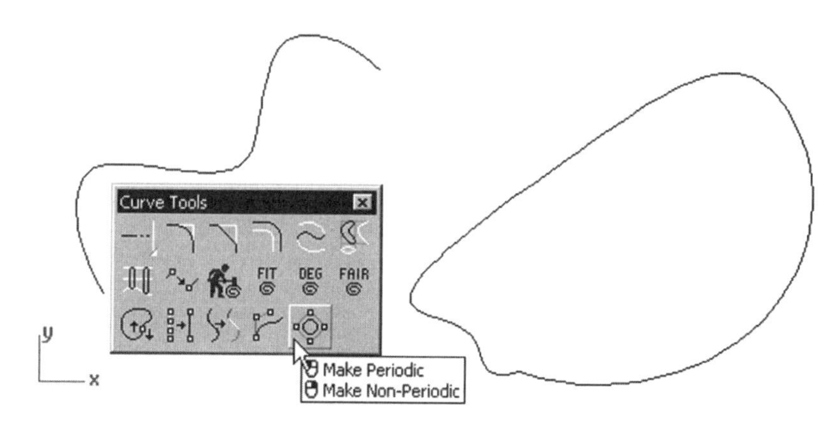

Fig. 3-52. Curve made periodic.

Adjusting Seam Location

A closed-loop curve, like an open-loop curve, has a start point and an end point. The difference between a closed-loop curve and an open-loop curve is that the start point and end point coincide. The coincident point is called the seam point. In the following, you will adjust the location of the seam point of a closed curve.

1. Select Curve > Curve Edit Tools > Adjust Closed Curve Seam, or click on the Adjust Closed Curve Seam button on the Curve Tools toolbar.
2. Select curve A (indicated in figure 3-53) and press the Enter key.
3. Select seam point B (indicated in figure 3-53).
4. Select a point along the curve to indicate a new position for the seam point.
5. Press the Enter key.

The seam is relocated.

6. If you want to save your file, save it as *CurveEdit.3dm*.

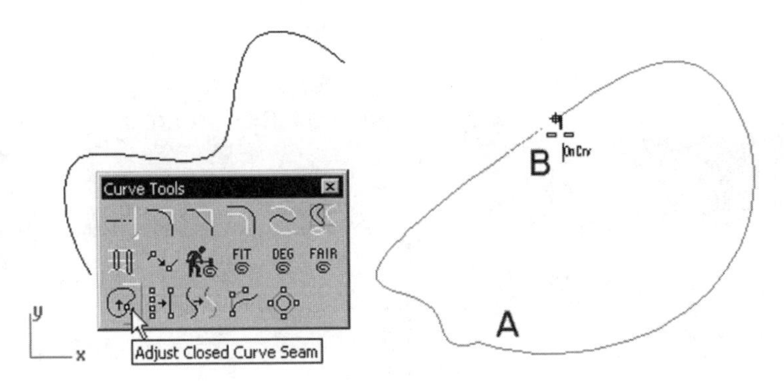

Fig. 3-53. Seam point being adjusted.

Transforming Curves

The transformation of curves involves translation and deformation. Transformation refers to changing the location of the curve without deforming it, and deformation refers to changing the shape of the curves. Some of the transform tools described here will apply not only to curves but to surfaces and solids. You perform transform operations from the Transform and Array toolbars, shown in figure 3-54. Drag mode affects how a selected object is moved when selected and dragged. If drag mode is on, selected objects are dragged on a plane parallel to the view plane. Otherwise, movement is parallel to the construction plane. Drag mode is accessed via the Drag Mode button on the Point Editing toolbar.

Table 3-5 outlines the transform commands and their locations in the main pull-down menu and respective toolbars.

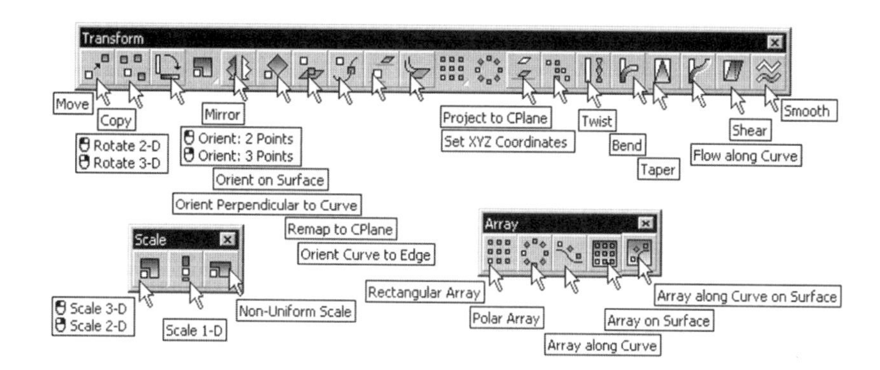

Fig. 3-54. Transform, Scale, and Array toolbars.

Table 3-5: Transform Commands and Their Functions

Toolbar Button	From Main Pull-down Menu	Function
Move	Transform > Move	Moving selected objects
Copy	Transform > Copy	Copying selected objects
Rotate 2-D	Transform > Rotate	Rotating selected objects about a point on the current construction plane
Rotate 3-D	Transform > Rotate 3-D	Rotating selected objects about a specified axis
Scale 3-D	Transform > Scale > Scale 3-D	Uniformly scaling the X, Y, and Z dimensions of selected objects
Scale 2-D	Transform > Scale > Scale 2-D	Uniformly scaling the X and Y dimensions (with Z dimension unchanged) of selected objects
Scale 1-D	Transform > Scale > Scale 1-D	Scaling selected objects in a specified direction
Non-uniform Scale	Transform > Scale > Non-uniform Scale	Scaling selected objects by specifying individual scale factors for the X, Y, and Z dimensions
Mirror	Transform > Mirror	Constructing a mirror copy of selected objects
Orient: 2 Points	Transform > Orient > 2 Points	Transforming selected objects to a specified location and orientation and uniformly scaling the selected objects by specifying two reference points and two target points
Orient: 3 Points	Transform > Orient > 3 Points	Translating selected objects to a specified location by defining target point, base direction, and orientation

Toolbar Button	From Main Pull-down Menu	Function
Orient on Surface	Transform > Orient > On Surface	Translating selected objects to specified locations on a surface
Orient Perpendicular to Curve	Transform > Orient > Perpendicular to Curve	Translating selected objects perpendicular to a curve
Remap to CPlane	Transform > Orient > Remap to CPlane	Translating selected objects to a specified construction plane
Orient Curve to Edge	Transform > Orient > Curve to Edge	Copying and aligning curves to a selected surface
Rectangular Array	Transform > Array > Rectangular	Constructing a rectangular pattern of selected objects
Polar Array	Transform > Array > Polar	Constructing a circular pattern of selected objects
Array along Curve	Transform > Array > Along Curve	Constructing a pattern of selected objects along a curve
Array on Surface	Transform > Array > Along Surface	Constructing a rectangular pattern of selected objects along the U and V directions of a surface
Array along Curve on Surface	Transform > Array > Along Curve on Surface	Constructing a pattern of select objects along a curve constructed on a surface
Project to CPlane	Transform > Project to CPlane	Projecting selected objects to a specified construction plane
Set XYZ Coordinates	Transform > Set Points	Deforming selected objects by aligning selected control points
Twist	Transform > Twist	Deforming selected curves or surfaces by twisting them about a specified axis
Bend	Transform > Bend	Deforming selected curves or surfaces by bending them about a bend axis
Taper	Transform > Taper	Deforming selected curves or surfaces by tapering objects
Flow along Curve	Transform > Flow along Curve	Deforming selected curves or surfaces by specifying an original backbone curve and a new backbone
Shear	Transform > Shear	Deforming selected objects by shearing them about a shear plane (the area of a sheared surface and the volume of a sheared box remain unchanged)

Toolbar Button	From Main Pull-down Menu	Function
Smooth	Transform > Smooth	Smoothing selected curves and surfaces by removing unwanted details and loops

The sections that follow explore the use of various transform tools on curves.

Move, Copy, and Rotate

The simplest methods of transforming a curve are to move or copy it to a new location or to make a mirror copy. After moving or copying, the orientation of the curve remains unchanged. To change the orientation of a curve, you rotate it about a point on the construction plane or about a 3D axis. To rotate a curve about a 3D axis, perform the following steps.

1. Start a new file. Use the metric (Millimeters) template.
2. With reference to figure 3-55, construct a free-form curve A in the Top viewport.

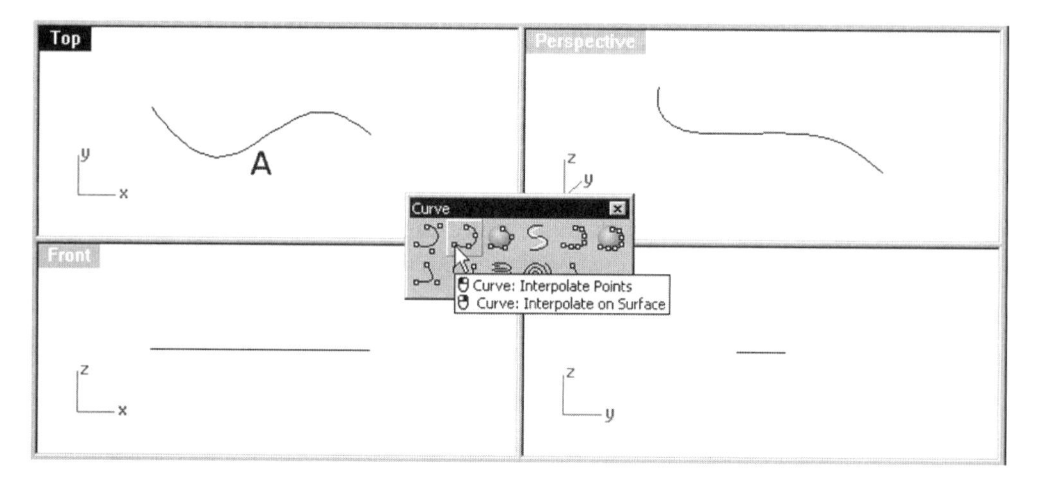

Fig. 3-55. Free-form curve constructed.

3. Select Transform > Rotate 3-D, or right-click on the *Rotate 2-D/ Rotate 3-D* button on the Transform toolbar.
4. Select curve A (indicated in figure 3-56) and press the Enter key.
5. Select location B (indicated in figure 3-56) in the Top viewport to indicate the start of the rotation axis.
6. Select location C (indicated in figure 3-56) in the Front viewport to indicate the end of the rotation axis.

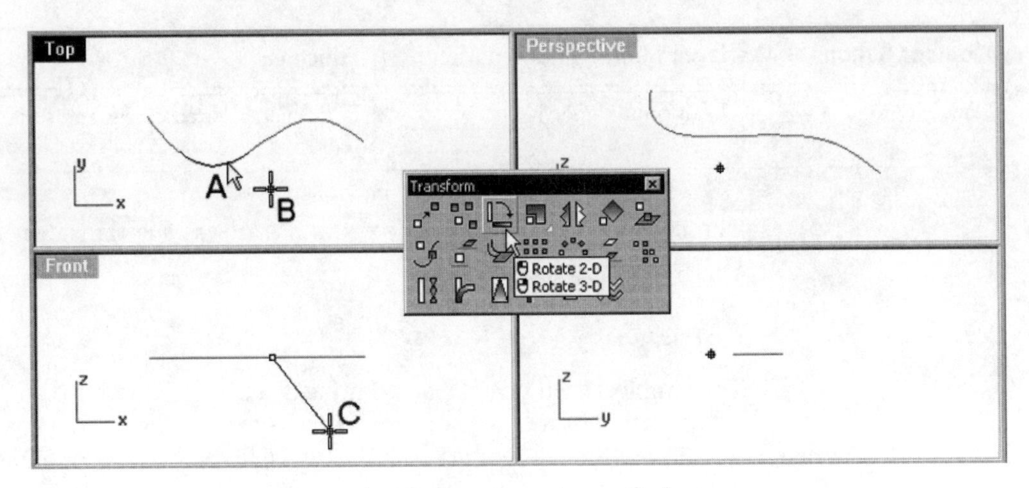

Fig. 3-56. Start and end of rotation axis specified.

7 Type *45* to specify the rotation angle.

The curve is rotated. To rotate a control point of the curve, continue with the following steps.

8 Turn on the control points of the curve.

9 Select Transform > Rotate 3-D, or right-click on the *Rotate 2-D/ Rotate 3-D* button on the Transform toolbar.

10 Select control point A (indicated in figure 3-57) and press the Enter key.

11 Select location B (indicated in figure 3-57) in the Top viewport to indicate the start of rotate axis.

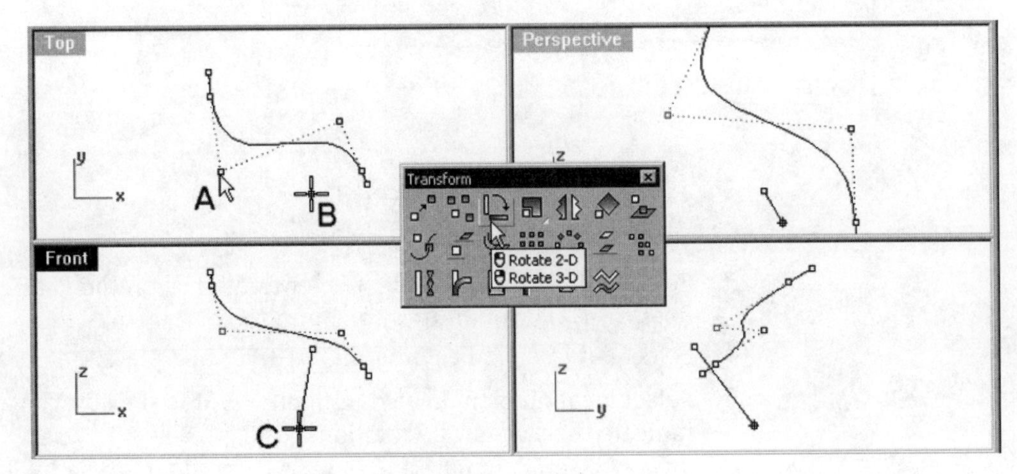

Fig. 3-57. Control point being rotated.

12 Select location C (indicated in figure 3-57) in the Front viewport to indicate the end of the rotation axis.

13 Type *45* to specify the rotation angle.

The selected control point is rotated.

14 Turn off the control points.

Scale

There are four ways to scale a curve. You scale the entire curve or selected control points of the curve uniformly in the X, Y, and Z dimensions; in two dimensions; in one dimension; or nonuniformly in three dimensions. When you scale the control points, you translate the locations of the selected control points in relation to a reference point. To scale a curve uniformly in the X, Y, and Z dimensions, perform the following steps.

1 Select Transform > Scale > Scale 3-D, or click on the *Scale 3-D/ Scale 2-D* button on the Scale toolbar.

2 Select curve A (indicated in figure 3-58) and press the Enter key.

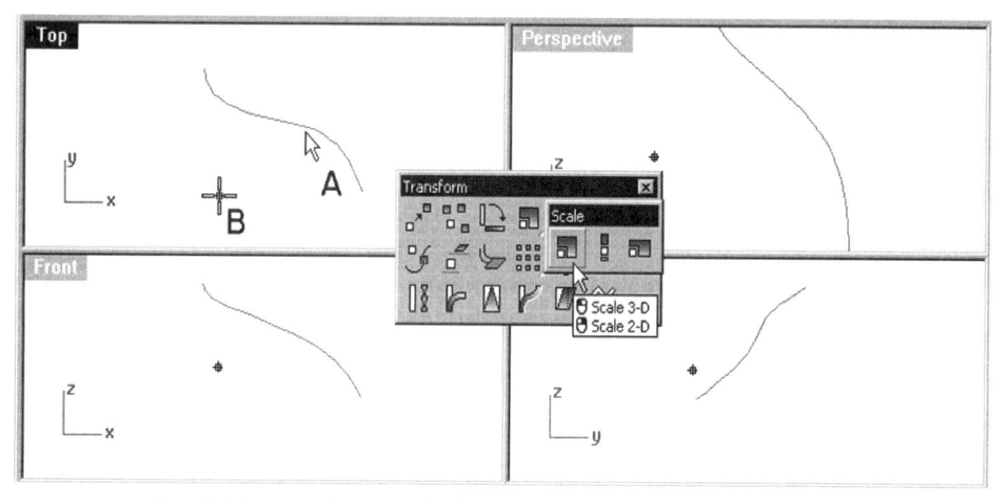

Fig. 3-58. Curve being scaled.

3 Select location B (indicated in figure 3-58) to specify the origin point.

4 Specify the scale factor of 0.8 at the command line interface or establish the first and second reference points.

The entire curve is scaled.

Mirror

The Mirror tool is used to mirror-copy a curve, as follows.

1. Select Transform > Mirror, or click on the Mirror button on the Transform toolbar.
2. Select curve A (indicated in figure 3-59) and press the Enter key.
3. Select location B (indicated in figure 3-59).
4. Select location C (indicated in figure 3-59).
5. If you want to save your file, save it as *Transform1.3dm*.

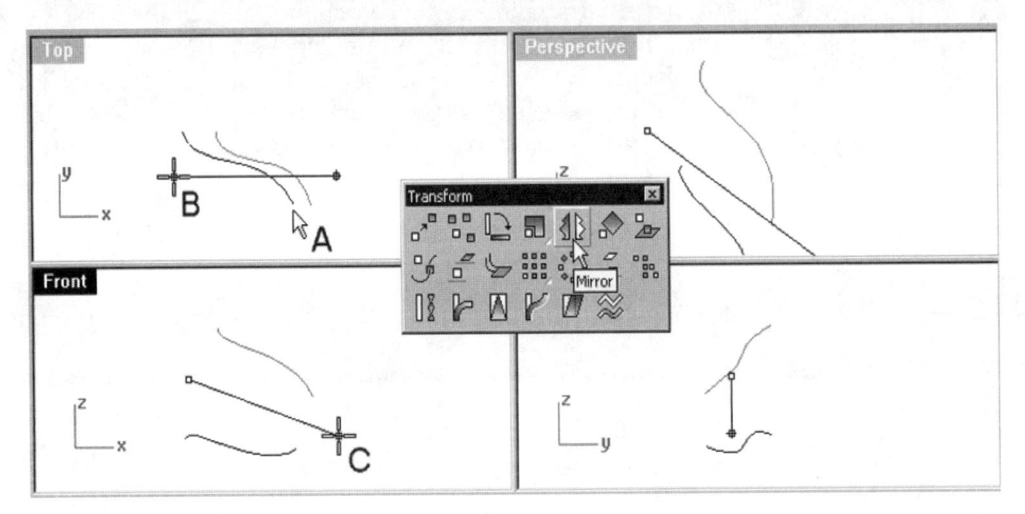

Fig. 3-59. Curve being mirrored.

Orient

The Orient tool is used to modify the orientation of selected objects. There are six ways to orient an object. These methods are discussed in the sections that follow.

Orient 2 Points

The Orient 2 Points option repositions and uniformly scales a curve with reference to two reference points and two target points. On a construction plane, you construct a curve that depicts your design intent, without having to bother about the accurate size of the curve, and then reposition and rescale it uniformly to a designated location, as follows.

1. Start a new file. Use the metric (Millimeters) template.
2. With reference to figure 3-60, construct two free-form curves A and B in the Top viewport, and a free-form curve C in the Right viewport.

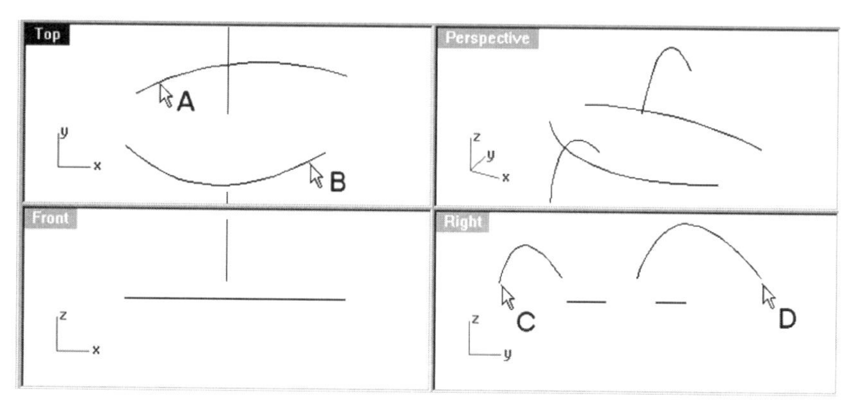

Fig. 3-60. Curves.

3 Check the End box on the Osnap dialog box.

4 Select Transform > Orient > 2 Points, or click on the *Orient: 2 Points/Orient: 3 Points* button on the Transform toolbar.

5 Select curve A (indicated in figure 3-61) and press the Enter key

6 Select end points B and C (indicated in figure 3-61) to specify the reference points.

7 Select end points D and E (indicated in figure 3-61) to specify the target points.

8 Repeat steps 4 through 7 to orient the other curve in the Right viewport, as shown in figure 3-62.

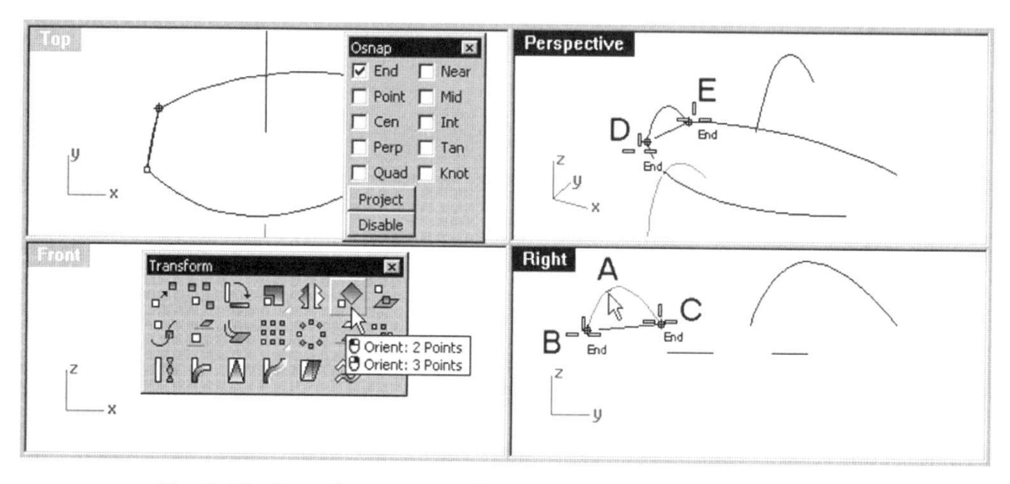

Fig. 3-61. Curve being oriented.

The curve is oriented to the circle. Note the change in the overall scale of the curve.

9 If you want to save your file, save it as *Orient2P.3dm*.

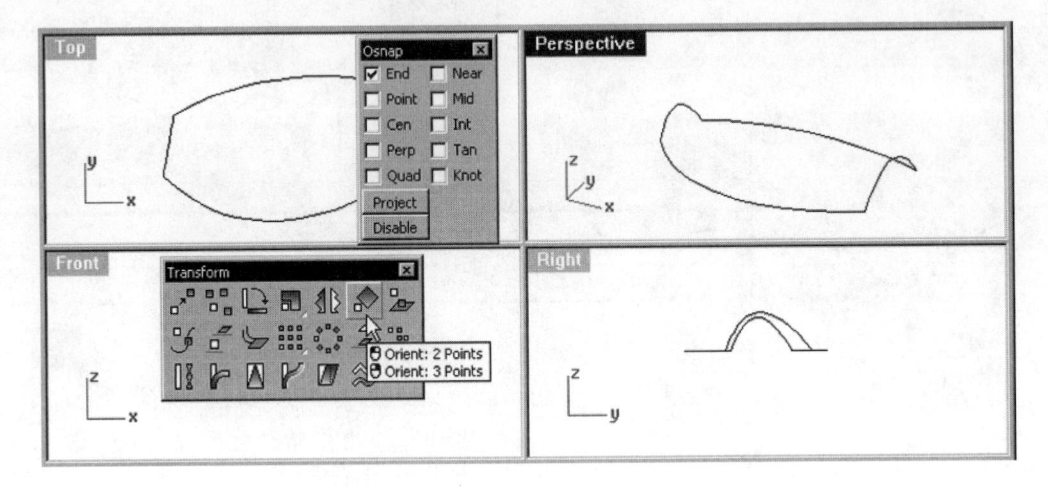

Fig. 3-62. Second curve oriented.

Orient 3 Points

The Orient 3 Points option repositions a curve to a new location and changes the curve's orientation, without scaling. Note that you can also do this with the Orient 2 Points command using the No Scale option. Perform the following steps.

1. Start a new file. Use the metric (Millimeters) template.
2. With reference to figure 3-63, construct a polyline by selecting location A in the Top viewport, location B in the Front viewport, and location C in the Right viewport.

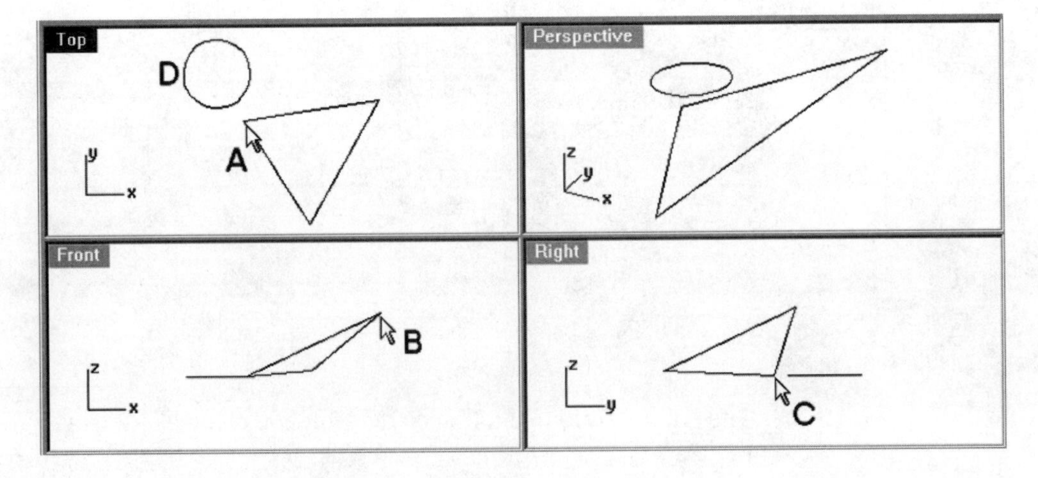

Fig. 3-63. Circle and polyline.

3 Construct a circle at location D in the Top viewport.
4 Check the Cen, Quad, and End boxes and clear all other boxes on the Osnap dialog box.
5 Select Transform > Orient > 2 Points, or right-click on the *Orient: 2 Points/Orient: 3 Points* button on the Transform toolbar.
6 Select the circle and press the Enter key.
7 Select the center of the circle A and two quadrant locations B and C (indicated in figure 3-64) as the reference points.
8 Select the end points of the polyline D, E, and F (indicated in figure 3-64) as the target points.

The circle is oriented, as shown in figure 3-65. Note that the oriented object does not change shape.

9 If you want to save your file, save it as *Orient3P.3dm*.

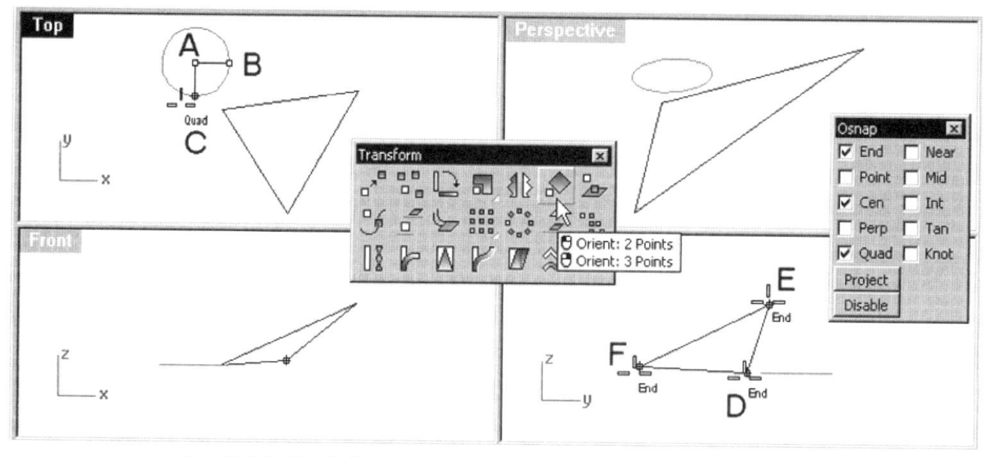

Fig. 3-64. Circle being oriented using Orient 3 Points.

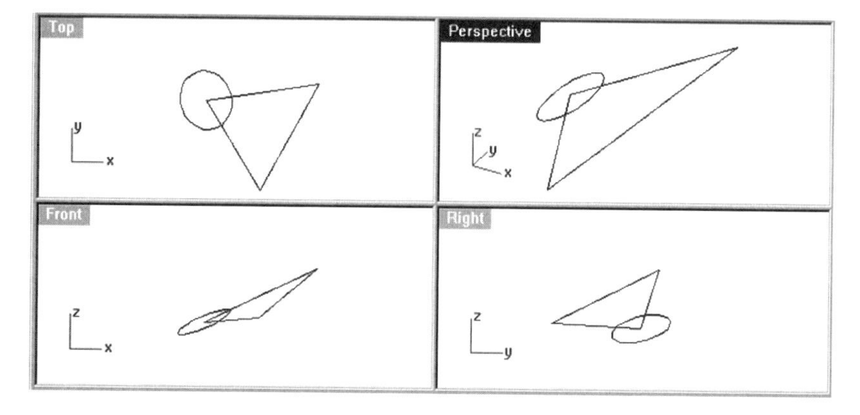

Fig. 3-65. Oriented circle using Orient 3 Points.

Orient Perpendicular to Curve

The Orient Perpendicular to Curve option repositions a curve to a point on a target curve with the plane of the repositioned curve perpendicular to the target curve. Note that this command is construction plane dependent. Perform the following steps.

1. Start a new file. Use the metric (Millimeters) template.
2. With reference to figure 3-66, construct a circle and a free-form curve in the Top viewport.
3. Select Transform > Orient > Perpendicular to Curve, or click on the Orient Perpendicular to Curve button on the Transform toolbar.
4. Select the circle and press the Enter key.
5. Select the center A of the circle as the base point.
6. Select the curve.
7. Select end point B of the curve.

The circle is oriented perpendicular to the curve, as shown in figure 3-67.

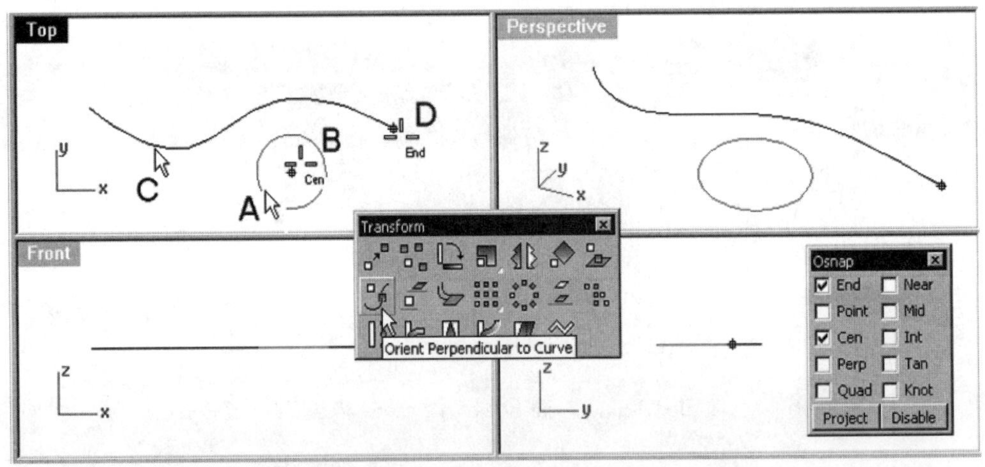

Fig. 3-66. Circle being oriented using Orient Perpendicular to Curve.

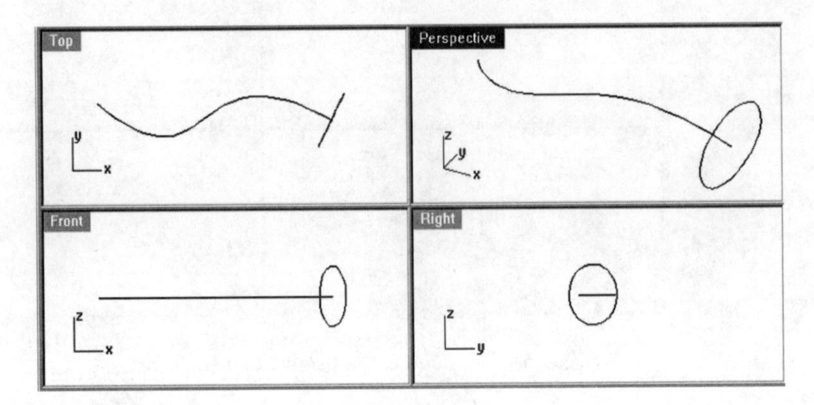

Fig. 3-67. Oriented circle using Orient Perpendicular to Curve.

8 If you want to save your file, save it as *OrientPC.3dm*.

Orient on Surface

To reposition an object on a surface with the plane of the curve tangent to the surface, perform the following steps.

1 Open the file *OrientonSurface.3dm* from the *Chapter 3* folder on the companion CD-ROM.
2 Select Transform > Orient > On Surface, or click on the Orient on Surface button on the Transform toolbar.
3 Select circle A (indicated in figure 3-68) and press the Enter key.
4 Select center B (indicted in figure 3-68) of the circle as the point to be oriented.
5 Select surface C (indicated in figure 3-68).
6 Select locations D, E, and F (indicated in figure 3-68) and press the Enter key.

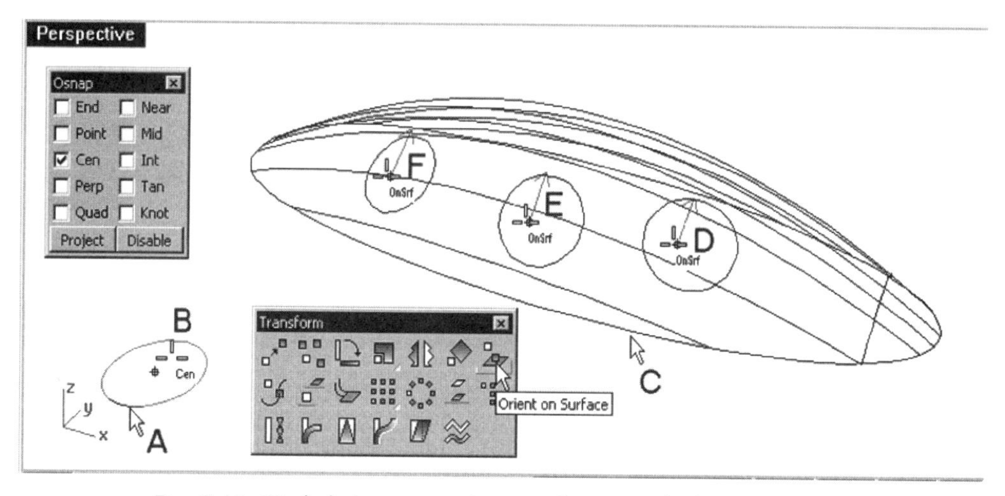

Fig. 3-68. Circle being oriented to a surface in multiple copies.

Orient Curve to Edge

The Orient Curve to Edge tool repositions selected objects on a point along the edge of a surface, as follows.

1 Open the file *Orient2Edge.3dm* from the *Chapter 3* folder on the companion CD-ROM.
2 Select Transform > Orient > Curve to Edge, or click on the Orient Curve to Edge button on the Transform toolbar.

3 Select circle A (indicated in figure 3-69) and press the Enter key.
4 Select surface edge B (indicated in figure 3-69).
5 Select locations C, D, and E (indicated in figure 3-69) and press the Enter key.

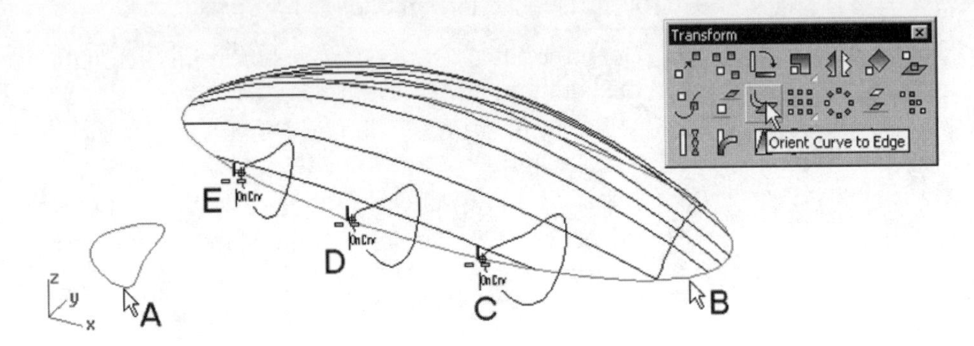

Fig. 3-69. Curve being oriented to the edge of a surface.

Remap to CPlane

To reposition a curve from a construction plane to another construction plane, perform the following steps. This is especially useful for orienting 2D drawings to 3D construction planes. You can map the front view to the Front viewport, the right view to the Right viewport, and so on. This is how most people use this command.

1 Start a new file. Use the metric (Millimeters) template.
2 Construct two free-form curves A and B, sharing an end point C, in the Front viewport.
3 Click on the Front viewport to set it as the current viewport.
4 Select Transform > Orient > Remap to CPlane, or click on the Remap to CPlane button on the Transform toolbar.
5 Select curve A (indicated in figure 3-70) and press the Enter key.
6 Click on the Right viewport.

A curve is oriented to the Right viewport. To see how remapping is used with the Curve from 2 Views option, continue with the following steps.

7 Select Curve > Curve from 2 Views, or click on the Curve from 2 Views button on the Curve Tools toolbar.
8 Select curves A and B (indicated in figure 3-71).
9 If you want to save your file, save it as *Remap.3dm*.

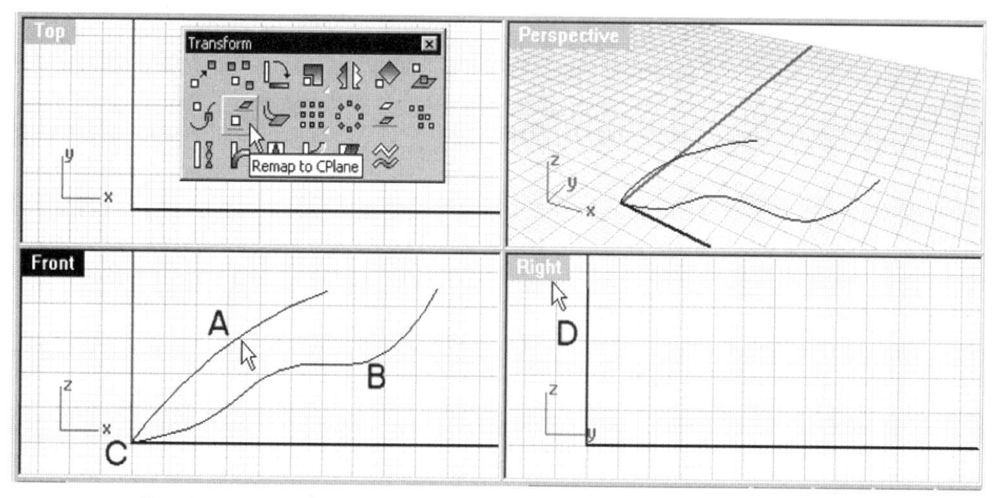

Fig. 3-70 Curve being remapped.

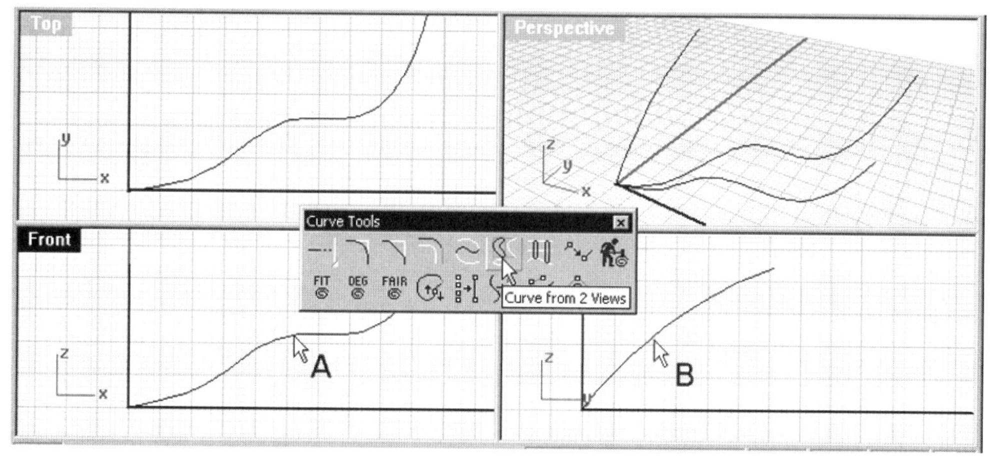

Fig. 3-71. 3D curve from two planar curves.

Arraying

An array is a set of repeated objects in a pattern. There are five ways to construct an array. These methods are discussed in the sections that follow.

Rectangular Array

A rectangular array repeats selected objects in a rectangular pattern. You specify the number of (and distance between) repeated objects in the X, Y, and Z directions. Perform the following steps.

1. Start a new file. Use the metric (Millimeters) template.
2. Maximize the Top viewport and construct a circle in accordance with figure 3-72.
3. Select Transform > Array > Rectangular, or click on the Rectangular Array button on the Array toolbar.
4. Select the circle and press the Enter key.
5. Type *3* to specify the number of objects in the X direction.
6. Type *2* to specify the number of objects in the Y direction.
7. Type *3* to specify the number of objects in the Z direction.
8. Type *10* to specify the X spacing.
9. Type *12* to specify the Y spacing.
10. Type *5* to specify the Z spacing. Note that negative values for distances change the direction of an array.

A rectangular array is constructed.

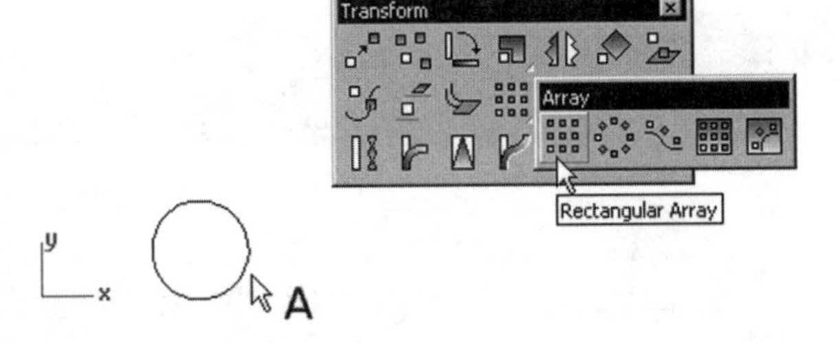

Fig. 3-72. Polyline constructed and being arrayed.

Polar Array

A polar array repeats selected objects in a circular pattern. You specify the center of the pattern, the number of repeated objects, and the angle to be filled by the repeated objects. Perform the following steps.

1. Set the display to a four-viewport configuration.
2. Select Transform > Array > Polar, or click on the Polar Array button on the Array toolbar.
3. Select circle A (indicated in figure 3-73) and press the Enter key.
4. Select location B (indicated in figure 3-73) to specify the center of the array.
5. Type *12* to specify the number of elements in the array.
6. Type *360* (or press the Enter key if the default is 360) to specify the angle to be filled.

A polar array is constructed.

7. If you want to save your file, save it as *Array.3dm*.

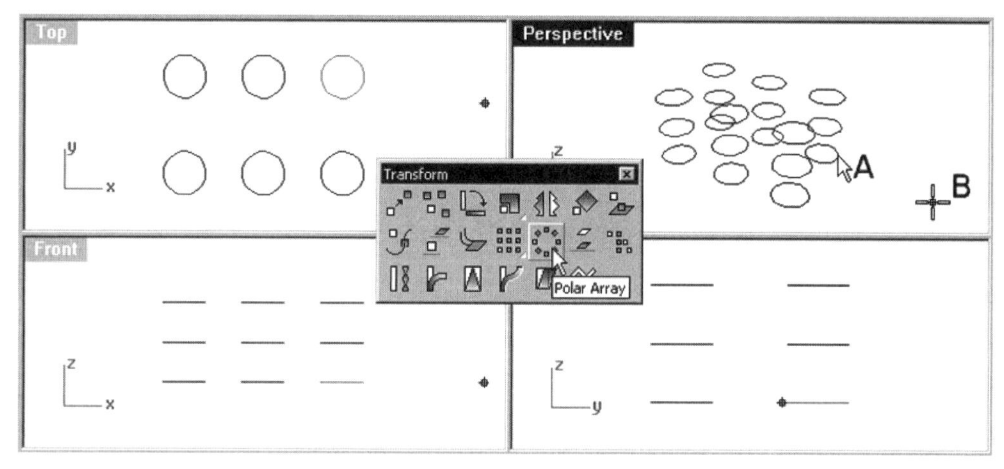

Fig. 3-73. Polar array being constructed.

Array Along a Curve

The Array Along Curve option repeats selected objects along a path curve. You specify the path curve, and the number of repeated objects or the distance between contiguous objects. Perform the following steps.

1. Start a new file. Use the metric (Millimeters) template.
2. Construct polyline A in the Front viewport and a free-form curve in the Top viewport, as indicated in figure 3-74.

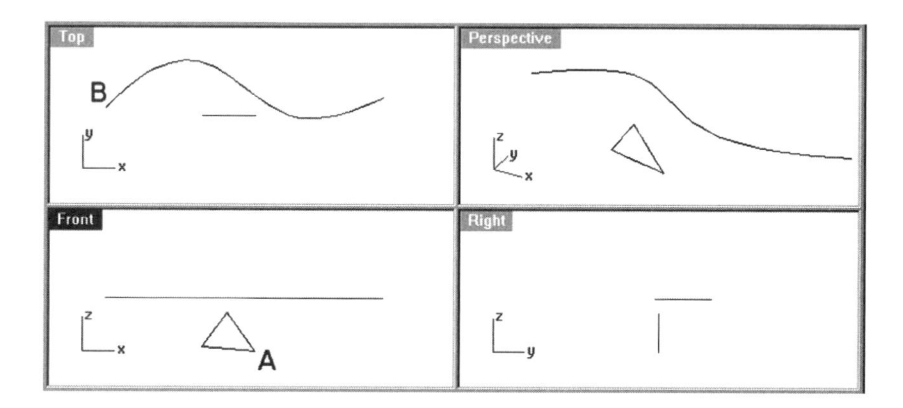

Fig. 3-74. Polyline and free-form curve.

3. Select Transform > Move, or click on the Move button on the Transform toolbar.
4. Select polyline A (indicated in figure 3-75) and press the Enter key.
5. Select end point B and then end point C (indicated in figure 3-75).

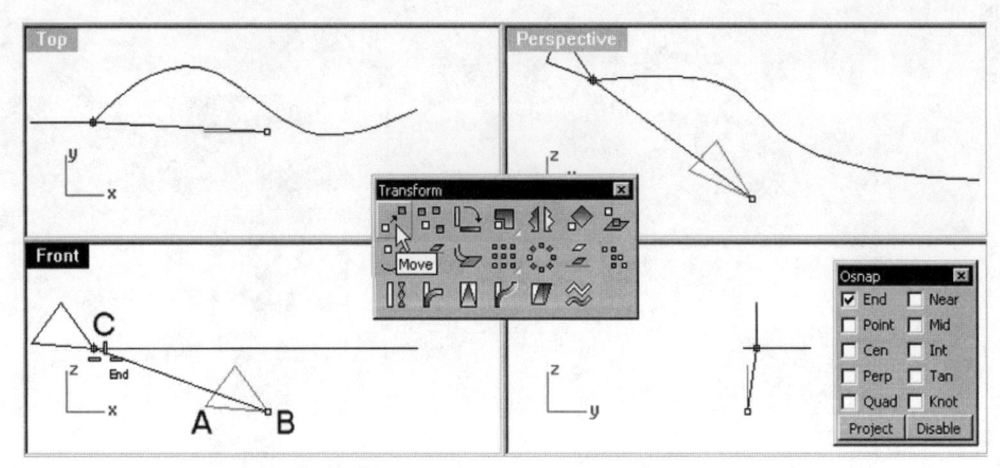

Fig. 3-75. Polyline being moved.

6 Select Transform > Array > Along Curve, or click on the Array Along Curve button on the Array toolbar.
7 Select polyline A (indicated in figure 3-76) and press the Enter key.
8 Select free-form curve B (indicated in figure 3-76) to use it as the path curve.
9 In the Array Along Curve Options dialog box, specify three (3) items in the *Number of items* box, select the *Freeform twisting* option from the list box, and then click on the OK button.

The polyline is arrayed along the curve, as shown in figure 3-77. Note that the triangle rotates about its own axis as it arrays along the curve.

10 If you want to save your file, save it as *ArrayCrv.3dm*.

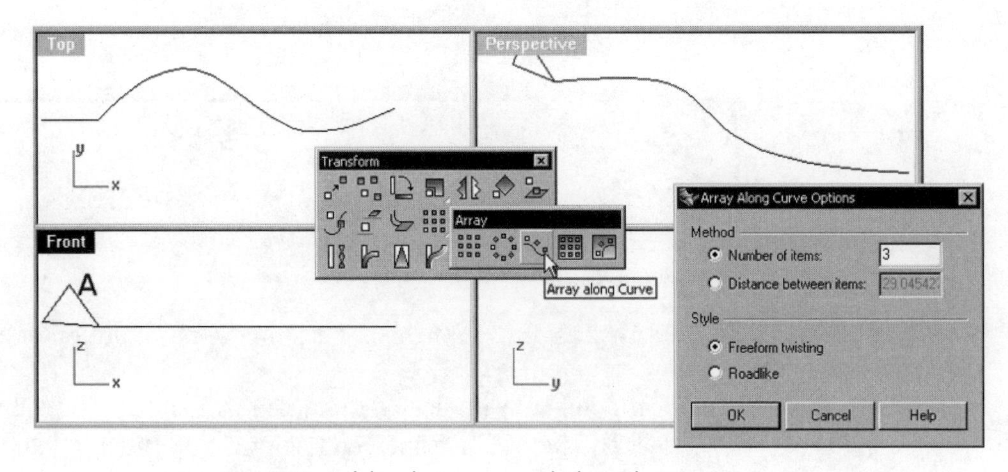

Fig. 3-76. Polyline being arrayed along the curve.

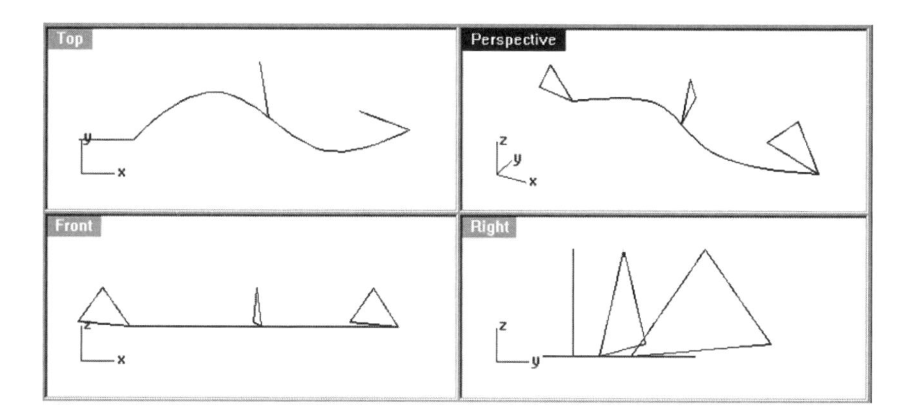

Fig. 3-77. Polyline arrayed along the curve.

Array on Surface

The Array on Surface option repeats selected objects on a surface along the surface's U and V directions, as follows.

1. Open the file *ArraySurface.3dm* from the *Chapter 3* folder on the companion CD-ROM.
2. Select Transform > Orient > On Surface, or click on the Orient on Surface button on the Transform toolbar.
3. Select the circle and line and press the Enter key.
4. Select end point A (indicated in figure 3-78) to orient from.

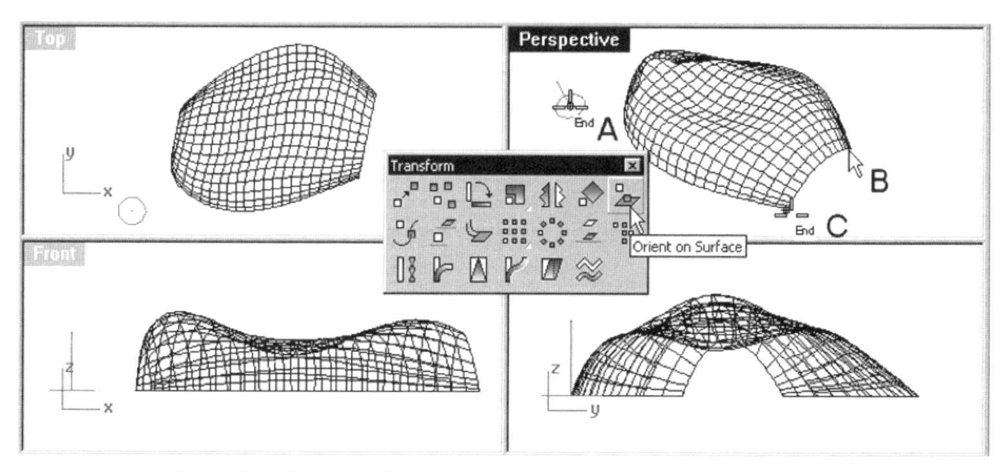

Fig. 3-78. Circle and line being oriented.

5. Select surface B (indicated in figure 3-78).
6. Select end point C (indicated in figure 3-78) and press the Enter key.

7 Maximize the Perspective viewport.

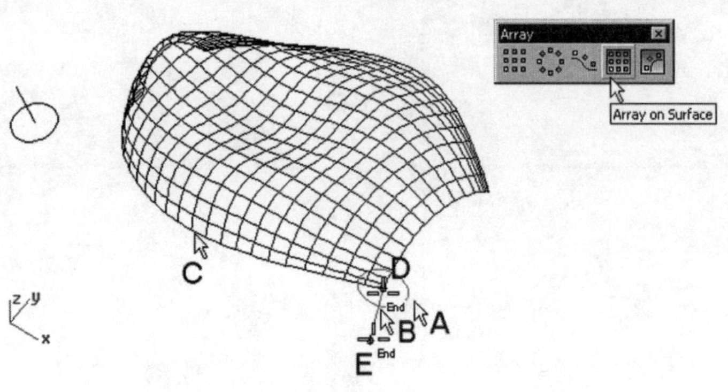

Fig. 3-79. Circle and line being arrayed.

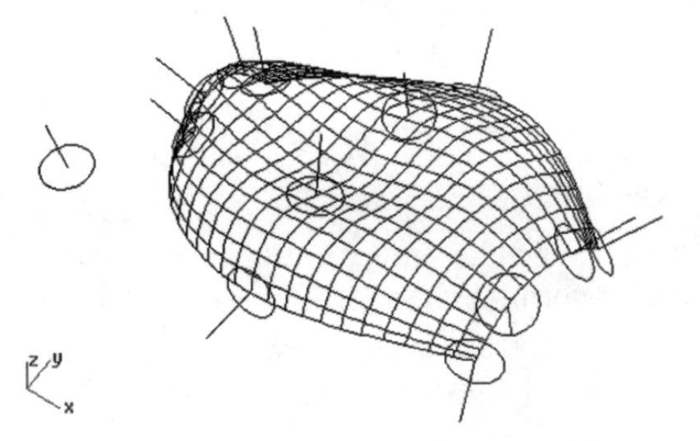

Fig. 3-80. Arrayed circle and line.

8 Select Transform > Array > Along Surface, or click on the Array on Surface button on the Array toolbar.
9 Select circle A and line B (indicated in figure 3-79) and press the Enter key.
10 Select end point C (shown in figure 3-79) to indicate the base point.
11 Select end point D (shown in figure 3-79) to indicate the normal direction.
12 Select surface E (indicated in figure 3-79).
13 Type *3* to specify the number of elements in the U direction.
14 Type *4* to specify the number of elements in the V direction.

An array is constructed, as shown in figure 3-80.

Array Along Curve on Surface

The Array Along Curve on Surface option repeats selected objects along a curve residing on a surface. Perform the following steps.

1 Open the file *ArrayCurveSurface.3dm* from the *Chapter 3* folder on the companion CD-ROM.
2 Select Transform > Array > Along Curve on Surface, or click on the Array Along Curve on Surface button on the Transform toolbar.
3 Select circle A (indicated in figure 3-81) and press the Enter key.
4 Select end point B (indicated in figure 3-81) as the base point.
5 Select curve C (indicated in figure 3-81).
6 Type *D* at the command line interface to divide the curve.
7 Type *4* at the command line interface.

The circle is arrayed along a curve on a surface, as shown in figure 3-82.

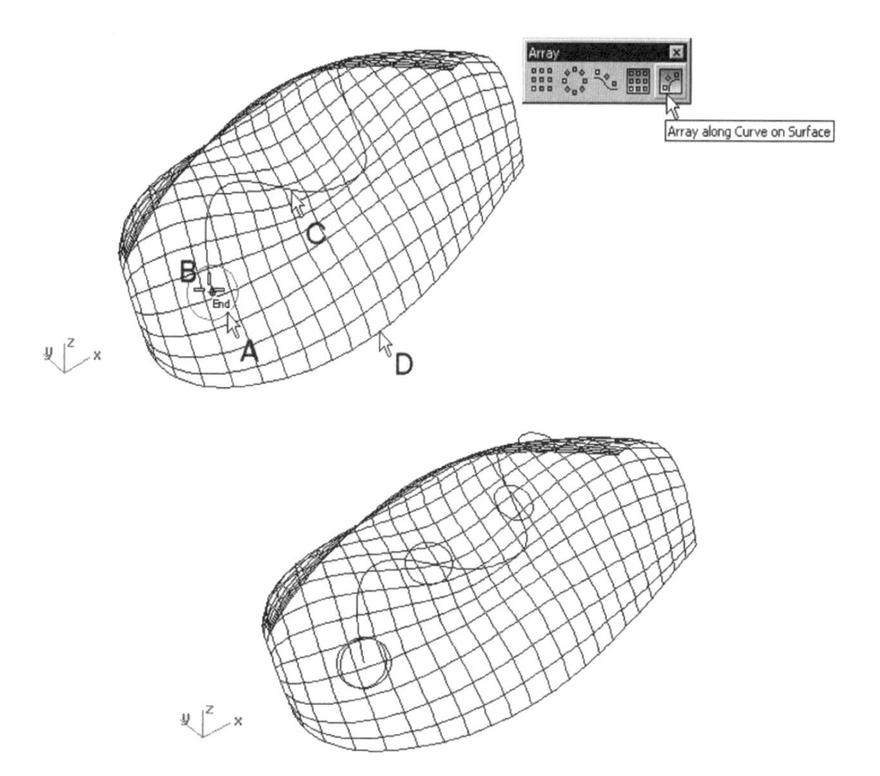

Fig. 3-81. Circle being arrayed along a curve on the surface.

Fig. 3-82. Arrayed circle.

Setting Points

You can change the shape of a curve by aligning the X, Y, and Z coordinates of selected control points or edit points of a curve. Perform the following steps.

1. Start a new file. Use the metric (Millimeters) template.
2. Maximize the Top viewport.
3. With reference to figure 3-83, construct a free-form curve and turn on its control points.
4. Select Transform > Set Points, or click on the Set XYZ Coordinates button on the Transform toolbar.
5. Select control points A and B (indicated in figure 3-83) and press the Enter key.
6. In the Set Points dialog box, check the Set Y and the Align to World boxes, and then click on the OK button.
7. Select location A (indicated in figure 3-84).
8. The Y coordinates of the selected control points are aligned.
9. If you want to save your file, save it as *SetPt.3dm*.

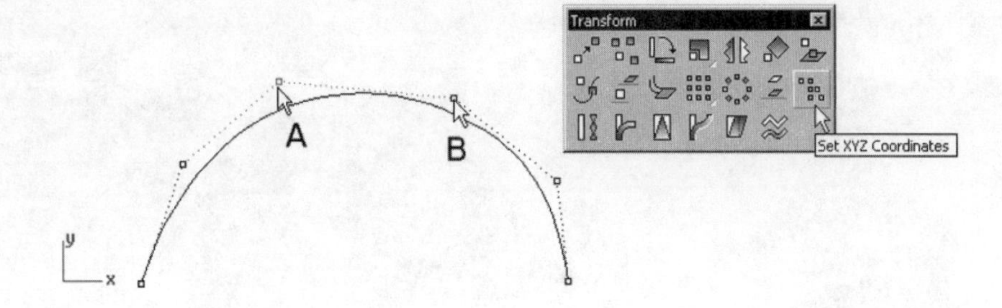

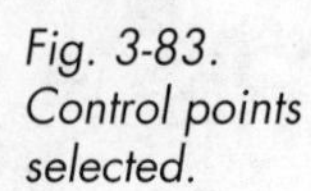
Fig. 3-83. Control points selected.

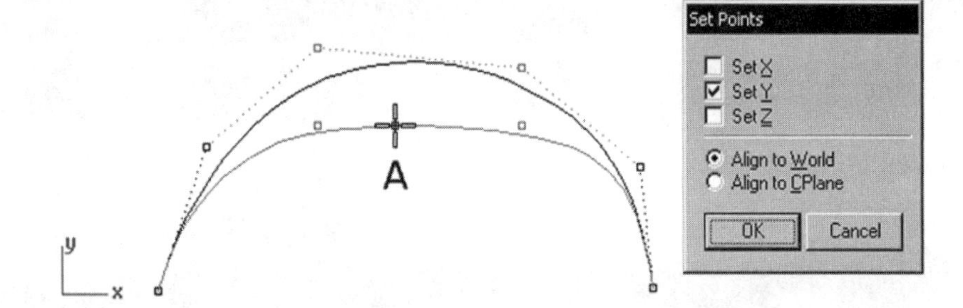

Fig. 3-84. Selected control points being set.

Smoothing

Smoothing transforms a curve by averaging its control points. To smooth a curve, perform the following steps.

1. Start a new file. Use the metric (Millimeters) template.
2. Maximize the Top viewport.
3. Construct a free-form curve in accordance with figure 3-85.
4. Select Transform > Smooth, or click on the Smooth button on the Transform toolbar.
5. Select curve A (indicated in figure 3-85) and press the Enter key.

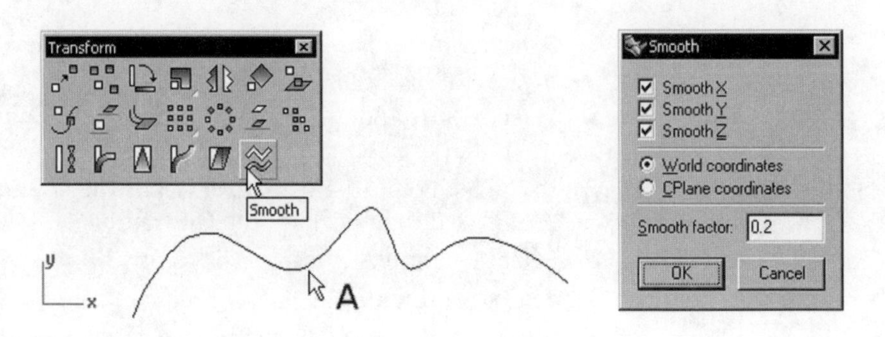

Fig. 3-85. Curve being smoothed.

6 In the Smooth dialog box, select Smooth X, Smooth Y, and Smooth Z, and then click on the OK button.

The curve is smoothed.

7 If you want to save your file, save it as *Smooth.3dm*.

Twisting

You can transform a curve by twisting the entire curve or selected control points of the curve. To twist a curve, perform the following steps.

1. Start a new file. Use the metric (Millimeters) template.
2. Construct a free-form curve and a line segment in the Top viewport.
3. Select Transform > Twist, or click on the Twist button on the Transform toolbar.
4. Select curve A (indicated in figure 3-86) and press the Enter key.
5. Select end points B and C (indicated in figure 3-86) to indicate the twist axis.
6. Type *90* at the command line interface to specify the twist angle.

The selected curve is twisted.

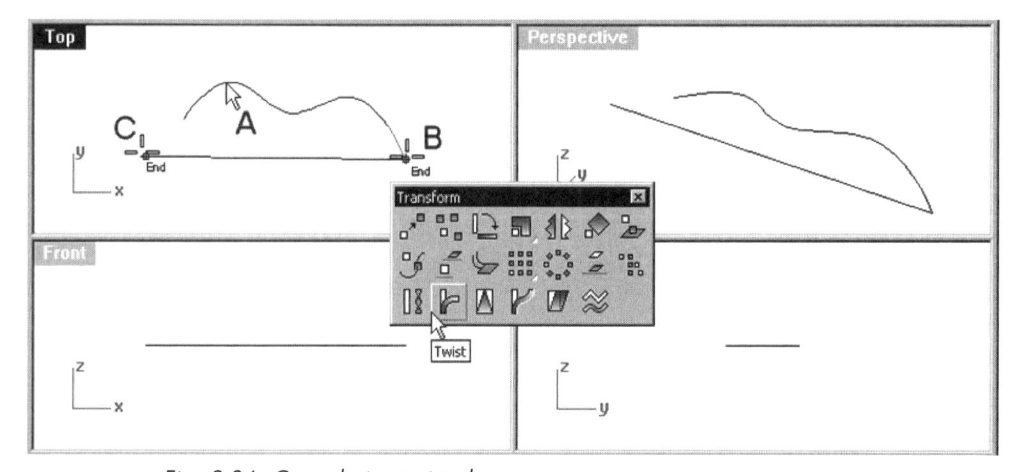

Fig. 3-86. Curve being twisted.

Bending

You can transform a curve by bending the entire curve or selected control points of the curve. To bend a curve, perform the following steps.

1. Select Transform > Bend, or click on the Bend button on the Transform toolbar.
2. Select curve A (indicated in figure 3-87) and press the Enter key.

3. Select end points B and C (indicated in figure 3-87) to specify a reference spline.
4. Select location D (indicated in figure 3-87) to specify the amount of bend.

The selected curve is bent.

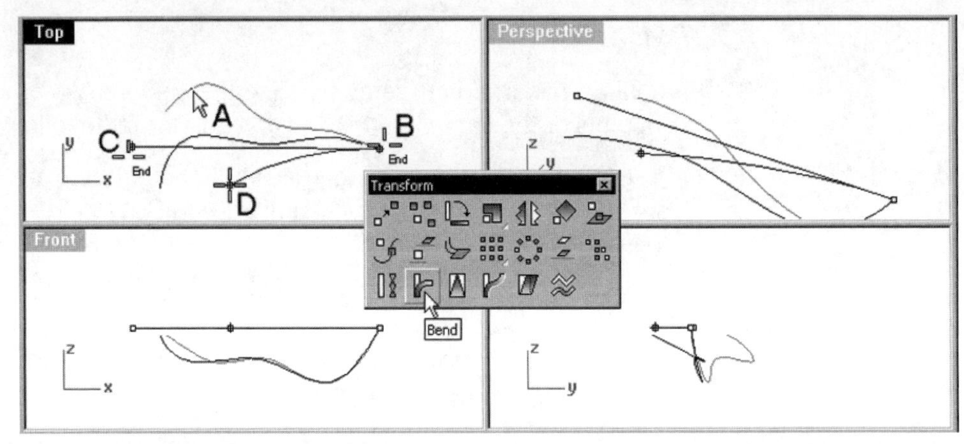

Fig. 3-87. Selected curve being bent.

Projection to CPlane

You can transform a curve by projecting the entire curve or selected control points of the curve onto a selected construction plane. To project a curve, perform the following steps.

1. Select Transform > Project to CPlane, or click on the Project to CPlane button on the Transform toolbar.
2. Select the Front viewport, select curve A (indicated in figure 3-88), and press the Enter key.

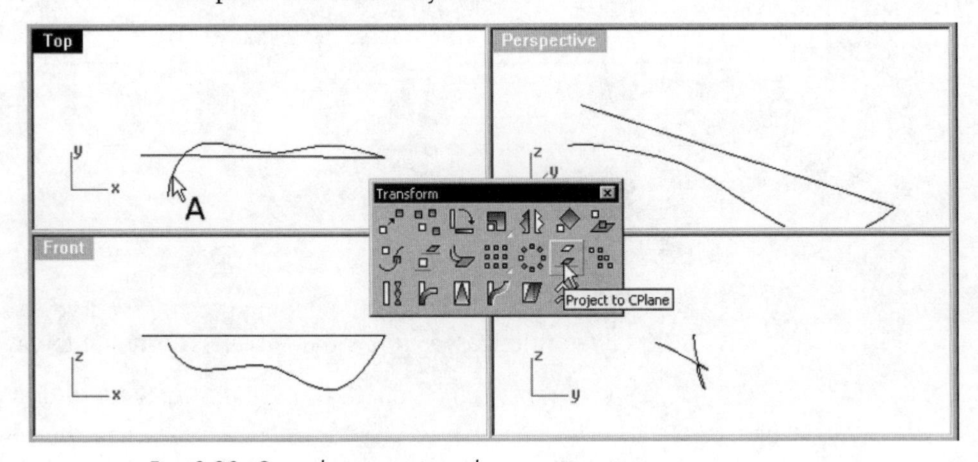

Fig. 3-88. Curve being projected.

3 The selected curve is projected onto the construction plane, corresponding to the Front viewport.

4 If you want to save your file, save it as *Project.3dm*.

Tapering

You can transform a curve by tapering the entire curve or selected control points of the curve. To taper a curve, perform the following steps.

1 Start a new file. Use the metric (Millimeters) template.

2 With reference to figure 3-89, construct a line and a helix curve in the Top viewport.

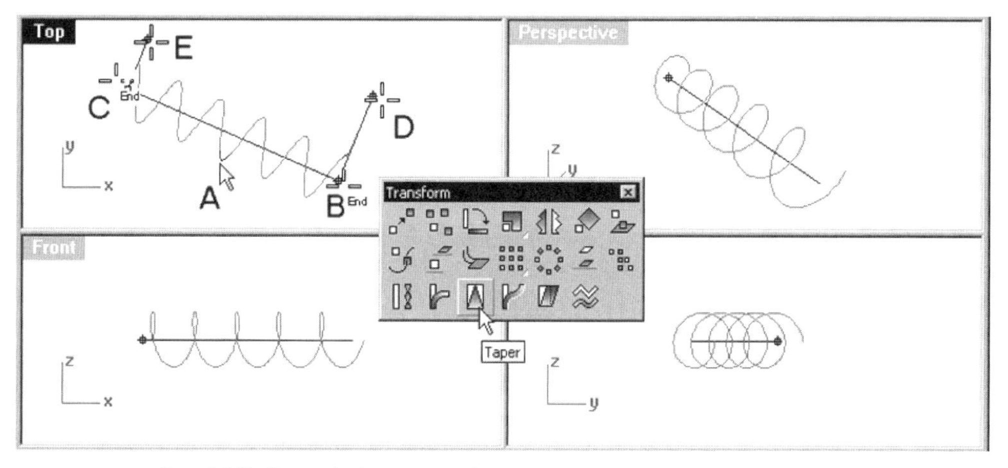

Fig. 3-89. Curve being tapered.

3 Select Transform > Taper, or click on the Taper button on the Transform toolbar.

4 Select helix A (indicated in figure 3-89) and press the Enter key.

5 Select end points B and C (shown in figure 3-89) to indicate the taper axis.

6 Select locations D and E (shown in figure 3-89) to indicate the start and end distances.

The selected curve is tapered.

Shear

You can shear an entire curve or selected control points of the curve. To shear a curve, perform the following steps.

1 Select Transform > Shear, or click on the Shear button on the Transform toolbar.

2 Select line A and curve B (indicated in figure 3-90) and press the Enter key.
3 Select end point C (indicated in figure 3-90) as the origin point.
4 Select end point D (indicated in figure 3-90) as the reference point.
5 Select location E (shown in figure 3-90) to indicate the shear angle.

The curve is sheared.

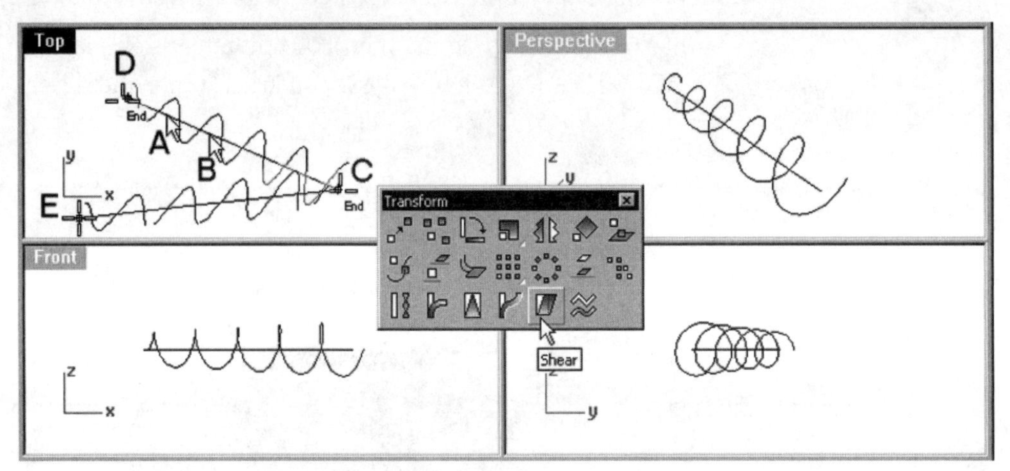

Fig. 3-90. Curves being sheared.

Flowing Along a Curve

You can transform a curve by setting the curve or its control points to flow along a selected curve, as follows. Flowing a curve along another curve involves an original backbone curve and a new backbone curve, in addition to the curve to be transformed. This process involves transforming the curve along its original backbone curve to the new backbone curve.

1 Construct a free-form curve A in the Top viewport.
2 Select Transform > Flow Along Curve, or click on the Flow Along Curve button on the Transform toolbar.
3 Select curve B (indicated in figure 3-91) and press the Enter key.
4 Select line C (indicated in figure 3-91) to use it as the original backbone curve.
5 Select curve A (indicated in figure 3-91) to use it as the new backbone curve.

A curve is transformed to flow along the new backbone curve, as shown in figure 3-92.

6 If you want to save your file, save it as *Flow.3dm*.

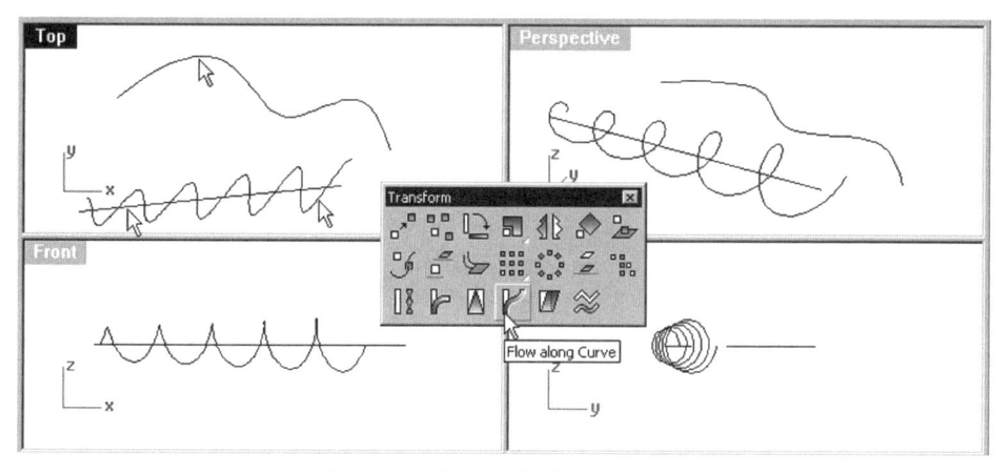

Fig. 3-91. Curve being transformed by flowing along another curve.

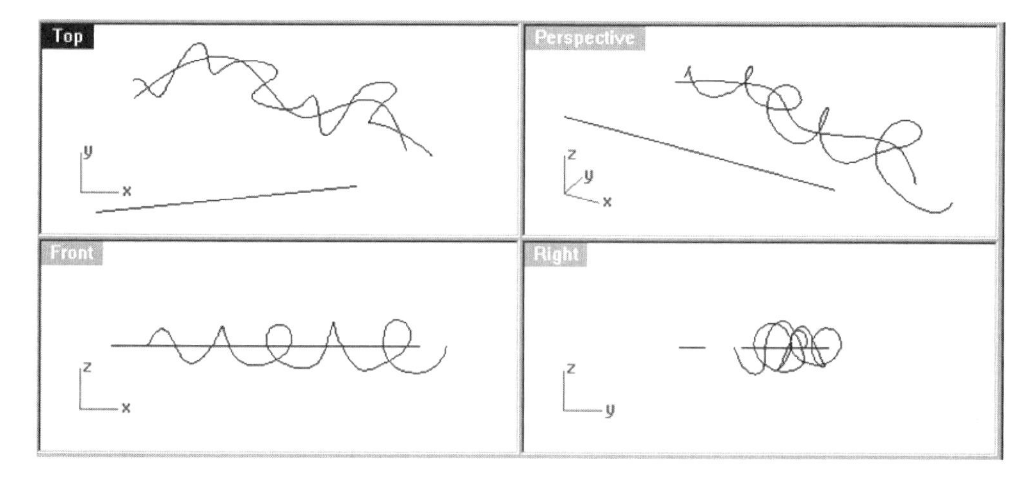

Fig. 3-92. Transformed curve.

Analyzing Curves

You have learned how to construct point objects, basic curves, and derived curves; to edit curves; and to transform curves. To find out information about point objects and curves, you use analysis tools from the Analyze menu and the Analyze toolbar, shown in figure 3-93.

Table 3-6 outlines the functions of the curve and point analysis commands and their locations in the main pull-down menu and respective toolbars.

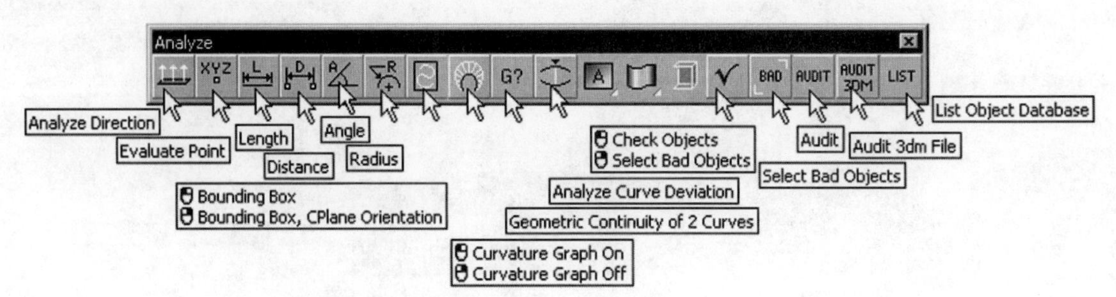

Fig. 3-93. Analyze toolbar.

Table 3-6: Curve and Point Analysis Commands and Their Functions

Toolbar Button	From Main Pull-down Menu	Function
Analyze Direction	Analyze > Direction	Displaying and flipping the direction of selected objects, including surfaces
Evaluate Point	Analyze > Point	Finding out the coordinates of selected points
Length	Analyze > Length	Measuring the length of a selected curve or edge of a surface
Distance	Analyze > Distance	Measuring the distance between two selected points on a curve or surface
Angle	Analyze > Angle	Measuring the angle between two lines defined by specifying the end points of the lines
Radius	Analyze > Radius	Measuring the radius of curvature at selected locations along a curve
	Analyze > Curve > Curvature, Circle	Constructing a tangent circle along a curve to depict the radius of curvature at a selected location
Bounding Box/ Bounding Box, CPlane Orientation	Analyze > Bounding Box	Constructing a bounding box to enclose a selected curve, surface, or polysurface
Curvature Graph On/ Curvature Graph Off	Analyze > Curve > Curvature Graph On Analyze > Curve > Curvature Graph Off	Displaying a graph along a curve to illustrate the radius of curvature along the curve; closing the curvature graph
Geometric Continuity of 2 Curves	Analyze > Curve > Geometric Continuity	Discerning the geometric continuity at a join between two curves

Toolbar Button	From Main Pull-down Menu	Function
Analyze Curve Deviation	Analyze > Curve > Deviation	Discerning the deviation between two curves
Check Objects/Select Bad Objects	Analyze > Diagnostics > Check	Diagnosing a selected object for error
Select Bad Objects	Analyze > Diagnostics > Select Bad Objects	Selecting any object that does not pass the check test
Audit	Analyze > Diagnostics > Audit	Displaying information on erroneous objects and their relationships to layers
Audit 3dm File	Analyze > Diagnostics > Audit 3dm File	Diagnosing Rhino files
List Object Database	Analyze > Diagnostics > List	Obtaining the data of selected objects

Curve Direction

Curves have direction. Sometimes, the direction affects the outcome of subsequent surface construction operations. To display and change the direction of a curve, perform the following steps.

1. Open the file *Analyze.3dm* from the *Chapter 3* folder on the companion CD-ROM.
2. Select Analyze > Direction, or click on the Analyze Direction button on the Analyze toolbar.
3. Select curve A (indicated in figure 3-94).

The direction of the curve is displayed.

Fig. 3-94. Direction of the curve displayed.

If you want to flip the direction of the curve, type *F* at the command line interface.

Evaluate Point

To discern the coordinates of selected points, perform the following steps.

1. Check the Point option in the Osnap dialog box.
2. Select Analyze > Point, or click on the Evaluate Point button on the Analyze toolbar.
3. Select point A (indicated in figure 3-95).

The coordinates of the selected point are displayed in the command line area.

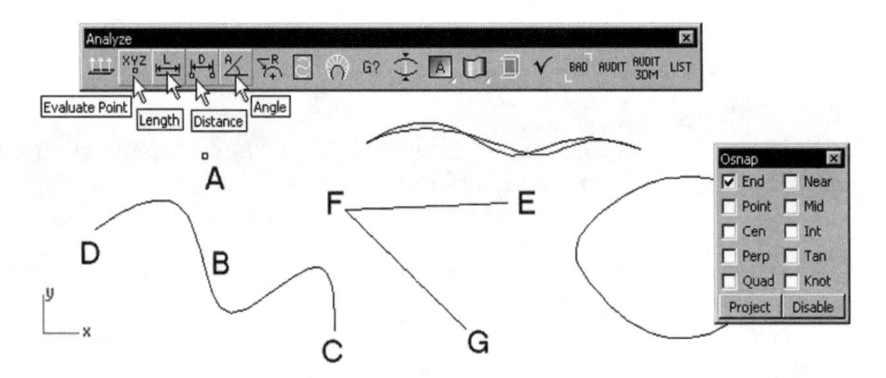

Fig. 3-95. Point and curves being evaluated.

Length Measurement

To measure the length of a curve, perform the following steps.

1. Select Analyze > Length, or click on the Length button on the Analyze toolbar.
2. Select curve B (indicated in figure 3-95).

The length of the selected curve is displayed.

Distance Measurement

To measure the distance between two selected points, perform the following steps.

1. Select Analyze > Distance, or click on the Distance button on the Analyze toolbar.

2 Select end points C and D (indicated in figure 3-95).

The distance between the selected points is displayed.

Angle Measurement

To measure the angle between two lines defined by four selected points, perform the following steps.

1 Select Analyze > Angle, or click on the Angle button on the Analyze toolbar.
2 Select points E and F and then points G and F (indicated in figure 3-95) to specify the first and second lines.

The angle between the first and second line is displayed.

Curvature Radius

To discern the radius of curvature of a selected point along a curve, perform the following steps.

1 Check the Near option of the Osnap dialog box.
2 Select Analyze > Radius, or click on the Radius button on the Analyze toolbar.
3 Select a point along the curve indicated in figure 3-96.

The radius of curvature of the selected location is displayed.

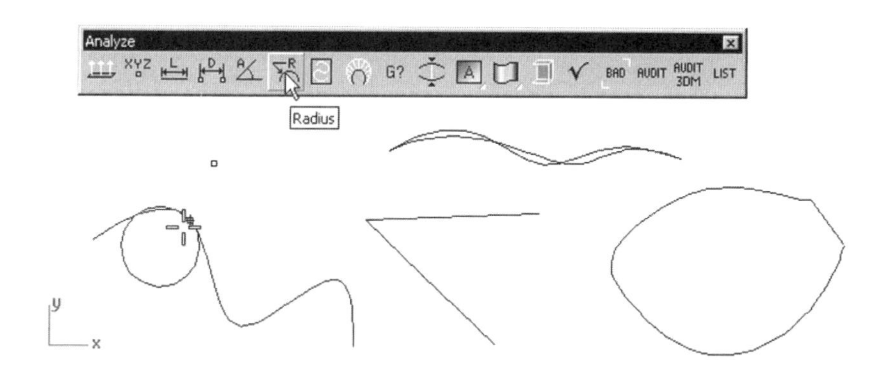

Fig.3-96. Radius of curvature being evaluated.

Curvature Graph

A curvature graph displays the radius of a curve along the curve in the form of a graph. It helps you to determine the smoothness of the curve and to discern any irregularities in the curve. To display and hide the curvature graph of a curve, perform the following steps.

1. Select Analyze > Curve > Curvature Graph On, or click on the *Curvature Graph On/Curvature Graph Off* button on the Analyze toolbar.
2. Select curve A (indicated in figure 3-97).

The curvature graph is displayed.

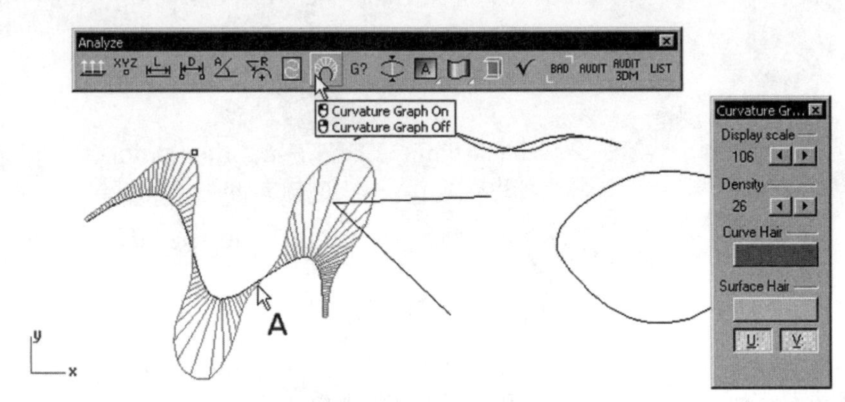

Fig. 3-97. Curvature graph displayed.

3. Select Analyze > Curve > Curvature Graph Off, or right-click on the *Curvature Graph On/Curvature Graph Off* button on the Analyze toolbar.

The curvature graph is turned off.

Geometric Continuity

As mentioned earlier in this chapter, there are three types of continuity in a connected curve: G0, G1, and G2. To discern the continuity of a joint in a curve, perform the following steps.

1. Select Analyze > Curve > Geometric Continuity, or click on the Geometric Continuity of 2 Curves button on the Analyze toolbar.
2. Select curves C and D (indicated in figure 3-98).

The geometric continuity is displayed.

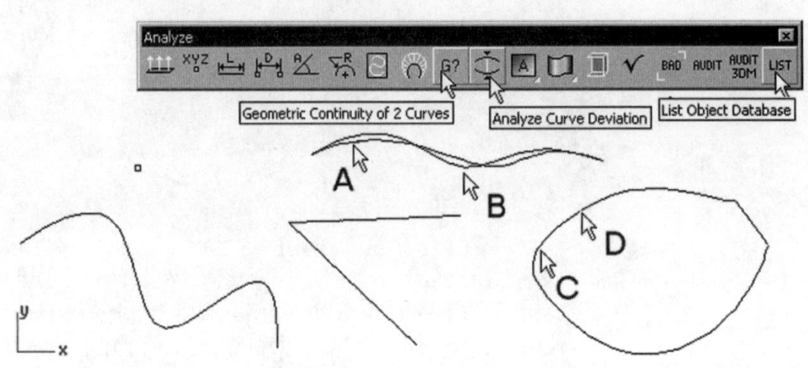

Fig. 3-98. Geometric continuity of curves being analyzed.

Deviation Between Curves

If you have two curves that closely resemble each other, you can find out their deviation by performing the following steps.

1. Select Analyze > Curve > Deviation, or click on the Analyze Curve Deviation button on the Analyze toolbar.
2. Select curves A and B (indicated in figure 3-98).

A report is displayed.

Listing Database

To display the database of a selected object in list form, perform the following steps.

1. Select Analyze > Diagnostics > List, or click on the List Object Database button on the Diagnostics toolbar.
2. Select curve A (indicated in figure 3-98) and press the Enter key.

The database of the selected objected is displayed as a list, as shown in figure 3-99.

3. Click on the Close button.

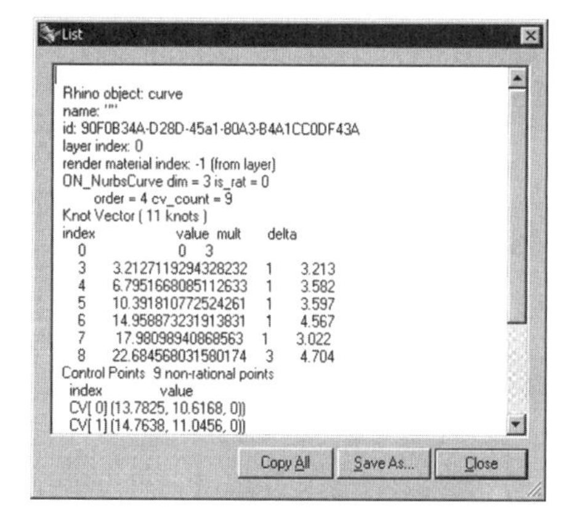

Fig. 3-99. Curve data listed.

Bounding Box Construction

A bounding box of a curve is the minimum size of a rectangular box that encloses the entire curve. To construct a bounding box, perform the following steps.

1. Select Analyze > Bounding Box, or click on the Curve Bounding Box button on the Analyze toolbar.
2. Select curve B (indicated in figure 3-100) and press the Enter key twice.

A bounding box is constructed.

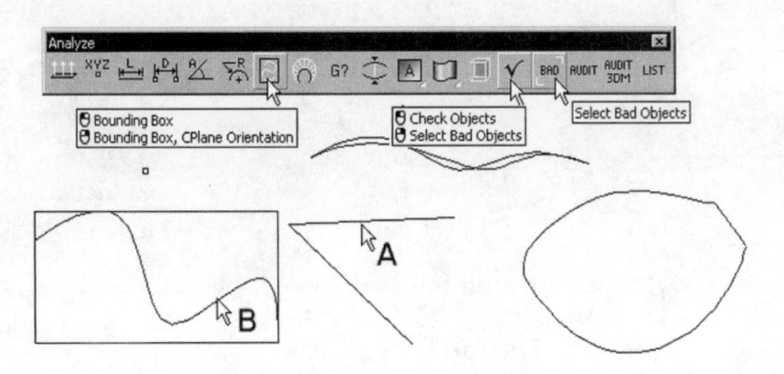

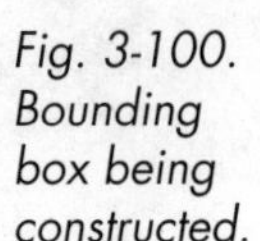

Fig. 3-100. Bounding box being constructed.

Bad Object Report

To check a curve for errors, perform the following steps.

1. Select Analyze > Diagnostics > Check, or click on the Check Object button on the Analyze toolbar.
2. Select curve A (indicated in figure 3-100).

The selected curve is checked.

3. Click on the Close button.

To check the entire file for bad objects, continue with the following.

4. Select Analyze > Diagnostics > Select Bad Objects, or click on the Bad Objects button on the Analyze toolbar.

A report is displayed.

Summary

Although it is logical to construct curves and/or points before creating surfaces and polysurfaces, detailing a 3D free-form model requires intertwined construction of curves, points, and surfaces. You construct surfaces from points and curves, and you construct points and curves from surfaces. Apart from constructing points and curves from scratch, you can construct them on existing objects, including surfaces and solids. Using these points and curves, you can develop and refine your design.

To develop and modify your design, you edit and transform curves. To evaluate curves you have constructed, edited, and transformed, you use the set of analysis tools. The process of analyzing, constructing, editing, and transforming curves should be iterative until you obtain a curve that meets your design needs. Curve editing includes the following three domains.

- Joining several contiguous curves into a single curve, exploding a set of joined curves into individual curves, trimming curves, and splitting curves
- Manipulation of the curve's control points, edit points, knot points, and selected locations along the curve
- Changing the curve's polynomial degree and fit tolerance

Curve transformation involves translation and deformation. You transform curves by moving, copying, rotating, and arraying. You deform curves by scaling, shearing, orienting, setting points, projecting, twisting, bending, tapering, smoothing, and flowing. Analysis procedures and tools include evaluating points, measuring length, measuring distance, measuring angle, evaluating curvature radius, constructing a bounding box, displaying a curvature graph, constructing a curvature circle, checking geometric continuity, finding out deviation between curves, displaying a database list, finding out a curve's direction, and checking bad objects in a file.

Review Questions

1. Outline the methods of constructing curves on existing objects.
2. Illustrate the methods of editing a curve by manipulating its control points, edit points, knots, and points along it.
3. Illustrate how a curve's polynomial degree and fit tolerance can be modified.
4. Explain methods of transforming a curve.
5. Explain how to find out the continuity of a joint in a curve.
6. What is the difference between displaying a database list and reporting bad objects in a file?

CHAPTER 4

Wireframe Modeling and Curves: Part 3

Introduction

This chapter further explores techniques involved in setting up construction planes in 3D space. The chapter also shows you how to construct points and curves by tracing a background image and using the MicroScribe digitizer.

Objectives

After studying this chapter you should be able to:

- ❒ Set up construction planes
- ❒ Trace a curve from background images
- ❒ Describe the function of the MicroScribe digitizer in regard to wireframe construction

Overview

One way to construct points and curves is to use the mouse as an input device. However, you are restricted to selecting locations on default construction planes unless you set up construction planes in various orientations. The ability to manipulate construction planes helps you construct points and curves in 3D space.

As a designer, you may wish to formulate your design ideas on a piece of paper. If you already have a design sketch and want to use the sketch as a template for making curves, you may incorporate it as a background image and trace a curve from it.

Sometimes, you may want to make a physical model before making a digital model in the computer. If you have a physical model, you can use a digitizer to obtain points and curves from the surface of your physical model.

This chapter is a continuation of Chapter 3, and is the final chapter on wireframe and curves. In this chapter, you will learn about construction planes, background images, and digitization.

The Construction Plane

In previous chapters you learned how to construct basic curves by selecting a command and inputting parameters of the curve. You input data by entering the coordinates at the command line area, or by using a pointing device or digitizer to select a set of points. Using a digitizer, you select a point on the surface of a physical object.

Using a pointing device, you select points on the construction plane. The construction plane is an imaginary plane. As explained in Chapter 1, there is a construction plane corresponding to each viewport. To facilitate point and curve construction using the pointing device, you set up construction planes in various ways using the commands on the Set CPlane toolbar, shown in figure 4-1.

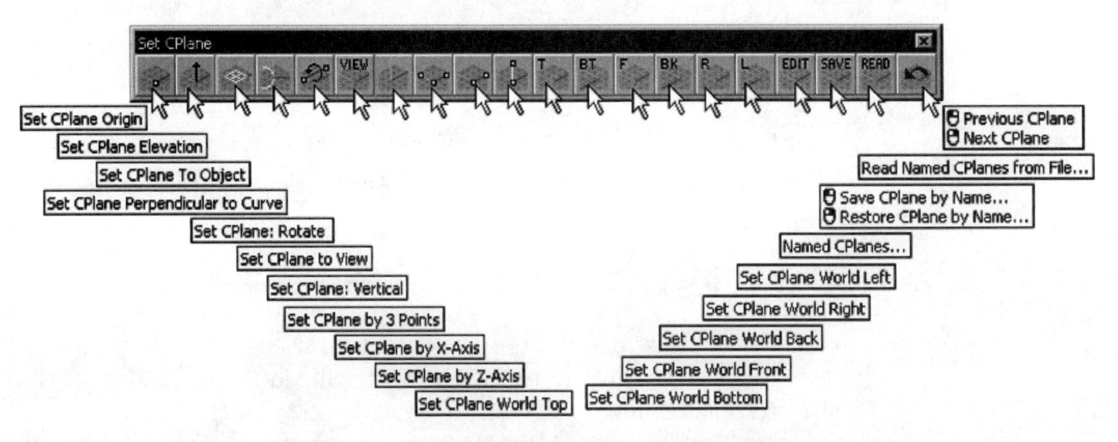

Fig. 4-1. Set CPlane toolbar.

Table 4-1 outlines the functions of the construction plane commands and their locations in the main pull-down menu and respective toolbars.

Table 4-1: Construction Plane Commands and Their Functions

Toolbar Button	From Main Pull-down Menu	Function
Set CPlane Origin	View > CPlane > Origin	Repositioning the origin to a new position
Set CPlane Elevation	View > CPlane > Elevation	Translating the construction plane in a perpendicular direction

Toolbar Button	From Main Pull-down Menu	Function
Set CPlane To Object	View > CPlane > To Object	Setting the construction plane in relation to a selected object
Set CPlane Perpendicular to Curve	View > CPlane > Perpendicular to Curve	Orienting the construction plane perpendicular to a curve at a selected point
Set CPlane: Rotate	View > CPlane > Rotate	Rotating the construction plane
Set CPlane To View	View > CPlane > To View	Setting the construction plane in relation to the display view
Set CPlane: Vertical	View > CPlane > Vertical	Repositioning the construction plane vertical to a line defined by two selected points
Set CPlane by 3 Points	View > CPlane > 3 Points	Defining a construction plane by specifying three points
Set CPlane by X-Axis	View > CPlane > X Axis	Orienting the construction plane by specifying the origin and the X axis
Set CPlane by Z-Axis	View > CPlane > Z Axis	Orienting the construction plane by specifying the origin and the Z axis
Set CPlane World Top	View > CPlane > World Top	Enabling use of the Top viewport construction plane
Set CPlane World Bottom	—	Enabling use of the Bottom viewport construction plane
Set CPlane World Front	View > CPlane > World Front	Enabling use of the Front viewport construction plane
Set CPlane World Back	—	Enabling use of the Back viewport construction plane
Set CPlane World Right	View > CPlane > World Right	Enabling use of the Right viewport construction plane
Set CPlane World Left	—	Enabling use of the Left viewport construction plane
Named CPlane	View > Named CPlane	Renaming or deleting saved construction planes
Save CPlane by Name/Restore CPlane by Name	—	Saving the current construction plane setting or retrieving a saved construction plane setting
Read Named CPlane from File	—	Enabling use of a saved construction plane from a Rhino file
Previous CPlane	View > CPlane > Undo CPlane Change	Enabling use of the previous construction plane

Toolbar Button	From Main Pull-down Menu	Function
Next CPlane	View > CPlane > Redo CPlane Change	Enabling use of the original construction plane after you use the previous construction plane

When working with construction plane commands, the active viewport when the command begins is the one that gets changed. If you are using parallel views, it is sometimes more effective to skew them a bit before changing the CPlane setting.

Setting the Construction Plane by Specifying Origin, X Axis, and Orientation

To construct a box and set the construction plane to the diagonal corners of the box, perform the following steps.

1. Open the file *CPlane.3dm* from the *Chapter 4* folder on the companion CD-ROM.
2. Maximize the Perspective viewport.

By default, the construction plane in the Perspective viewport is the same as that of the Top viewport. To set the construction plane by specifying three points, continue with the following steps.

3. Select View > Set CPlane > 3 Points, or click on the Set CPlane by 3 Points button on the Set CPlane toolbar.
4. Check the End box on the Osnap dialog box.
5. Select end points A, B, and C (indicated in figure 4-2).

The construction plane in the Perspective viewport is set. Note that the construction planes in the Top, Front, and Right viewports remain unchanged.

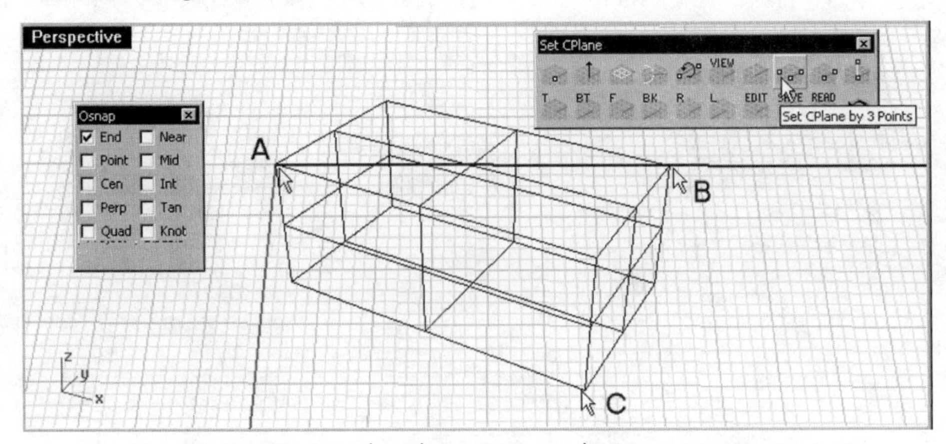

Fig. 4-2. Construction plane being set up in the Perspective viewport.

Construction Plane Origin

To set the origin of the construction plane without changing its orientation, perform the following steps.

1. Select View > Set CPlane > Origin, or click on the Set CPlane Origin button on the Set CPlane toolbar.
2. Select end point A (indicated in figure 4-3).

The origin of the construction plane in the Perspective viewport is relocated.

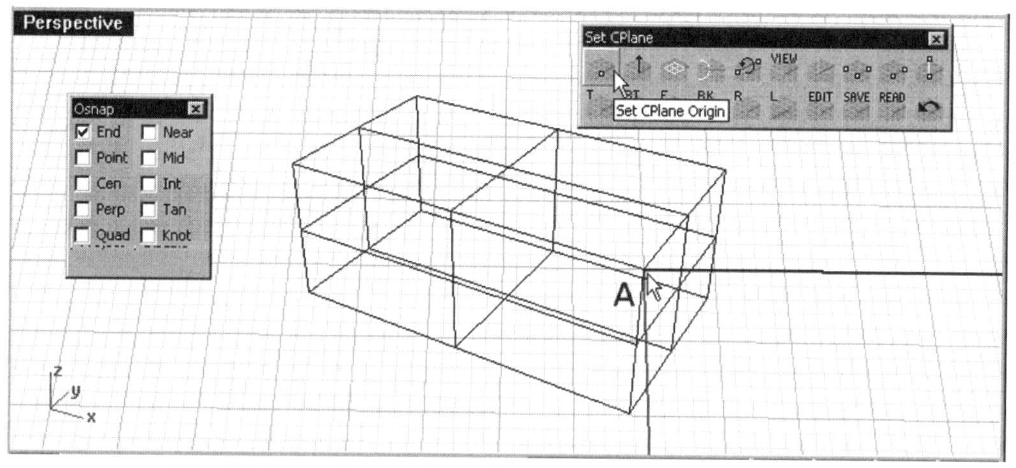

Fig. 4-3. Origin being relocated.

Previous Construction Plane

To return to the previous construction plane setting, perform the following.

1. Select View > Set CPlane > Undo CPlane Change, or click on the *Previous CPlane/Next CPlane* button on the Set CPlane toolbar.

The construction plane is reverted to the previous setting.

Next Construction Plane

After reverting to the previous construction plane, you need to change back to the *next* construction plane. *Next* refers to the next construction plane of the *previous* construction. Perform the following.

1. Select View > Set CPlane > Redo CPlane Change, or right-click on the *Previous CPlane/Next CPlane* button on the Set CPlane toolbar.

The construction plane setting in the Perspective viewport is changed.

Setting Construction Plane Elevation

In the following you will construct a curve in the Top viewport, change the construction plane elevation in the Top viewport, and construct another curve there.

1. With reference to figure 4-4, maximize the Top viewport and construct a curve.
2. Select View > Set CPlane > Elevation, or click on the Set CPlane Elevation button on the Set CPlane toolbar.
3. Type *10* at the command line interface to change the elevation.
4. Construct another curve in the Top viewport.

Return the display to a four-viewport configuration. Note in the other viewports the vertical distance between the two curves, as shown in figure 4-5.

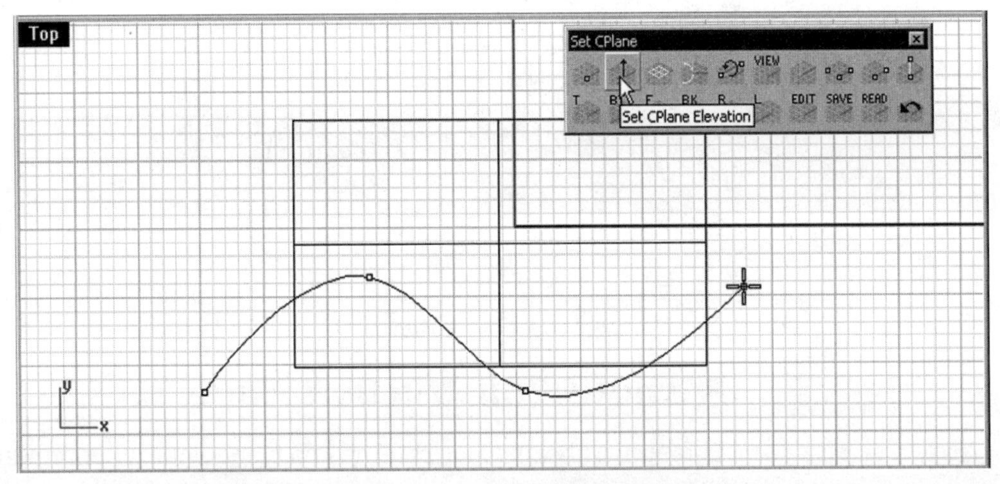

Fig. 4-4. Circle constructed in the Top viewport.

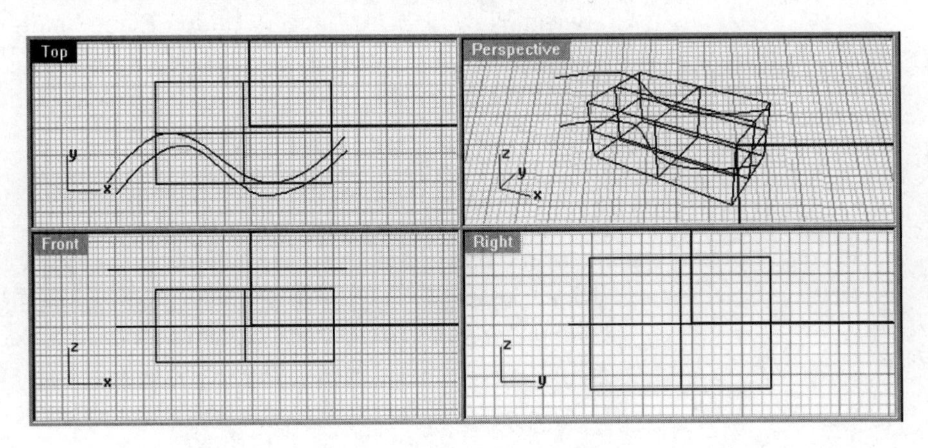

Fig. 4-5. Construction plane's elevation changed and second circle constructed in the Top viewport.

Rotating a Construction Plane

To rotate a construction plane in the Right viewport, perform the following steps.

1. Maximize the Right viewport.
2. Select View > Set CPlane > Rotate, or click on the *Set CPlane: Rotate* button on the Set CPlane toolbar.
3. Type *Z* at the command line interface to use the current construction plane Z axis as the rotation axis.
4. Type *25* to rotate the construction plane 25 degrees.

The construction plane is rotated, as shown in figure 4-6.

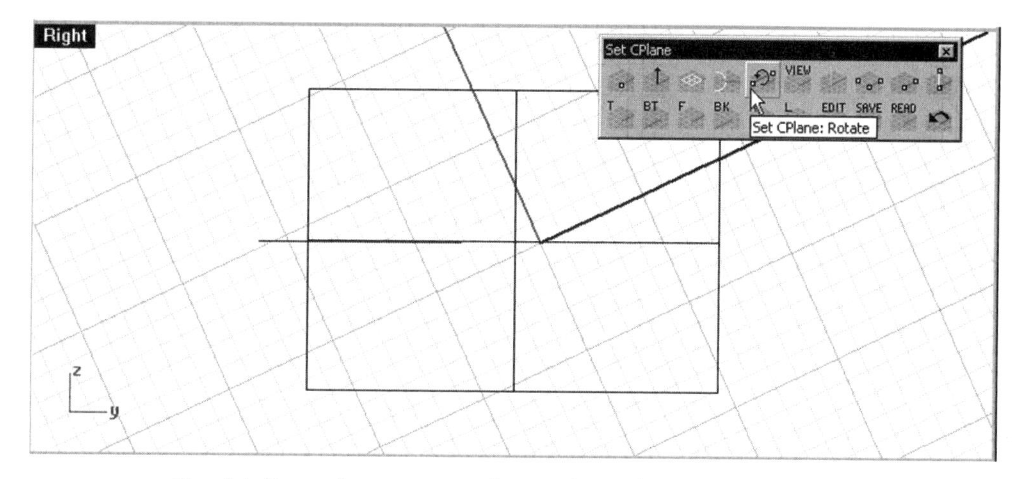

Fig. 4-6. Rotated construction plane in the Right viewport.

Object Construction Plane

To set a construction plane to a selected object, perform the following steps.

1. Maximize the Perspective viewport.
2. Select View > Set CPlane > To Object, or click on the Set CPlane To Object button on the Set CPlane toolbar.
3. Select curve A (indicated in figure 4-7).

The construction plane in the Front viewport is set to the selected object.

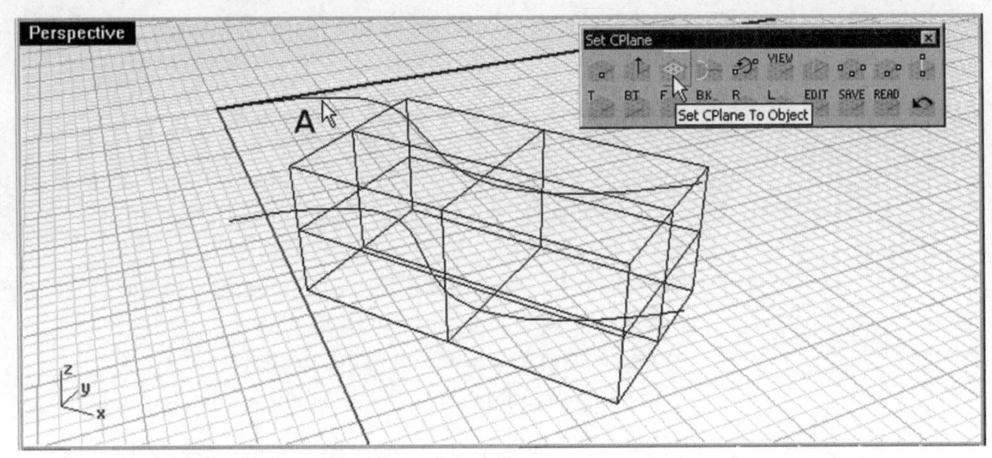

Fig. 4-7. Construction plane in the Perspective viewport set to an object.

Vertical Construction Plane

To set the construction plane perpendicular to the current construction plane, perform the following steps.

1. Maximize the Front viewport.
2. Select View > Set CPlane > Vertical, or click on the *Set CPlane: Vertical* button on the Set CPlane toolbar.
3. Select location A (indicated in figure 4-8) to specify the location of the origin.

The construction plane is vertical to two selected points.

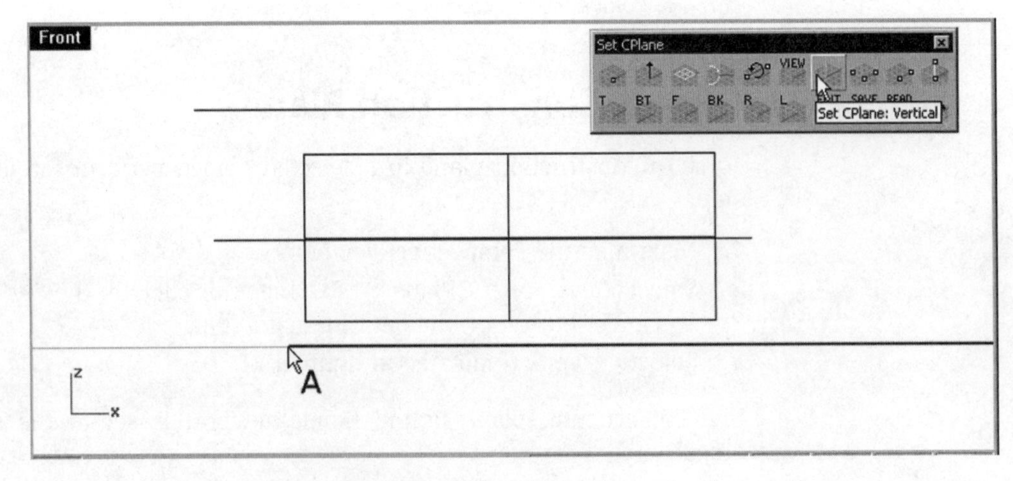

Fig. 4-8. Construction plane in the Front viewport being set vertical to current construction plane.

World Top Construction Plane

To align a construction plane to the "Top" view of the World coordinate system, perform the following steps.

1. Return to a four-viewport display and click on the Top viewport.
2. Select View > Set CPlane > World Top, or click on the Set CPlane World Top button on the Set CPlane toolbar.

The construction plane is set to World Top, as shown in figure 4-9.

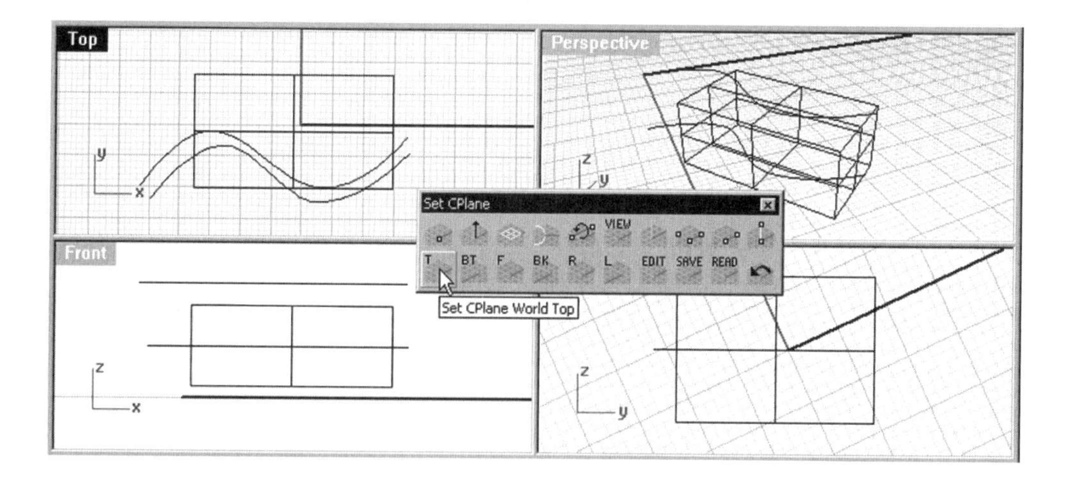

Fig. 4-9. Construction plane of the Top viewport set to World Top.

World Front Construction Plane

To align a construction plane to the "Front" view of the World coordinate system, perform the following steps.

1. Click on the Front viewport.
2. Select View > Set CPlane > World Front, or click on the Set CPlane World Front button on the Set CPlane toolbar.

The construction plane is set to World Front, as shown in figure 4-10.

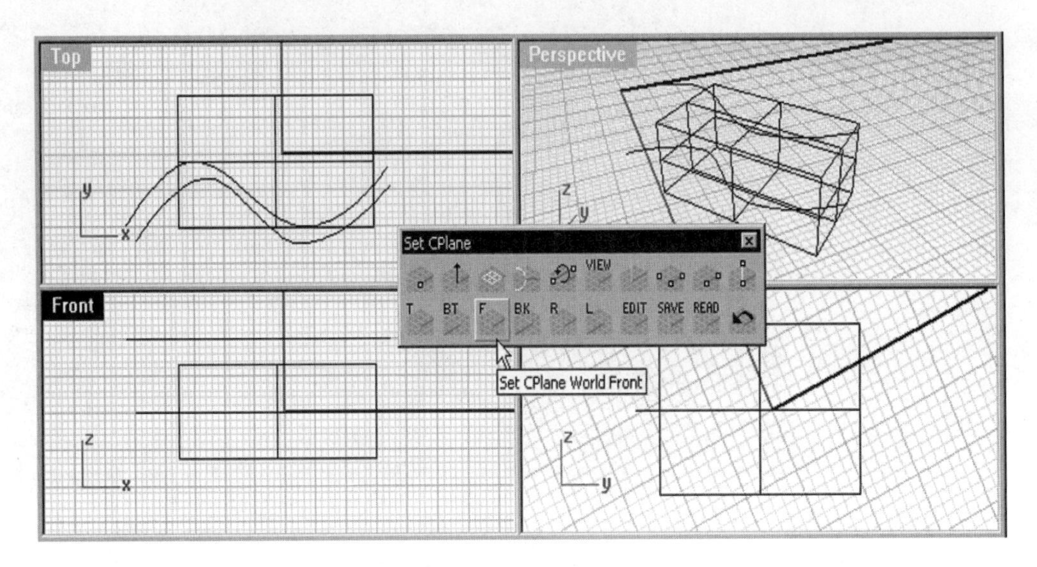

Fig. 4-10. Construction plane of the Front viewport set to World Front.

World Right Construction Plane

To align a construction plane to the "Right" view of the World coordinate system, perform the following steps.

1. Click on the Right viewport.
2. Select View > Set CPlane > World Right, or click on the Set CPlane World Right button on the Set CPlane toolbar.

The construction plane is set to World Right, as shown in figure 4-11.

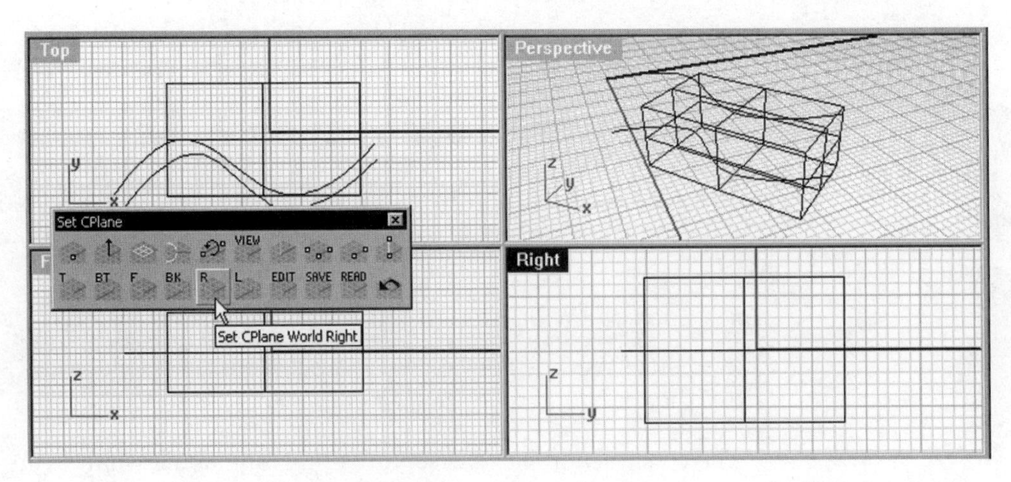

Fig. 4-11. Construction plane of the Right viewport set to World Right.

View Construction Plane

To set a construction plane parallel to the viewport, perform the following steps.

1. Click on the Perspective viewport.
2. Select View > Set CPlane > To View, or click on the Set CPlane To View button on the Set CPlane toolbar.

The construction planes in the perspective viewports are set parallel to the viewport, as shown in figure 4-12.

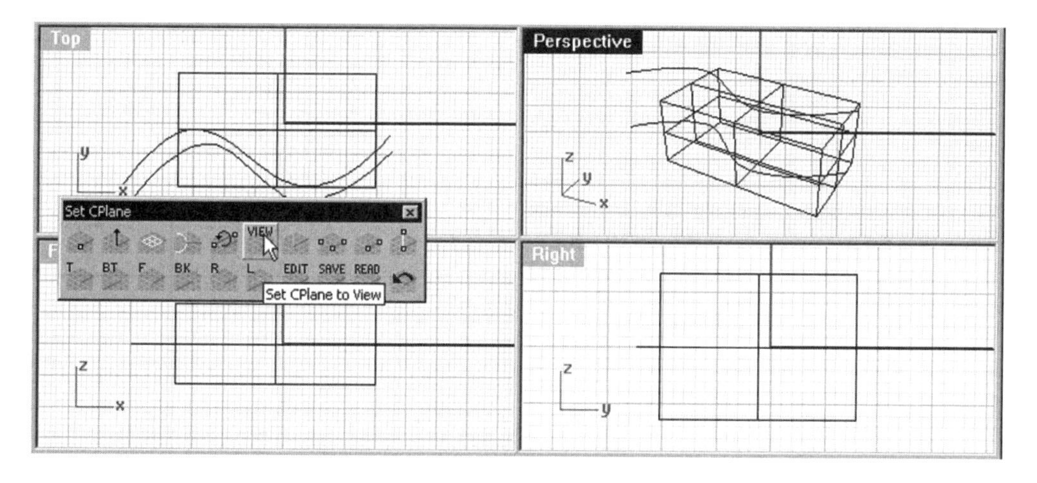

Fig. 4-12. Construction plane in the Perspective viewport set.

Viewport Label's Right-click Menu

To use the right-click menu of the Perspective viewport to set the construction plane, perform the following steps.

1. Select the Perspective viewport title and right-click.
2. In the right-click menu, shown in figure 4-13, select Set View > Perspective.

The display is set and the construction plane of the viewport follows the default setting. In this case, the default construction plane setting of the Perspective viewport is the Top-viewport construction plane.

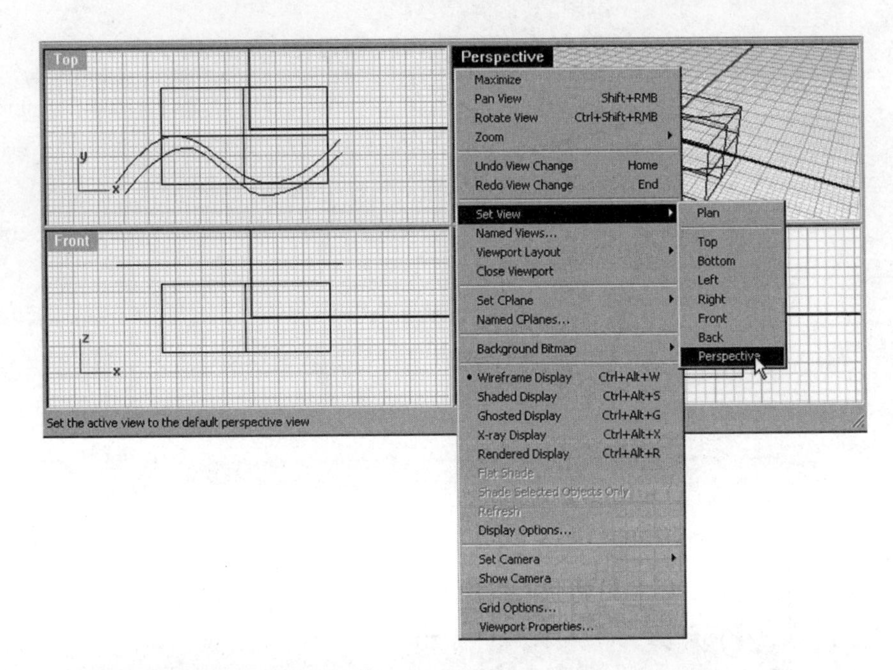

Fig. 4-13. Right-click menu.

Construction Plane Perpendicular to a Curve

To set a construction plane perpendicular to a curve, perform the following steps.

1. Maximize the Perspective viewport.
2. Select View > Set CPlane > Perpendicular to Curve, or click on the Set CPlane Perpendicular to Curve button on the Set CPlane toolbar.
3. Select curve A (indicated in figure 4-14).
4. Select end point B (indicated in figure 4-14).

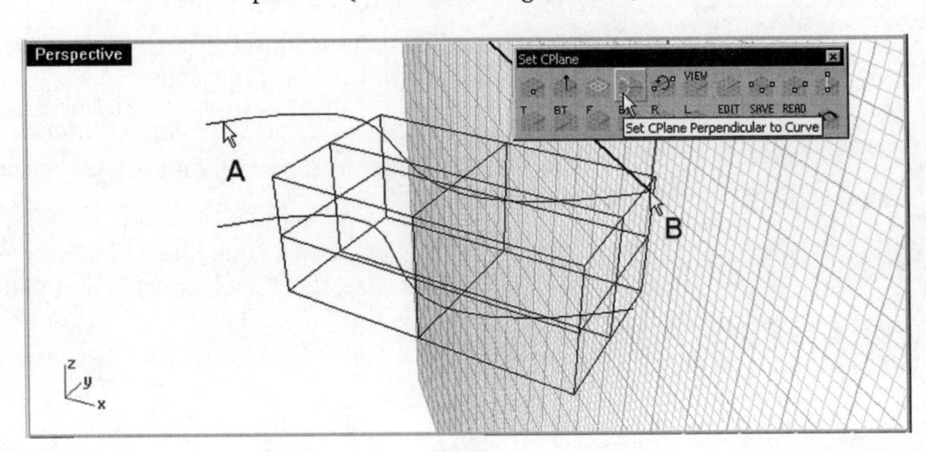

Fig. 4-14. Construction plane being set perpendicular to a curve.

The construction plane is set perpendicular to the curve. Note that constructing a curve on this construction plane is equivalent to constructing a curve on one of the viewport's default construction planes and then orienting the curve perpendicular to another curve.

Specifying a Construction Plane via Its X Axis

To specify a construction plane via its X axis, perform the following steps.

1. Maximize the Top viewport.
2. Select View > Set CPlane > X Axis, or click on the Set CPlane by X-Axis button on the Set CPlane toolbar.
3. Select locations A and B (indicated in figure 4-15).

The construction plane is specified per its X axis.

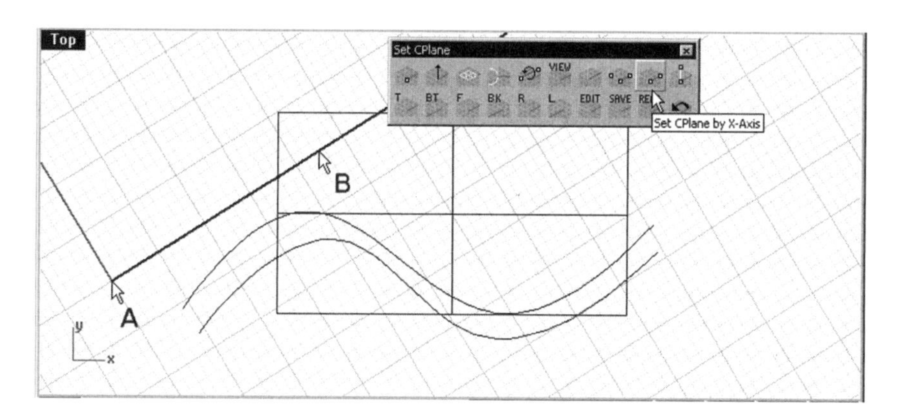

Fig. 4-15. Construction plane being specified per its X axis.

Specifying a Construction Plane via Its Z Axis

To specify a construction plane via its Z axis, perform the following steps.

1. Maximize the Front viewport.
2. Select View > Set CPlane > Z Axis, or click on the Set CPlane by Z-Axis button on the Set CPlane toolbar.
3. Select locations A and B (indicated in figure 4-16).

The construction plane is specified per its Z axis, as shown in figure 4-17.

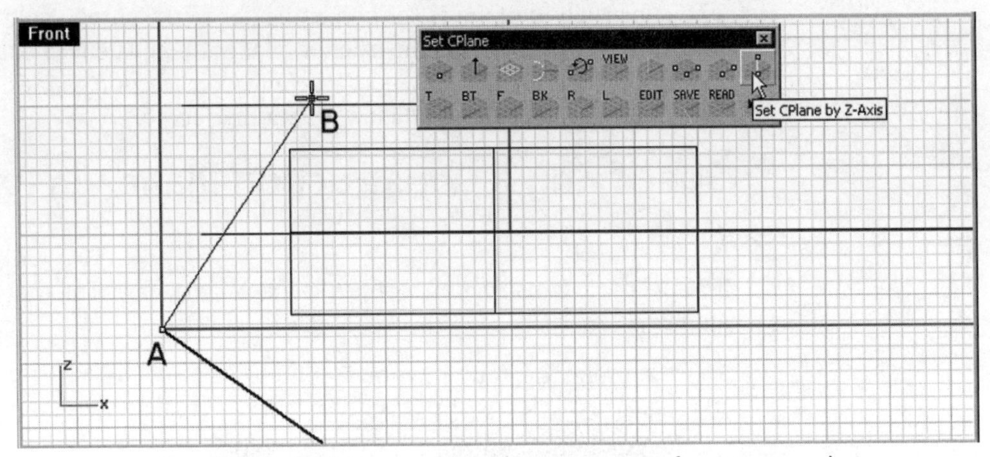

Fig. 4-16. Z axis being established as determiner of construction plane.

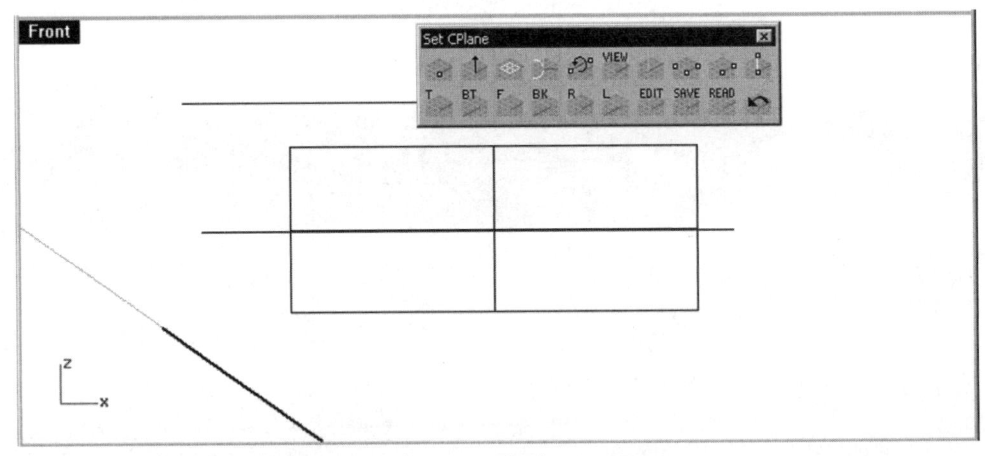

Fig. 4-17. Construction plane specified per its Z axis.

Saving a Construction Plane Configuration

To save a construction plane configuration so that you can retrieve it later, perform the following steps.

1. Select View > Named CPlane, or click on the *Save CPlane by Name/Restore CPlane by Name* button on the Set CPlane toolbar.
2. Select View > Named CPlane, and click on the Save button in the Named CPlane dialog box. Specify a name in the Save CPlane dialog box, click on the OK button, and then click on the Close button. Alternatively, click on the *Save CPlane by Name/Restore CPlane by Name* button on the Set CPlane toolbar, specify a name at the command line interface, and press the Enter key. The construction plane being saved is shown in figure 4-18.

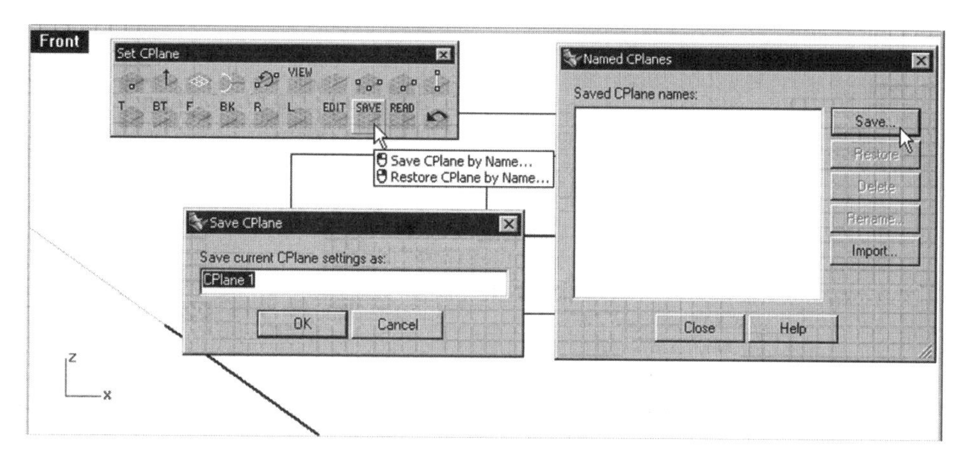

Fig. 4-18. Construction plane configuration being saved.

Retrieving a Construction Plane Configuration

To retrieve a saved construction plane configuration, perform the following steps.

1. Click on the Perspective viewport.
2. Select View > Named CPlane, or right-click on the *Save CPlane by Name/Restore CPlane by Name* button on the Set CPlane toolbar.
3. Select View > Named CPlane, select a saved CPlane, click on the Restore button, and click on the Close button. Alternatively, right-click on the *Save CPlane by Name/Restore CPlane by Name* button on the Set CPlane toolbar, specify a saved name at the command line interface, and press the Enter key.

The saved construction plane configuration is applied to the Perspective viewport, as shown in figure 4-19.

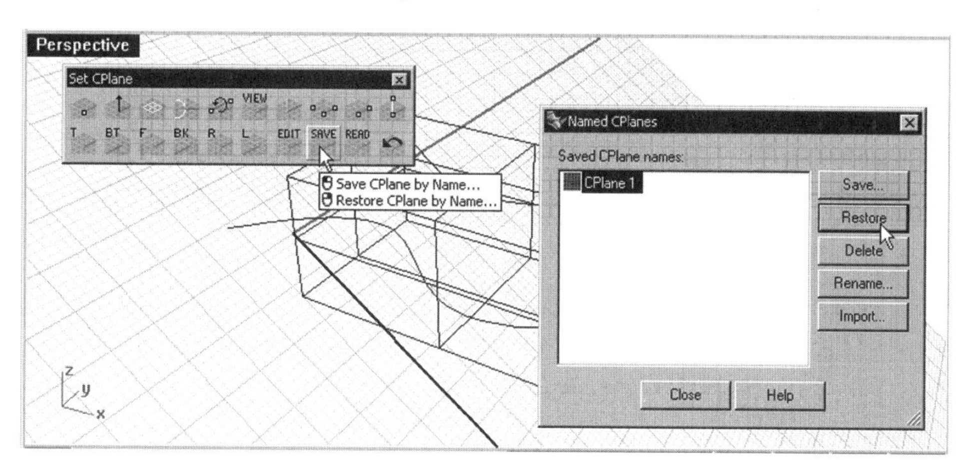

Fig. 4-19. Saved configuration being retrieved.

Editing a Saved Construction Plane Configuration

1. Select View > Named CPlane, or click on the Named CPlanes button on the Set CPlane toolbar.
2. In the Named CPlane dialog box, select a saved CPlane, and click on the Rename button. Specify a name in the Rename CPlane dialog box, click on the OK button, and click on the Close button.

A saved CPlane is renamed, as indicated in figure 4-20.

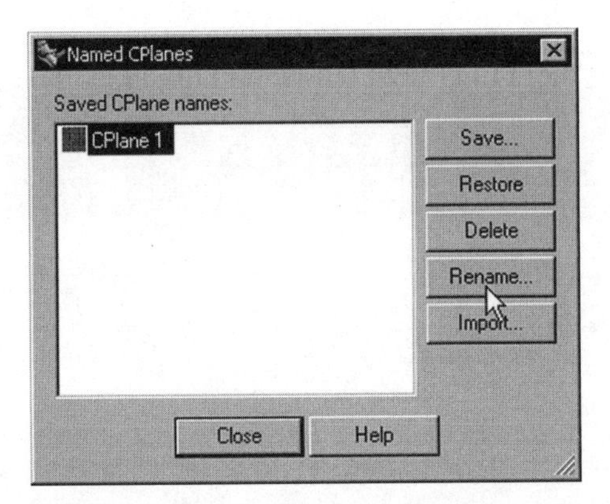

Fig. 4-20. CPlane being renamed.

Reading a Construction Plane Configuration from a File

If a construction plane configuration has been saved in a file, you can retrieve the configuration from the saved file, as follows.

1. Start a new file. Use the metric (Millimeters) template.
2. Select View > Named CPlane, or click on the Read Named CPlanes from File button on the Set CPlane toolbar.
3. In the Named CPlane dialog box, click on the Import button. Select a file in the Import dialog box, click on the Open button, and then click on the Close button.

The construction plane configuration is imported, as indicated in figure 4-21. Now you can retrieve the saved CPlane configuration. Do not save the file.

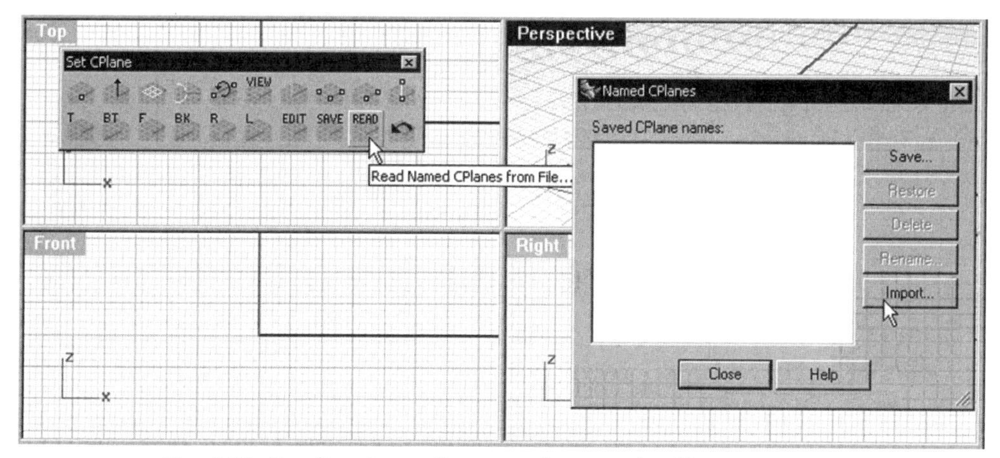

Fig. 4-21. Reading the configuration from another file.

Tracing Background Images

Knowing that the fundamental issue of point and basic curve construction is to specify a location, you need to be able to use the pointing device to select a point or input the coordinates at the command line interface. To help determine the locations, you might sketch your design idea on graph paper. From the graph paper, you extract point locations. If you already have a physical object, you can use a digitizer to obtain the required coordinates. (You will learn how to use a digitizer in the next section.)

Another way of using the sketch is to scan it as a digital image and insert it in the viewport as a background image. If the image is properly scaled, you can directly sketch in the viewport. Commands related to using an image as a background for a sketch are accessed from the Background Bitmap toolbar, shown in figure 4-22.

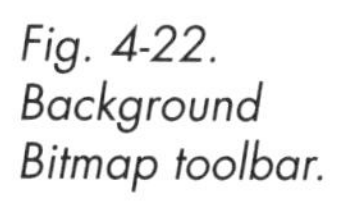

Fig. 4-22. Background Bitmap toolbar.

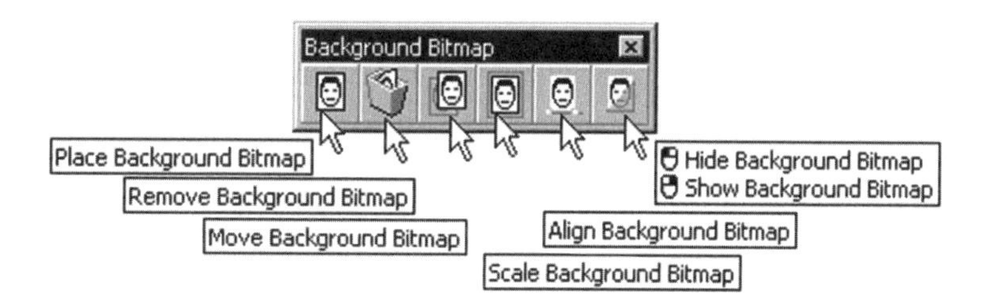

Table 4-2 outlines the functions of the background image commands and their locations in the main pull-down menu and respective toolbars.

Table 4-2: Background Image Commands and Their Functions

Toolbar Button	From Main Pull-down Menu	Function
Place Background Image	View > Background Bitmap > Place	Placing background image in viewport
Remove Background Bitmap	View > Background Bitmap > Remove	Removing background image placed in viewport
Move Background Bitmap	View > Background Bitmap > Move	Moving background image placed in viewport
Scale Background Bitmap	View > Background Bitmap > Scale	Scaling background image placed in viewport
Align Background Bitmap	View > Background Bitmap > Align	Aligning background image
Hide Background Bitmap/Show Background Bitmap	View > Background Bitmap > Hide View > Background Bitmap > Show	Hiding and showing background image in viewport

To try this way of sketching, perform the following steps.

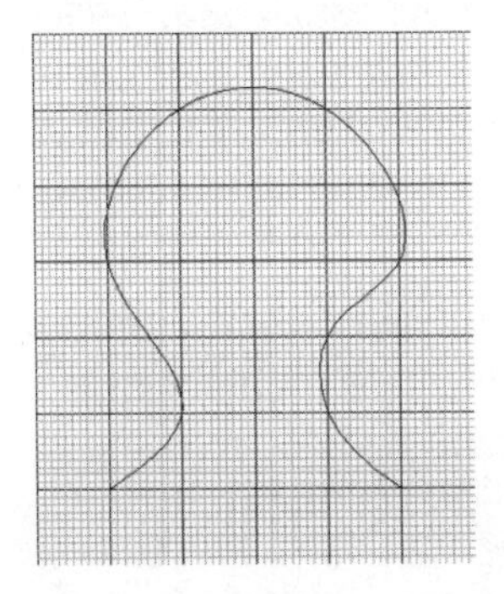

Fig. 4-23. Bitmap image.

1. Obtain a piece of graph paper.
2. Use a pencil to construct a free-hand sketch depicting the curve you want to construct in the computer.
3. Scan the sketch as a digital image.
4. After scanning, use an image processing application to trim away the unwanted portion of the image, and write down the dimensions of the grid lines in the image. For example, the bitmap image shown in figure 4-23 is 30 mm wide and 35 mm tall.

Incorporate the digital image as a background image in the Front viewport by continuing with the following steps.

5. Start a new file. Use the metric (Millimeters) template.
6. Maximize the Top viewport.
7. Check the Snap option on the status bar to turn on snap mode.
8. Select View > Background Bitmap > Place, or click on the Place Background Bitmap button on the Background Bitmap toolbar.
9. In the Open Bitmap dialog box, select the bitmap file *Background-Image.tga* from the *Chapter 4* folder on the companion CD-ROM, and then click on the Open button.

10 Select a point on the Front viewport, as shown in figure 4-24. Select a point 30 mm to the right of the first selected point, because the width of the bitmap used in this example is 30 mm wide.

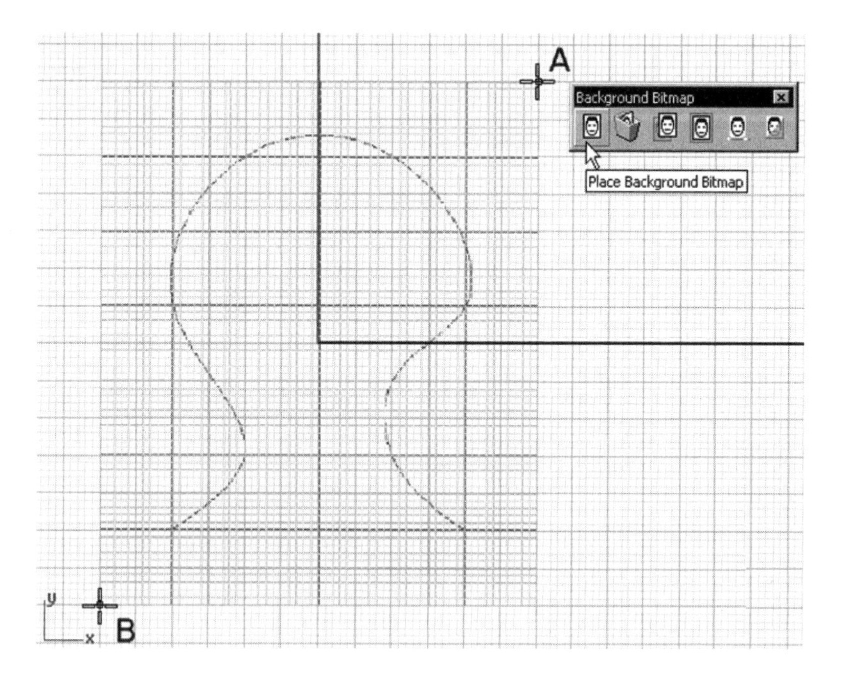

Fig. 4-24. Image placed on the background of Front viewport.

11 Select Curve > Free-form > Sketch, or click on the Sketch button on the Curve toolbar. (Alternatively, you can use the Curve from Control Points option or the Interpolated Curve option to trace the images.)

12 Hold down the left mouse button to sketch a free-form curve.

13 Select View > Background Bitmap > Hide, or click on the Hide Background Bitmap button on the Background Bitmap toolbar.

14 Turn off grid display.

A sketch is constructed and the bitmap is hidden, as shown in figure 4-25. You should note that the free-hand sketch constructed this way is not smooth enough and you may need to use the editing tools and transformation tools to modify it.

15 If you want to save your file, save it as *Sketch.3dm*.

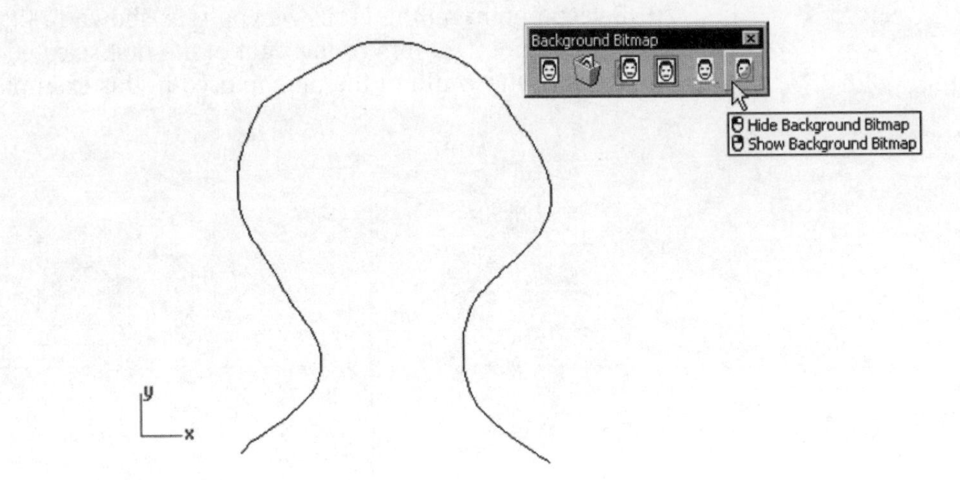

Fig. 4-25. Sketch constructed, image hidden, and grid mesh turned off.

Digitizing Physical Objects and Using the MicroScribe Digitizer

By using a digitizer, you obtain coordinates of selected points on the surface of a physical object. If the physical object is a hand-made mock-up of the real object, chances are the surfaces are not smooth enough. For example, the silhouette and profiles of a physical model you digitize may deviate from the model's ideal shape and may even be wavy.

If the waviness and irregularities on the mock-up are not intended in the final computer model, you need to edit the curves you obtained from the digitizer and/or the basic free-form surfaces you construct from the digitized data. (In this chapter, you already learned curve editing. You will learn surface editing in Chapter 5.)

There are many types of digitizers. The following section discusses digitizing using the MicroScribe digitizer. If you do not have a MicroScribe digitizer connected to your computer, you may skip the following exercise and proceed to the summary to this chapter. However, the following has general value in terms of understanding how digitizers operate and how they are employed. Figure 4-26 shows the MicroScribe digitizer. The MicroScribe digitizer has eight key components: base, shoulder, counterweight, upper arm, elbow, lower arm, wrist, and stylus.

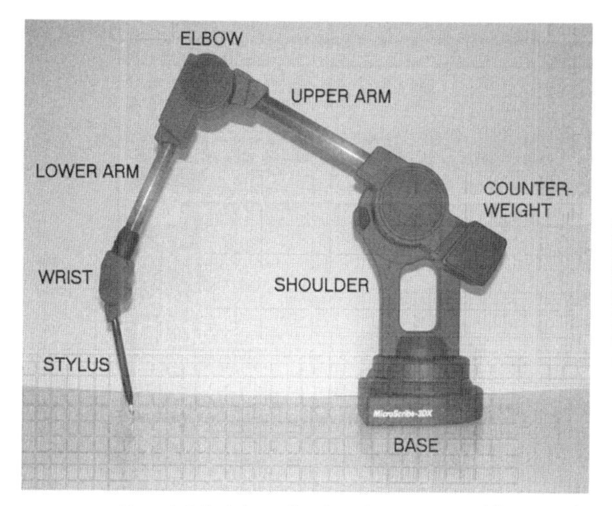

Fig. 4-26. MicroScribe digitizer and foot pedal.

The base unit is the foundation of the digitizer. You place it on any flat surface. At the back of the base unit are connections to the foot pedal, electric power supply, and data transmission cord to the computer. On the pedal unit are two buttons that function similarly to the buttons of a mouse.

On top of the base unit is a swiveling, upright shoulder unit that can freely rotate horizontally. At the upper end of the shoulder unit is a pivot joint that connects to the upper arm of the digitizer. The upper arm then connects to the elbow unit that, in turn, connects to the lower arm. At the other end of the lower arm is a wrist unit that links to the stylus. To balance the weight of the elbow and other objects connected to it, a counterweight is placed at the lower end of the upper arm.

To operate the digitizer, you select a command, hold the stylus, and place the stylus tip on the surface of the object you want to digitize. You then depress the right-hand foot pedal. Information regarding the position of the stylus tip relative to the base unit is transmitted to the computer.

MicroScribe Digitizer Commands

A set of built-in commands is available if your computer is connected to the 3D MicroScribe digitizer. These commands are found in the 3-D Digitizing toolbar, shown in figure 4-27.

Table 4-3 outlines the functions of MicroScribe digitizer commands and their locations in the main pull-down menu and respective toolbars.

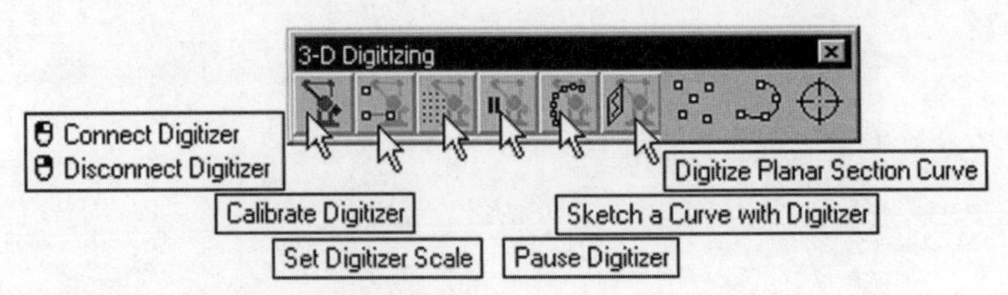

Fig. 4-27. 3D Digitizing toolbar.

Table 4-3: Digitizer Commands and Their Functions

Toolbar Button	From Main Pull-down Menu	Function
Connect Digitizer	Tools > 3D Digitizer > Connect	Activating the connected MicroScribe digitizer
Disconnect Digitizer	Tools > 3D Digitizer > Disconnect	Deactivating the digitizer
Calibrate Digitizer	Tools > 3D Digitizer > Calibrate	Calibrating the origin and the X, Y, and Z axes of the digitizer
Set Digitizer Scale	Tools > 3D Digitizer > Set Scale	Setting the scale of digitized output
Pause Digitizer	Tools > 3D Digitizer > Pause	Pausing the digitizer
Sketch a Curve with Digitizer	Tools > 3D Digitizer > Sketch Curve	Sketching a curve
Digitize Planar Section Curve	Tools > 3D Digitizer > Planar Section Curve	Constructing a series of cross-section curves

Activating and Calibrating the MicroScribe Digitizer

Prior to using the MicroScribe digitizer, you need to connect it to an electric power supply and to your computer. You also need a flat surface on which to place the base unit of the digitizer and the physical object you want to digitize. To activate the digitizer, perform the following steps.

1. Start a new file. Use the metric (Millimeters) template.
2. Select Tools > 3-D Digitize > Connect, or click on the Connect Digitizer button (see figure 4-28) on the MicroScribe toolbar. This activates the Select Digitizer dialog box.
3. Type *D* at the command line interface to select a digitizer.
4. Type *M* to select the MicroScribe digitizer.

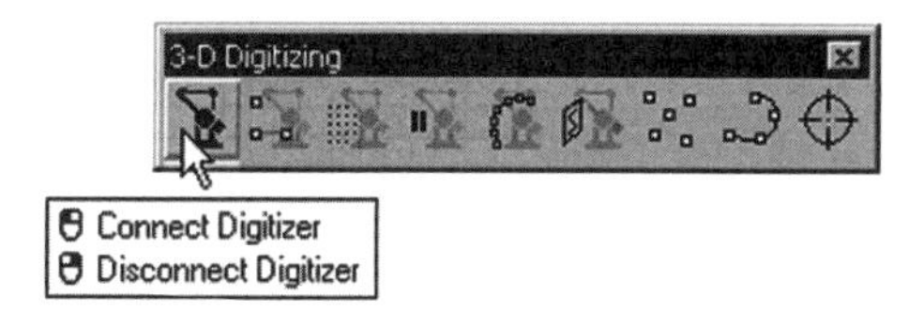

Fig. 4-28. Select Digitizer dialog box.

5 In the MicroScribe Port and Baud dialog box, shown in figure 4-29, accept the default settings and then click on the OK button.

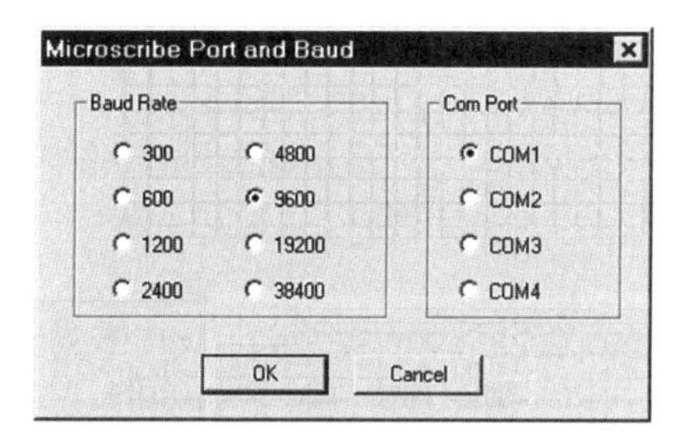

Fig. 4-29. MicroScribe Port and Baud dialog box.

The digitizer is activated. After activating the digitizer, you need to calibrate it. First, you will specify three points to depict a reference plane. The first point is the origin, the second point (together with the first point) defines the X axis, and the third point (together with the first and second points) defines a plane and the Y axis. Continue with the following steps.

6 Hold the stylus, move the stylus tip on a point on the flat surface to use the point as the reference origin, and depress the right-hand foot pedal. This operation is depicted in figure 4-30.

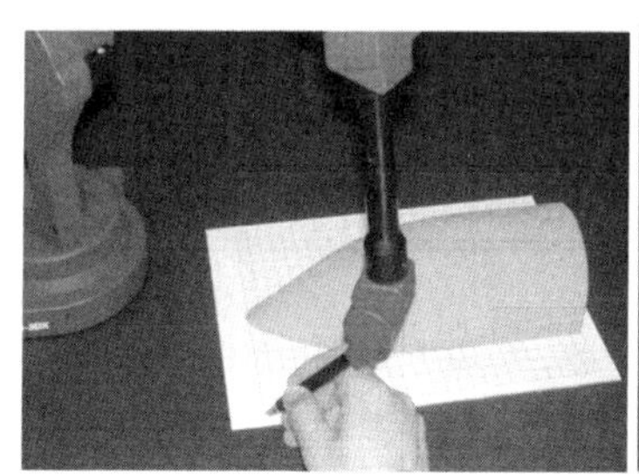
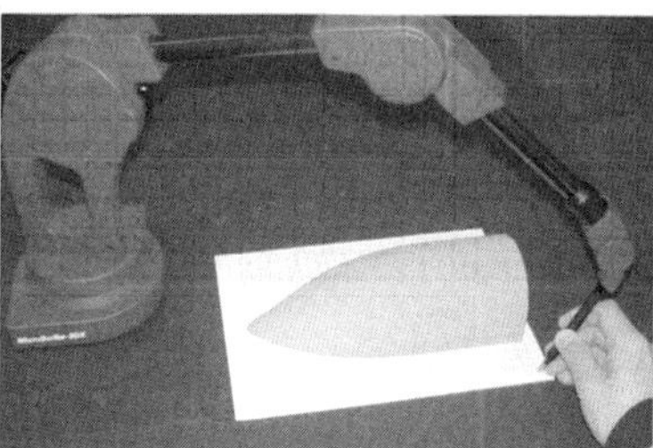
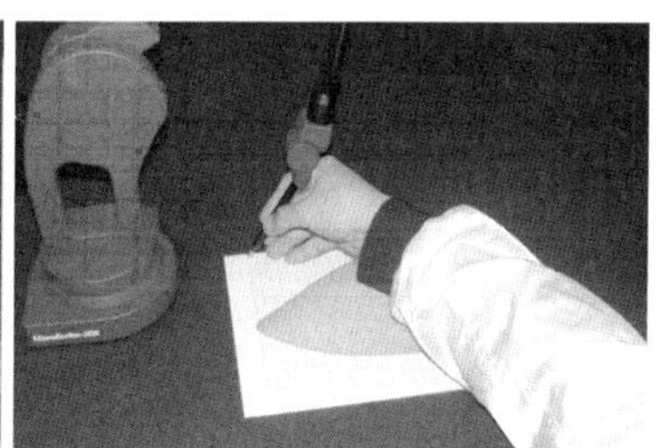

Fig. 4-30. Specifying the origin, X axis, and Y axis.

7 Bring the stylus tip to a second point on the flat surface to define the X axis direction and depress the right-hand foot pedal.

8 Bring the stylus tip to a third point to define a plane and depress the right-hand foot pedal.

9 Press the Enter key to use the World origin.

The MicroScribe digitizer is connected to your computer and is calibrated. If you wish to recalibrate the digitizer, you use the Calibrate command, as follows.

10 Select Tools > 3-D Digitize > Calibrate, or click on the Calibrate Digitizer button on the MicroScribe toolbar.

11 Select three points to indicate the origin, X axis, and reference plane, and press the Enter key.

Digitizing Sketched Curves

In essence, you use the digitizer as a 3D mouse. In contrast to an ordinary mouse, with which you can select only 2D points on a construction plane, the digitizer enables you to select points in 3D, regardless of the current construction plane. Using the digitizer with appropriate commands, you construct point objects and curves.

Before you start to digitize, you have to decide the types of surfaces to be constructed, because they need different patterns of points and curves. (See Chapter 5, on basic free-form surfaces.) To construct a curve, perform the following steps.

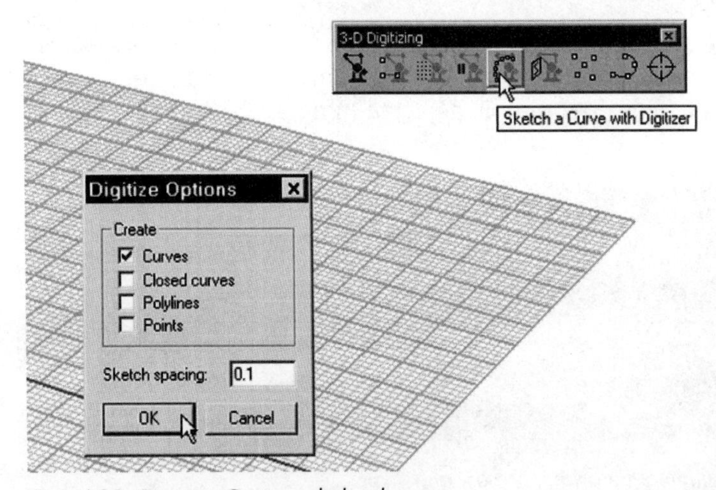

Fig. 4-31. Digitize Options dialog box.

1 Select Tools > 3-D Digitize > Sketch Curve, or click on the Sketch a Curve with Digitizer button on the MicroScribe toolbar.

2 In the Digitize Options dialog box, shown in figure 4-31, check only the Curves box and then click on the OK button. This way, open-loop curves will be constructed.

3 Place the stylus tip on the surface of the model at the location you want to start sketching, and depress and hold down the right-hand foot pedal.

4 Move the stylus tip across the face of the physical model and then release the foot pedal.

A free-form curve is constructed along the path of the stylus movement.

5 Repeat the command by depressing the left-hand foot pedal, and repeat steps 3 and 4 to digitize.

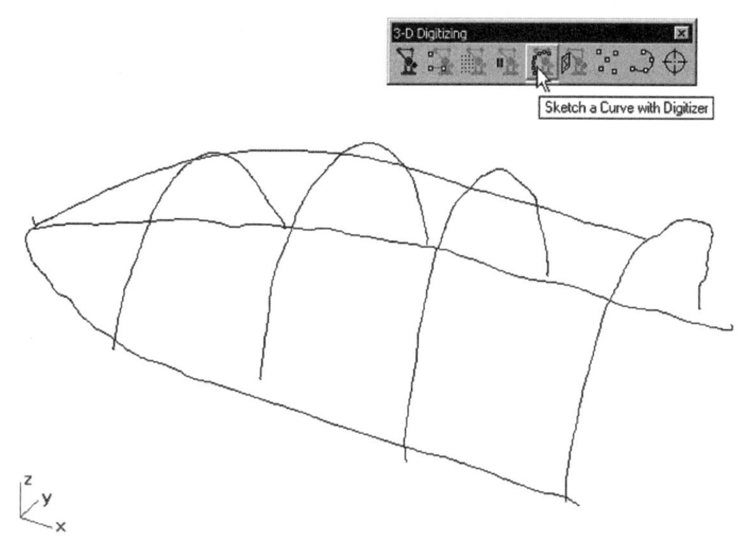

Fig. 4-32. Free-form sketch curves.

When you have sufficient curves, you may use a free-form surface command to construct a surface. Figure 4-32 shows a set of free-form curves constructed for making a patch surface. (You will learn about various types of surfaces in Chapter 5.)

Because the curves may not be smooth enough for a making a smooth surface, you need to edit the curve, as follows.

6 Select Curve > Curve Edit Tools > Fair, or click on the Fair Curve button on the Curve Tools toolbar.

7 Select all curves and press the Enter key. The result is shown in figure 4-33.

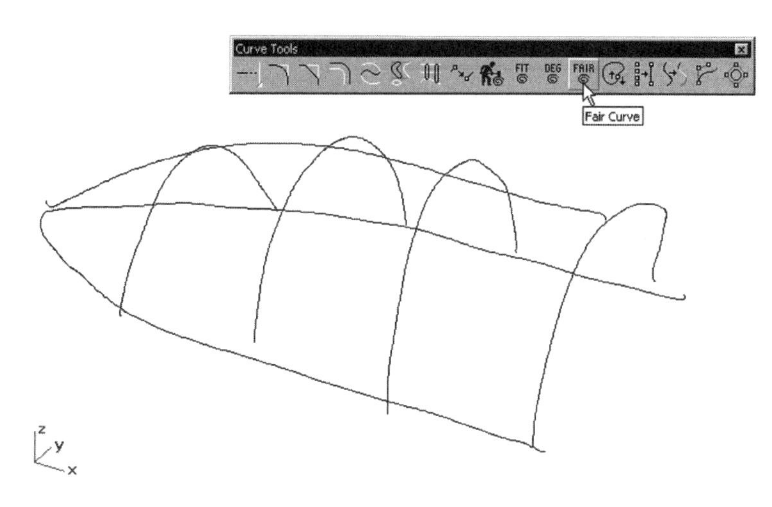

Fig. 4-33. Curves faired.

8 Select Curve > Curve Edit Tools > Match, or click on the Match Curve button on the Curve Tools toolbar. Select the curves indicated in figure 4-34 to match the end points.

9 Repeat the Match command to match the other end points.

To construct a patch surface, continue with the following steps.

10 Select Surface > Patch, or click on the Patch button on the Sur face toolbar.

11 Select the curves and press the Enter key.

A patch surface is constructed, as shown in figure 4-35.

12 Save your file as *DigitizePatch.3dm.*

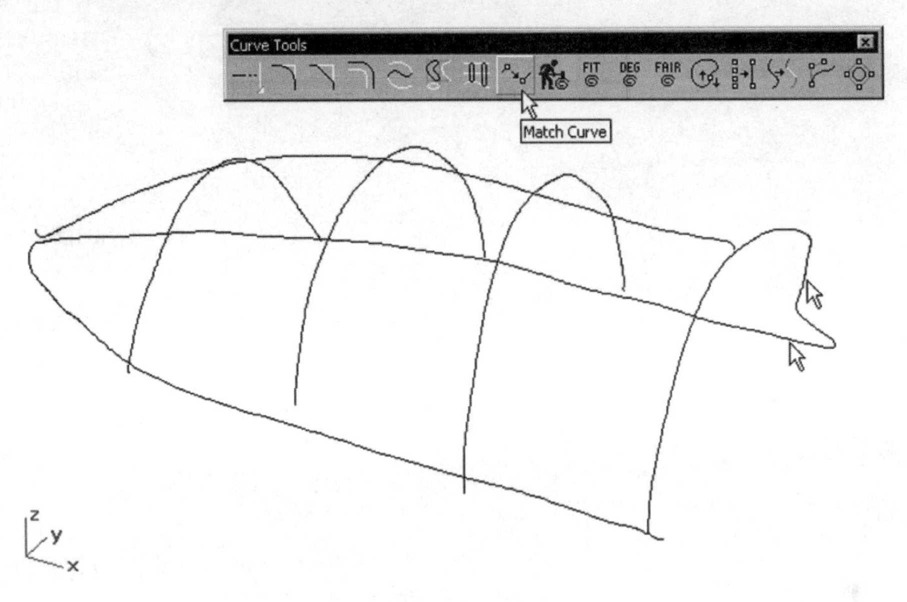

Fig. 4-34. Curves being matched.

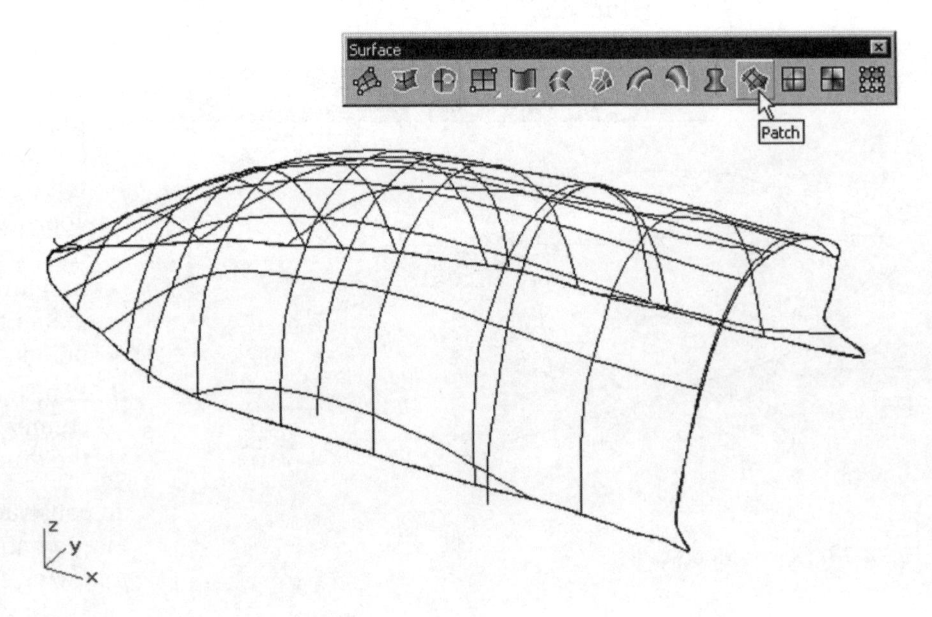

Fig. 4-35. Patch surface constructed.

Digitizing Cross-section Profiles

In the following, you will construct a set of cross-section curves using the digitizer. Figure 4-36 shows the physical object placed on a flat horizontal board with two vertical boards attached that along with the base board serve as digitizing reference planes.

Fig. 4-36. Physical object to be digitized.

1. Start a new file. Use the metric (Millimeters) template.
2. Maximize the Perspective viewport.
3. Select Tools > 3-D Digitize > Planar Section Curve, or click on the Digitize Planar Section Curve button on the MicroScribe toolbar.
4. In the Digitize Options dialog box, shown in figure 4-37, select Curve and then click on the OK button.

To define a vertical plane and the X axis, continue with the following steps.

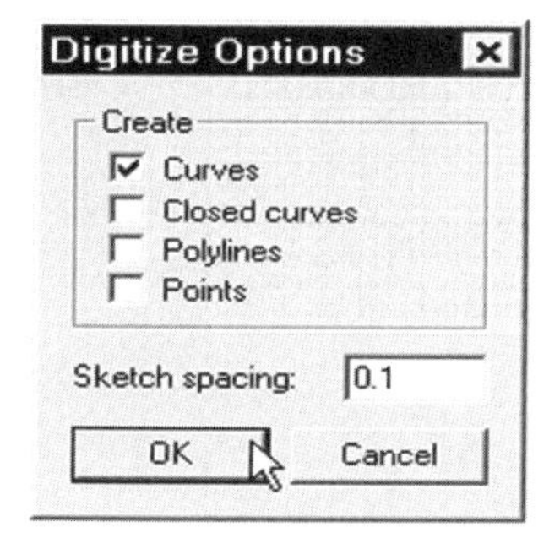

Fig. 4-37. Digitize Options dialog box.

5. Select a point and depress the right-hand foot pedal to define the origin of the vertical plane, shown in figure 4-38.
6. Select a point and depress the right-hand foot pedal to define the second point of the vertical plane.
7. Select a point and depress the right-hand foot pedal to define the third point of the vertical plane.

A vertical plane is defined, as shown in figure 4-39.

8. Select a point and depress the right-hand foot pedal to define the start point of the X axis, shown in figure 4-40.
9. Select a point and depress the right-hand foot pedal to define the end point of the X axis.

The X axis is defined, as shown in figure 4-41.

10. In the Section Plane Spacing dialog box, set the number of planes to 10 and then click on the OK button.

Ten section planes are defined. They are parallel to the vertical plane and evenly spaced along the X axis.

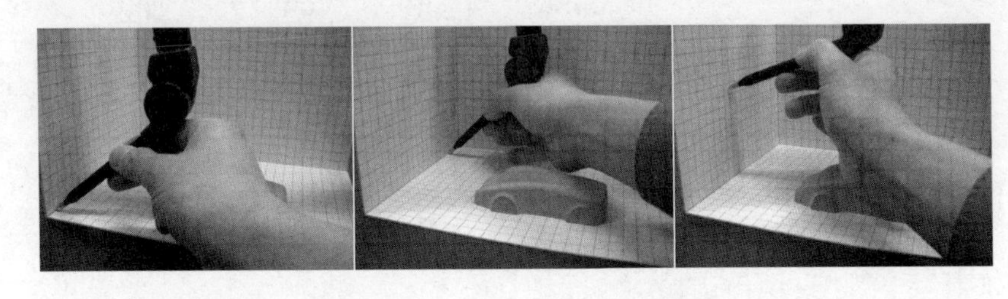

Fig. 4-38. Defining a vertical plane by selecting three points on a vertical board.

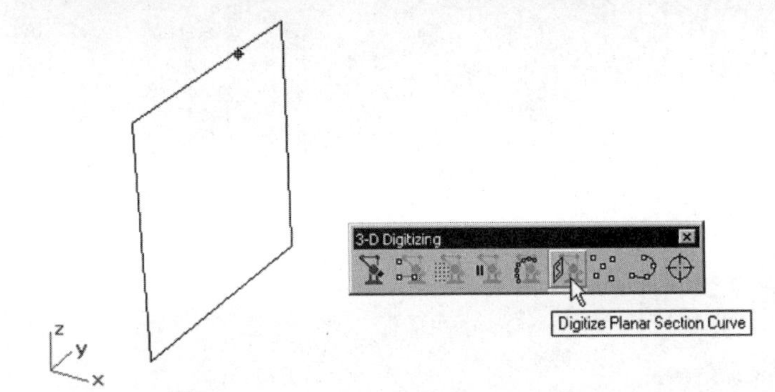

Fig. 4-39. Vertical plane defined.

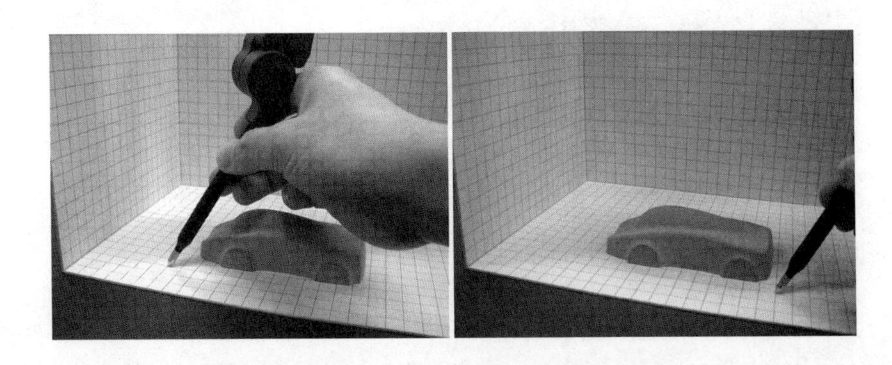

Fig. 4-40. Defining the X axis.

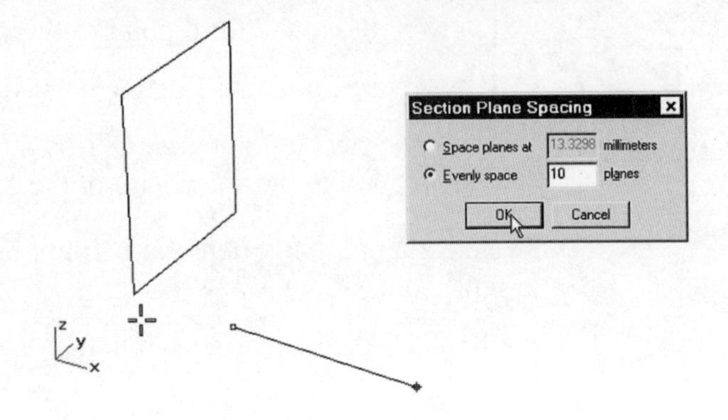

Fig. 4-41. X axis defined.

To construct a set of section curves, continue with the following steps.

11 Bring the stylus onto the surface of the physical object, and depress and hold down the right-hand foot pedal, as indicated in figure 4-42.

12 Move the stylus along the surface to describe a curve, and then release the right-hand foot pedal.

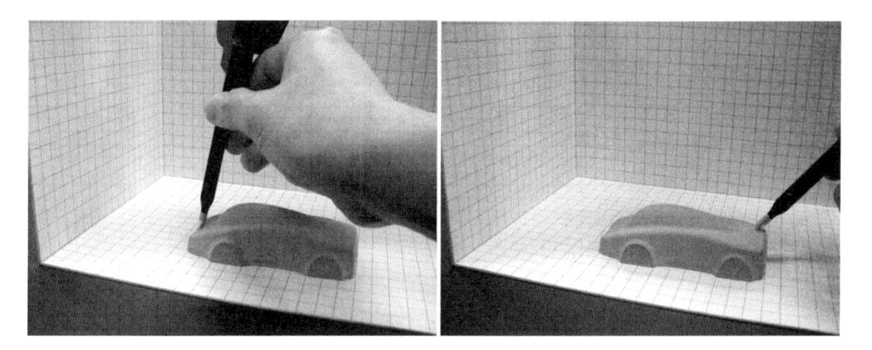

Fig. 4-42. Describing a curve along the surface.

Point objects are displayed at the intersection between this curve and the vertical planes, as shown in figure 4-43.

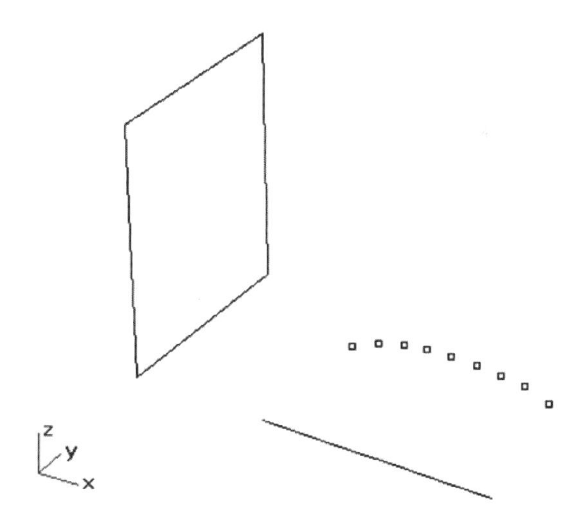

Fig. 4-43. Intersection points displayed.

13 Repeat steps 11 and 12 to construct more points, as shown in figure 4-44.

14 Press the Enter key.

15 Fair the curves, as shown in figure 4-45.

Figure 4-46 shows a lofted surface constructed from the curves.

16 Save your file as *DigitLoft.3dm*.

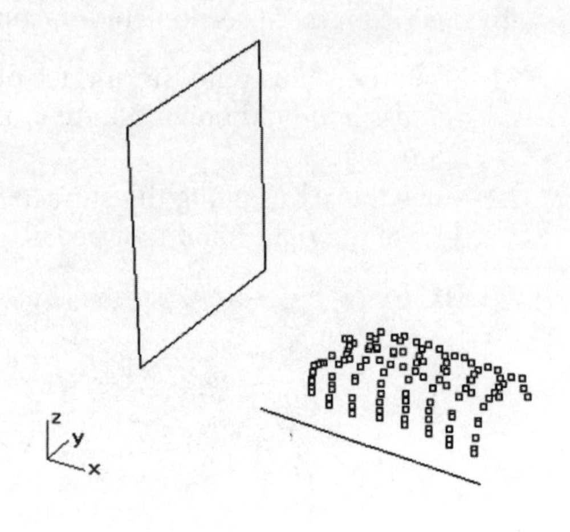

Fig. 4-44. More intersection points displayed.

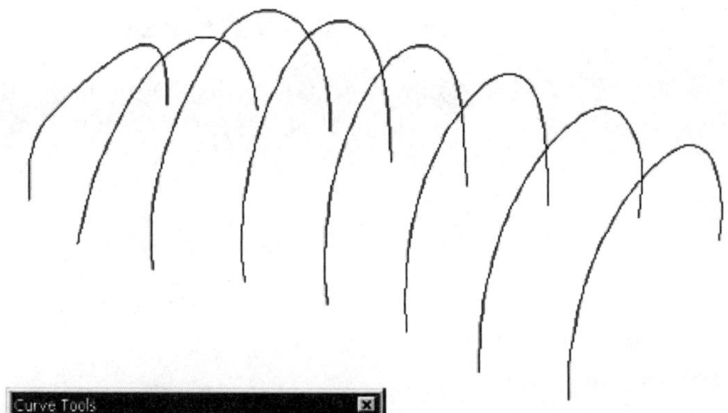

Fig. 4-45. Fairing curves.

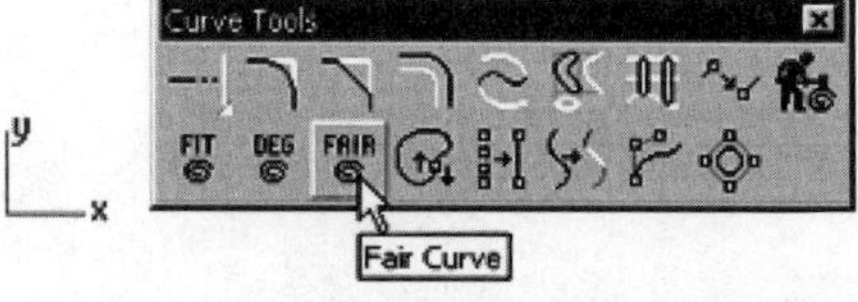

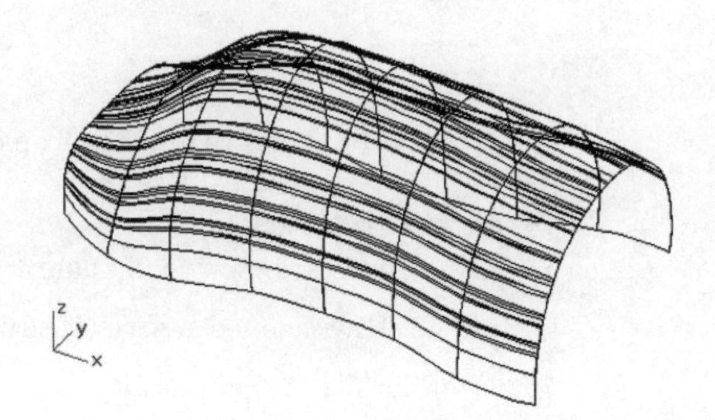

Fig. 4-46. Surface constructed.

Digitizing Point Objects and Curves

Apart from the DigSketch and DigSection commands, you can use the MicroScribe digitizer to replace your mouse to select 3D points for constructing the point objects and curves you learned about in Chapter 2.

Summary

To construct objects in 3D space, you input a set of coordinates at the command line interface or use the input device to select points in the viewports. There is an imaginary construction plane in each of the viewports, and thus selecting a point in a viewport specifies a point in the corresponding viewport construction plane. To construct points and curves in 3D, you manipulate construction planes in various ways.

To reiterate, you construct points and/or curves prior to constructing a basic free-form surface. In making the point objects and curves, you have to specify locations. To specify locations, you can extract coordinate information from sketches that depict your design idea, use the grid mesh background or bitmap background and construct free-form sketches directly by selecting locations in the viewport, or use a digitizer if you already have a physical object. In the case of tracing a background image or digitizing a physical object you may need to analyze, edit, and transform the curves before using them for surface construction.

Review Questions

1. Give a brief account of methods of setting up the construction plane in the viewport.
2. Explain how to use free-hand sketches to help construct curves in the computer.
3. Give a brief account on how you will use a digitizer to help construct points and curves in the computer.

CHAPTER 5

Surface Modeling: Part 1

Introduction

This chapter explores key concepts involved in surface modeling, to explain the basic ways of representing a surface in the computer, to outline an approach to surface modeling, to delineate the advantages and limitations of surface modeling, and to explain the use of Rhino NURBS surface modeling tools in the construction of planar surfaces, basic free-form surfaces, and derived surfaces. This chapter also explores the three types of surface continuity.

Objectives

After studying this chapter you should be able to:

- ❒ Explain the key concepts involved in surface modeling
- ❒ State the two types of surface representation methods
- ❒ Explain the advantages and limitations of surface modeling
- ❒ Use Rhino as a tool to construct planar surfaces, basic free-form NURBS surfaces, and derived surfaces
- ❒ Explain the three types of surface continuity

Overview

In our daily lives, we encounter many objects of free-form shape. Some examples are the handle of a razor, the blade of an electric fan, the casing of a computer pointing device (mouse), the casing of a mobile phone, the handle of a joystick, and the body panels of an automobile. An example of free-form shapes is shown in figure 5-1.

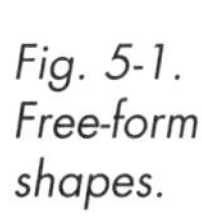

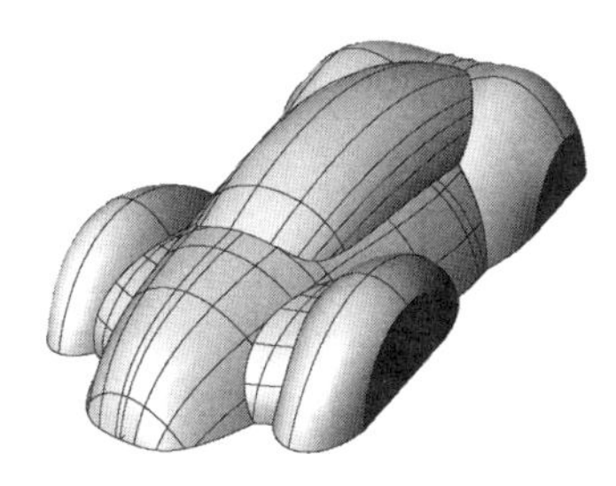

Fig. 5-1. Free-form shapes.

Free-form surfaces are used in objects to meet two basic design needs: aesthetic and functional. Aesthetically, an object has to be eye-pleasing and eye-catching to attract customers. Hence, various types of free-form shapes are used in many consumer products.

Functionally, a surface needs to comply with a certain form and shape to serve specific purposes. For example, the blade of an electric fan should conform to aerodynamic requirements, the profile and silhouette of a shoe should match the shape of a human foot, and the car body shown in figure 5-1 has to allow for wheels.

In chapters 2 through 4 you learned methods of constructing point objects, as well as how to construct, edit, transform, and analyze curves. In this chapter you will learn various surface construction methods.

Surface Modeling Concepts

Surface modeling is a means of representing the geometric shape of a 3D object in the computer by using a set of surfaces put together to resemble the boundary faces of the object. Each surface is a mathematical expression that represents a 3D shape with no thickness. There are two basic ways of representing a surface in the computer: using polygon meshes to approximate a surface and using complex mathematics to obtain an exact representation of the surface.

Polygon Mesh

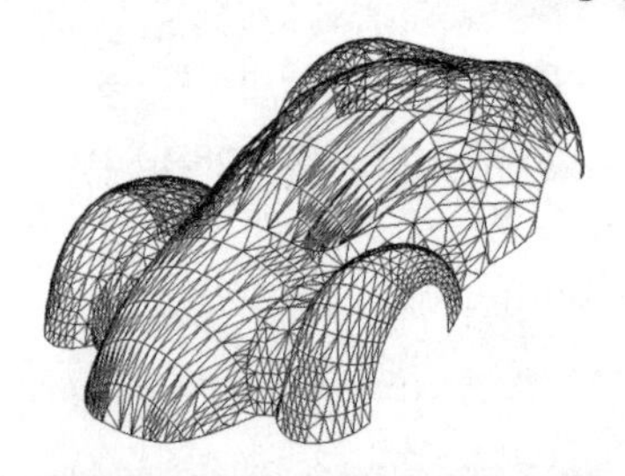

Fig. 5-2. Polygon mesh.

A polygon mesh is a method of approximately representing a surface. It reduces a smooth, free-form surface to a set of planar polygonal faces and curved edges, and reduces silhouettes to sets of straight line segments. Accuracy of representation is inversely proportional to the size of the polygon faces and line segments. A mesh with smaller polygon size represents a surface better, but the memory required to store the mesh is larger. Figure 5-2 shows the polygon mesh of a scale model car body.

A severe drawback of the polygon mesh method is that despite using a very small polygon it can never represent a surface accurately because the surface is always faceted. As a result, this method can be used only for visualization of the real object, and cannot be used in most downstream computerized manufacturing systems.

NURBS Surfaces

To accurately represent free-form smooth surfaces in 3D design applications and computerized manufacturing systems, a higher-order spline surface is used. This is a surface that uses NURBS (Non-Uniform rational basis spline) mathematics to define a set of control vertices and a set of parameters (knots).

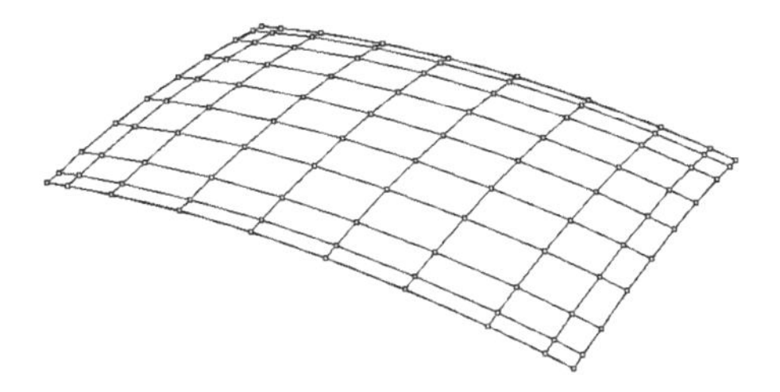

Fig. 5-3. NURBS surface and control vertices.

The distribution of control vertices together with the values of the parameters controls the shape of the surface. The use of NURBS mathematics allows the implementation of multi-patch surfaces with cubic surface mathematics and maintains full continuity control even with trimmed surfaces. Figure 5-3 shows a NURBS surface and control vertices.

Isocurves

Because a NURBS surface is an accurate representation of a smooth surface, there is no facet but only boundary edges and silhouettes. To help visualize a smooth NURBS surface in the computer display, isocurves distributed in two orthogonal directions are placed on the surface. To avoid confusion with the X and Y axes of the coordinate system, these directions are called U and V, examples of which are shown in figure 5-4. It must be stressed that the isocurves are not physical curves. You can set isocurve density by selecting Edit > Properties and changing the density value in the Properties dialog box.

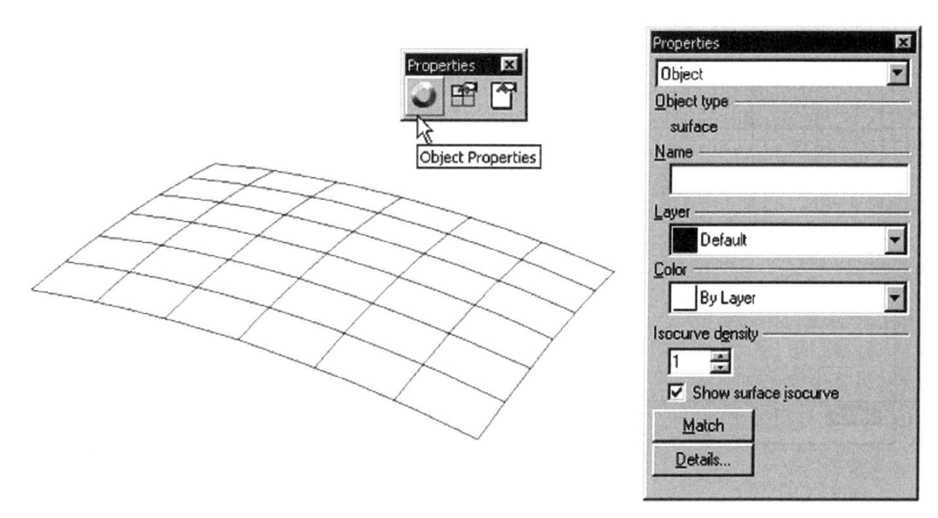

Fig. 5-4. U and V isocurves.

Surface Modeling Approaches

There are many ways to construct NURBS surfaces in the computer. If you want to construct a surface of basic primitive geometric shape, you simply select a primitive surface from the menu and specify the parameters. To meet aesthetic and functional needs, free-form surfaces are required.

The fundamental method of creating a free-form surface is to construct a framework of points and/or curves and let the computer generate the surfaces. If you already have a physical object or mock-up of the physical object, you use the digitizer to obtain appropriate data from the object in constructing the framework of points and/or curves.

Framework of Points and/or Curves

To construct a surface from a framework of points and/or curves, you construct the framework, select a command, and let the computer construct the surface. The outcome of the surface is determined by the location and pattern of the points and/or curves and the command applied. To prepare yourself for making free-form models, you will learn surface construction using various types of points and curves. Figure 5-5 shows a set of curves, a surface constructed from the curves, and a rendered image of the surface.

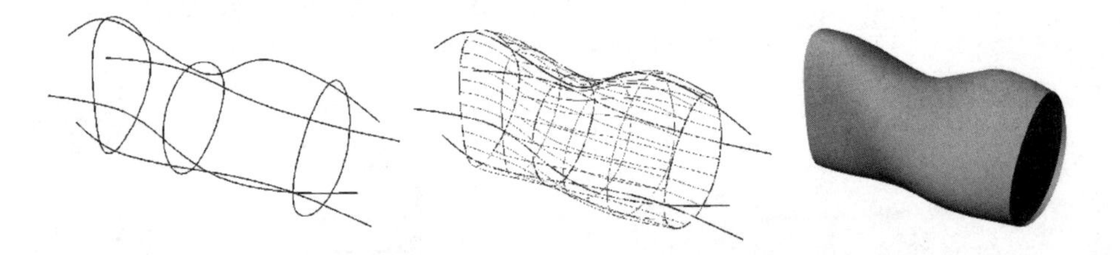

Fig. 5-5. Curve framework and surface constructed from the framework.

Thinking About Surface Modeling

Because a surface model represents an object by using a set of surfaces, surface modeling concerns making a set of surfaces in accordance with the appearance of the object. The sections that follow discuss various aspects of surface modeling with which you need to be familiar.

Deconstruction

To make a surface model, you start thinking about the boundary faces of the object. With careful analysis of the shape of the object, you deconstruct the object's surface into a number of discrete regions of surfaces, with each region following a particular geometric pattern. Then you think about the shape of each individual surface and identify ways of constructing those surfaces.

After this initial thinking process, you construct the surfaces accordingly. Thus, the process of surface modeling consists of two major steps: reduction of the complex 3D object into simple objects (individual surfaces) and making the simple objects to form the complex object. Figure 5-6 shows the surface model of a car body and the individual surfaces exploded.

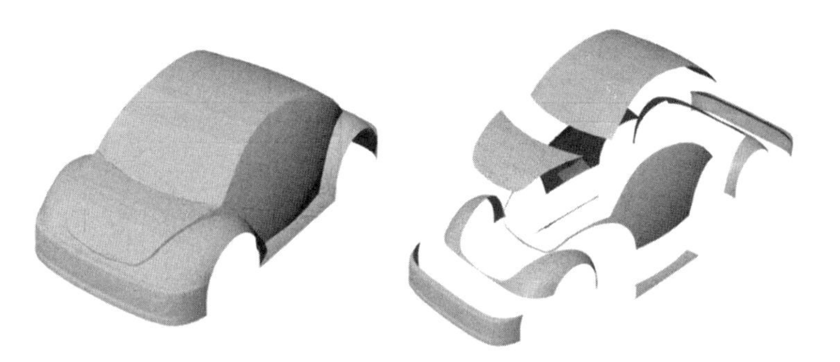

Fig. 5-6. Surface model and individual surfaces exploded.

After you deconstruct the surface of a complex 3D object into a set of surfaces, you start thinking about how to construct each of the surfaces. Among the many ways to construct an individual surface, the fundamental method is to construct 3D curves and/or points as a framework defining the profile and silhouette of the surface and apply NURBS surface modeling tools on the framework.

Thinking About Curves and Points

To construct a surface from a framework of curves and/or points, you should think about the curves and points and reduce the surface to its defining curves and/or points. Because the curves and points of a surface are not readily perceived, reduction of a surface into its defining curves and points requires in-depth analysis of what curves and/or points are needed, where the curves and/or points should be located, and what surface modeling commands are to be applied on them to obtain the required surface.

To be able to identify the locations and types of curves and/or points and the command to apply on them, you need to master various surface construction methods. You also need a good understanding of the shapes of surfaces generated from these methods.

In any design project, you will find that the most tedious job in surface modeling is creating curves and points, and that making surfaces from curves is simple. You need only use the appropriate surface construction commands. Hence, the prime focus of surface modeling is thinking about the curves and points, designing the curves and points, and making the curves and points. Try to determine the curves and/or points for making the surface shown in figure 5-7.

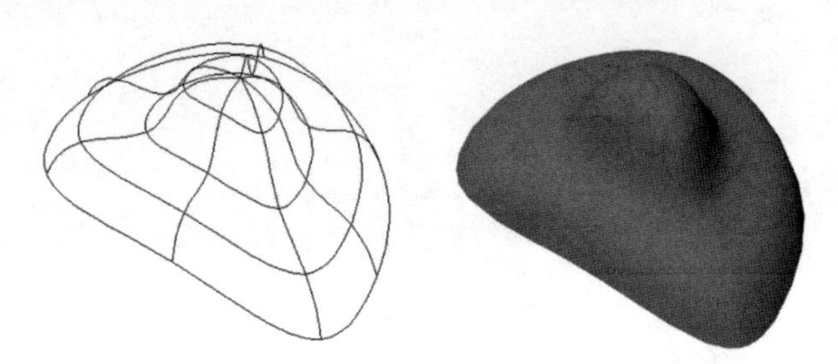

Fig. 5-7. Rail revolve surface.

Curve Data

In Chapter 2 you learned basic techniques of curve construction, editing, and transformation. To guide you through various curve and surface construction techniques, some of the data for the curves are given to you in the tutorials. However, you must realize that this type of data is not readily available when you design a surface. You need to think about the curves and determine the data for the curves on your own.

One way to determine the curves is to use graph paper and sketches. (See Chapter 2.) After you have discerned the relationships among surface shapes, the defining curves, and the applicable surface modeling commands, you construct freehand sketches on graph paper that depict the curves.

From these sketches, you obtain the coordinates of interpolation points. Using these interpolation points, you construct preliminary curves in the computer. From the preliminary curves, you improvise and construct curves to better represent the surface according to your design intent.

Digitizing a Physical Model

Another way to define curves is to make a physical model of the object and use a digitizer to obtain the coordinates of the control points of the surface. Using the coordinates of the markings, you construct curves. Figure 5-8 shows the 3D digitizer discussed in Chapter 3.

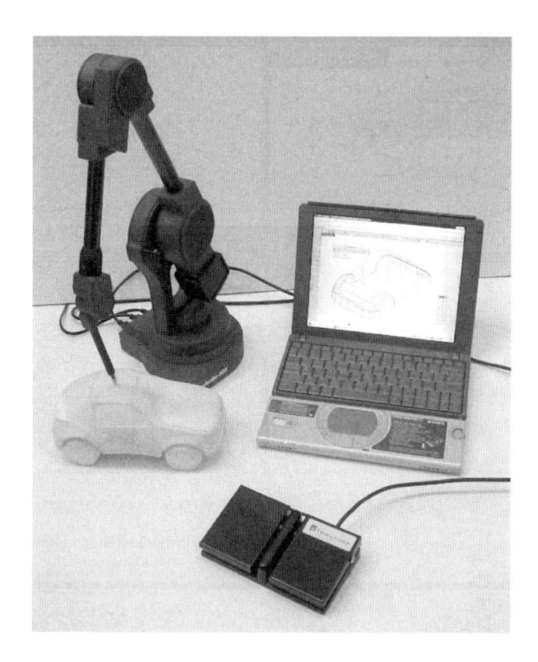

Fig. 5-8. Digitizing a physical model.

Prerequisite Knowledge and Thinking

Bearing in mind that the curves and points for the surface are implicit and imaginary, you may not find them on the object. Thus, a good understanding of the characteristics of various surface modeling commands and the types of curves and points required for these commands is essential. Insight into the types of curves and points needed for a particular type of surface is advantageous.

Modification

There are times when you are not satisfied with the surface constructed from a set of curves. It is probable that the surface you construct does not match the surface you perceive in your mind. To improvise, you have two options: (1) think about the curves and points again, reconstruct the curves and points, and make a new surface from the curves and points, or (2) modify the surface by editing and transforming.

One way to modify a surface is to manipulate control points. It is important to note that when you translate a control point on a surface not only is the control point moved but quite a large area of the surface is affected. Figure 5-9 shows a surface modified by translating two control points.

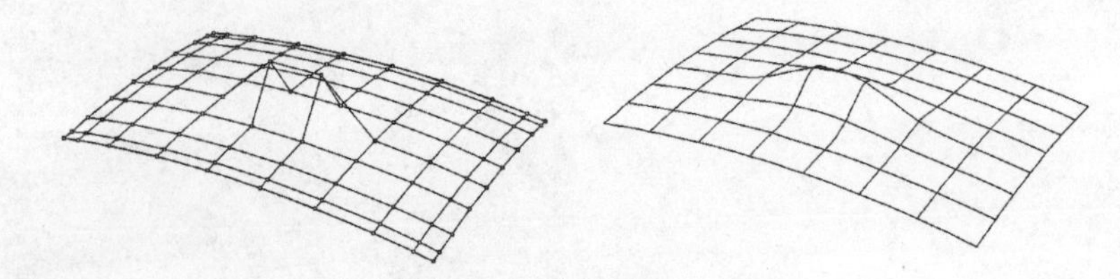

Fig. 5-9. Surface modified by moving its control points.

Advantages of Surface Models

Because a NURBS surface accurately represents smooth, free-form surfaces in the computer, it is the most appropriate tool for aesthetic and engineering design. A surface model contains all surface data of a 3D object, and you can retrieve the coordinates of any point on the surface of the object. Hence, you can use a surface model in most downstream computerized manufacturing operations, as well as in visualization of the object.

For example, you generate stereolithographic data for making rapid prototypes and cross sections for CNC machining. In visualization, you generate hidden-line projection views, shaded images, and photorealistic rendered images. Figure 5-10 shows a rapid prototype constructed by using a 3D printing machine.

Fig. 5-10. Rapid prototype constructed from a surface model.

Limitations of Surface Models

A surface has no thickness. Hence, a surface model is simply a thin shell with no volume, and therefore represents no information other than surface data. As a result, you cannot evaluate the mass properties of a surface model. This produces difficulties such as detecting collision between two surface models in situations in which one surface model lies completely inside another. Figure 5-11 shows the surface model of a joystick lying partly inside the surface model of a joypad.

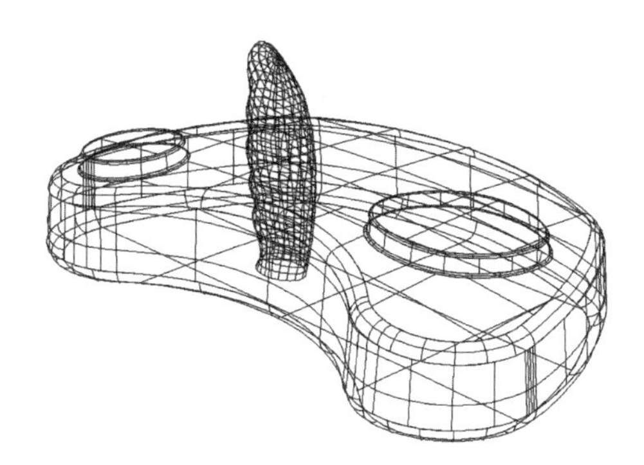

Fig. 5-11. One surface model lying partly inside another surface model.

Rhino Surface Modeling Tools

Basically, Rhino uses NURBS mathematics to accurately represent surfaces in the computer. In addition, Rhino enables you to construct and manipulate polygon meshes. (You will learn about methods of manipulating polygon meshes in Chapter 6.) By joining a set of surfaces, you obtain a polysurface. If the set of polysurfaces encloses a volume without any gap or opening, a solid is implied. (You will learn about solid modeling in Chapter 6.) Figure 5-12 shows a set of surfaces joined to form a polysurface, and figure 5-13 shows an implied solid.

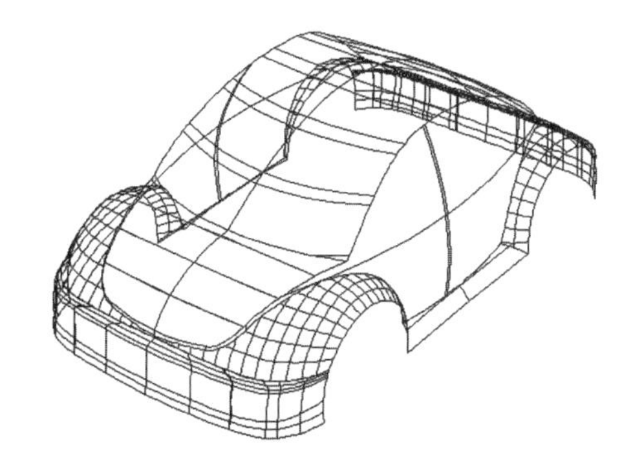

Fig. 5-12. Polysurface.

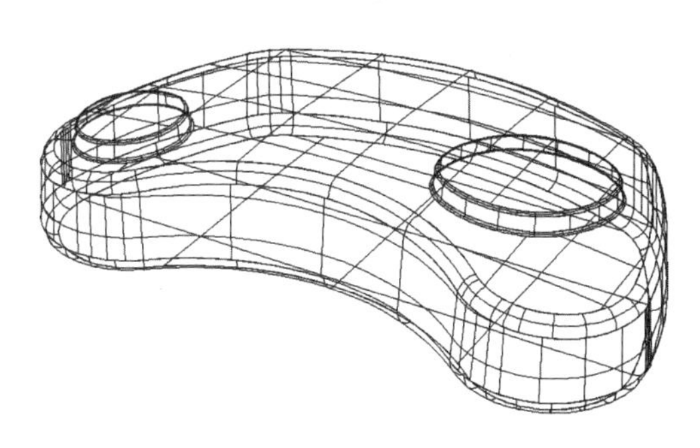

Fig. 5-13. Polysurface enclosing a volume.

Command Menu

Using Rhino as a surface modeling tool, you construct, edit, transform, and analyze NURBS surfaces as well as manipulate polygon meshes. The NURBS surface construction tools are available from the Surface pull-down menu. To edit and transform a NURBS surface, you use the

Edit and Transform pull-down menu. Figure 5-14 shows Surface pull-down menu items for NURBS surface modeling, the Polygon Mesh cascading menu of the Tools pull-down menu, and the Surface, Surface Tools, and Mesh toolbars.

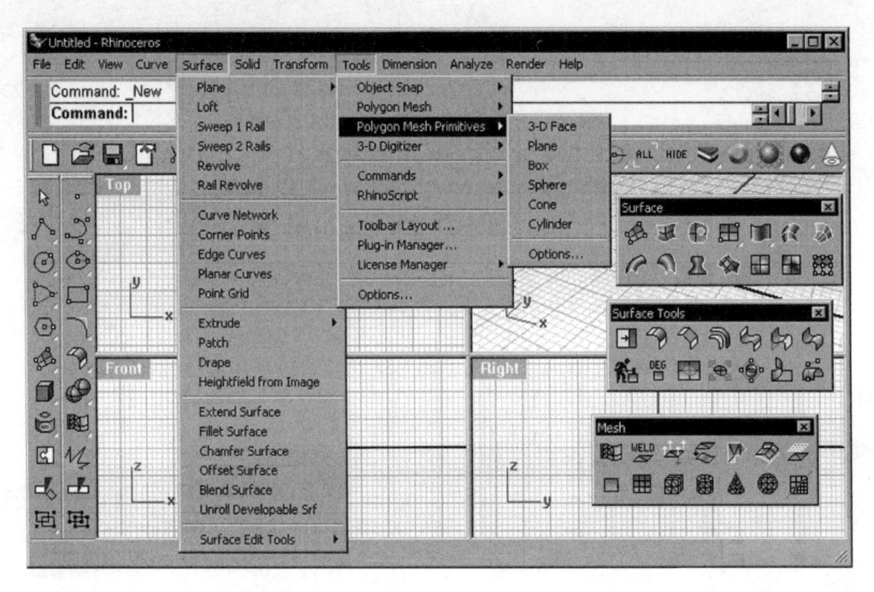

Fig. 5-14. NURBS surface modeling pull-down menu items.

NURBS Surfaces

Because of their importance in design and manufacture, this chapter focuses mainly on NURBS surfaces. Basically, you can classify NURBS surfaces into three main categories in accordance with how they are constructed. The first category of surfaces is the primitives. To construct these types of surfaces, you select a surface type from the menu and specify the parameters of the surface.

The second category of surfaces is the basic free-form surfaces constructed from a framework of curves. The third category derives from existing surfaces. Their shape bears a relationship to surfaces you have already constructed. After you have constructed these surfaces, you can modify them in various ways and analyze them to determine various surface data.

Constructing Planar Surfaces

Primitive surfaces are surfaces you specify without having to make a framework of points or curves. To produce a primitive surface, you select a command and specify the parameters. Corner-point surfaces and rectangular surfaces are primitive surfaces. A corner-point surface is a surface constructed by specifying four corner points. A rectangular surface, as the name implies, is a planar rectangular surface. In essence, both are quadrilateral planar surfaces.

Figure 5-15 shows the commands for constructing planar surfaces in the Surface and Plane toolbars. Apart from these planar surfaces, you may construct other types of primitive surfaces, such as boxes, cylinders, and spheres. Because these primitive objects are closed polysurfaces (i.e., with no gap or opening), they are treated as solids (which you will learn about in Chapter 6).

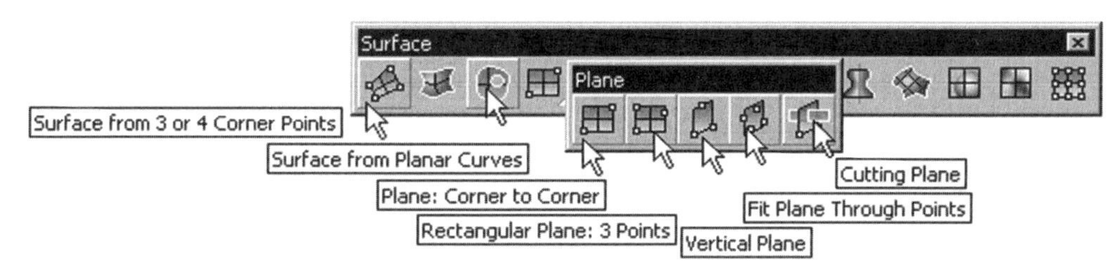

Fig. 5-15. Primitive surface construction commands.

Table 5-1 outlines the functions of primitive surface commands and their locations in the main pull-down menu and respective toolbars.

Table 5-1: Primitive Surface Commands and Their Functions

Toolbar Button	From Main Pull-down Menu	Function
Surface from 3 or 4 Corner Points	Surface > Corner Points	Constructing a planar surface by specifying three or four corner points
Surface from Planar Curves	Surface > Planar Curves	Constructing a trimmed planar surface from planar closed curves
Plane: Corner to Corner	Surface > Plane > Corner to Corner	Constructing a rectangle by specifying its diagonal corners
Rectangular Plane: 3 Points	Surface > Plane > 3 Points	Constructing a rectangle by specifying two end points of an edge and a point on the opposite edge
Vertical Plane	Surface > Plane > Vertical	Constructing a rectangle perpendicular to the construction plane
Fit Plane Through Points	Surface > Plane > Through Points	Constructing a rectangle through a set of selected points
Cutting Plane	Surface > Plane > Cutting Plane	Constructing a rectangle large enough to intersect selected objects

Primitive Planar Surfaces

Perform the following to construct four types of primitive planar surfaces.

4 Start a new file. Use the metric (Millimeters) template.

5 Maximize the Top viewport.

6 Select Surface > Corner Points, or click on the Surface from 3 or 4 Corner Points button on the Surface toolbar.

7 Select locations A, B, and C (indicated in figure 5-16) and press the Enter key.

A triangular planar surface is constructed.

8 Select Surface > Corner Points, or click on the Surface from 3 or 4 Corner Points button on the Surface toolbar.

9 Select locations A, B, C, and D (indicated in figure 5-16).

A quadrilateral planar surface is constructed.

10 Select Surface > Plane > Corner to Corner, or click on the *Plane: Corner to Corner* button on the Plane toolbar.

11 Select locations J and K (indicated in figure 5-16).

A rectangular planar surface is constructed.

12 Select Surface > Plane > 3 Points, or click on the *Rectangular Plane: 3 Points* button on the Plane toolbar.

13 Select locations L, M, and N (indicated in figure 5-16).

A rectangular planar surface inclined at an angle is constructed.

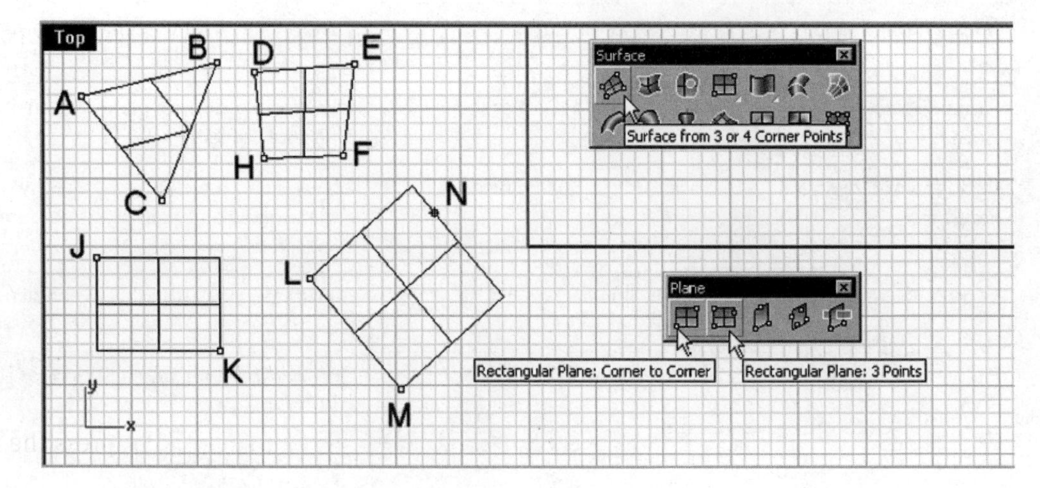

Fig. 5-16. Planar surfaces.

14 Set the display to a four-viewport configuration.

15 Select Surface > Plane > Vertical, or click on the Vertical Plane button on the Plane toolbar.

16 Select locations A and B (indicated in figure 5-17) in the Top viewport.

17 Select location C (indicated in figure 5-17) in the Front viewport.

A vertical rectangular surface is constructed.

18 If you want to save your file, save it as *PlanarSurface1.3dm*. (If you are using the evaluation version, do not save your file.)

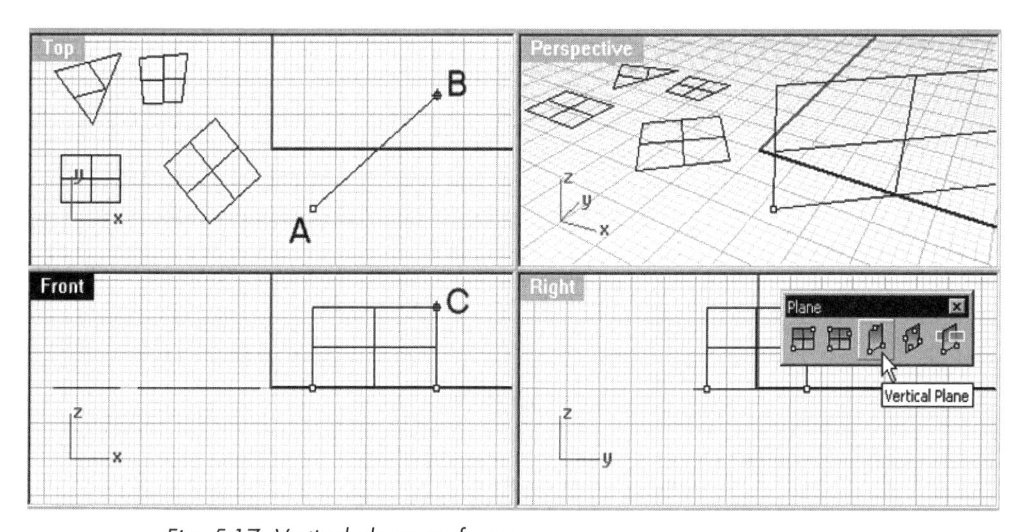

Fig. 5-17. Vertical planar surfaces.

Planar Surfaces Constructed from Points and Curves

Perform the following to construct planar surfaces from point and curve objects.

1 Start a new file. Use the metric (Millimeters) template.

2 Maximize the Top viewport.

3 With reference to figure 5-18, construct a point cloud and three closed-loop free-form curves.

4 Select Surface > Plane > Through Points, or click on the Fit Plane Through Points button on the Plane toolbar.

5 Select point cloud A (indicated in figure 5-19) and press the Enter key.

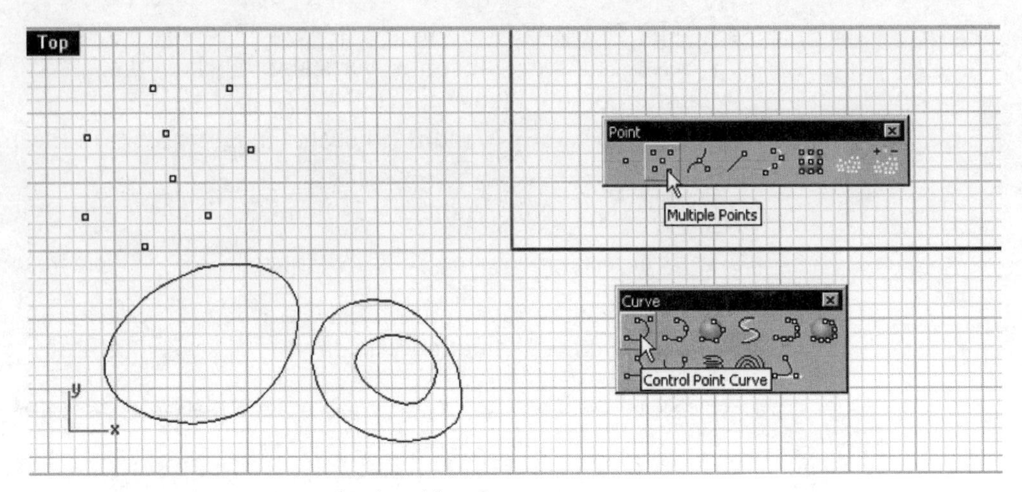

Fig. 5-18. Point cloud and free-form curves.

A rectangular surface passing through all points is constructed.

6 Select Surface > Planar Curves, or click on the Surface from Planar Curves button on the Surface toolbar.
7 Select curves B, C, and D (indicated in figure 5-19) and press the Enter key.

Two trimmed planar surfaces are constructed from the curves.

8 Set the display to a four-viewport configuration.
9 With reference to figure 5-20, construct two free-form curves in the Top viewport and two free-form curves in the Front viewport.

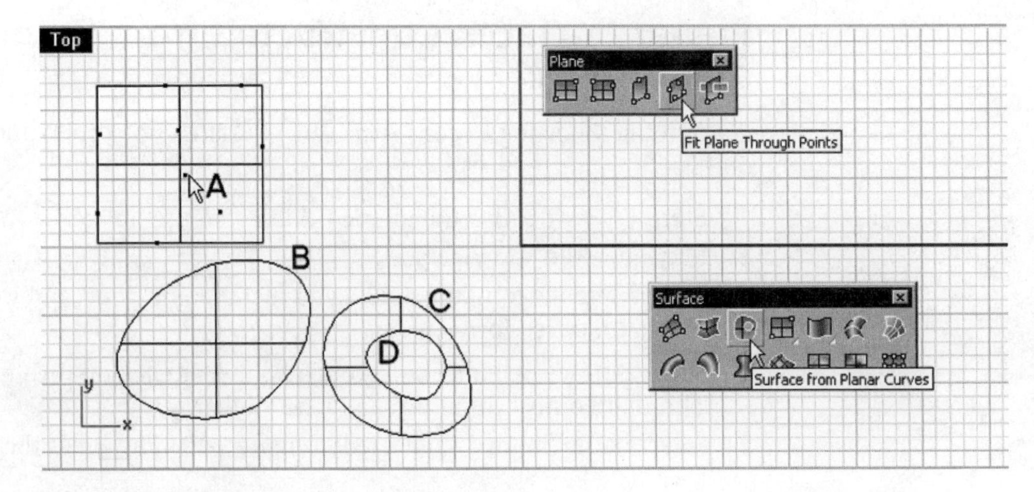

Fig. 5-19. Three planar surfaces.

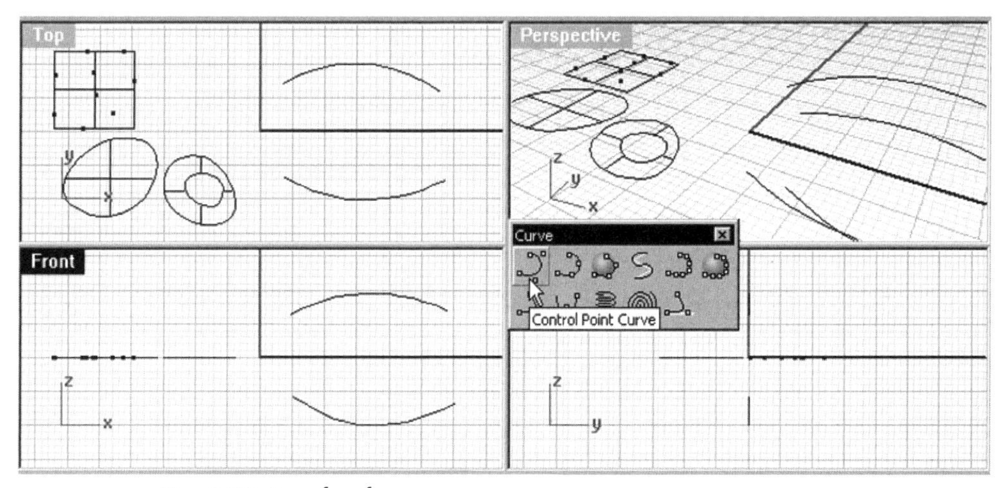

Fig. 5-20. Four free-form curves.

10 Select Surface > Plane > Cutting Plane, or click on the Cutting Plane button on the Plane toolbar.

11 Select curves A, B, C, and D (indicated in figure 5-21) and press the Enter key.

12 Select locations E and F (indicated in figure 5-21). A planar surface is constructed.

13 Select locations G and H (indicated in figure 5-21). Another planar surface is constructed.

14 Press the Enter key to terminate the command.

Two planar surfaces cutting across the curves are constructed.

15 If you want to save your file, save it as *PlanarSurface2.3dm.*

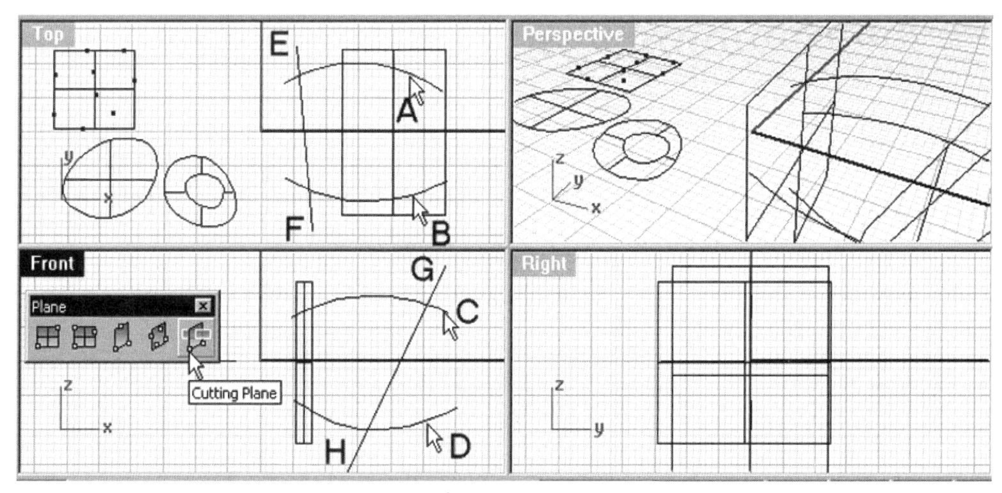

Fig. 5-21. Two cutting planes.

Perform the following to construct planar surfaces from points *not* lying on the same plane.

1. Start a new file. Use the metric (Millimeters) template.
2. With reference to figure 5-22, construct three point objects A, B, and C in the Top viewport and three point objects D, E, and F in the Front viewport.
3. Construct a point cloud that includes all point objects.

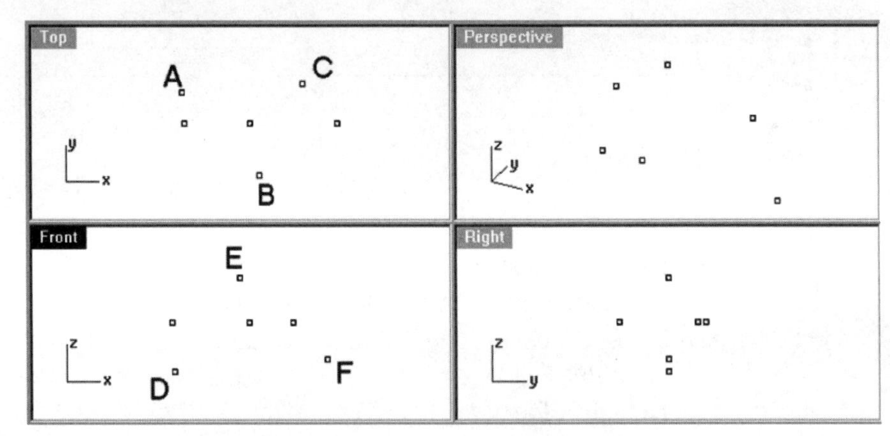

Fig. 5-22. Point cloud.

4. Select Surface > Plane > Through Points, or click on the Fit Plane Through Points button on the Plane toolbar.
5. Select the point cloud and press the Enter key.

A rectangular surface approximating the midplane of the points is constructed, as shown in figure 5-23.

6. If you want to save your file, save it as *PlanarSurface3.3dm*.

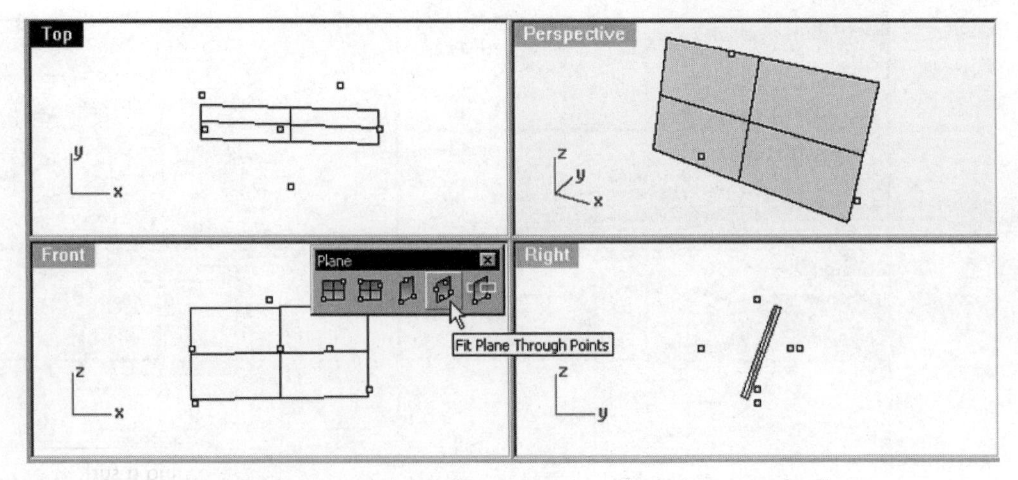

Fig. 5-23. Rectangular surface constructed from the point cloud.

Constructing Basic Free-form Surfaces

As explained earlier, the fundamental method of constructing a free-form surface is to construct a set of curves and/or point objects and let the computer make the surface. The surface profile is determined by the location and pattern of the curves and/or points and the method used to construct the surface. Figure 5-24 shows the free-form surface construction commands of the Surface toolbar.

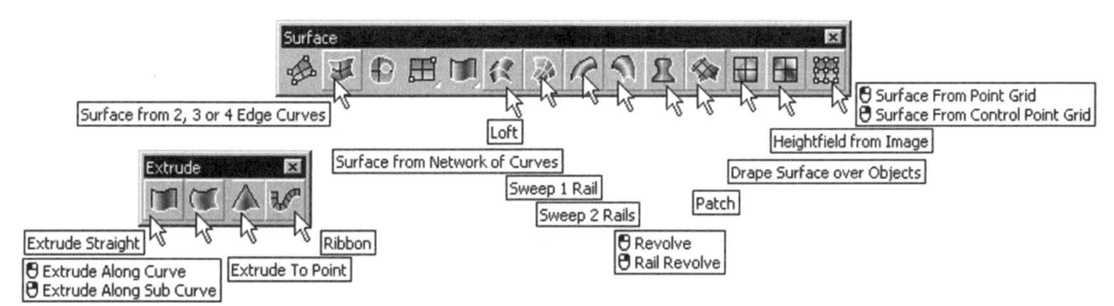

Fig. 5-24. Free-form surface commands.

Table 5-2 outlines the functions of basic free-form surface commands and their locations in the main pull-down menu and respective toolbars.

Table 5-2: Basic Free-form Surface Commands and Their Functions

Toolbar Button	From Main Pull-down Menu	Function
Surface from 2, 3 or 4 Edge Curves	Surface > Edge Curves	Constructing a surface from two, three, or four edge curves
Extrude Straight	Surface > Extrude > Straight	Constructing a surface by extruding a curve perpendicular to the construction plane
Extrude Along Curve	Surface > Extrude > Along Curve	Constructing a surface by extruding a curve along a curve
Extrude Along Sub Curve	—	Constructing a surface by extruding a curve along part of a curve
Extrude To Point	Surface > Extrude > To Point	Constructing a surface by extruding a curve toward a point
Ribbon	Surface > Ribbon	Constructing a surface by offsetting a curve
Loft	Surface > Loft	Constructing a surface that passes through a set of selected curves

Toolbar Button	From Main Pull-down Menu	Function
Surface from Network of Curves	Surface > Curve Network	Constructing a surface from a network of curves
Sweep 1 Rail	Surface > Sweep 1 Rail	Constructing a surface by sweeping a set of section curves along a rail curve
Sweep 2 Rails	Surface > Sweep 2 Rails	Constructing a surface by sweeping a set of section curves along two rail curves
Revolve/Rail Revolve	Surface > Revolve Surface > Rail Revolve	Constructing surfaces by revolving curves about an axis, and constructing a surface by revolving a curve along a path curve and about an axis
Patch	Surface > Patch	Constructing a patch surface from selected curves and/or points
Drape Surface over Objects	Surface > Drape	Constructing a surface by dropping a rectangular sheet over selected objects
Heightfield from Image	Surface > Heightfield from Image	Constructing a surface from a bitmap in accordance with the color value of the bitmap
Surface From Point Grid	Surface > Point Grid	Constructing a surface on a matrix of point objects
Surface From Control Point Grid	—	Constructing a surface by using a matrix of point objects as control points of the surface

Surface from Edge Curves

The simplest way to construct a free-form surface is to define two, three, or four edges of the surface. Perform the following steps to construct surfaces from edge curves.

1. Start a new file. Use the metric (Millimeters) template.
2. With reference to figure 5-25, construct two free-form curves A and B in the Front viewport and two free-form curves C and D in the Right viewport.
3. Select the curves one by one and drag them to new positions in accordance with figure 5-26.

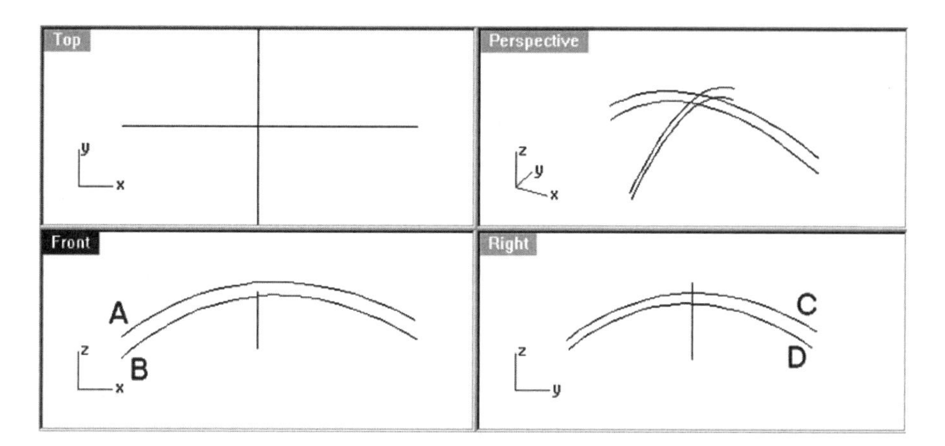

Fig. 5-25. Free-form curves constructed in the Front and Right viewports.

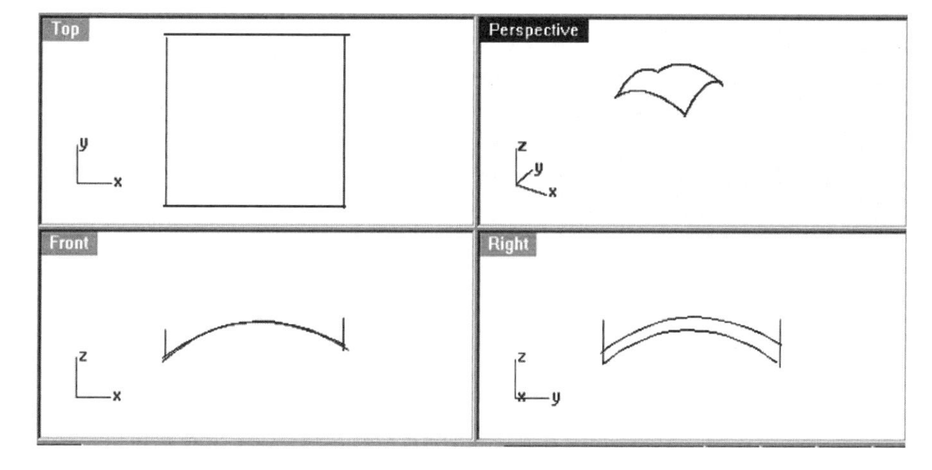

Fig. 5-26. Locations of curves moved.

4 Maximize the Perspective viewport.

5 Select Surface > Edge Curves, or click on the Surface from 2, 3, or 4 Edge Curves button on the Surface toolbar.

6 Select curves A and B (indicated in figure 5-27) and press the Enter key.

A surface is constructed from two edge curves.

7 Select the surface and drag the surface to a new position.

8 Select Surface > Edge Curves, or click on the Surface from 2, 3, or 4 Edge Curves button on the Surface toolbar.

9 Select curves A, B, and C (indicated in figure 5-27) and press the Enter key.

A surface is constructed from three edge curves.

10 Select the surface and drag it to a new position.

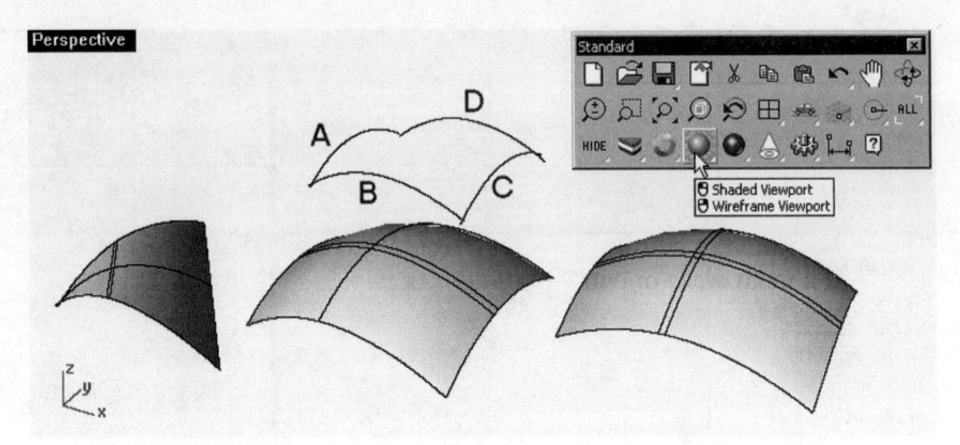

Fig. 5-27. Surfaces constructed from two edge curves, three edge curves, and four edge curves.

11 Select Surface > Edge Curves, or click on the Surface from 2, 3, or 4 Edge Curves button on the Surface toolbar.

12 Select curves A, B, C, and D (indicated in figure 5-27).

13 Select the surface and drag it to a new position.

14 Select Shaded Viewport on the Standard toolbar.

A surface is constructed from four edge curves.

15 If you want to save your file, save it as *EdgeSurface.3dm*.

Extruded Surfaces

By translating a curve along a straight line or curve, you construct an extruded surface. Perform the following steps to construct five types of extruded surfaces.

1 Start a new file. Use the metric (Millimeters) template.

2 With reference to figure 5-28, construct four free-form curves in the Front viewport and two free-form curves in the Top viewport.

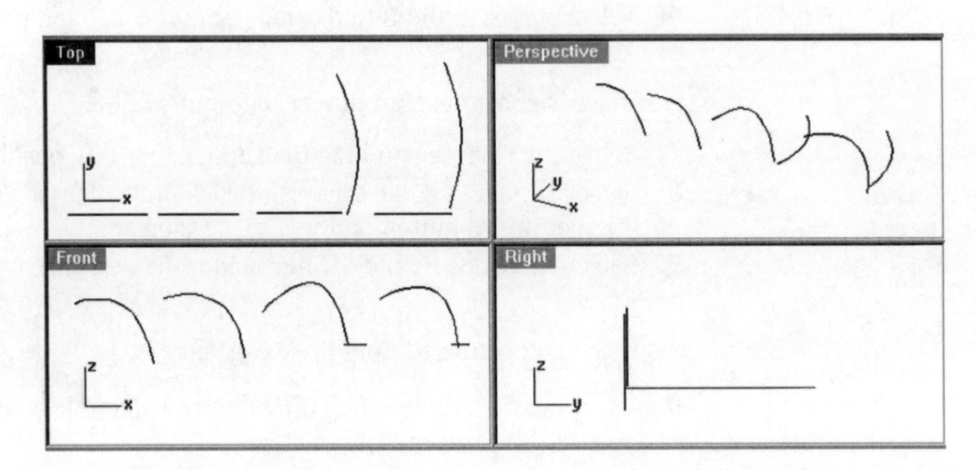

Fig. 5-28. Six free-form curves.

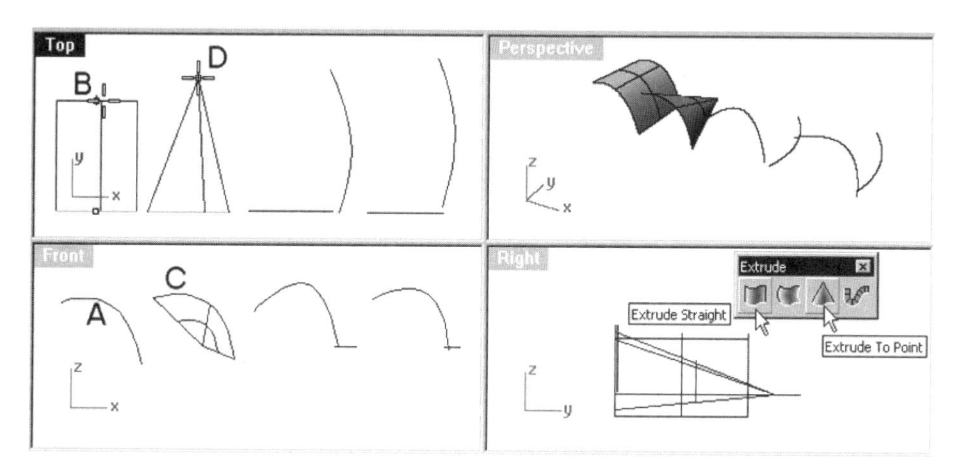

Fig. 5-29. Two extruded surfaces.

3 Select Surface > Extrude > Straight, or click on the Extrude Straight button on the Extrude toolbar.

4 Select curve A (indicated in figure 5-29) and press the Enter key.

5 Select location B (indicated in figure 5-29).

A curve is extruded in a straight line.

6 Select Surface > Extrude > To Point, or click on the Extrude to Point button on the Extrude toolbar.

7 Select curve C (indicated in figure 5-29) and press the Enter key.

8 Select location D (indicated in figure 5-29).

A curve is extruded to a point.

9 Select Surface > Extrude > Along Curve, or click on the *Extrude Along Curve/Extrude Along Sub Curve* button on the Extrude toolbar.

10 Select curve A (indicated in figure 5-30) and press the Enter key.

11 Select location B (indicated in figure 5-30).

A curve is extruded along a curve.

12 Select Surface > Extrude > Along Curve, or right-click on the *Extrude Along Curve/Extrude Along Sub Curve* button on the Extrude toolbar.

13 Select curve C (indicated in figure 5-30) and press the Enter key.

Skip steps 14 and 15 if you run this command from the toolbar.

14 Type *M* to set the mode.

15 Type *L* to use the AlongSubCurve option.

16 Select curve D (indicated in figure 5-30).

17 Select location E along curve D (indicated in figure 5-30).

18 Select location F along curve D (indicated in figure 5-30).

A curve is extruded along a path defined by two points along a path curve.

19 If you want to save your file, save it as *ExtrudeSurface1.3dm*.

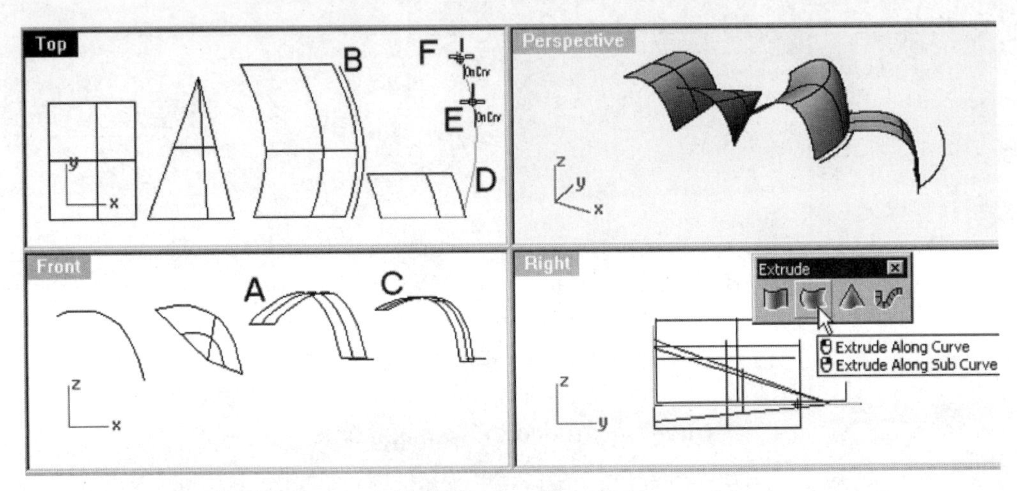

Fig. 5-30. Curves extruded along paths.

Ribbon Surfaces

Perform the following steps to construct two ribbon surfaces.

1 Start a new file. Use the metric (Millimeters) template.

2 With reference to figure 5-31, construct a free-form curve in the Front viewport.

3 Select Surface > Extrude > Ribbon, or click on the Ribbon button on the Extrude toolbar.

4 Select curve A (indicated in figure 5-31).

5 Type *D* to set the distance of the ribbon.

6 Type a value to assign the ribbon's width.

7 Select location B (indicated in figure 5-31) in the Top viewport.

8 Select Surface > Extrude > Ribbon, or click on the Ribbon button on the Extrude toolbar.

9 Select curve A (indicated in figure 5-31) in the Front viewport.

10 Because there are two edges of the ribbon surface you just constructed and a curve located in the same position, you need to select one of them from the pop-up menu that opens. Select Curve from the pop-up menu.

11 Select location C in the Front viewport.

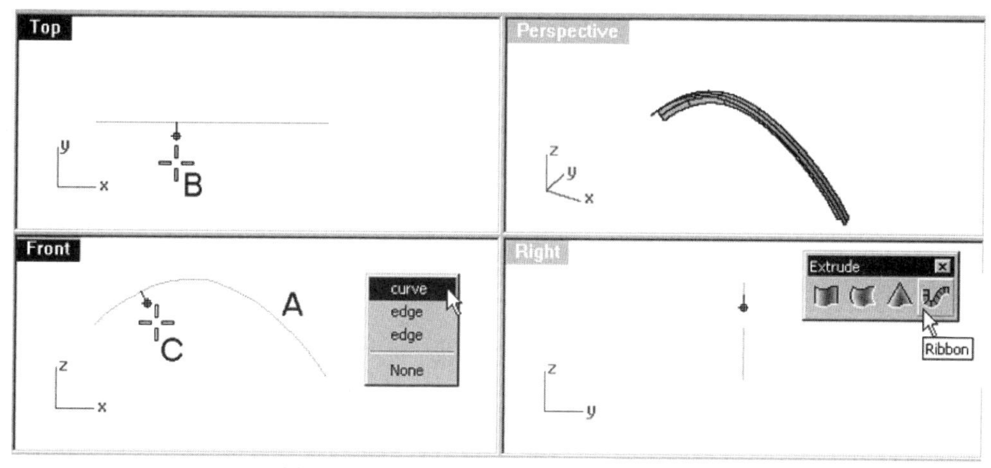

Fig. 5-31. Ribbon surfaces being constructed.

Two ribbon surfaces are constructed.

12 If you want to save your file, save it as *Extrude Surface2.3dm*.

Lofted Surface

You can construct a surface by defining a series of cross sections. This type of surface is called a lofted surface. It interpolates from the first section curve to the second, and to the next curve. To construct a lofted surface, perform the following steps.

1. Start a new file. Use the metric (Millimeters) template.
2. With reference to figure 5-32, construct three free-form curves in the Top viewport.
3. Select and drag the curves one by one in the Front viewport to move them to the locations indicated in figure 5-33.
4. Select Surface > Loft, or click on the Loft button on the Surface toolbar.
5. Select curves A, B, and C (indicated in figure 5-33) and press the Enter key.

When you move the curve over seam points D, E, and F (indicated in figure 5-33) of the curve, you will find a direction arrow. If the directions of the arrows are not congruent, type *F* and select the noncongruent seam points.

6. Press the Enter key.
7. In the Loft Options dialog box, shown in figure 5-34, click on the OK button.

A lofted surface is constructed, as shown in figure 5-35.

8 If you want to save your file, save it as *Loft.3dm*.

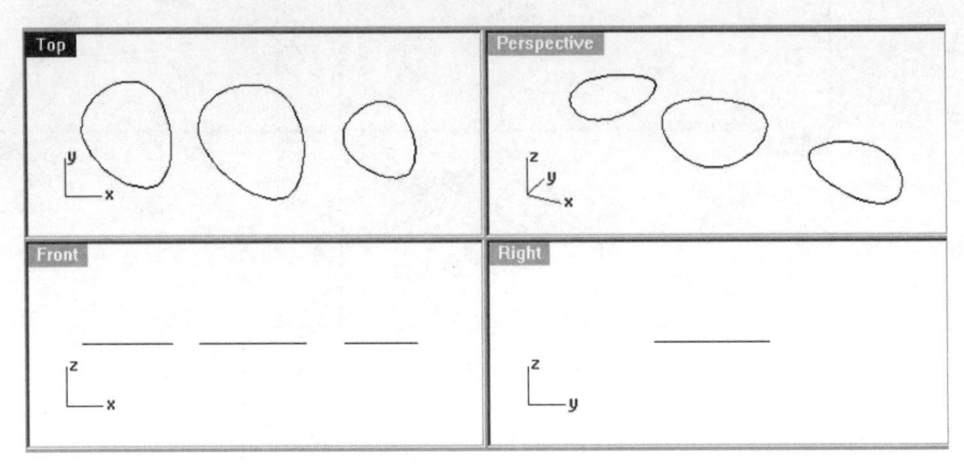

Fig. 5-32. Free-form curves.

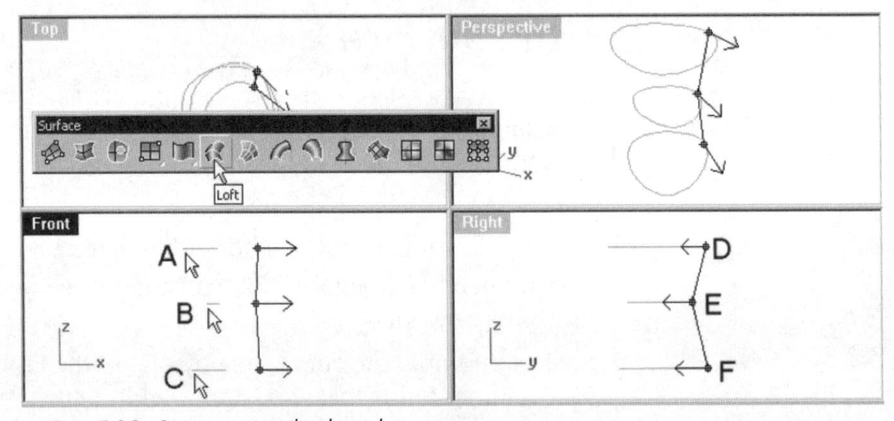

Fig. 5-33. Seam points displayed.

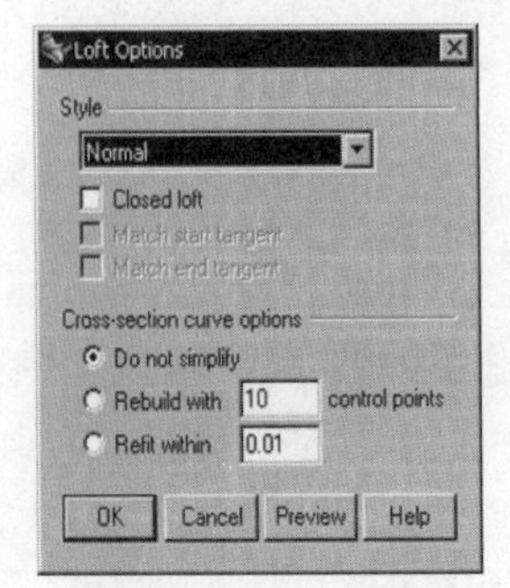

Fig. 5-34. Loft Options dialog box.

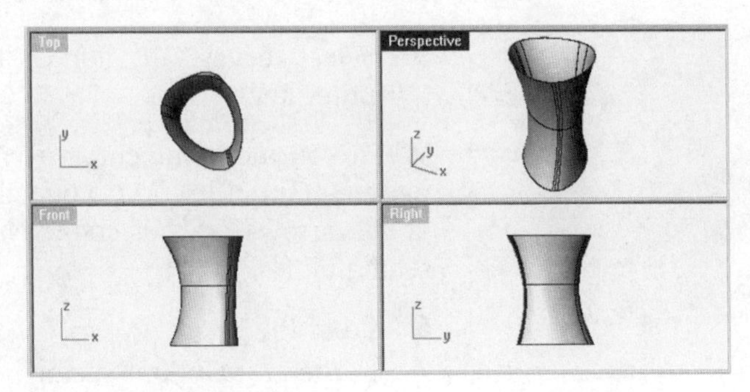

Fig. 5-35. Lofted surface.

Surface from Curve Network

A surface constructed from a curve network is similar to a surface-from-edge curve, with additional curves in two orthogonal directions. In essence, valid curves will be automatically sorted into two sets of orthogonal curves, regardless of the sequence of selection. However, if one of the curves is invalid or ambiguously defined, the system will prompt you to select the curve sets manually. Then you should select one set of curves, press the Enter key, continue to select the second set of orthogonal curves, and press the Enter key again. To construct this type of surface, perform the following steps.

1. Start a new file. Use the metric (Millimeters) template.
2. Construct three free-form curves in the Front viewport and three free-form curves in the Right viewport in accordance with figure 5-36.

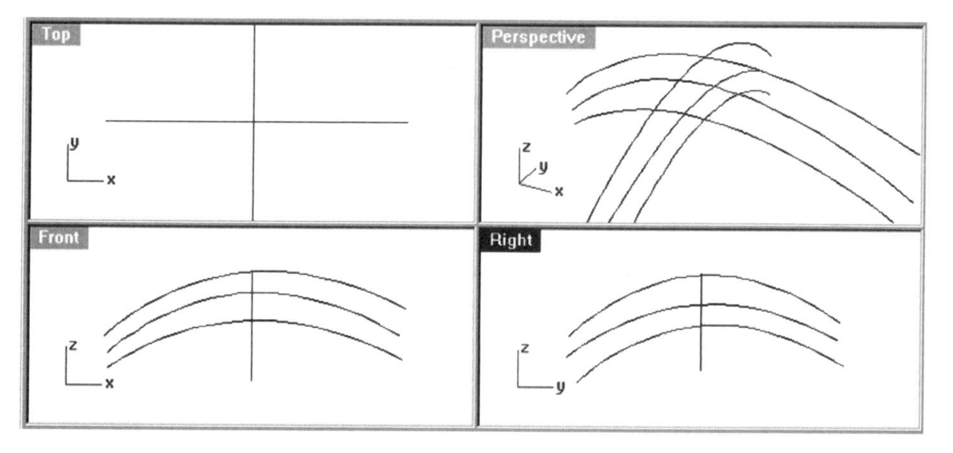

Fig. 5-36. Free-form curves.

3. Translate the curves so that they form a network of curves in two orthogonal directions, as shown in figure 5-37.
4. Select Surface > Curve Network, or click on the Surface from Network of Curves button on the Surface toolbar.
5. Select all curves and press the Enter key.

After selecting the curves, the Surface From Curve Network dialog box displays. Here, you specify edge tolerance and continuity of matching edges, as follows.

6. In the Surface From Curve Network dialog box, shown in figure 5-38, accept the default and then click on the OK button.

A surface is constructed from a curve network, as shown in figure 5-39.

7. If you want to save your file, save it as *NetworkSurface.3dm*.

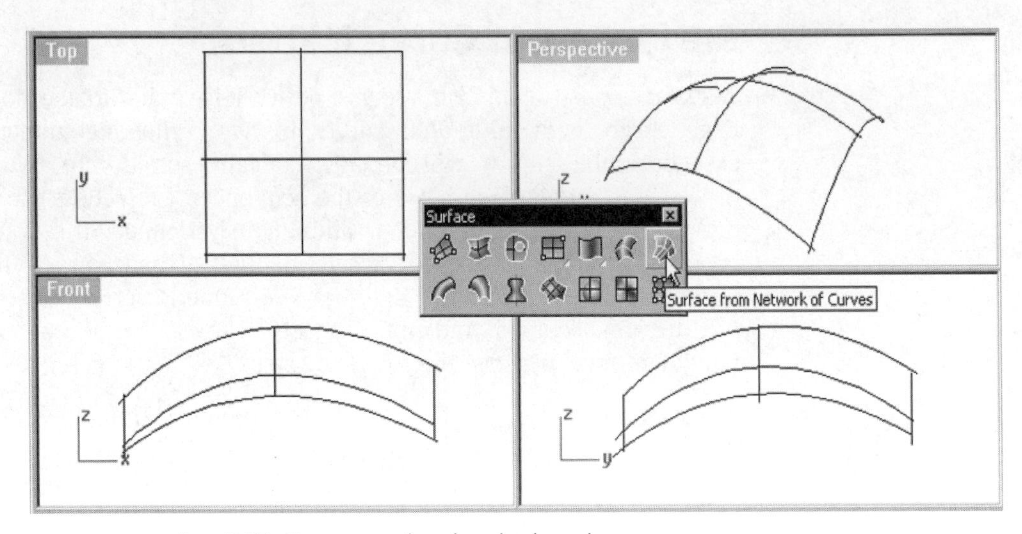

Fig. 5-37. Curves translated and selected.

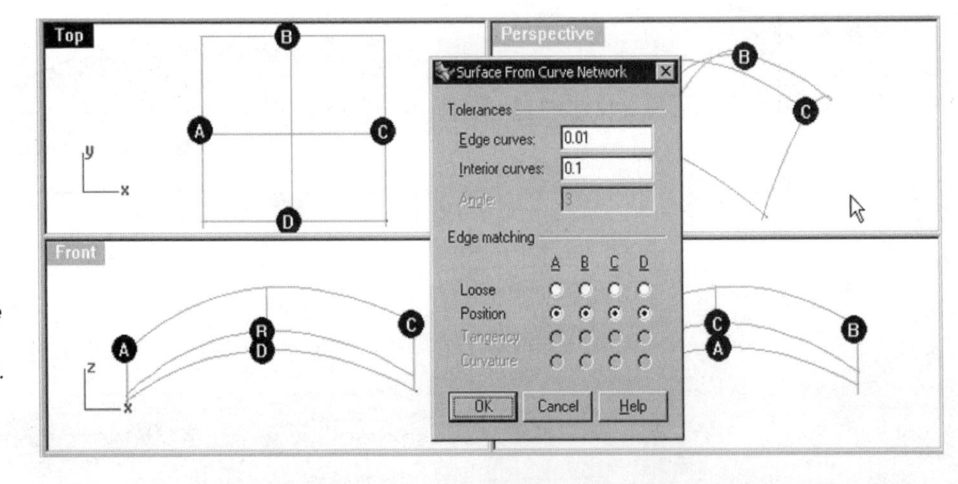

Fig. 5-38. Surface From Curve Network dialog box.

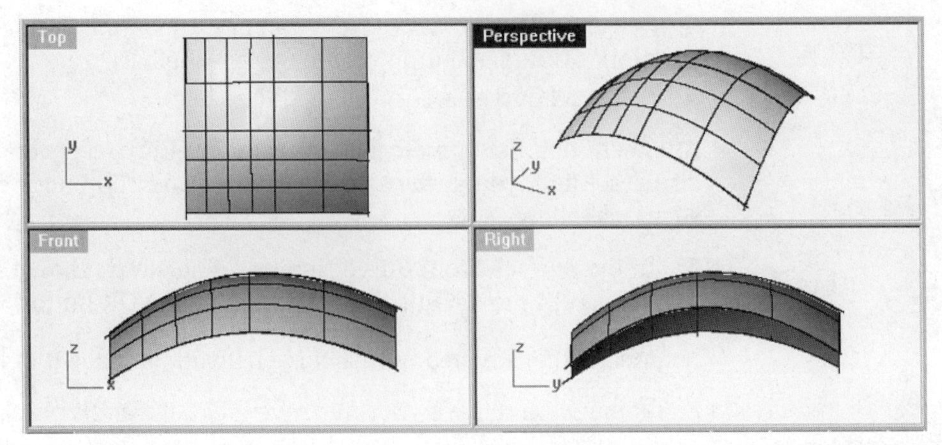

Fig. 5-39. Surface from curve network.

Sweeping Along a Rail

In creating a lofted surface, the surface profile interpolates smoothly from the first curve to the second curve, and to the next curve. To impose control on the flow of a surface profile from a section curve to another section curve, you may add a guiding rail. The surface is called a sweep surface. To construct a surface by sweeping cross-section curves along a rail, perform the following steps.

1. Start a new file. Use the metric (Millimeters) template.
2. Construct a free-form curve, a circle around the free-form curve, and an ellipse around the free-form curve in accordance with figure 5-40.
3. Select Surface > Sweep 1 Rail, or click on the Sweep 1 Rail button on the Surface toolbar.
4. Select free-form curve A (indicated in figure 5-40) as the rail curve, select circle B and ellipse C (indicated in figure 5-40) as the cross-section curves, and press the Enter key.

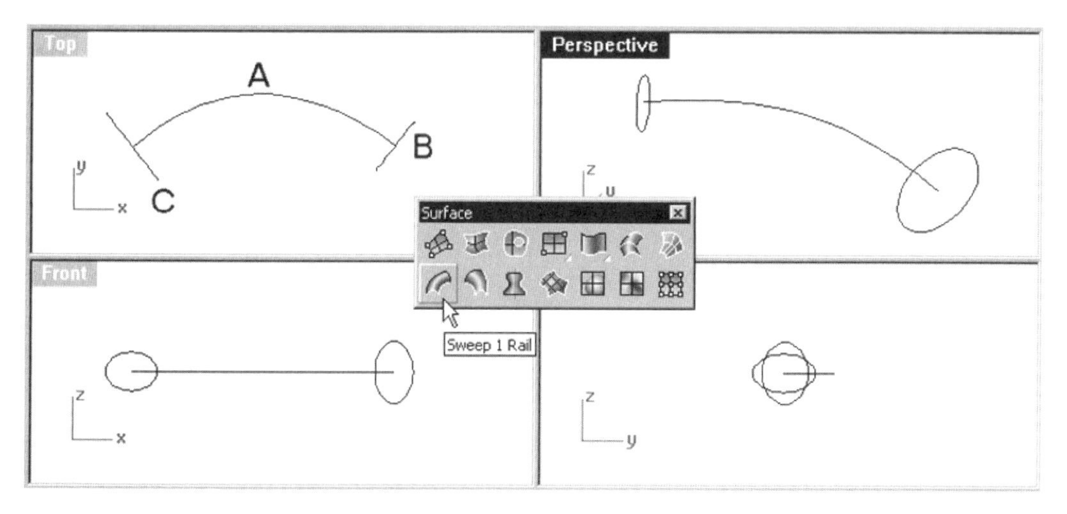

Fig. 5-40. Free-form curve, and circle and ellipse around the free-form curve.

5. In the Sweep 1 Rail Options dialog box, shown in figure 5-41, accept the default and then click on the OK button.

A surface is constructed, as shown in figure 5-42.

6. If you want to save your file, save it as *Sweep1.3dm*.

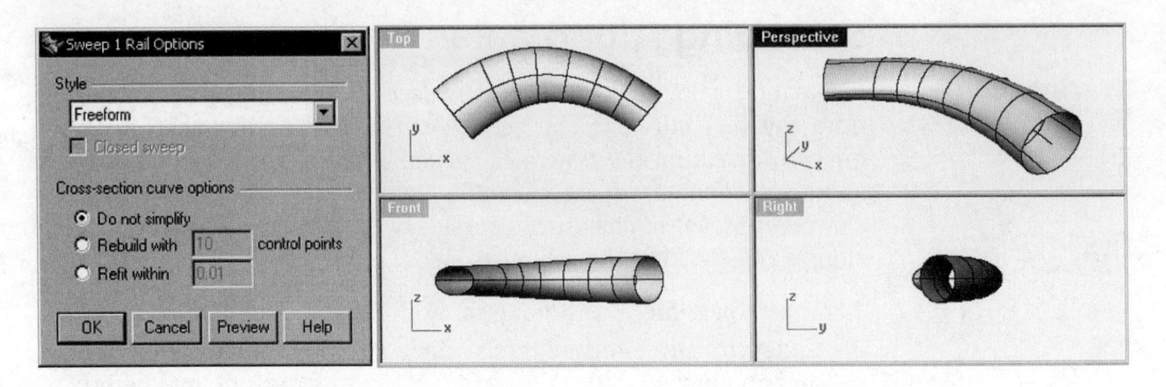

Fig. 5-41. Sweep 1 Rail Options dialog box.

Fig. 5-42. Surface constructed by sweeping curves along two rails.

Sweeping Along Two Rails

To add further control to the surface profile while sweeping, you may use two guiding rails. To construct a surface by sweeping cross-section curves along two rails, perform the following steps.

1. Start a new file. Use the metric (Millimeters) template.
2. With reference to figure 5-43, construct two free-form curves in the Top viewport and two free-form curves in the Right viewport.
3. Check the End box in the Osnap dialog box.

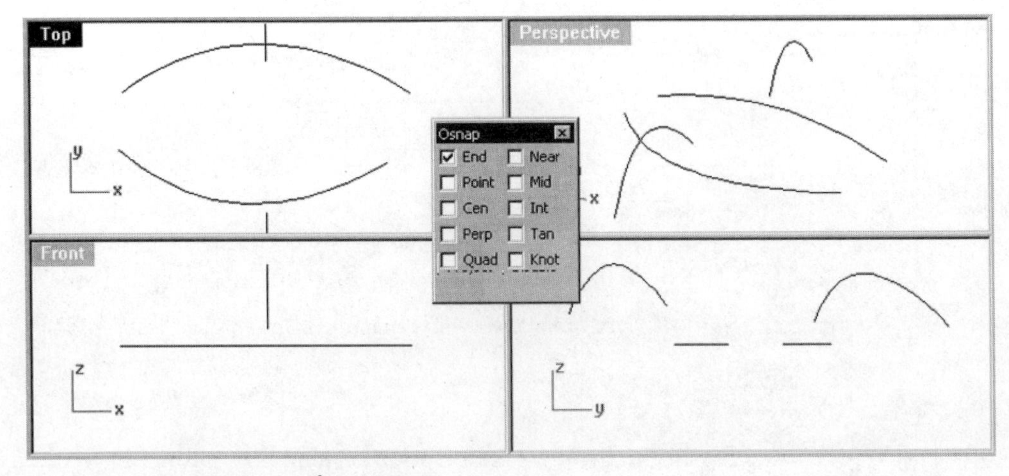

Fig. 5-43. Free-form curves.

4. Select Transform > Orient > 2 Points, or click on the *Orient: 2 Points/Orient: 3 Points* button on the Transform toolbar.

5 Select curve A (indicated in figure 5-44) and press the Enter key.
6 Select points B and C (indicated in figure 5-44) as the reference points.
7 Select points D and E (indicated in figure 5-44) as the target points.
8 Repeat steps 4 through 7 to orient curve F (indicated in figure 5-44).

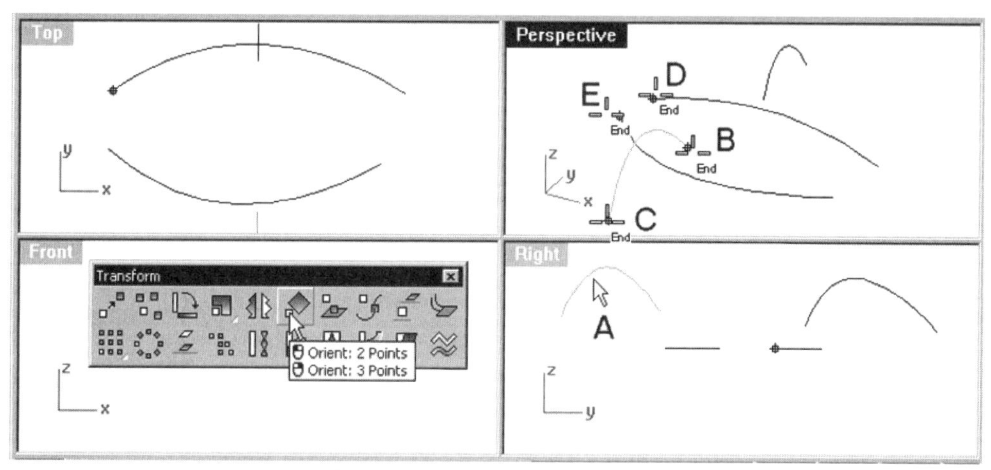

Fig. 5-44. Curves being oriented.

9 Select Surface > Sweep 2 Rail, or click on the Sweep 2 Rails button on the Surface toolbar.
10 Select curves A, B, C, and D (indicated in figure 5-45) and press the Enter key.

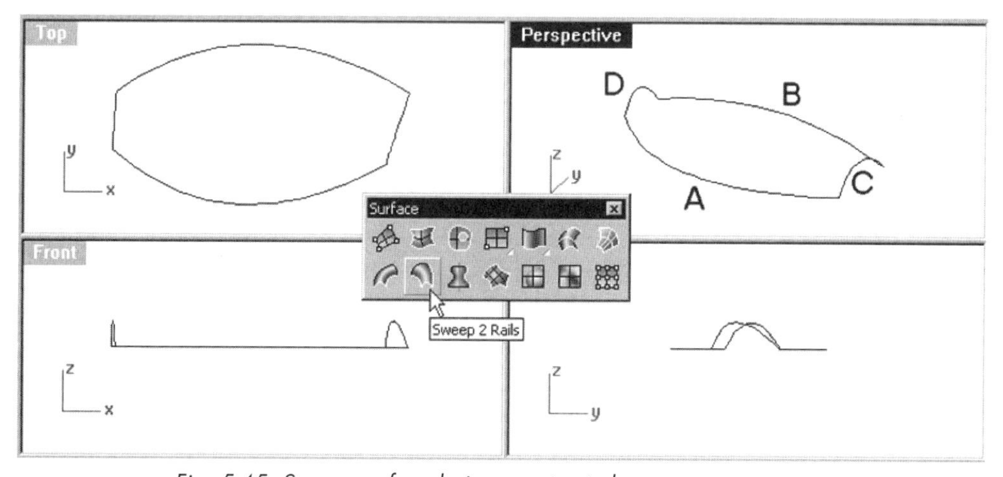

Fig. 5-45. Sweep surface being constructed.

11 In the Sweep 2 Rails Options dialog box, shown in figure 5-46, click on the OK button.

A surface is constructed by sweeping two curves along two rails, as shown in figure 5-47.

12 If you want to save your file, save it as *Sweep2.3dm*.

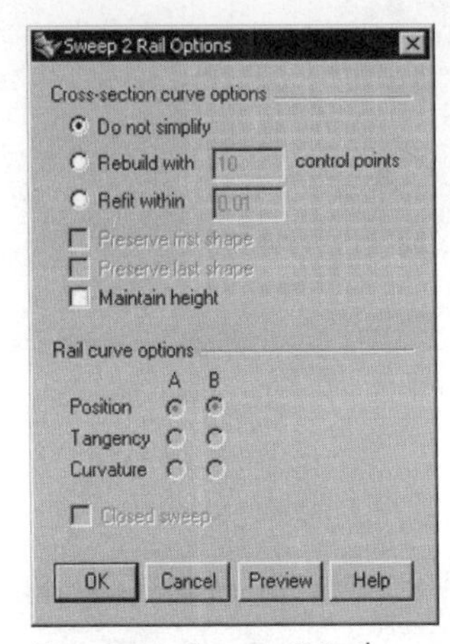

Fig. 5-46. Sweep 2 Rail Options dialog box.

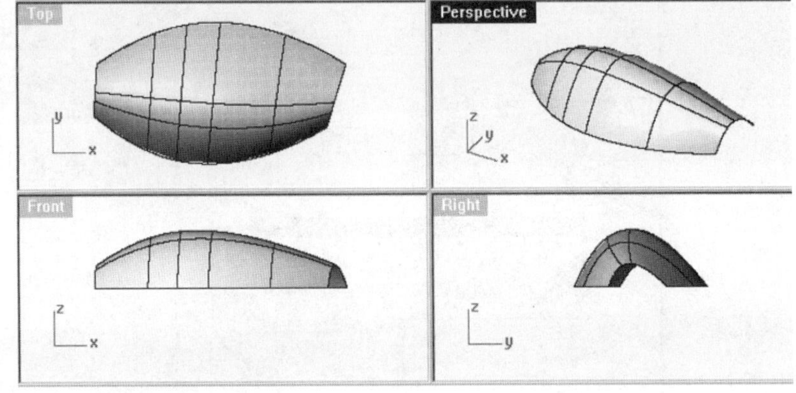

Fig. 5-47. Surface constructed by sweeping curves along two rails.

Revolved Surface and Revolved Surface Guided by a Rail

A revolved surface is constructed by revolving a curve about an axis. To control the surface profile while it is being revolved about an axis, you can add a guide rail. To construct a revolved surface and a revolved surface guided by a rail, perform the following steps.

1. Start a new file. Use the metric (Millimeters) template.
2. Construct free-form curve A in the Top viewport.
3. Construct straight lines B and C and free-form curves D and E in the Front viewport, as shown in figure 5-48.
4. Check the End box in the Osnap dialog box.
5. Select Surface > Revolve, or click on the *Revolve/Rail Revolve* button on the Surface toolbar.
6. Select curve A (indicated in figure 5-49) and press the Enter key.
7. Select end points C and D (indicated in figure 5-49).
8. In the Revolve Options dialog box, click on the OK button.
9. Select Surface > Rail Revolve, or right-click on the *Revolve/Rail Revolve* button on the Surface toolbar.

10 Select curve E (indicated in figure 5-49) as the profile curve.

11 Select curve F (indicated in figure 5-49) as the rail curve.

12 Select end points G and H (indicated in figure 5-49) to define the axis of revolution.

A revolved surface is constructed.

13 If you want to save your file, save it as *Revolve.3dm*.

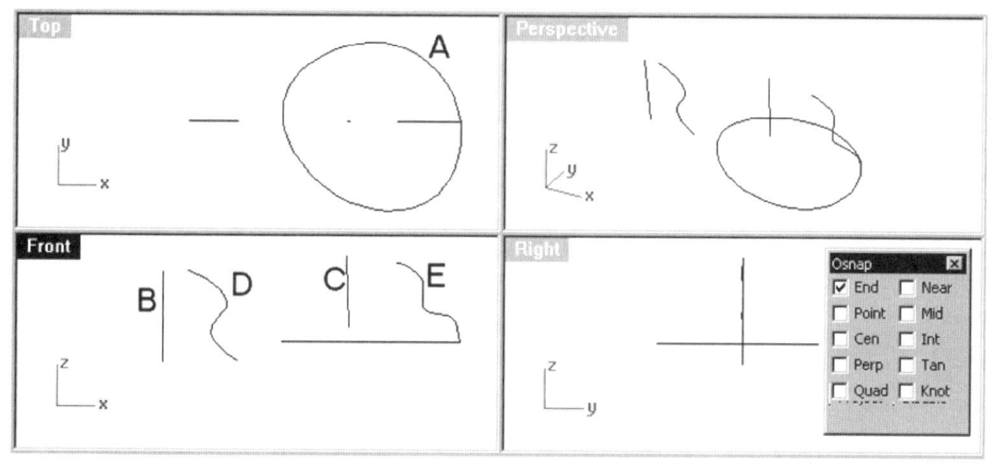

Fig. 5-48. Lines and curves.

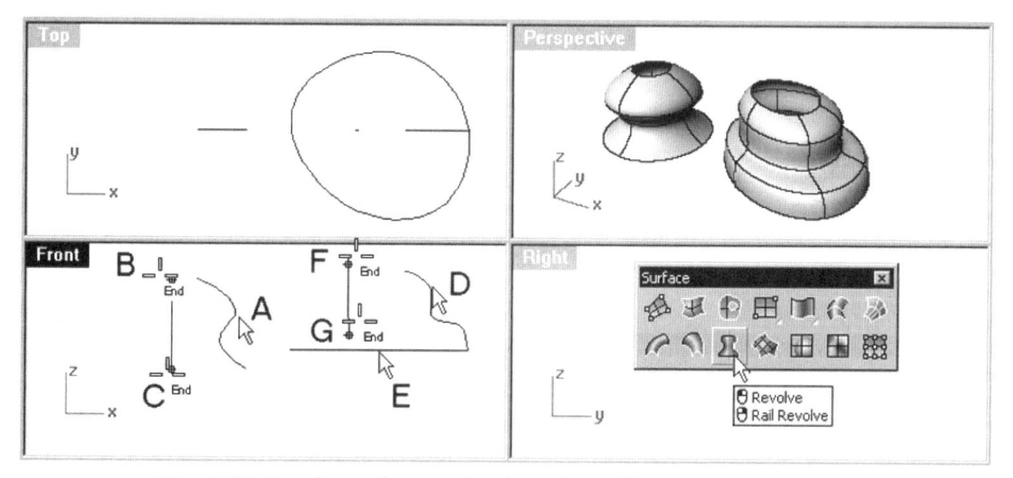

Fig. 5-49. Revolve surface and rail revolve surface.

Patch Surfaces

The input data for construction of a patch surface can be curves and/or points. In the following you will construct five types of patch surfaces from a combination of points and curves: a patch surface from a closed-loop curve, a patch surface from point objects, a patch surface from a closed-loop curve and point objects, a patch surface from an open-loop curve and point objects, and a patch surface from a number of curves. To construct two curves and a set of point objects, perform the following steps.

1. Start a new file. Use the metric (Millimeters) template.
2. With reference to figure 5-50, construct two free-form curves (an open-loop curve and a closed-loop curve) in the Top viewport, three point objects in the Front viewport, and three point objects in the Right viewport.

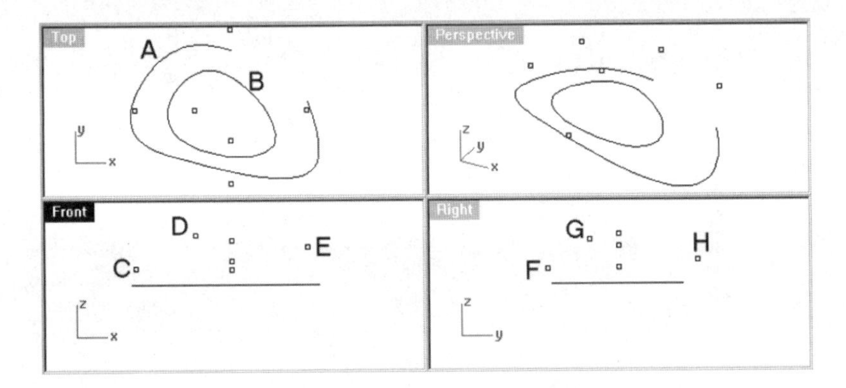

Fig. 5-50. Curve and point objects.

Patch Surface from Closed-loop Curve

To construct a patch surface from a curve, perform the following steps.

1. Set the current layer to *Layer 01*.
2. Select Surface > Patch, or click on the Patch button on the Surface toolbar.
3. Select closed-loop curve A (indicated in figure 5-51) and press the Enter key.

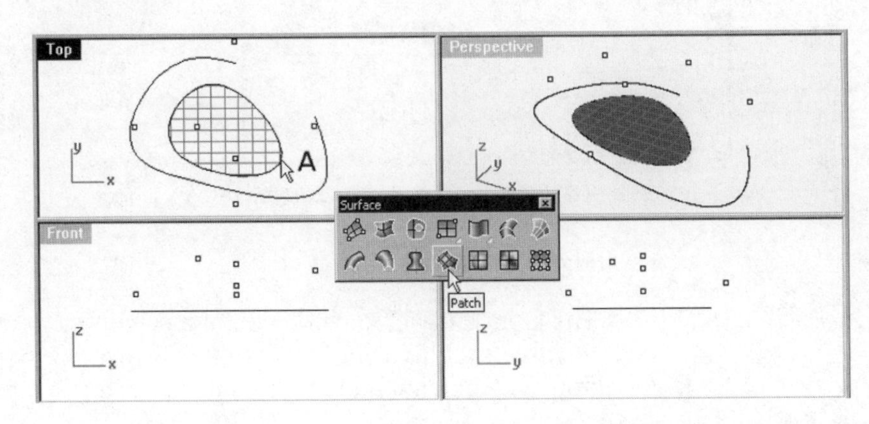

Fig. 5-51. Patch surface constructed from a closed-loop curve.

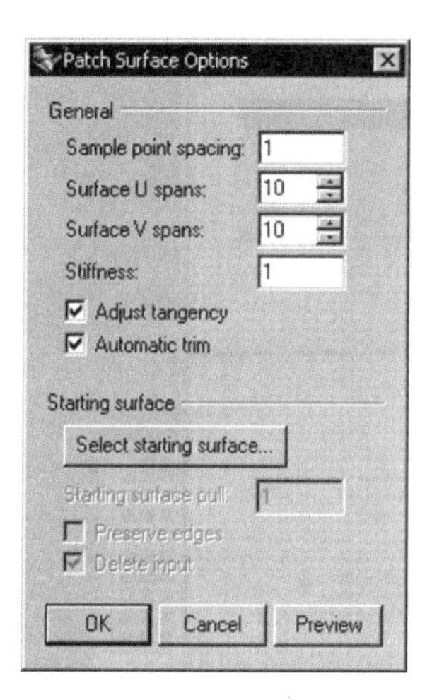

4 In the Patch Surface Options dialog box, shown in figure 5-52), accept the default and then click on the OK button.

A patch surface is constructed from the closed-loop curve.

NOTE: *Do not confuse this command with the PlanarSrf command (accessed by selecting Surface > Planar Curves, or by clicking on the Surface from Planar Curves button on the Surface toolbar) in which the curves for making a surface have to be planar.*

Fig. 5-52. Patch Surface Options dialog box.

Patch Surface from Point Objects

To construct a patch surface from point objects, perform the following steps.

1 Turn off *Layer 01* and set the current layer to *Layer 02*.

2 Select Surface > Patch, or click on the Patch button on the Surface toolbar.

3 Select points A, B, C, D, E, and F and press the Enter key.

4 Click on the OK button in the Patch Options dialog box.

5 A patch surface is constructed from the points, as shown in figure 5-53.

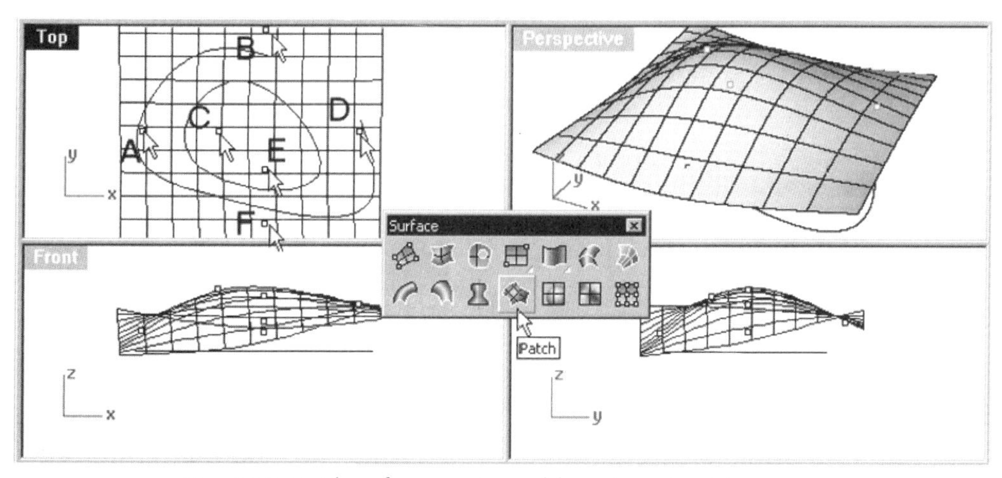

Fig. 5-53. Patch surface constructed from the point objects.

Patch Surface from Closed-loop Curve and Point Objects

To construct a patch surface from the closed-loop curve and point objects, perform the following steps.

1. Turn off *Layer 02* and set the current layer to *Layer 03*.
2. Select Surface > Patch, or click on the Patch button on the Surface toolbar.
3. Select the closed-loop curve and the points, and then press the Enter key.
4. Click on the OK button of the Patch Options dialog box.

A patch surface from a closed-loop curve and point objects is constructed, as shown in figure 5-54.

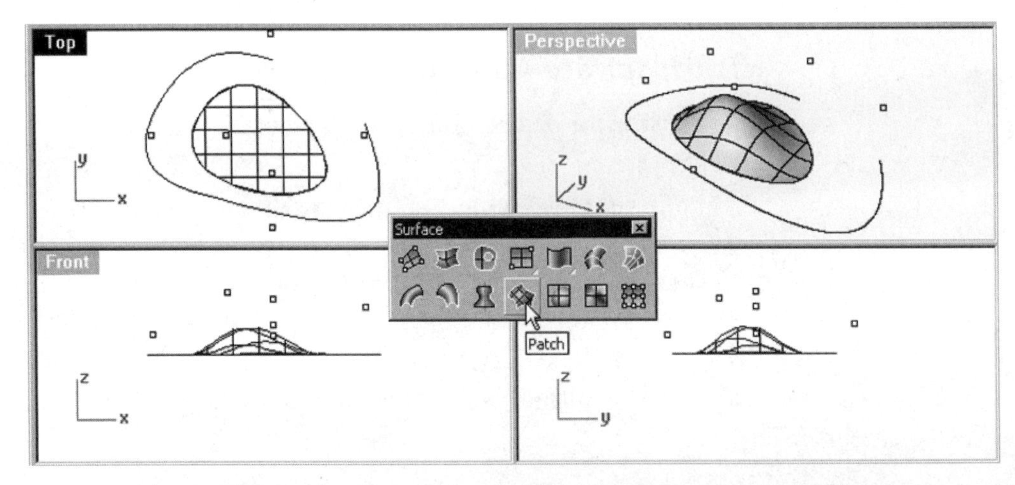

Fig. 5-54. Patch surface constructed from point objects and closed-loop curve.

Patch Surface from Open-loop Curve and Point Objects

To construct a patch surface from an open-loop curve and point objects, perform the following steps.

1. Turn off *Layer 03* and set the current layer to *Layer 04*.
2. Select Surface > Patch, or click on the Patch button on the Surface toolbar.
3. Select the open-loop curve and the points, and then press the Enter key.
4. Click on the OK button of the Patch Options dialog box.

A patch surface is constructed, as shown in figure 5-55.

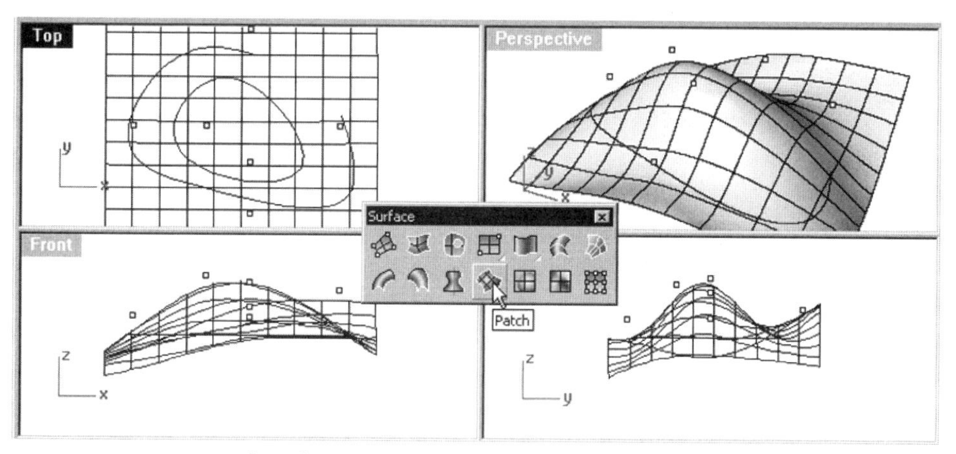

Fig. 5-55. Patch surface constructed from point objects and open-loop curve.

Patch Surface from Curves

To construct two curves and a patch surface from these curves, perform the following steps.

1 Turn off *Layer 04* and set the current layer to *Layer 05*.
2 Construct free-form curve A in the Front viewport and free-form curve B in the Right viewport, as shown in figure 5-56.

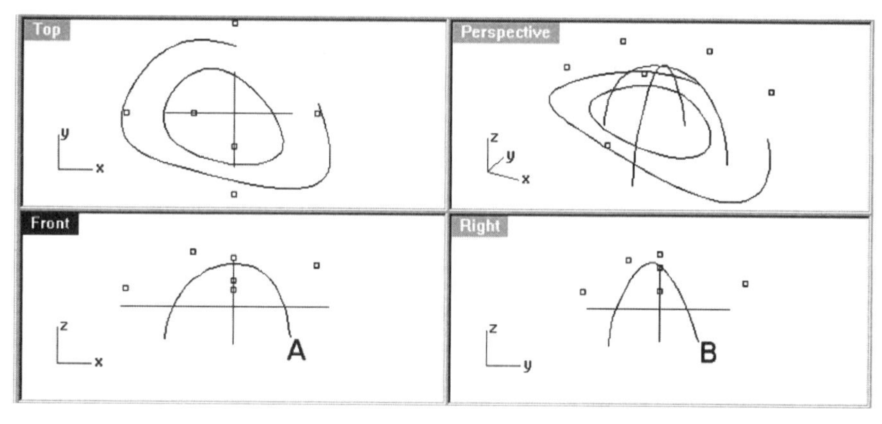

Fig. 5-56. Free-form curves.

3 Select Surface > Patch, or click on the Patch button on the Surface toolbar.
4 Select the curves indicated in figure 5-57 and the closed-loop curve, and then press the Enter key.
5 Click on the OK button of the Patch Options dialog box.

A patch surface from the curves is constructed, as shown in figure 5-57. Five patch surfaces are complete. Try to relate the results with the input points and/or curves.

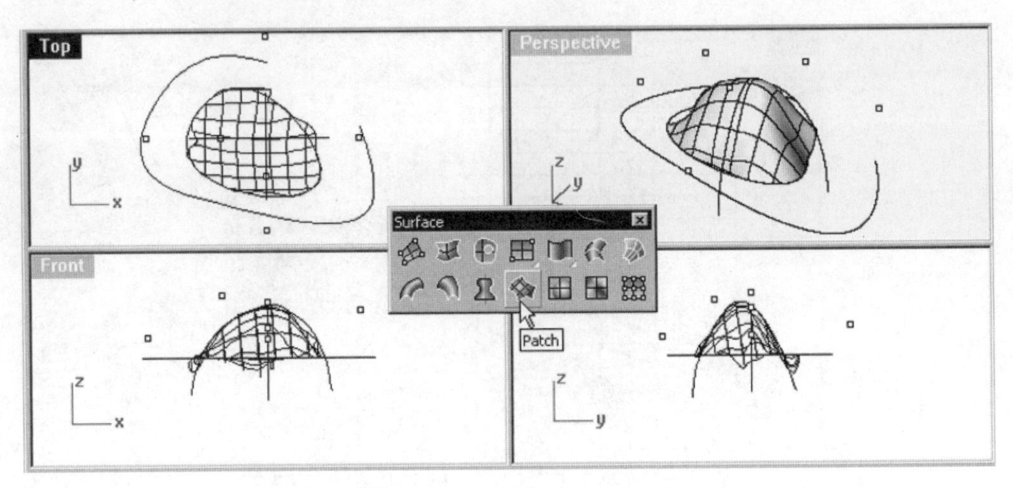

Fig. 5-57. Patch surface constructed from curves.

6 If you want to save your file, save it as *Patch.3dm*.

Drape

Making a draped surface is analogous to warping a rectangular piece of plastic sheeting on a set of 3D objects in a way similar to vacuum forming. Vacuum forming is a type of plastics forming process. You heat up plastic sheeting, warp the sheet onto a 3D object (the mold), and apply a vacuum to deform the sheet. Figure 5-58 shows a vacuum-form machine.

Fig. 5-58. Vacuum-form machine.

1 To construct draped surfaces from a cone and a sphere, perform the following steps.
2 Select File > Open.
3 In the *Files of type* box of the Open dialog box, select IGES (*.igs*; *.iges*) and select the file *Drape.igs* from the *Chapter 5* folder on the companion CD-ROM.
4 Select Surface > Drape, or click on the Drape Surface over Objects button on the Surface toolbar.
5 Select locations A and B (indicated in figure 5-59).
6 A draped surface is constructed, as shown in figure 5-60.
7 If you want to save your file, save it as *Drape.3dm*.

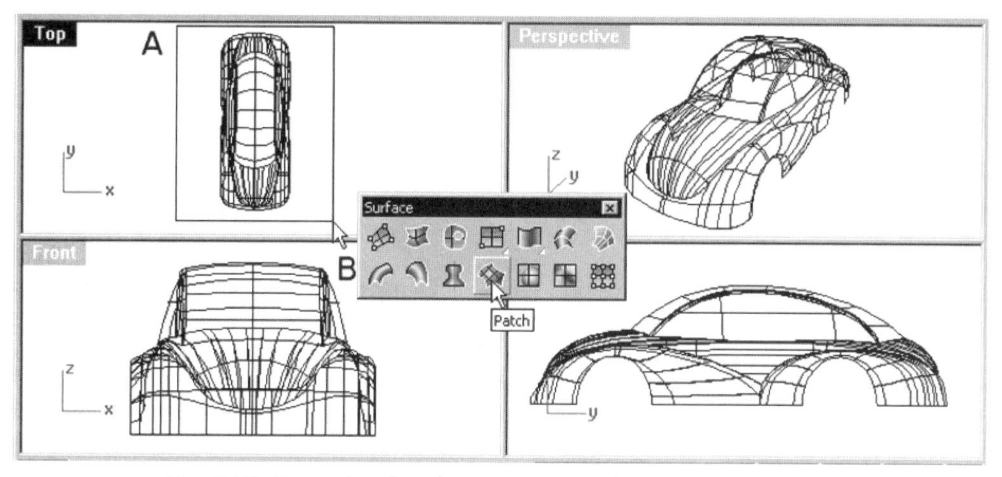

Fig. 5-59. Draped surface being constructed.

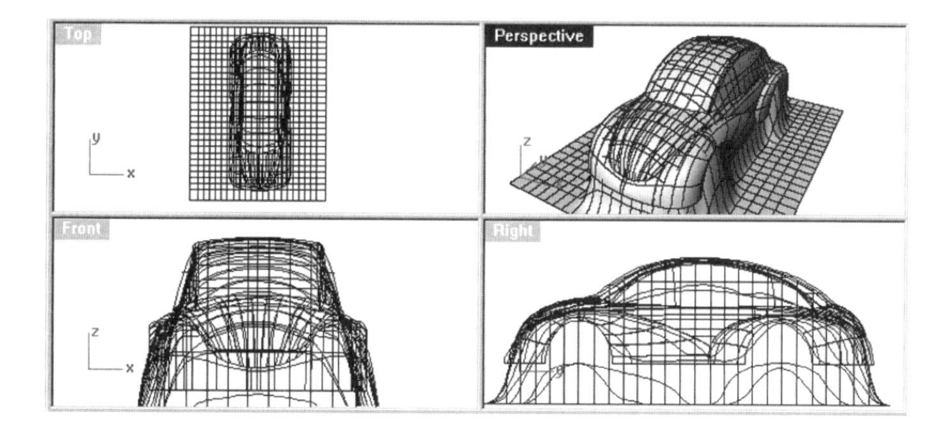

Fig. 5-60. Draped surface.

Surfaces from Bitmaps

Using the color value of a bitmap, you can derive a surface. Basically, the bitmap can be color or black and white. However, it is more appropriate to use a black-and-white image because only the brightness value is considered when a surface is derived. If you use a black-and-white image, you can better perceive the outcome before making the surface.

Naturally, you need a bitmap to make a surface of this type. Figure 5-61 shows digital black-and-white images being taken via digital camera. To derive a surface from a bitmap, perform the following steps.

1. Start a new file. Use the metric (Millimeters) template.
2. Maximize the Top viewport.

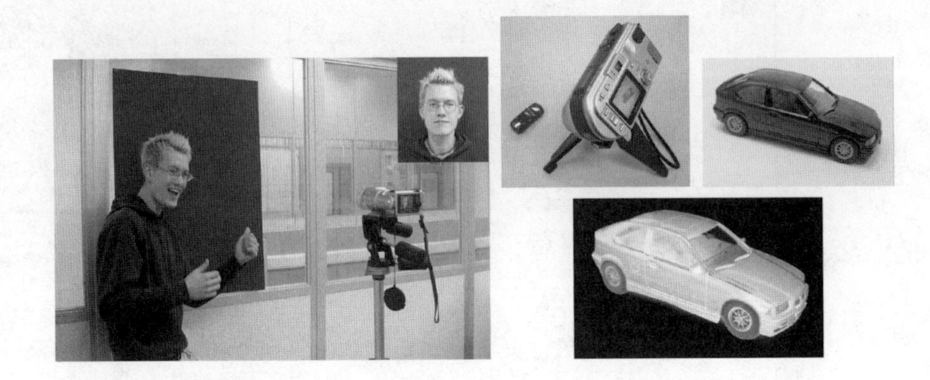

Fig. 5-61. Digital black-and-white images being taken.

3 Select Surface > Heightfield from Image, or click on the Heightfield from Image button on the Surface toolbar.

4 In the Select Bitmap dialog box, select the image file *Car.tga* from the *Chapter 5* folder on the companion CD-ROM.

5 Pick two points A and B in the Top viewport to indicate the size of the surface to be derived from the bitmap, as shown in figure 5-62.

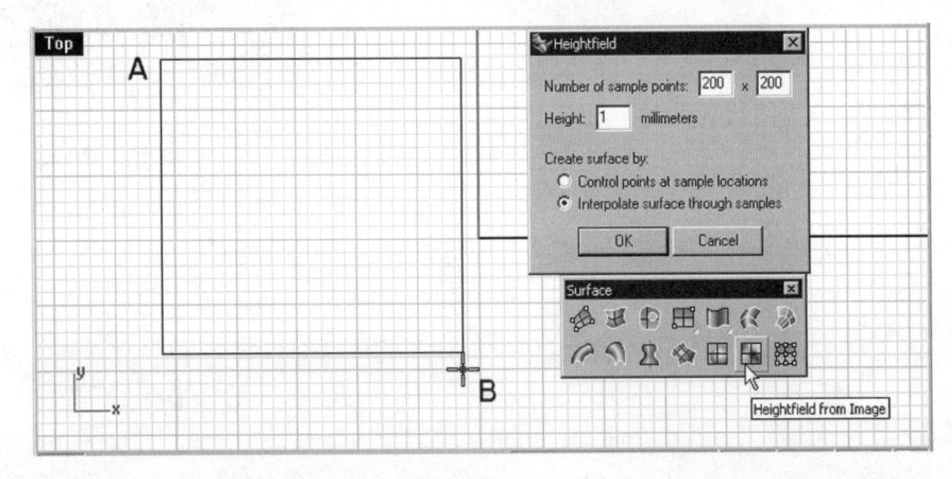

Fig. 5-62. Heightfield surface being constructed.

6 In the Heightfield dialog box, set the number of sample points to 200 times 200, set Height to 1, select *Interpolate surface through samples,* and then click on the OK button.

A surface is constructed from the bitmap image, as shown in figure 5-63. The surface is complete.

7 If you want to save your file, save it as *Heightfield.3dm*.

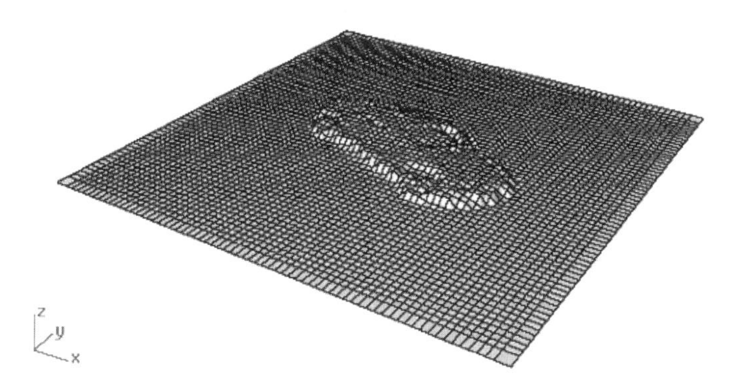

Fig. 5-63. Heightfield surface.

Surface from a Point Matrix

The point matrix used to construct a surface must be a regular rectangle. To construct this type of surface, perform the following steps.

1. Select File > Open.
2. Select the file *PointGrid.3dm* from the *Chapter 5* folder on the companion CD-ROM.
3. Check the Point box on the Osnap dialog box.
4. Select Surface > Point Grid, or click on the *Surface From Point Grid/Surface From Control Point Grid* button on the Surface toolbar.
5. Type *4* at the command line interface twice to specify the number of points in rows and columns.
6. Select points A1, A2, A3, A4, B1, B2, B3, B4, C1, C2, C3, C4, D1, D2, D3, and D4, indicated in figure 5-64.

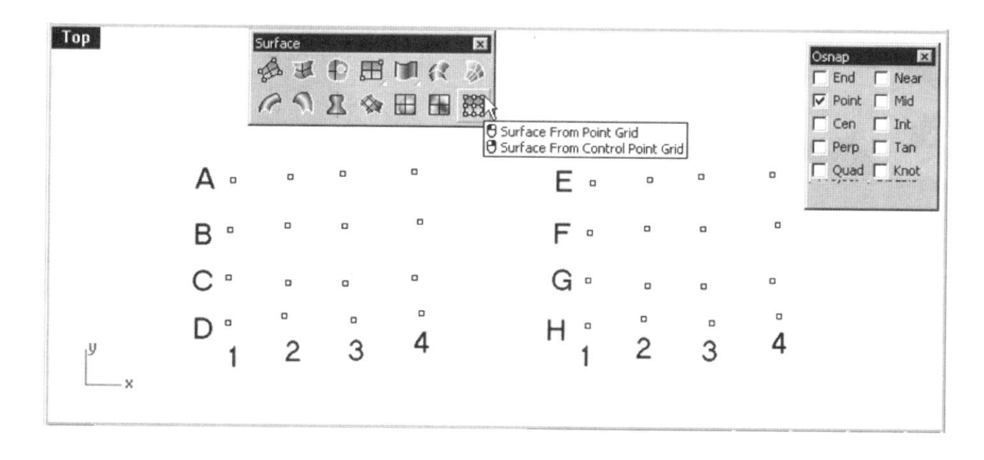

Fig. 5-64. Point grid surfaces being constructed.

A surface passing through the point matrix is constructed.

7 Right-click on the *Surface From Point Grid/Surface From Control Point Grid* button on the Surface toolbar.
8 Press the Enter key twice to accept the number of rows and columns, which should be 4 for both.
9 Select points E1, E2, E3, E4, F1, F2, F3, F3, G1, G2, G3, G4, H1, H2, H3, and H4, indicated in figure 5-64.

A surface using the selected point matrix as control points is constructed, as shown in figure 5-65.

10 Set the display to a four-viewport configuration.

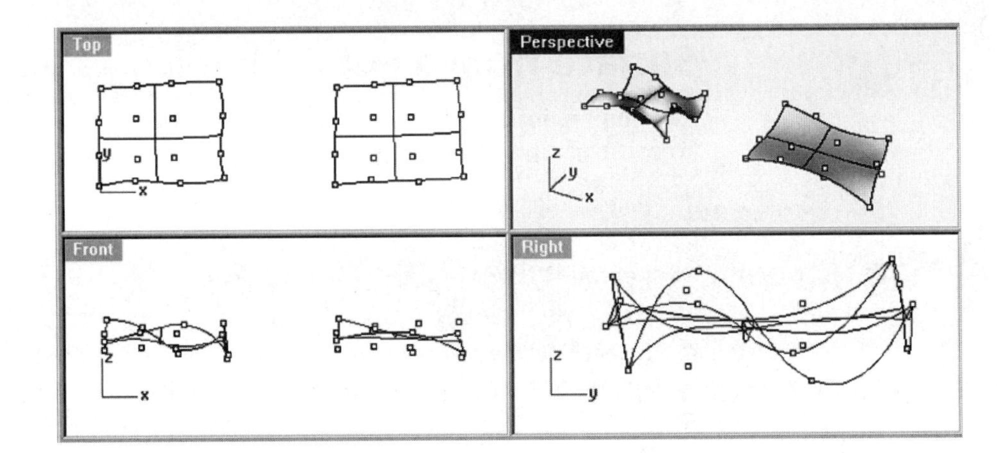

Fig. 5-65. Point grid surface.

Constructing Derived Surfaces

Derived surfaces, as the name implies, are surfaces derived from existing objects. To construct a derived surface, you do not need to construct any point objects or curves. Instead, you use existing objects to derive new surfaces. Extended, filleted, chamfered, offset, and blended derived surfaces are explored in the sections that follow. Figure 5-66 shows the derived surface commands found on the Surface Tools toolbar.

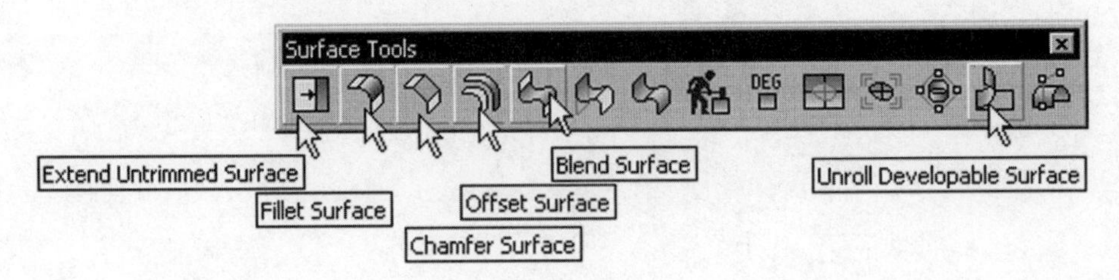

Fig. 5-66. Surface Tools toolbar commands.

Table 5-3 outlines the functions of derived surface commands and their locations in the main pull-down menu and respective toolbars.

Table 5-3: Derived Surface Commands and Their Functions

Toolbar Button	From Main Pull-down Menu	Function
Extend Untrimmed Surface	Surface > Extend Surface	Extending the untrimmed edge of a surface
Fillet Surface	Surface > Fillet Surface	Constructing a filleted surface at the intersection of two surfaces
Chamfer Surface	Surface > Chamfer Surface	Constructing a chamfer surface at the intersection of two surfaces
Offset Surface	Surface > Offset Surface	Deriving surfaces offset from selected surfaces
Blend Surface	Surface > Blend Surface	Constructing a blended surface connecting two sets of surface edges
Unroll Developable Surface	Surface > Unroll Developable Srf	Constructing a flat pattern of a developable surface

Unrolling Developable Surfaces

Before you make a sheet metal object, you need a development (flat) pattern of the object as a 2D sheet. You then roll or fold the 2D sheet to create the 3D object. An object constructed this way can be unrolled into a flat sheet. A cylindrical surface, for example, can be unrolled to a rectangle. To unroll a developable surface, perform the following steps.

1. Start a new file. Use the metric (Millimeters) template.
2. Construct a free-form curve in accordance with figure 5-67, and then construct an extruded surface from the curve.
3. Select Surface > Unroll Developable Srf, or click on the Unroll Developable Surface button on the Surface Tools toolbar.
4. Select surface A and then press the Enter key.

The surface is unrolled, as shown in figure 5-68.

5. If you want to save your file, save it as *UnrollSurface.3dm*.

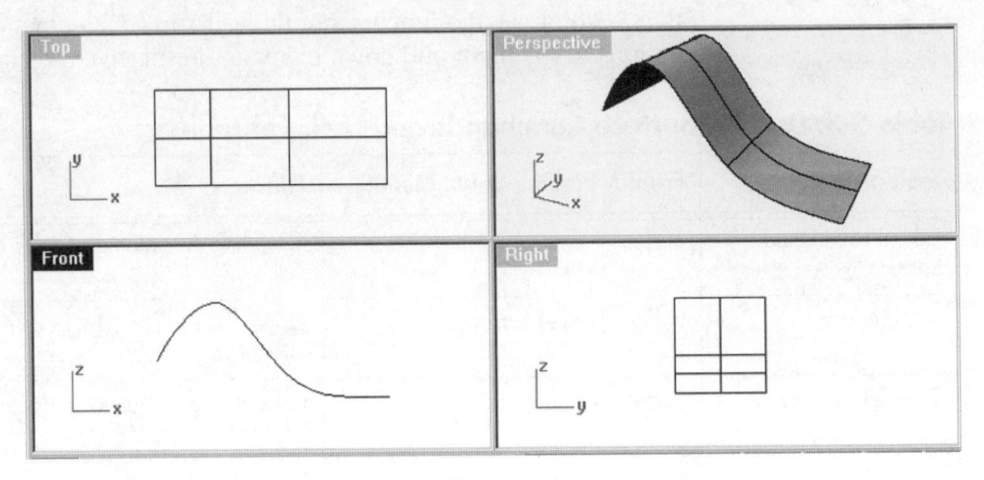

Fig. 5-67. Curve constructed and extruded.

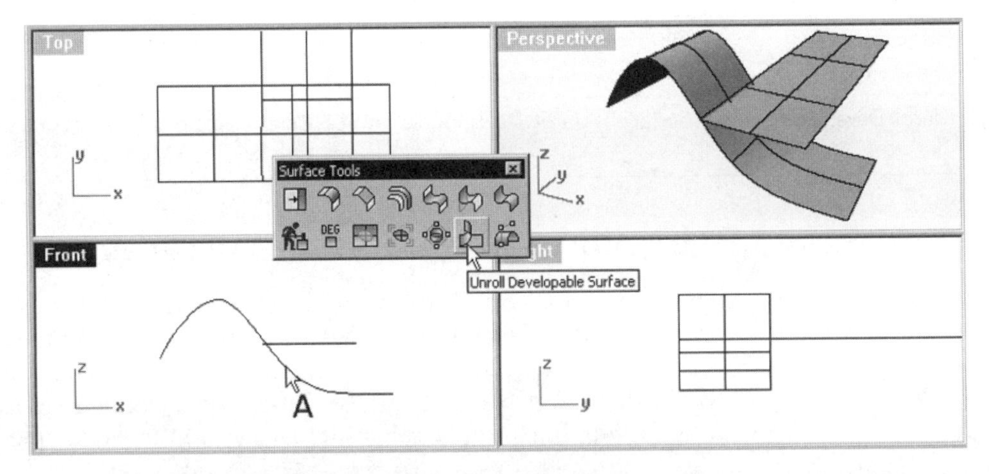

Fig. 5-68. Surface unrolled.

Extended, Filleted, Chamfered, Offset, and Blended Derived Surfaces

In the following, you will construct a set of surfaces. Based on these surfaces, you will derive extended, filleted, chamfered, offset, and blended surfaces.

1. Start a new file. Use the metric (Millimeters) template.
2. Maximize the Front viewport.
3. Construct two free-form curves in accordance with figure 5-69, and then construct two extruded surfaces from the curves.
4. If you want to save your file, save it as *DerivedSurfaces.3dm*.

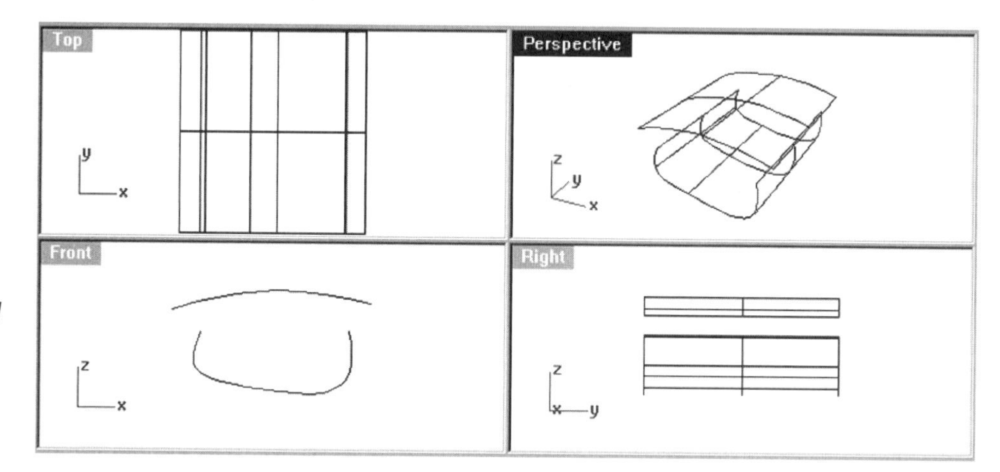

Fig. 5-69. Free-form curves and extruded surfaces.

Extend

You can enlarge a surface by extending untrimmed edges, as follows.

1. Maximize the Perspective viewport.
2. Select Surface > Extend Surface, or click on the Extend Untrimmed Surface button on the Surface Tools toolbar.
3. Type *T* at the command line interface to change the extend mode to Smooth, if the default is Line.
4. Select edge A (indicated in figure 5-70).

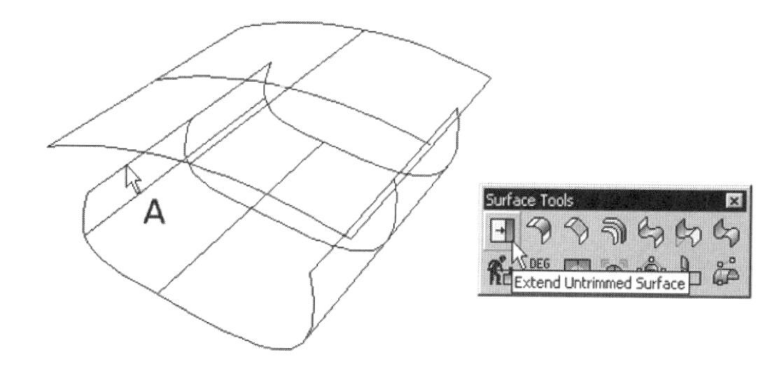

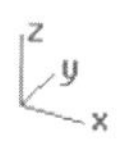

Fig. 5-70. Surface being extended.

5. Type *20* at the command line interface to set the extension factor.
6. Select Surface > Extend Surface, or click on the Extend Untrimmed Surface button on the Surface Tools toolbar.
7. Select edge A (indicated in figure 5-71).

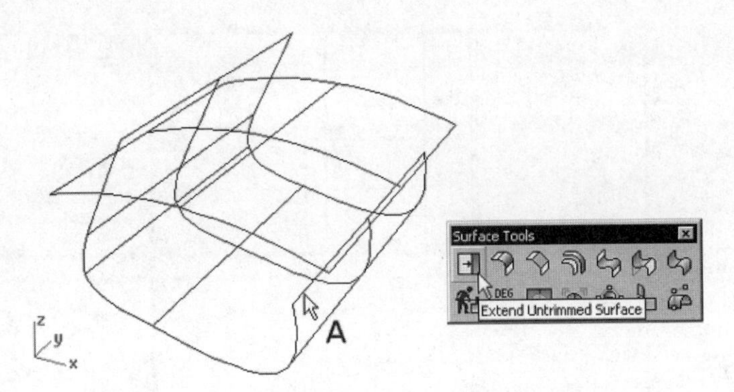

Fig. 5-71. Second edge being extended.

8 Type *25* at the command line interface to set the extension factor.

Two untrimmed edges of a surface are extended.

Fillet

A filleted surface has an arc-shaped cross section. It derives from two intersecting surfaces. For any pair of intersecting surfaces there are four possible filleted surfaces. The location at which you select the original surface determines the location of the filleted surface. To derive a filleted surface, perform the following steps.

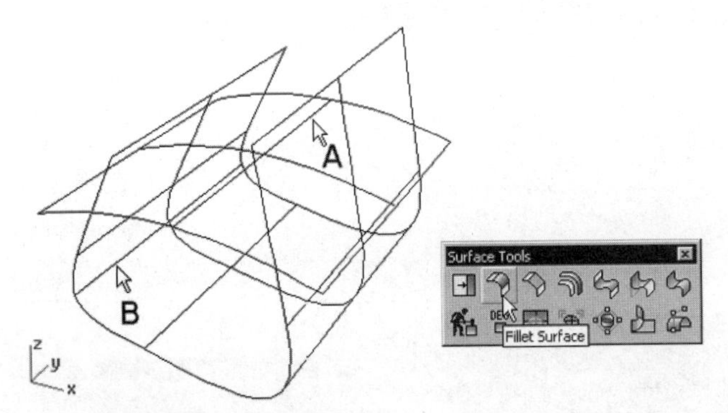

Fig. 5-72. Surface being filleted.

1. Select Surface > Fillet Surface, or click on the Fillet Surface button on the Surface Tools toolbar.
2. Type R at the command line interface.
3. Type 7 at the command line interface.
4. Select surfaces A and B (indicated in figure 5-72).

Note that you have the option of not trimming the surfaces.

Chamfer

A chamfered surface also derives from two intersecting surfaces. It forms a beveled edge between two surfaces. Similar to filleting, there are four possible chamfered surfaces on the intersecting surfaces. To create a chamfered surface, perform the following steps.

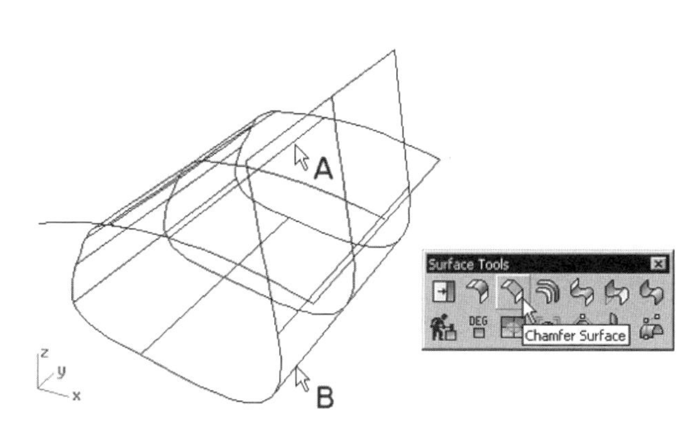

Fig. 5-73. Chamfered surface being constructed.

1. Select Surface > Chamfer Surface, or click on the Chamfer Surface button on the Surface Tools toolbar.
2. Type *D* at the command line interface.
3. Type *4* to set the first chamfer distance to 4 units. The first chamfer distance applies to the first selected surface.
4. Type *6* to set the second chamfer distance.
5. Select surfaces A and B (indicated in figure 5-73).

Offset

An offset surface is a surface that offsets from an existing surface. Every point on the offset surface is equal in distance from the original surface. To create an offset surface, perform the following steps.

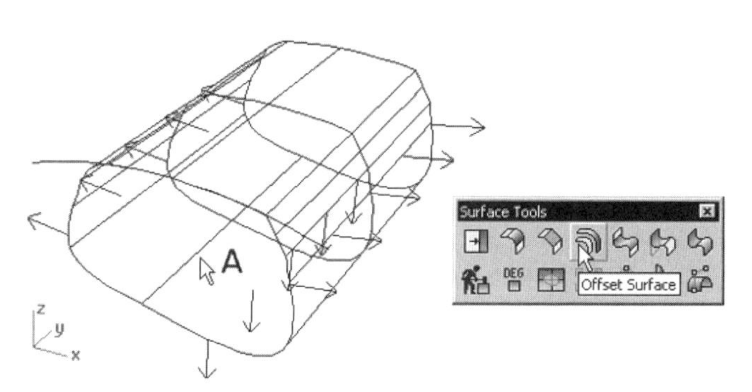

Fig. 5-74. Surface being offset.

1. Select Surface > Offset Surface, or click on the Offset Surface button on the Surface Tools toolbar.
2. Select surface A (indicated in figure 5-74). If you do not find the tracking lines indicating the direction of normal of the surface, set the color of the Tracking Lines option on the Color tab of the Rhino Options dialog box to black.
3. If the arrow direction is not the same as that indicated in figure 5-74, type *F* at the command line interface. (Note: You can also click on the surface to flip the normal.)
4. Type *4* to specify the offset distance.
5. Repeat steps 1 through 4 to construct another offset surface, shown in figure 5-75.
6. Select File > Incremental Save.

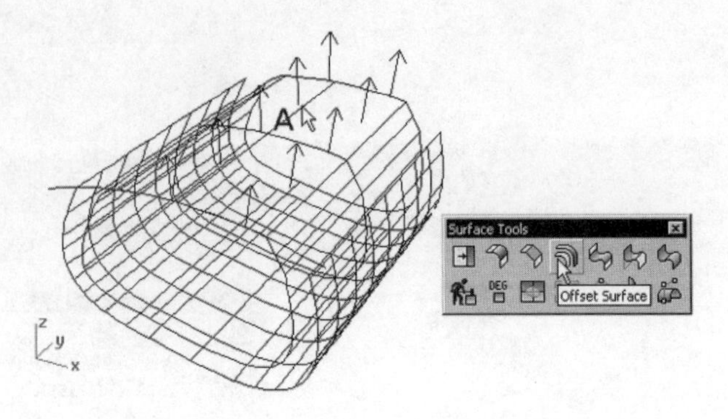

Fig. 5-75.
Second offset surface being constructed.

Blend

A blended surface joins two nonintersecting surfaces. It is a smooth transition between the two original surfaces. To construct a blended surface, perform the following steps.

1. Select Surface > Blend Surface, or click on the Blend Surface button on the Surface Tools toolbar.
2. Select edges A and B (indicated in figure 5-76) and press the Enter key.
3. In the Adjust Blend Bulge dialog box, click on the OK button.

A blend surface is constructed, as shown in figure 5-77.

4. If you want to save your file, save it as *DerivedSurfaces.3dm.*

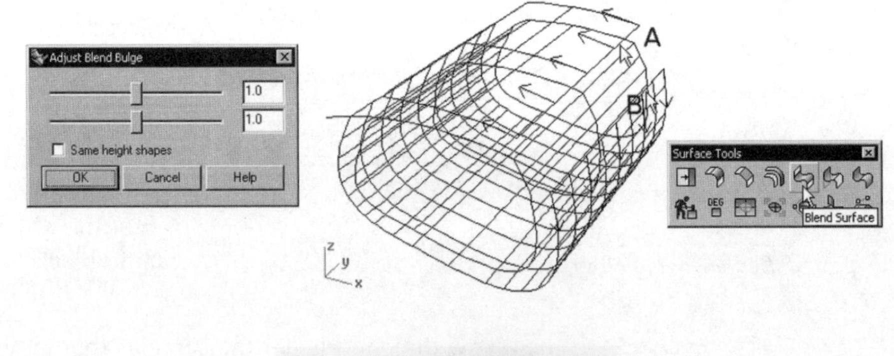

Fig. 5-76.
Blended surface being constructed.

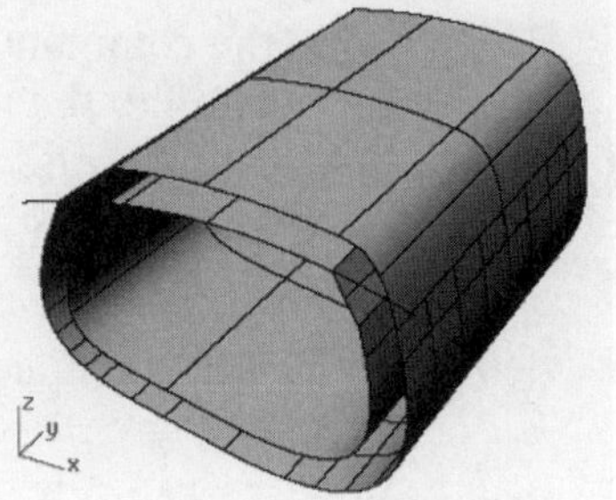

Fig. 5-77.
Blended surface.

Continuity of Contiguous Surfaces

Chamfered, filleted, and blended surfaces have one thing in common: they bridge two surfaces. Chamfered and filleted surfaces bridge two intersecting surfaces, and the blended surface bridges two nonintersecting surfaces. At the "joint" between the surfaces, three types of surface continuity are manifested, which are discussed in the sections that follow.

Positional Continuity (G0 Continuity)

In a G0 (positional) continuity joint, the control points at the edges of contiguous surfaces coincide. The chamfered surface edge A indicated in figure 5-78 has G0 continuity.

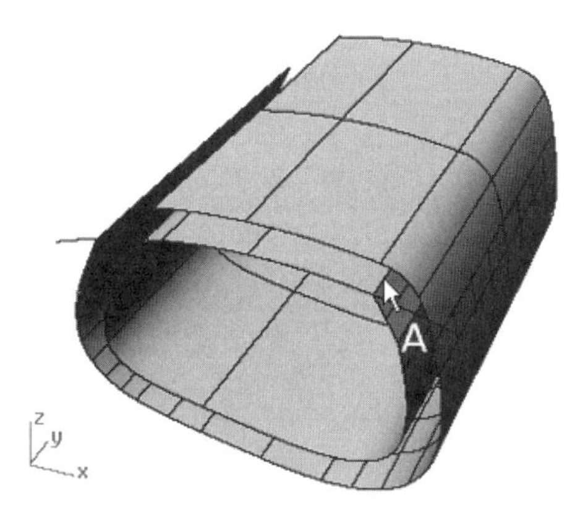

Fig. 5-78. G0 (positional) continuity.

Tangent Continuity (G1 Continuity)

In a G1 (tangent) continuity joint, the tangent directions of the control points at the edges of the contiguous surfaces are the same. The filleted surface edges A and B indicated in figure 5-79 have G1 continuity.

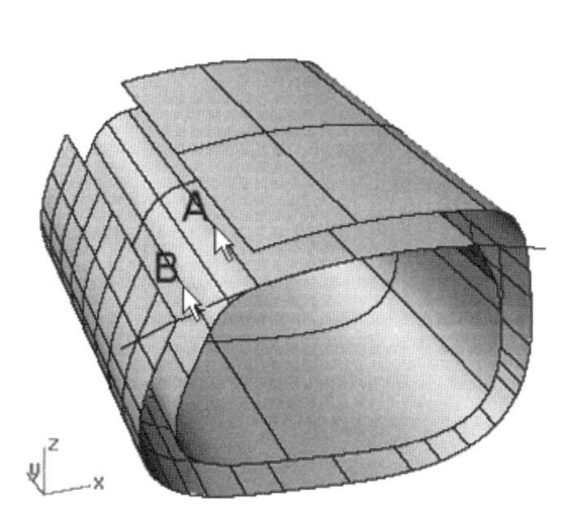

Fig. 5-79. G1 (tangent) continuity.

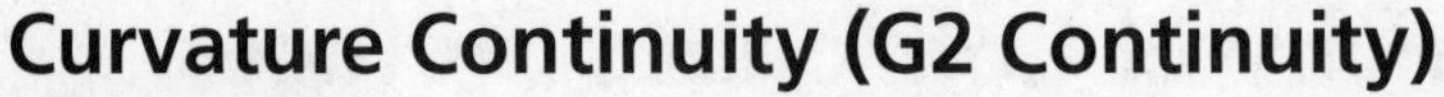

Curvature Continuity (G2 Continuity)

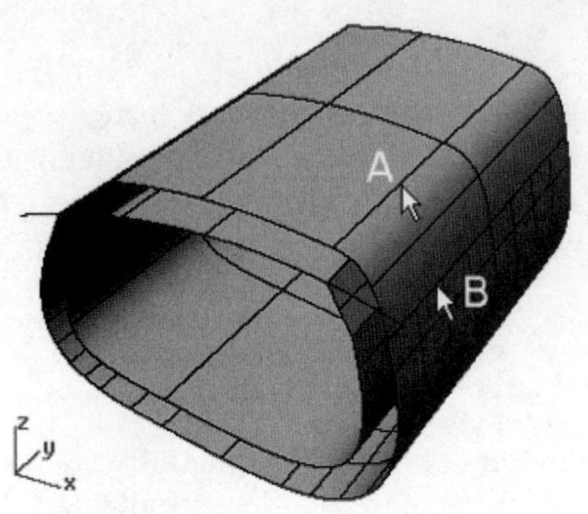

In a G2 (curvature) continuity joint, the curvature and tangent direction of the control points at the edges of contiguous surfaces are the same. The blended surface edges A and B indicated in figure 5-80 have G2 continuity.

Fig. 5-80. G2 (curvature) continuity.

Summary

A surface model is a set of 3D surfaces joined in 3D space to represent a 3D object. Each surface of a surface model is a mathematical expression depicting the profile and silhouette of the surface. There are two basic methods of representing a surface in the computer: using planar polygon meshes to approximate a surface and using complex spline surfaces to exactly represent a surface. Using Rhino, you construct and manipulate both polygon meshes and accurate surfaces.

A polygon mesh approximates a smooth surface by employing a set of small planar polygonal faces. Accuracy of representation is inversely proportional to the size of the polygon. The smaller the polygon size the more accurate the surface. One major disadvantage of the polygon mesh is that file size increases tremendously with a decrease in polygon size. Another disadvantage is that a surface can never be accurately represented with a mesh, no matter how small the polygon size. Hence, the polygon mesh is only useful in visualization and is not appropriate for downstream computerized manufacturing operations.

A better way to represent a continuously smooth surface is to use spline mathematics. In most contemporary computer-aided design applications, NURBS surface mathematics is used. From points and curves, you can construct planar surfaces and basic free-form surfaces. From existing objects, you can construct derived surfaces.

Basic free-form surfaces are those you use most often in constructing 3D surface models. This category of surface includes the following: extruded surfaces, revolved surfaces, rail revolved surfaces, surfaces derived from edge curves, lofted surfaces, one-rail swept surfaces, two-rail swept surfaces, curve network surfaces, patch surfaces, draped surfaces, surfaces derived from bitmaps, and point grid surfaces. To construct free-form surfaces, you define a set of curves and/or points. Creating free-form surfaces is simple after the curves and points are constructed. However, making the curves and points can be a tedious job.

Regarding derived surfaces, you can use the Extend tool to enlarge an existing surface, use the Chamfer tool to bevel the edges between two surfaces, use the Fillet tool to construct rounded edges, use the Offset tool to construct a surface with a uniform distance from an existing surface, and use the Blend tool to construct a surface connecting two surfaces. In a surface model consisting of a number of surfaces, continuity of contiguous surfaces is classified into three categories: G0 (positional), G1 (tangent), and G2 (curvature).

Review Questions

1. Explain the concept of surface modeling.
2. Differentiate the NURBS surface and the polygon mesh.
3. Outline the advantages and limitations of surface modeling.
4. List NURBS planar surfaces and basic free-form surfaces.
5. Use simple sketches to illustrate the types of points and/or curves required to construct various types of basic free-form surfaces.
6. List of the various derived surfaces.
7. What are the three types of continuity between contiguous surfaces?

CHAPTER 6

Surface Modeling: Part 2

Introduction

This chapter explores editing and transforming surfaces, as well as methods of analyzing surfaces. This chapter also introduces the use of polygon meshes.

Objectives

After studying this chapter you should be able to:

- ❐ Edit surfaces
- ❐ Transform surfaces
- ❐ Use surface analysis tools
- ❐ Use polygon meshes

Overview

To compose a set of surfaces to form a surface model, you may have to perform editing and transforming operations such as those you have learned in dealing with curves. To evaluate a surface, you use analysis tools. As mentioned in Chapter 5, there are two ways to represent a surface in the computer: using NURBS surfaces to exactly represent surfaces or using polygon meshes to approximate surfaces. To cope with other upstream and downstream computer applications that use polygon meshes to represent surfaces, you need to know how to manipulate polygon meshes.

This chapter is a continuation of Chapter 5. Here you will learn how to modify surfaces in several ways, transform the shapes of surfaces, and analyze surfaces you have constructed. You will also work on polygon meshes.

Surface Editing Tools

Similar to editing curves, there are three categories of surface editing, as follows. These types of surface editing are discussed in the sections that follow.

- Edit by joining, exploding joined surfaces, trimming, and splitting
- Edit by manipulating points of a surface
- More advanced editing

Joining, Exploding, Trimming, and Splitting

You learned how to use the Join, Explode, Trim, and Split commands in Chapter 3 (see figure 3-17 and Table 3-2). These commands, explored in the sections that follow, also apply to surfaces.

Joining and Exploding

Joining two or more contiguous surfaces, you get a polysurface. Note that the surfaces will not join unless they meet edge to edge within tolerance. (Later in this chapter, you will learn how to merge edges before joining.) Contrary to joining, exploding a polysurface renders a set of individual surfaces. Perform the following steps.

1. Open the file *JoinExplode.3dm* from the *Chapter 6* folder on the companion CD-ROM.
2. Select Edit > Join, or click on the Join button on the Main1 toolbar.
3. Select surfaces A, B, C, and D (indicated in figure 6-1) and press the Enter key.

The surfaces are joined to become a polysurface. Continue with the following steps to explode the polysurface to revert it into a set of individual surfaces.

4. Select Edit > Explode, or click on the *Explode/Extract Surfaces* button on the Main2 toolbar.
5. Select polysurface A (indicated in figure 6-1), or any part of the polysurface, and press the Enter key.

The polysurface is exploded. In Chapter 7 you will learn how to extract individual surfaces from a polysurface without exploding it.

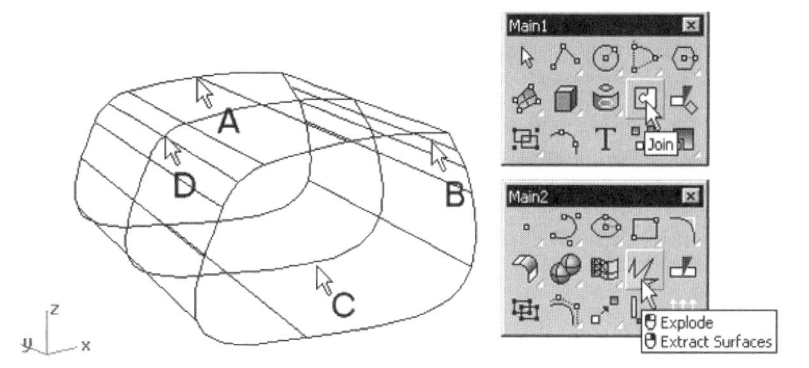

Fig. 6-1. Surfaces.

Trimming and Splitting Surfaces

To produce a smooth surface, it is necessary to use smooth defining wires and smooth boundary lines. However, most of the surfaces you will use to compose a design do not necessarily have smooth boundaries, although they have smooth profiles. To construct a smooth surface with an irregular boundary, you construct a large, smooth surface and then trim the smooth surface with a cutting object. The resulting surface is a smooth, trimmed surface.

The splitting operation is very similar to trimming in that you need a cutting object. The difference is that the cutting object cuts the target object in two. Cutting objects can be curves or surfaces. However, it is important to know that you can use curves for trimming or splitting surfaces, but not polysurfaces. If you want to trim or split a polysurface using a curve, you have to first explode it into individual surfaces. Another point to note when using curves as cutting objects is that cutting is viewport dependent.

Trimming

To trim a surface, perform the following steps.

1. Open the file *TrimSurface1.3dm* from the *Chapter 6* folder on the companion CD-ROM.
2. Select Edit > Trim, or click on the *Trim/Untrim Surface* button on the main toolbar.
3. Select surfaces A and B (indicated in figure 6-2) as cutting objects and press the Enter key.
4. Select locations D and E (shown in figure 6-2) to indicate the portion to be trimmed away, and then press the Enter key.

You will find that both surfaces are not trimmed, owing to the fact that both A and B (the cutting objects) are not wide enough to pass through the objects to be trimmed. Continue with the following steps.

5 Select Edit > Trim, or click on the *Trim/Untrim Surface* button on the main toolbar.

6 Click on the Top viewport to make it the active viewport.

7 Select surfaces B and C (indicated in figure 6-2) as cutting objects and press the Enter key.

8 Select locations F and G (shown in figure 6-2) to indicate the portion to be trimmed away, and then press the Enter key.

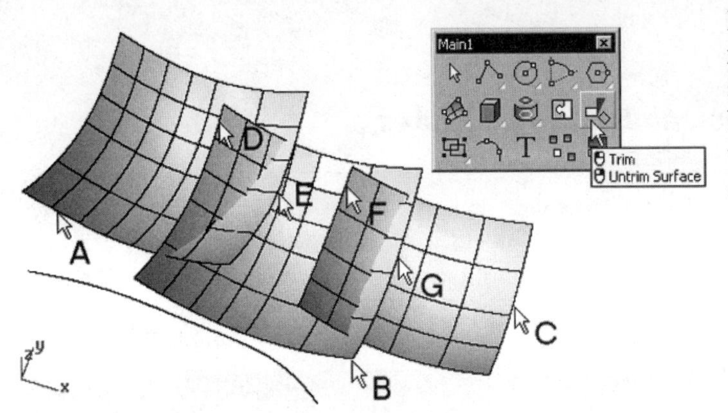

Fig. 6-2. Surfaces being trimmed.

Note that surface C (the narrower surface) is trimmed by surface B (the wider surface), but not vice versa. Continue with the following steps to use curves as cutting edges.

9 Double click on the Perspective viewport label to return to a four-viewport display.

10 Maximize the Top viewport.

11 Select Edit > Trim, or click on the *Trim/Untrim Surface* button on the main toolbar.

12 Select curves A and B (indicated in figure 6-3) as cutting objects and press the Enter key.

13 Select locations C and D (shown in figure 6-3) to indicate the portion to be trimmed away, and then press the Enter key.

Both surfaces are trimmed by the open-loop curves having end points extending outside the boundaries of the surfaces.

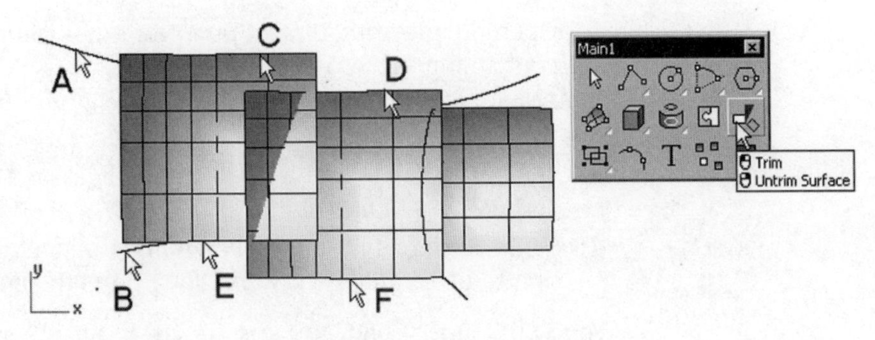

Fig. 6-3. Surfaces being trimmed by curves.

Continue with the following steps to trim surfaces.

14 Maximize the Perspective viewport.

15 Select Edit > Trim, or click on the *Trim/Untrim Surface* button on the main toolbar.

16 Select surfaces A and B (indicated in figure 6-4) as cutting objects and press the Enter key.

17 Select locations C and D (shown in figure 6-4) to indicate the portion to be trimmed away, and then press the Enter key.

The surfaces are trimmed.

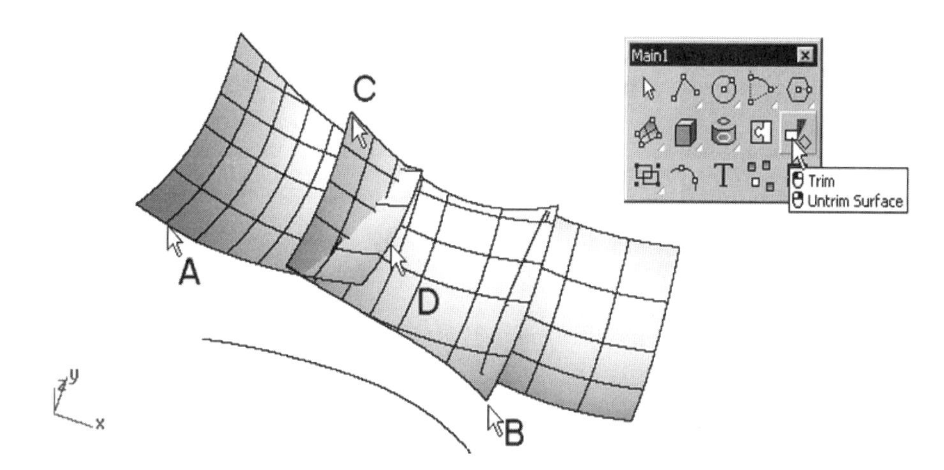

Fig. 6-4. Surfaces being trimmed.

Splitting

Perform the following steps to split a surface using another surface and a curve.

1 Open the file *SplitSurface.3dm* from the *Chapter 6* folder on the companion CD-ROM.

2 Select Edit > Split, or click on the Split button on the Main2 toolbar.

3 Select surface A (indicated in figure 6-5) as the object to be split, and then press the Enter key.

4 Select curve A and surface B (indicated in figure 6-5) as splitting objects, and then press the Enter key.

To realize that the surface is split, continue with the following.

5 Select Edit > Object Properties, or click on the Object Properties button on the Properties toolbar.

6 Select surface A (indicated in figure 6-6) and press the Enter key.

7 In the Properties dialog box, set the color to green and close the dialog box by clicking on the checkmark in the upper right-hand corner.

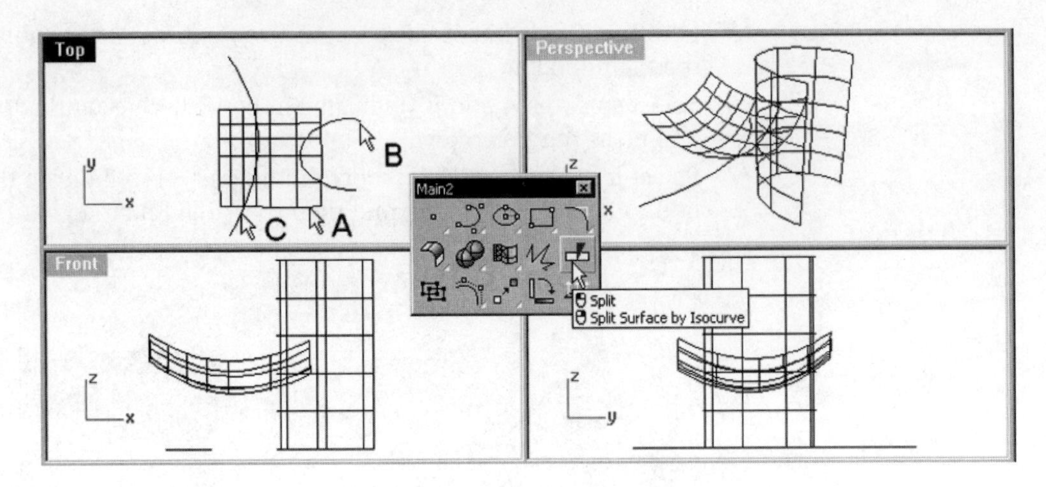

Fig. 6-5. Surfaces being split.

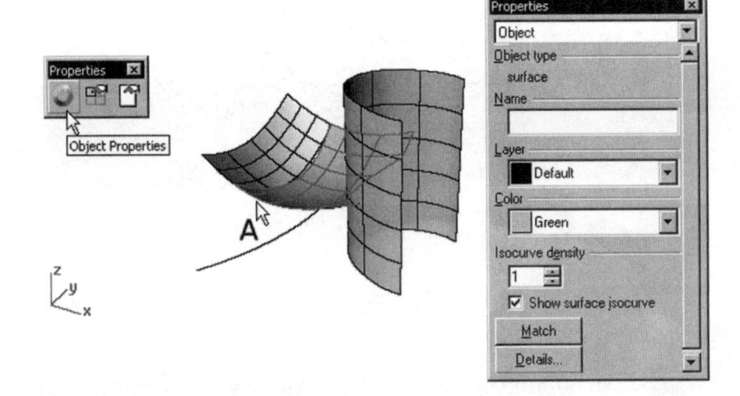

Fig. 6-6. Surface's color changed.

Trimming in Two Viewports

Perform the following steps to trim a surface in two different viewports.

1. Open the file *TrimSurface2.3dm* from the *Chapter 6* folder on the companion CD-ROM.
2. Select Edit > Trim, or click on the *Trim/Untrim Surface* button on the main toolbar.
3. Click Select curves A and B (indicated in figure 6-7) as cutting objects, and then press the Enter key.
4. Click on the Top viewport and select location C (indicated in figure 6-7).
5. Click on the Front viewport, select location D (indicated in figure 6-7), and press the Enter key.

The surfaces are trimmed, as shown in figure 6-8.

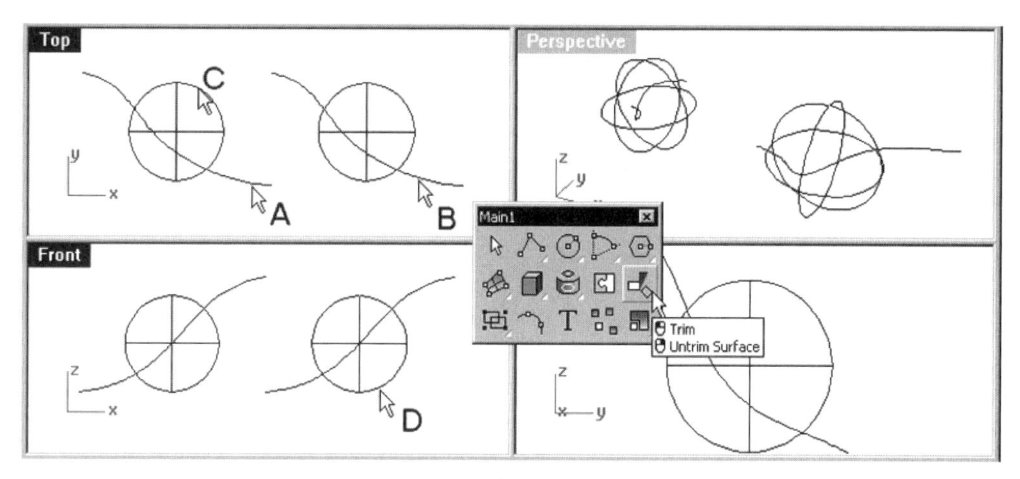

Fig. 6-7. Surfaces being trimmed by curves.

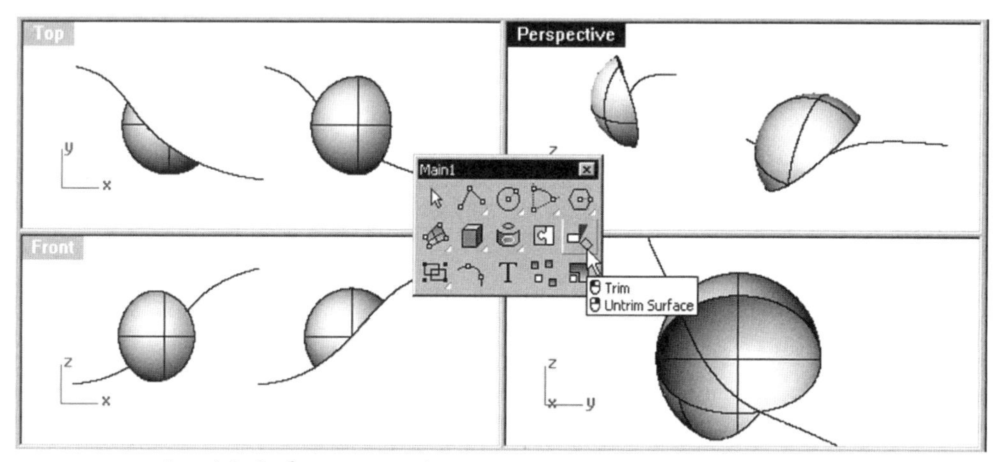

Fig. 6-8. Surfaces trimmed.

Later in this chapter, you will learn how to untrim a trimmed surface.

Point Editing

Similar to curve editing, you edit a surface by manipulating its control point location and weight, using a handlebar, and adding or removing knots. However, there is no edit point available on a surface, and there is no kink point because any kink point on a curve will result in multiple surfaces. Related menus, toolbars, and their descriptions are shown in figure 3-23 and outlined in Table 3-3 of Chapter 3.

Control Point Manipulation

One method of modifying the shape of a surface is to edit its control points. There are four ways to move a control point, as follow.

- Select the control point, hold down the mouse button, and drag it to a new position.
- Select the control point and use the nudge keys. (By default, the nudge keys are the Alt key plus the arrow keys.)
- Use the MoveUVN command.
- Use any transform command.

To modify a control point of a surface using the nudge keys, perform the following steps.

1. Open the file *ControlPoint.3dm* from the *Chapter 6* folder on the companion CD-ROM.
2. Select Edit > Control Points > Control Points On, or click on the *Control Points On/Points Off* button on the Point Editing toolbar.
3. Select surface A (indicated in figure 6-9) and press the Enter key.
4. Select control point B (indicated in figure 6-9) in the Top viewport.
5. Hold down the Alt key and press the Up arrow key.

The selected control point is moved.

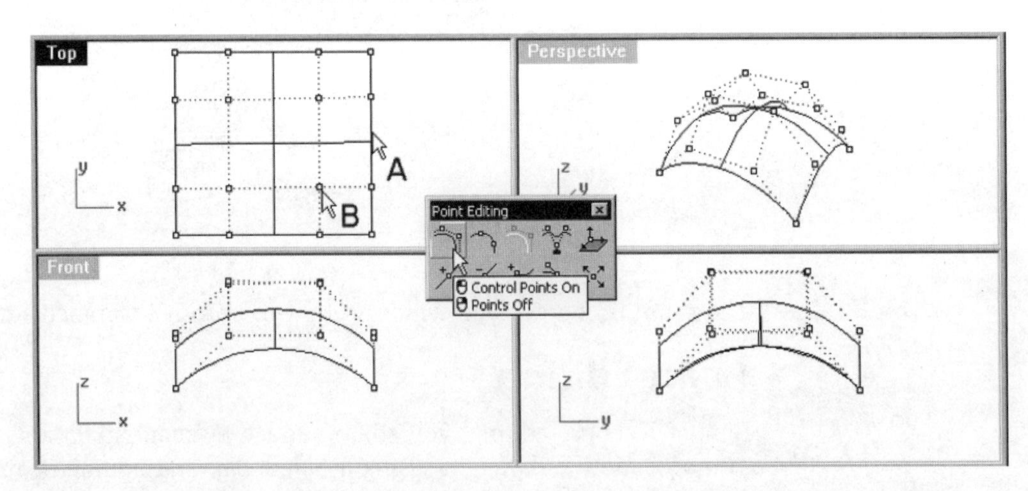

Fig. 6-9. Control points of a surface being modified.

Continue with the following to use the MoveUVN command to move the control point's location.

6. Maximize the Top viewport.

7 Select Transform > Move UVN, or click on the *MoveUVN/Turn MoveUVN Off* button on the Point Editing toolbar.
8 Select control point A (indicated in figure 6-10).
9 In the MoveUVN dialog box, drag the U, V, and N slider bar to move the control point along the U, V, and N directions.

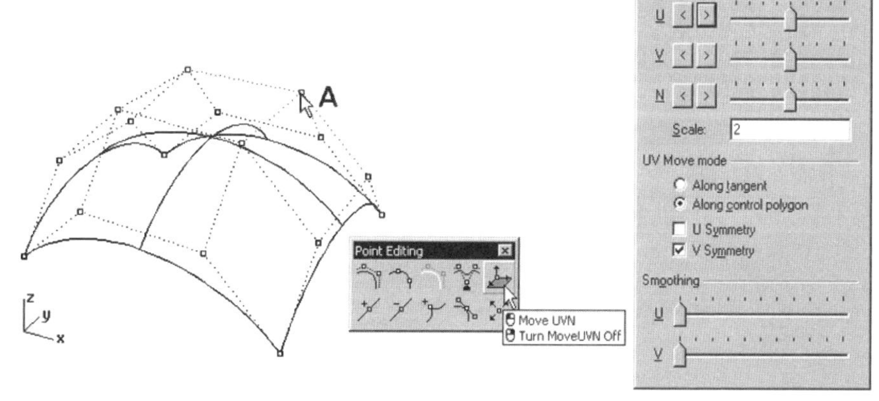

Fig. 6-10. Control point being moved along U, V, and N directions.

10 Continue with the following to change the control point's weight. Like the control points of a spline, the higher the weight the closer the surface will be pulled to the control point. Select Edit > Control Points > Edit Weight, or click on the Edit Control Point Weight button on the Point Editing toolbar.
11 Select control point A (indicated in figure 6-11) and press the Enter key.
12 In the Set Control Point Weight dialog box, drag and move the slider bar to increase or decrease the weight of the selected control point.

The surface is modified.

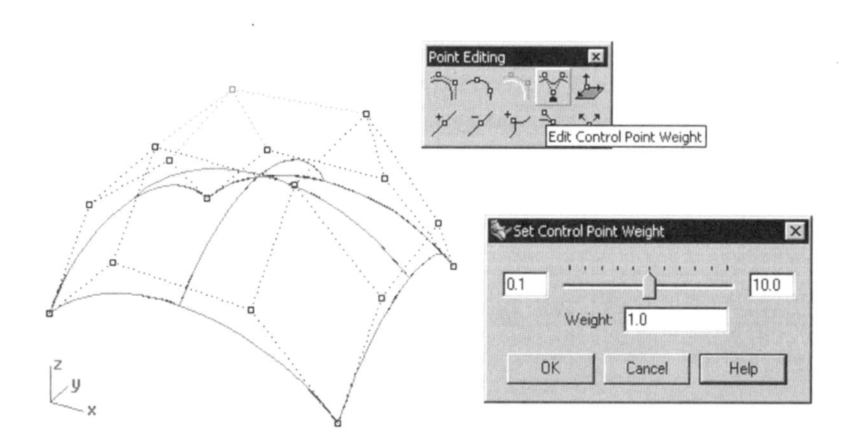

Fig. 6-11. Control point's weight being modified.

Handlebar Editor

The Handlebar editor enables you to modify the shape of a surface by adjusting the tangency of selected locations on the surface. To facilitate selection of specific locations on a surface, you may increase the isocurve density. To reiterate, isocurves are aids to help you visualize and select a surface. Changing the density does not affect the number of control points and the degree of polynomial. To edit a surface using the handlebar, perform the following steps.

1. Open the file *HandleBar.3dm* from the *Chapter 6* folder on the companion CD-ROM.
2. Select Edit > Object Properties, or click on the Object Properties button on the Properties toolbar.
3. Select surface A (indicated in figure 6-12).

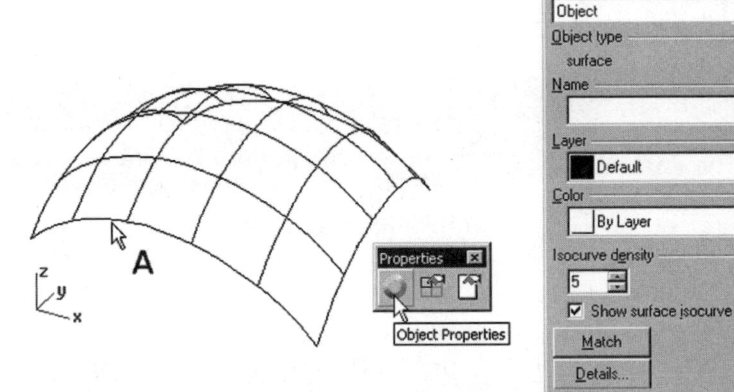

Fig. 6-12. Isocurve density being modified.

4. In the Properties dialog box, change the value of the isocurve density to 5, and then close the dialog box by clicking on the X in the upper right-hand corner of the dialog box.
5. Select Edit > Control Points > Handlebar Editor, or click on the Handlebar Editor button on the Point Editing toolbar.
6. Select location A (indicated in figure 6-13).

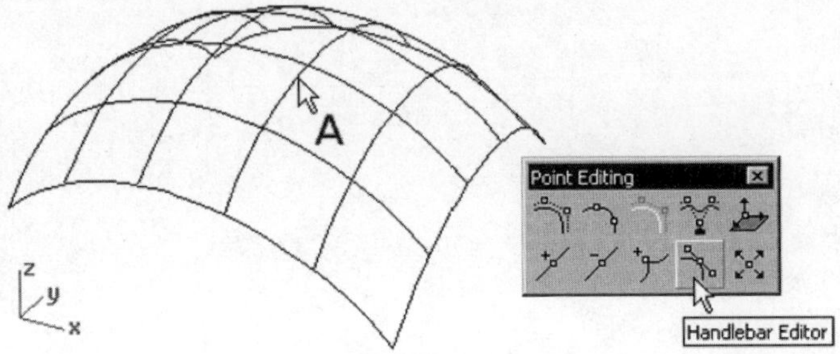

Fig. 6-13. Handlebar editor being activated at selected location of a surface.

7 Select and drag one of the five handles of the handlebar to modify the shape of the surface at the selected location, as shown in figure 6-14. Press the Enter key when finished.

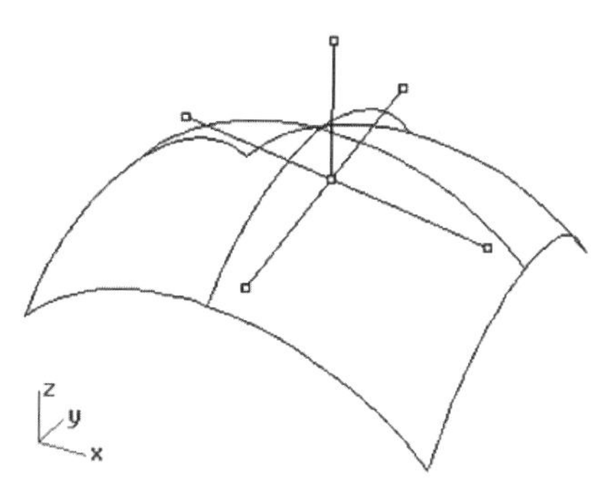

Fig. 6-14. Handlebar being manipulated.

Adding and Removing Knots

The number of control points of a surface has a significant effect on how the surface will change in shape if a control point is moved. Adding knots increases the number of control points. The surface profile will not change until the next time you manipulate its control points. In the following, you will add knots to the surface.

1. Open the file *Knot.3dm* from the *Chapter 6* folder on the companion CD-ROM.
2. Select Edit > Control Points > Control Points On, or click on the *Control Points On/Points Off* button on the Point Editing toolbar.
3. Select surface A (indicated in figure 6-15) and press the Enter key.
4. Select Edit > Control Points > Insert Knot, or click on the *Insert Knot/Insert Edit Point* button on the Point Editing toolbar.
5. Select surface A and then location B (indicated in figure 6-15) to add a control point.
6. Continue to add as many control points as may be required. Press the Enter key when finished.

The number of control points is increased. To remove a knot from the surface, continue with the following steps.

7. Select Edit > Control Points > Remove Knot, or click on the Remove Knot button on the Point Editing toolbar.
8. Select surface A and then control point B (indicated in figure 6-16).
9. Press the Enter key when finished.

The control points are removed.

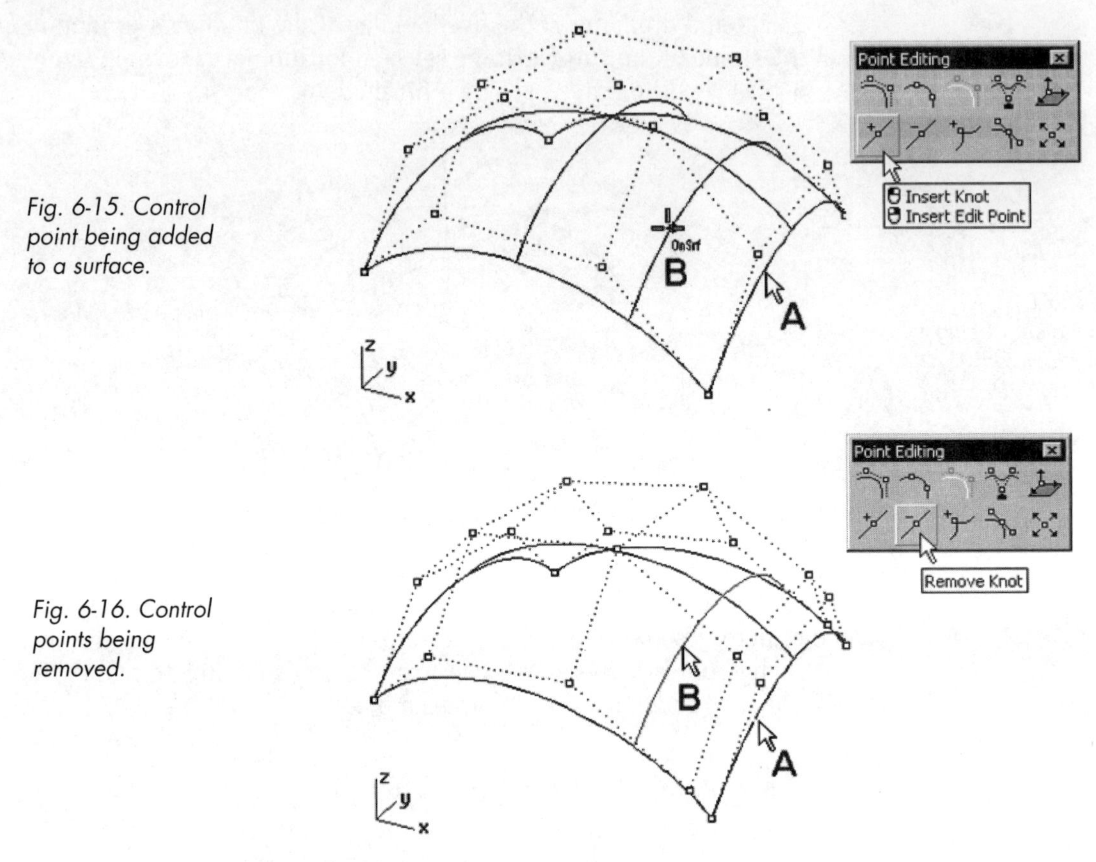

Fig. 6-15. Control point being added to a surface.

Fig. 6-16. Control points being removed.

Advanced Editing

The more advanced commands for editing a surface are indicated in the Surface Tools, Main2, and Edge Tools toolbars, shown in figure 6-17.

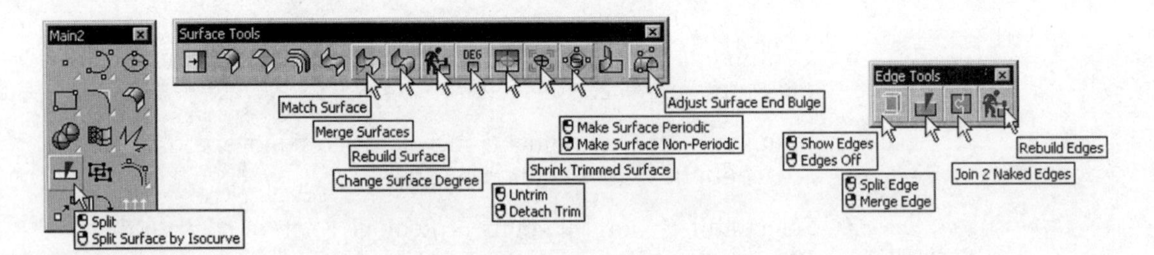

Fig. 6-17. Surface editing tools.

Table 6-1 outlines the functions of surface editing commands and their locations in the main pull-down menu and respective toolbars.

Table 6-1: Surface Editing Commands and Their Functions

Toolbar Button	From Main Pull-down Menu	Function
Match Surface	Surface > Surface Edit Tools > Match	Matching an untrimmed edge of a surface with an untrimmed surface of another surface by modifying the position, tangency, and/or curvature of the selected surfaces
Merge Surfaces	Surface > Surface Edit Tools > Merge	Merging and joining the untrimmed edges of two surfaces
Rebuild Surface	Edit > Rebuild	Reconstructing a selected surface with a new degree setting and number of control points in the U and V directions
Change Surface Degree	Edit > Change Degree	Modifying a surface by changing its degree setting
Untrim	Surface > Surface Edit Tools > Untrim	Removing a selected trimmed edge of a surface to restore it to its untrimmed state
Detach Trim	Surface > Surface Edit Tools > Detach Trim	Removing a selected trimmed edge of a surface and converting the trimmed edge to a curve
Split	Edit > Split	Splitting a surface into two surfaces
Split Surface by Isocurve	Surface > Surface Edit Tools > Split at Isocurve	Splitting a surface into multiple surfaces along an isocurve
Shrink Trimmed Surface	Surface > Surface Edit Tools > Shrink Trimmed Surface	Reducing the size of the underlying untrimmed surface of a trimmed surface
Adjust Surface End Bulge	Surface > Surface Edit Tools > Adjust End Bulge	Modifying a surface at the edge without changing the tangency direction
Make Surface Periodic	Surface > Surface Edit Tools > Make Periodic	Constructing a periodic surface from a surface
Make Surface Non-Periodic	—	Making a periodic surface become non-periodic
—	Surface > Surface Edit Tools > Divide Surfaces on Creases	Splitting a surface with creases or a polysurface into individual surfaces
Show Edges/Edges Off	Analyze > Edge Tools > Show Edges	Displaying edges of selected surfaces, including the edges of a closed-loop surface (such as a cylinder or sphere)
Split Edge	Analyze > Edge Tools > Split Edge	Splitting an edge in two
Merge Edge	Analyze > Edge Tools > Merge Edge	Merging split edges

Toolbar Button	From Main Pull-down Menu	Function
Join 2 Naked Edges	Analyze > Edge Tools > Join 2 Naked Edges	Joining edges of contiguous surfaces
Rebuild Edges	Analyze > Edge Tools > Rebuild Edges	Restoring the original edges of healed surfaces after naked edges are joined

Matching a Surface to Another Surface

Matching modifies the shape of an untrimmed edge of a surface for it to match the edge (either trimmed or untrimmed) of another surface to yield one of the three types of continuity (G0, G1, or G2) between the surfaces. Perform the following steps.

1. Open the file *EditSurfaces1.3dm* from the *Chapter 6* folder on the companion CD-ROM.
2. Select Surface > Surface Edit Tools > Match, or click on the Match Surface button on the Surface Tools toolbar.
3. Select edge A (indicated in figure 6-18) as the surface edge to be modified.
4. Select edge B (indicated in figure 6-18) as the target surface. Note that where you pick the surfaces is important in terms of eliminating twisted surfaces.
5. In the Match Surface dialog box, check the Position and Preserve opposite end check boxes and then click on the OK button.

A surface's edge is changed to match the other, as shown in figure 6-18.

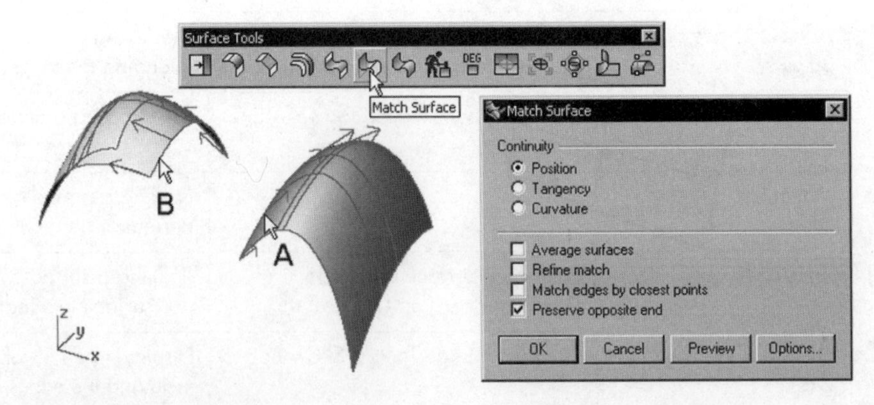

Fig. 6-18. Surface being modified to match the other surface.

Merging Surfaces

Contiguous surfaces sharing a common untrimmed edge can be merged to become a single surface. Note that unlike joining merging results in a single surface, not a polysurface. Therefore, you cannot explode it into the surfaces you merged. To merge surfaces, perform the following steps.

1. Select Surface > Surface Edit Tools > Merge, or click on the Merge Surfaces button on the Surface Tools toolbar.
2. Select surfaces A and B (indicated in figure 6-19) and press the Enter key.

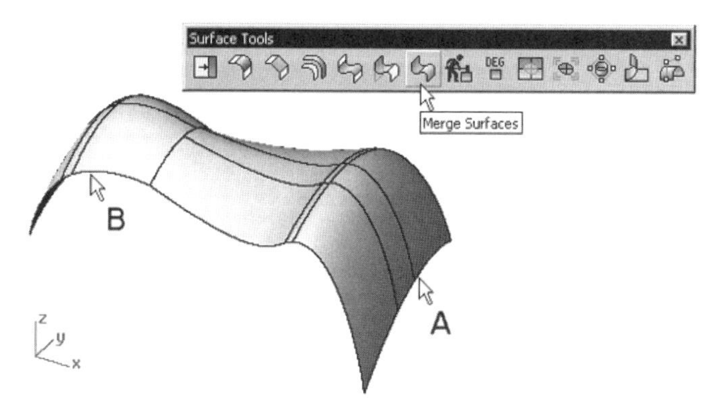

Fig. 6-19. Two surfaces being merged at their untrimmed edges.

The surfaces are merged into one, as shown in figure 6-19. If the surfaces failed to merge, repeat the command, type *T* at the command line interface, specify a larger joining tolerance value, and select the surfaces again.

Rebuilding

To improve the smoothness of a surface, by rebuilding you can change its number of control points and/or the degree of polynomials along the U and V directions of the surface. To rebuild a surface, perform the following steps.

1. Select Edit > Rebuild, or click on the Rebuild Surface button on the Surface Tools toolbar.
2. Select surface A (indicated in figure 6-20) and press the Enter key.
3. In the Rebuild Surface dialog box, set the U and V point count to 6 and degree of polynomial to 4, and then click on the OK button.

The surface is rebuilt, as shown in figure 6-20.

Changing Polynomial Degree

If you increase the degree of polynomial, the surface's shape will not change. However, if you reduce the degree of polynomial, the surface's shape is simplified. To change the polynomial degree of a surface along its U and V directions, perform the following steps.

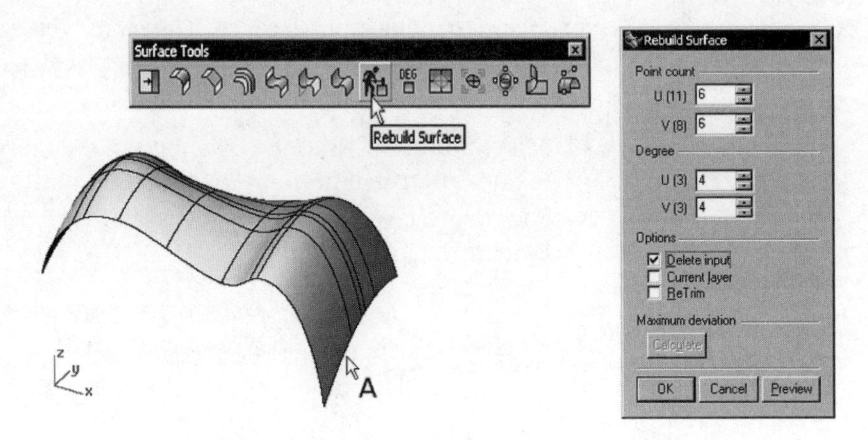

Fig. 6-20. Surface being rebuilt.

1. Select Edit > Change Degree, or click on the Change Surface Degree button on the Surface Tools toolbar.
2. Select surface A (indicated in figure 6-21) and press the Enter key.
3. Type *3* at the command line interface to set the U degree.
4. Type *3* at the command line interface to set the V degree.

The surface's degree of polynomial is changed, as shown in figure 6-21.

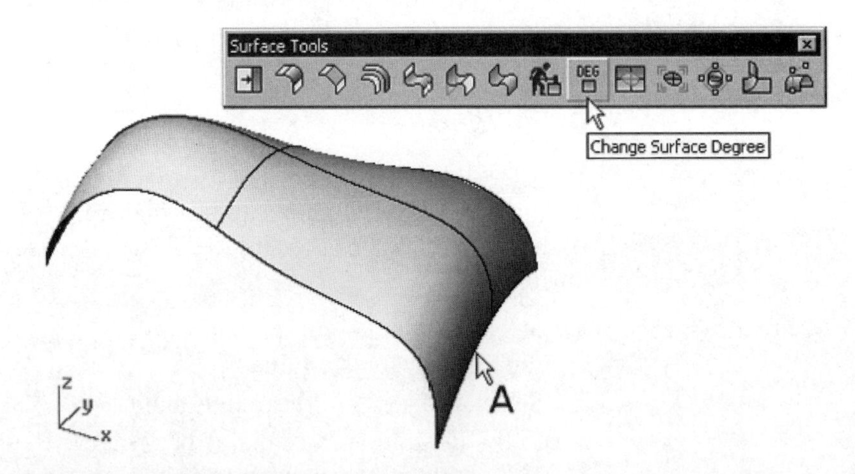

Fig. 6-21. Degree of polynomial being changed.

Adjusting End Bulge

Similar to editing the end bulge of a curve, the tangency direction of a surface's edge is not changed when adjusting end bulge. To adjust end bulge, perform the following steps.

1. Select Surface > Surface Edit Tools > Adjust End Bulge, or click on the Adjust Surface End Bulge button on the Surface Tools toolbar.

2 Select surface edge A (indicated in figure 6-22).

3 Select location B (indicated in figure 6-22).

4 Select control point C (indicated in figure 6-22) and press the Enter key.

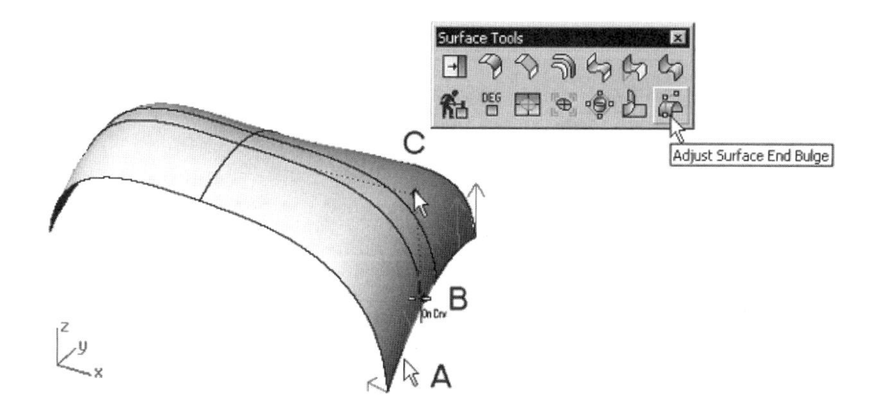

Fig. 6-22. Surface edge selected.

5 Drag the control point to location A (indicated in figure 6-23).

The surface is modified, as shown in figure 6-24.

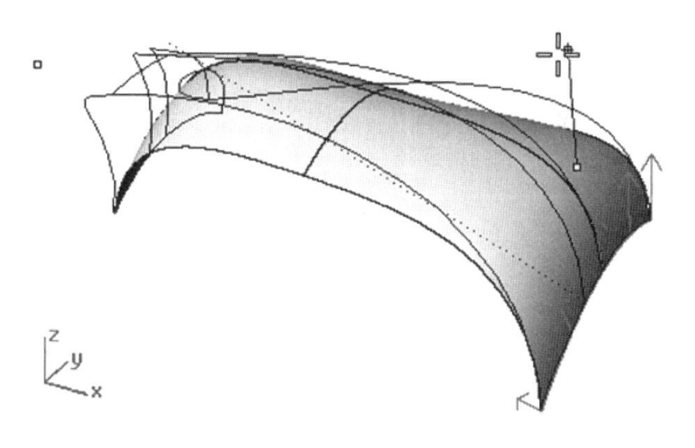

Fig. 6-23. Control point being manipulated.

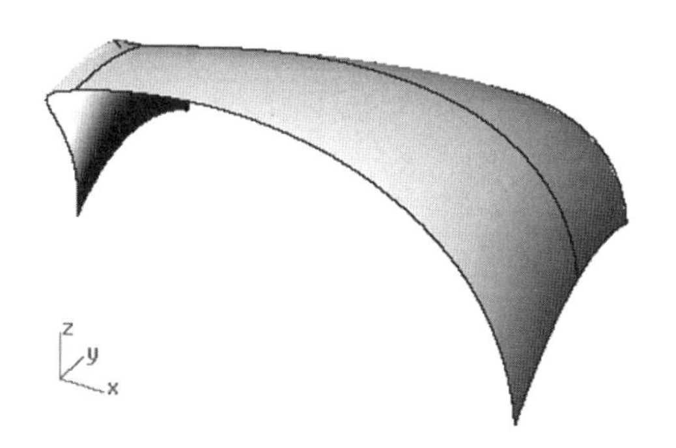

Fig. 6-24. Modified surface.

Making a Surface Periodic

A periodic surface is a smooth, closed-loop surface. To make a surface periodic, perform the following steps.

1. Open the file *EditSurfaces2.3dm* from the *Chapter 6* folder on the companion CD-ROM.

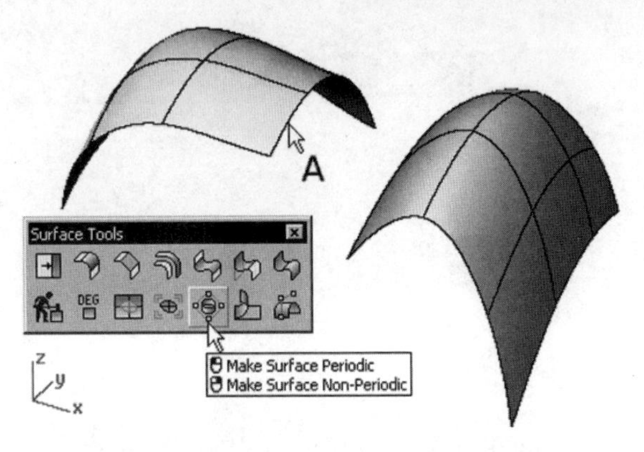

Fig. 6-25. Surface being changed to a periodic surface.

2. Select Surface > Surface Edit Tools > Make Periodic, or click on the *Make Surface Periodic/Make Surface Non-Periodic* button on the Surface Tools toolbar.
3. Select edge A (indicated in figure 6-25) and press the Enter key.
4. If the default is yes for delete input, press the Enter key. Otherwise, type *Y*.

The surface is changed to a periodic surface.

Splitting Along Isocurve and Divide Along Creases

To split a surface along an isocurve, perform the following steps.

1. Surface > Surface Edit Tools > Split at Isocurve, or right-click on the *Split/Split Surface by Isocurve* button on the Main2 toolbar.
2. Select surface A (indicated in figure 6-26) and press the Enter key.
3. Select location B (indicated in figure 6-26) and press the Enter key.

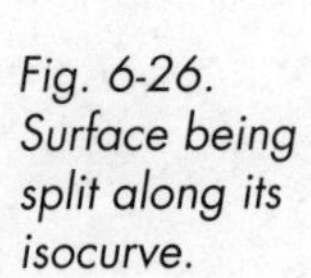

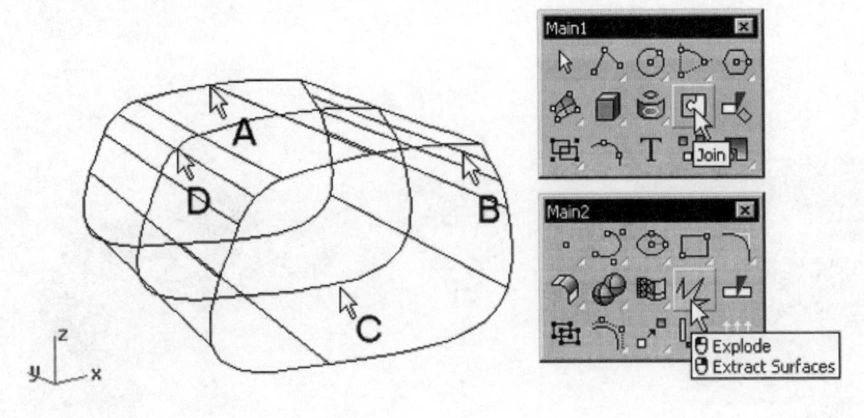

Fig. 6-26. Surface being split along its isocurve.

The surface is split.

If a surface or polysurface has creases, you may divide it into individual smooth surfaces by selecting Surface > Surface Edit Tools > Divide Surfaces on Creases.

Untrimming

Because the original untrimmed boundary of a surface is still retained in memory after the surface is trimmed, you can revert a trimmed surface to its original untrimmed state, as follows.

1. Open the file *EditSurfaces3.3dm* from the *Chapter 6* folder on the companion CD-ROM.
2. Select Surface > Surface Edit Tools > Untrim, or click on the *Untrim/Detach Trim* button on the Surface Tools toolbar.
3. Select edge A (indicated in figure 6-26).

The surface is untrimmed. Now undo the last command, as follows.

4. Select Edit > Undo.

Detaching a Trimmed Boundary

To untrim a trimmed surface and retain the trimmed boundary curve, perform the following steps.

1. Surface > Surface Edit Tools > Detach Trim, or right-click on the *Untrim/Detach Trim* button on the Surface Tools toolbar.
2. Select edge A (indicated in figure 6-27).

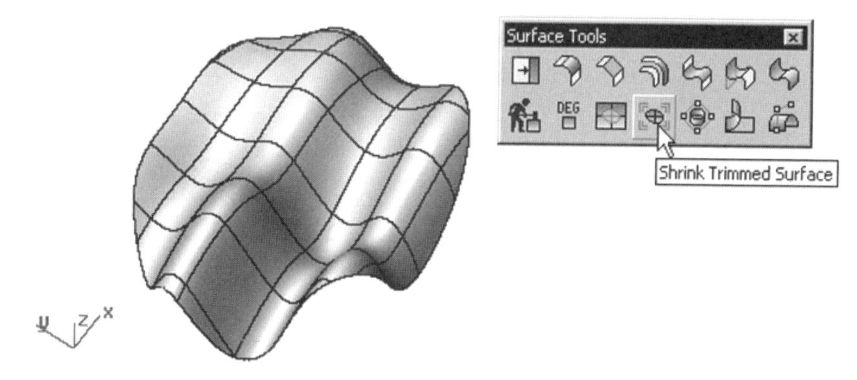

Fig. 6-27. Surface being untrimmed.

The surface is untrimmed and the trimmed boundary is retained, as shown in figure 6-28. Again, undo this command, as follows.

3. Select Edit > Undo.

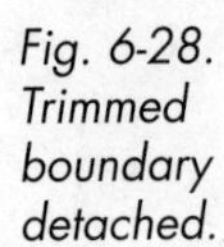

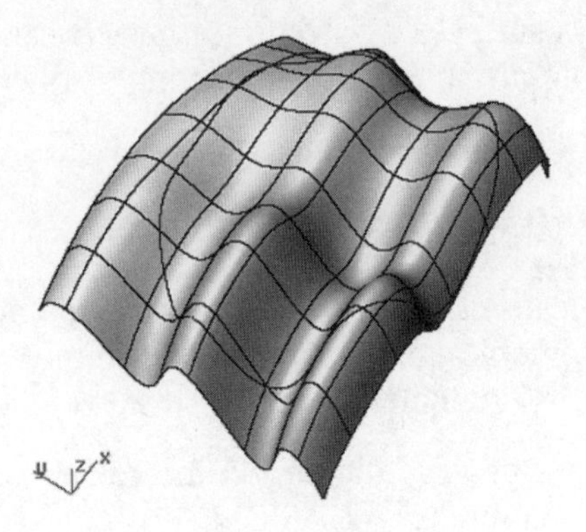

Fig. 6-28. Trimmed boundary detached.

Shrinking a Trimmed Surface

After you untrim or detach a trimmed boundary, the trimmed surface returns to its original shape because the original surface is kept in the database. To reduce the memory required to store a trimmed surface, you can shrink the trimmed surface, as follows.

1. Select Surface > Surface Edit Tools > Shrink Trimmed Surface, or click on the Shrink Trimmed Surface button on the Surface Tools toolbar.
2. Select edge A (indicated in figure 6-29) and press the Enter key. Note that shrinking a trimmed surface also makes subsequent fillet, blend, and sweep commands work better.

To appreciate the change, detach the trimmed boundary of the shrunk surface, as shown in figure 6-30. Save your file. Apart from a reduction in memory space, the topology of the control points of a shrunk, trimmed surface is also different from that of the same trimmed surface before it is shrunk.

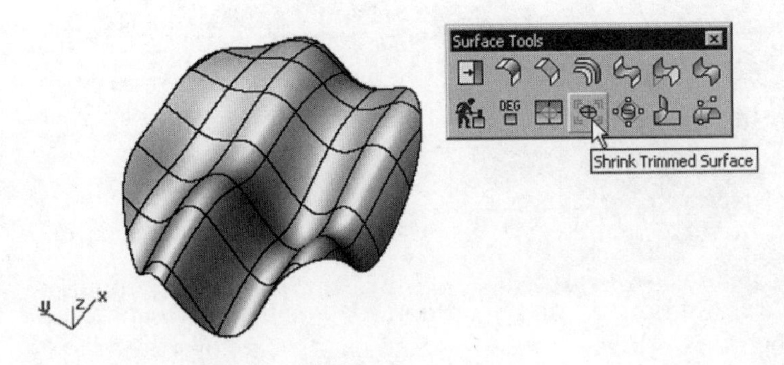

Fig. 6-29. Untrimmed surface being shrunk.

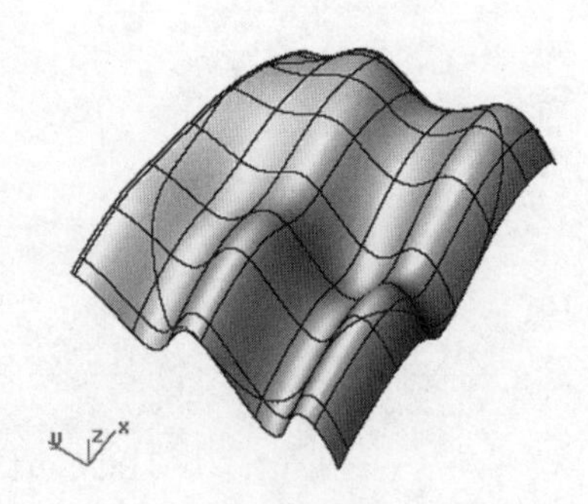

Fig. 6-30. Shrunk surface boundary detached.

Manipulation of Surface Edge of Contiguous Surfaces

Edges of a surface can be manipulated in four ways: splitting, merging, joining, and rebuilding.

Show Edges

In the computer display, both surface edges and isocurves are displayed as curves. To see clearly the edges of a surface, you highlight them, as follows.

1. Open the file *EditSurfaces4. 3dm* from the *Chapter 6* folder on the companion CD-ROM.

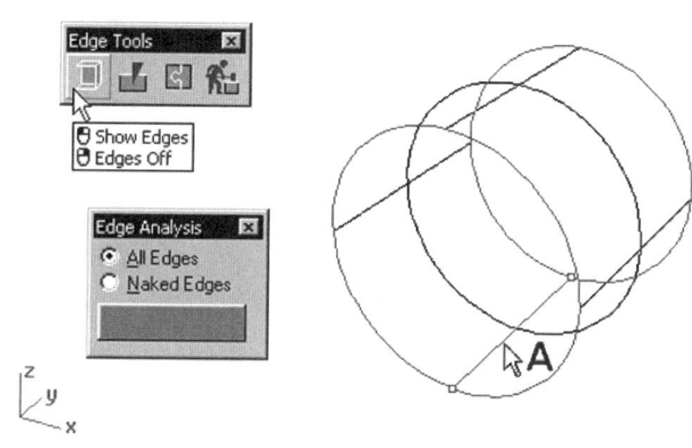

Fig. 6-31. All edges highlighted.

2. Select Analyze > Edge Tools > Show Edges, or click on the *Show Edges/Edges Off* button on the Edge Tools toolbar.
3. Select the cylindrical surface and press the Enter key.
4. In the Edge Analysis dialog box, activate the All Edges box.
5. Edge A, which is not a naked edge, is highlighted in addition to the other curves, as shown in figure 6-31.
6. Click on Naked Edges in the Edge Analysis dialog box. Edge A is not highlighted.

Close the Edge Analysis dialog box.

Splitting an Edge

By splitting an edge of a surface in two, you can make use of one of the edges to further construct surfaces. Perform the following steps to split an edge.

1. Select Analyze > Edge Tools > Split Edge, or click on the *Split Edge/Merge Edge* button on the Edge Tools toolbar.
2. Select edge A (indicated in figure 6-32).
3. Select location B (indicated in figure 6-32).

The selected edge is split. To extrude one of the split edges, continue with the following steps.

4. Select Surface > Extrude > Straight, or click on the Extrude Straight button on the Extrude toolbar.

5 Select edge A (indicated in figure 6-33) and press the Enter key.
6 Select location B (shown in figure 6-33) to indicate the extrusion height.

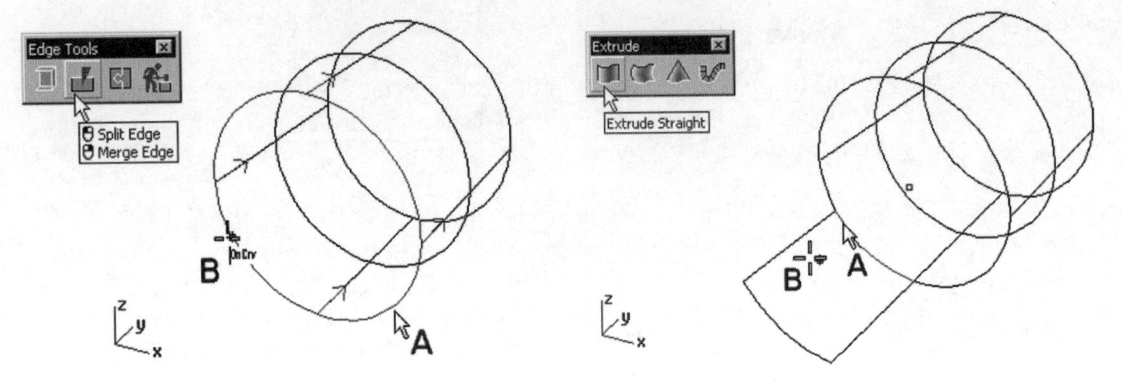

Fig. 6-32. Edge split.

Fig. 6-33. An edge being extruded.

An edge is extruded. To see how surfaces separated a small distance can be joined, you need to move the extruded surface, as follows.

7 Move the extruded surface as indicated in figure 6-34.

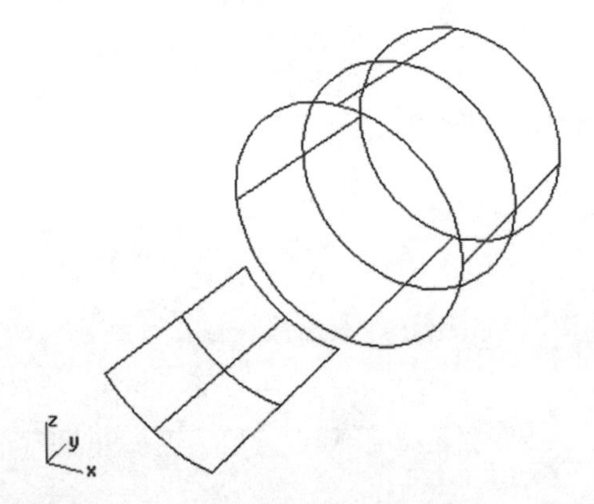

Fig. 6-34. Extruded surface moved.

Merging Edges

The opposite of splitting an edge is merging split edges, as follows.

1 Select Analyze > Edge Tools > Merge Edge, or right-click on the *Split Edge/Merge Edge* button on the Edge Tools toolbar.
2 Select split edge A (indicated in figure 6-35).
3 Select All in the pop-up dialog box.

The split edges are merged into a single edge.

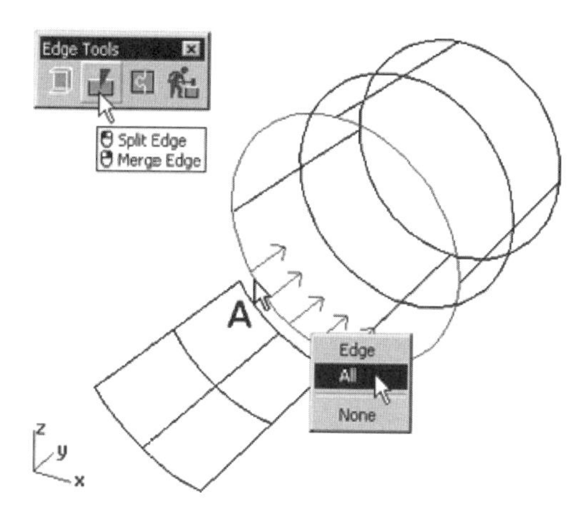

Fig. 6-35. Split edges being merged.

Joining Edges

Naked edges of two surfaces, although separated by a small distance, can be merged to become a single edge. Joining naked edges repairs any small gaps between contiguous surfaces. The display of the edges is forced closed. The geometry does not change. This command may work for models that are for rendering, but generally is not good enough for manufacturing. To join the naked edges of two surfaces, perform the following steps.

1. Select Analyze > Edge Tools > Join 2 Naked Edges, or click on the Join 2 Naked Edges button on the Edge Tools toolbar.
2. Select edges A and B (indicated in figure 6-36).

A dialog box informing you of the tolerances between the edges is displayed. Continue with the following.

3. In the Edge Joining dialog box, click on the OK button.

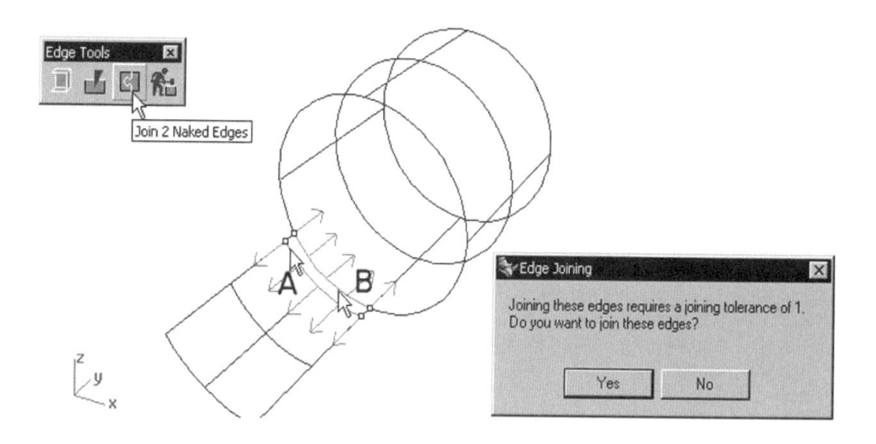

Fig. 6-36. Edges being joined.

Rebuilding Edges

To restore the original edge after you have joined edges, you explode the joined surface into individual surfaces and rebuild the edges, as follows.

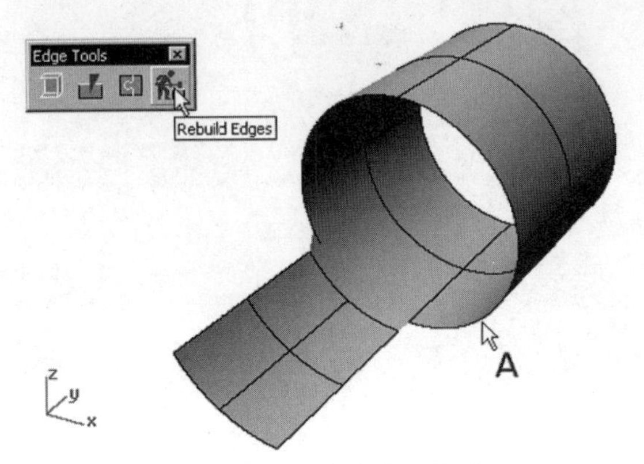

Fig. 6-37. Polysurface exploded and edges being restored.

1. Select Edit > Explode, or click on the Explode button on the Main2 toolbar.
2. Select the surface and press the Enter key.
3. Select Analyze > Edge Tools > Rebuild Edges, or click on the Rebuild Edges button on the Edge Tools toolbar.
4. Select edge A (indicated in figure 6-37).

Note that this command is also one of the first things to apply if your surface shows as a bad object.

Transforming Surfaces

Transformation of surfaces involves translation and deformation. That is, in the transformation of surfaces you translate surfaces and deform surfaces and control points of surfaces. The transform tools you learned about in Chapter 3 (see figure 3-54 and Table 3-5) in regard to curves also apply to the transformation of surfaces. The sections that follow discuss various aspects and procedures related to the transformation of surfaces.

Move, Copy, Rotate, Scale, Mirror, Orient, and Array

The operation of these commands as applied to surfaces is similar to that for curves. See the techniques delineated for these commands in Chapter 2.

Setting Points

Like aligning the control points of curves, you can align the control points of a surface. To align the Y coordinates of the control points of a surface, perform the following steps.

1. Open the file *TransformSurface1.3dm* from the *Chapter 6* folder on the companion CD-ROM.

2 Turn on the control points.

3 Select Transform > Set Points, or click on the Set XYZ Coordinates button on the Transform toolbar.

4 Select control points A and B (indicated in figure 6-38) and press the Enter key.

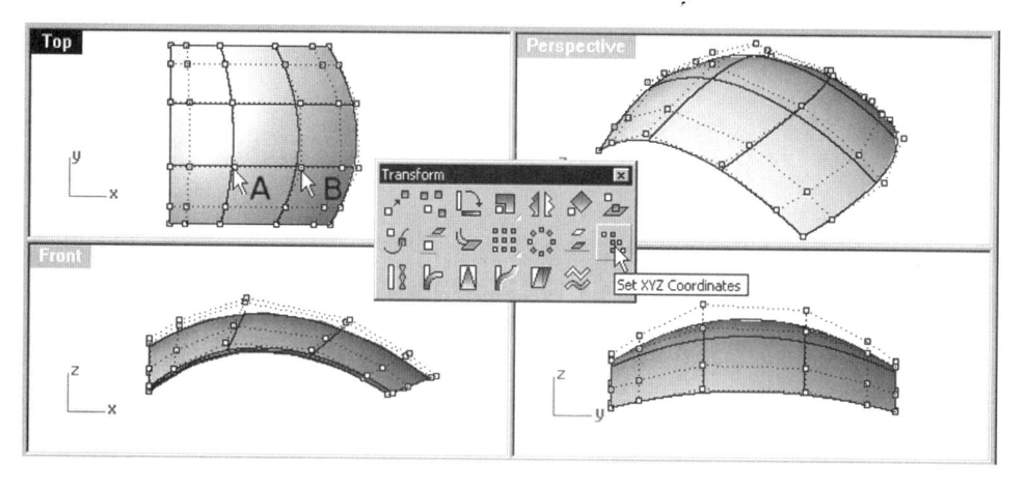

Fig. 6-38. Control points being selected.

5 In the Set Points dialog box, check the Set Z box and Align to World box and then click on the OK button.

6 Select location A (indicated in figure 6-39) in the Front viewport.

The Y coordinates of the selected control points are aligned, as shown in figure 6-40.

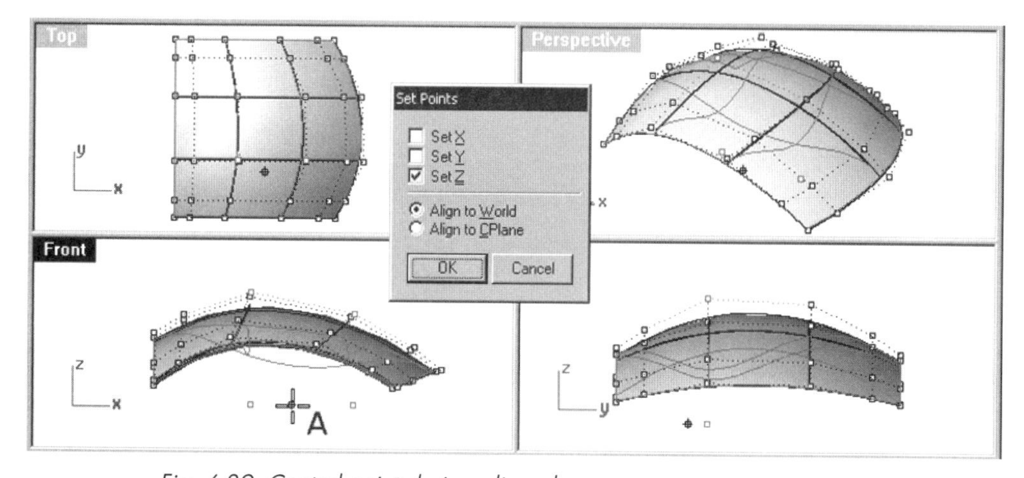

Fig. 6-39. Control points being aligned.

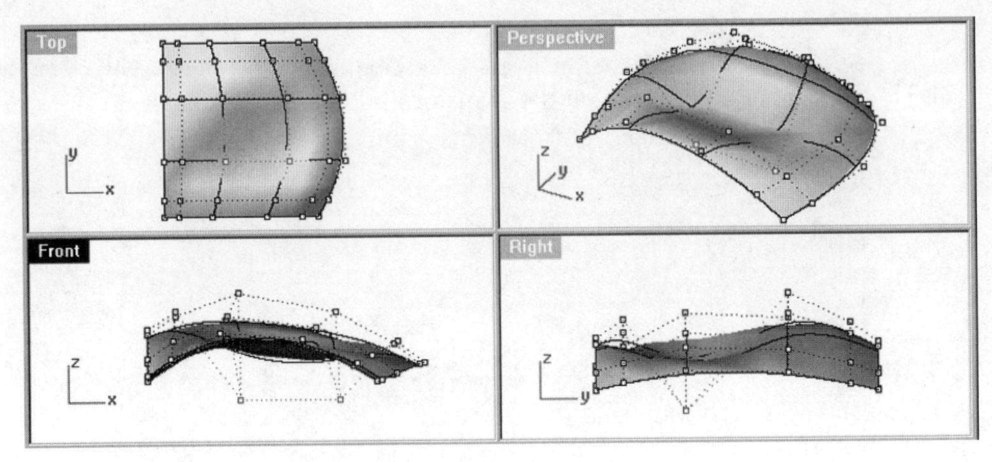

Fig. 6-40. Aligned control points.

Projection

By projecting the control points of a surface or the entire surface to the current construction plane, you transform the surface, as follows.

1. Select Transform > Project to CPlane, or click on the Project to CPlane button on Transform toolbar.
2. Select control points A and B (indicated in figure 6-41) and press the Enter key.
3. Press the Enter key if the default deletes input objects.

The control points are projected onto the construction plane. This command is also very useful when making curves with the Osnap option on. Continue with the following steps.

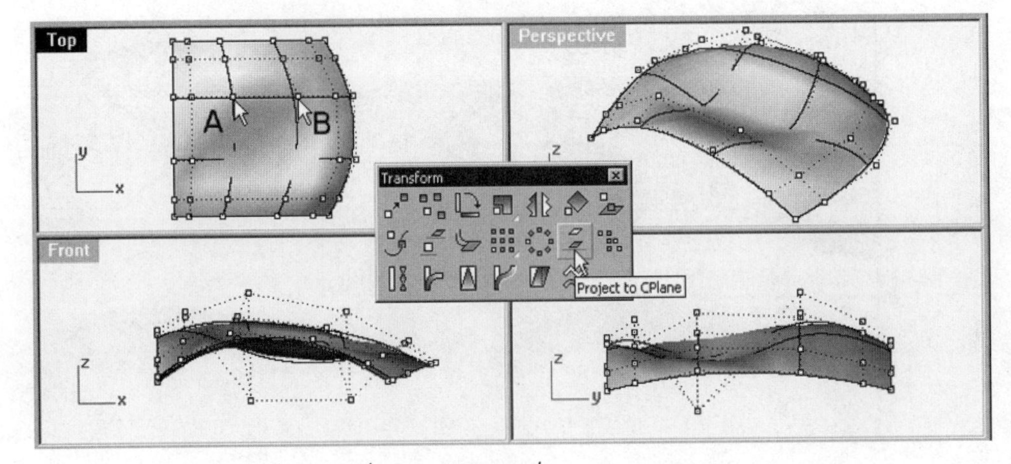

Fig. 6-41. Control points projected.

4 Turn off the control points.
5 Select Transform > Project to CPlane, or click on the Project to CPlane button on Transform toolbar.
6 Select surface A (indicated in figure 6-42) and press the Enter key.

The entire surface is projected, as shown in figure 6-43.

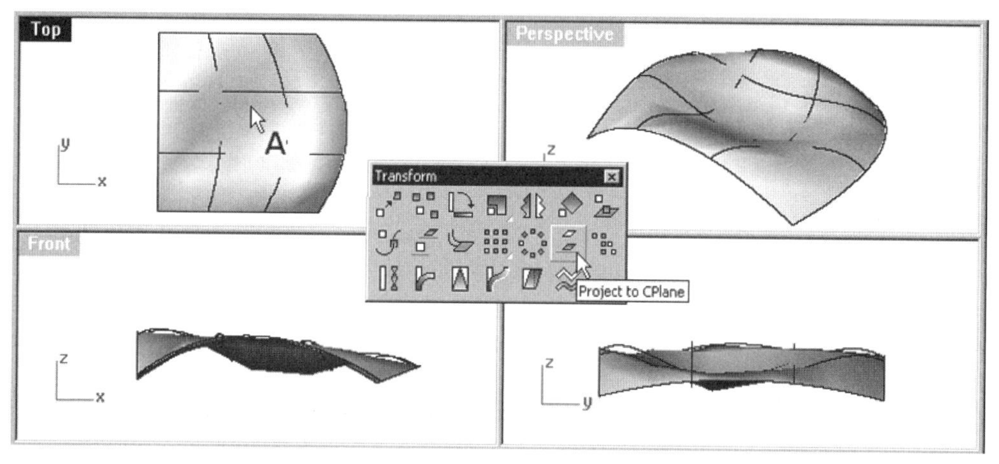

Fig. 6-42. Surface being projected.

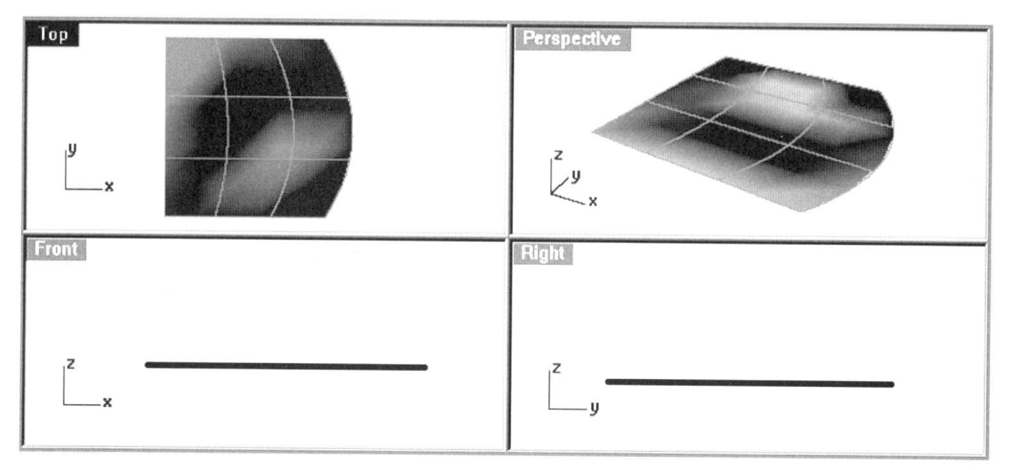

Fig. 6-43. Entire surface projected.

Twisting

Twisting transforms the shape of a surface by twisting a selected portion of the surface around an axis. Perform the following steps.

1. Open the file *TransformSurface2.3dm* from the *Chapter 6* folder on the companion CD-ROM.
2. Select Transform > Twist, or click on the Twist button on the Transform toolbar.
3. Select surface A (indicated in figure 6-44) and press the Enter key.
4. Select end point B (indicated in figure 6-44) to specify the start of the twist.
5. Select end point C (indicated in figure 6-44) to specify the end of the twist.
6. Select location D (indicated in figure 6-44) to specify the first reference point. (You may type an angular value at the command line interface to specify an angle, and skip step 7.)
7. Select location E (indicated in figure 6-44) to specify the second reference point.

Because the start of the twist and the end of the twist extend beyond the surface's boundary, the entire surface is twisted.

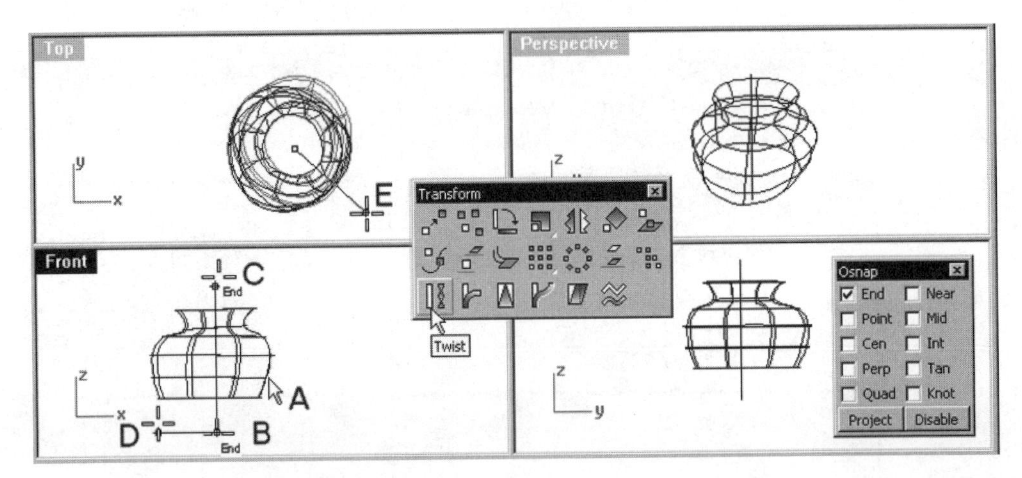

Fig. 6-44. Surface being twisted.

Tapering

Tapering transforms the shape of a surface by beveling the surface. To taper a surface, perform the following steps.

1. Select Transform > Taper, or click on the Taper button on the Transform toolbar.
2. Select surface A (indicated in figure 6-45) and press the Enter key.
3. Select end point B (indicated in figure 6-45) to specify the start of the taper axis.

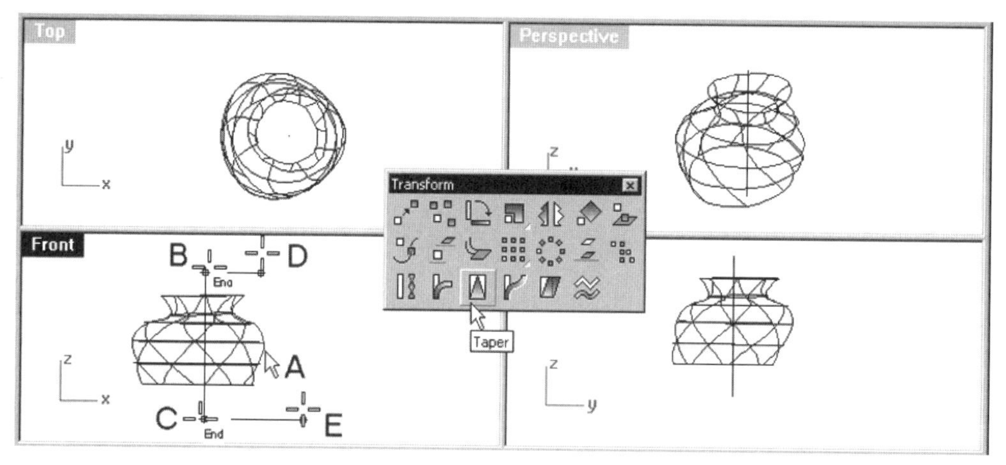

Fig. 6-45. Surface being tapered.

4 Select end point C (in figure 6-45) to specify the end of the taper axis.
5 Select location D (in figure 6-45) to specify the start distance.
6 Select location E (in figure 6-45) to specify the end distance.

The surface is tapered.

Shearing

Shearing transforms the shape of a surface by shearing it along a shear plane. To shear a surface, perform the following steps.

1 Select Transform > Shear, or click on the Shear button on the Transform toolbar.
2 Select surface A (indicated in figure 6-46) and press the Enter key.

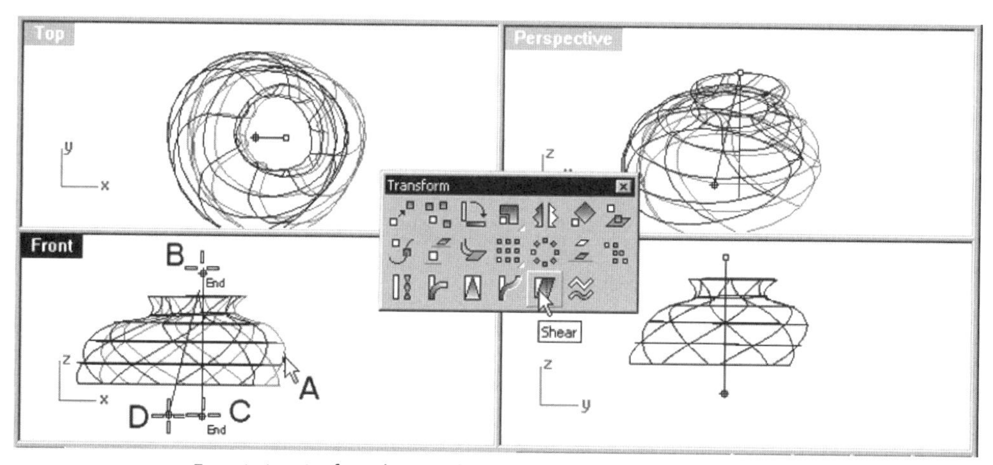

Fig. 6-46. Surface being sheared.

3 Select end point B (indicated in figure 6-46) as the origin point.
4 Select end point C (indicated in figure 6-46) as the reference point.
5 Select location D (indicated in figure 6-46) to specify the shear angle.

The surface is sheared.

Bending

Bending transforms a surface by bending the surface along an imaginary spline. To bend a surface, perform the following steps.

1 Select Transform > Bend, or click on the Bend button on the Transform toolbar.
2 Select surface A (indicated in figure 6-47) and press the Enter key.

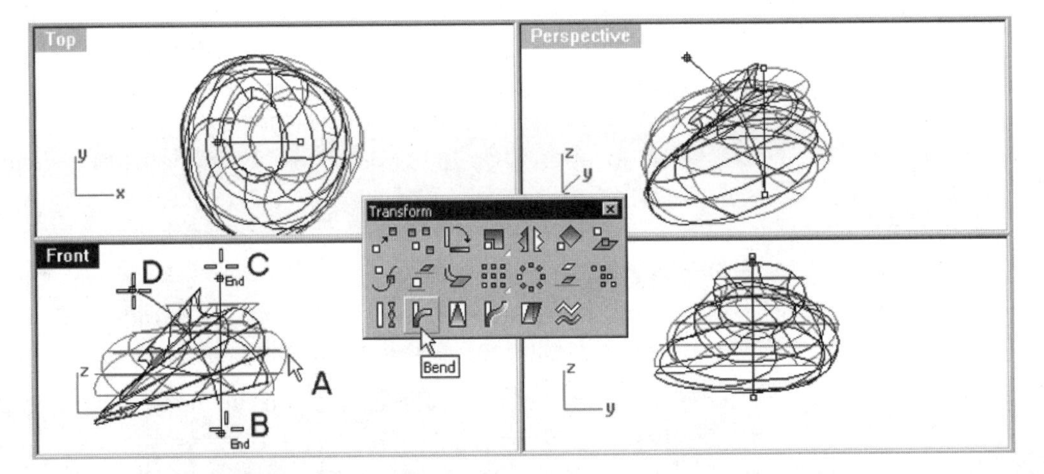

Fig. 6-47. Surface being bent.

3 Select end point B (indicated in figure 6-47) to specify the start of the spine.
4 Select end point C (indicated in figure 6-47) to specify the end of the spine.
5 Select location D (indicated in figure 6-47) to indicate the point to bend through.

The surface is bent, as shown in figure 6-48. This command also works with selected points on the surface for more subtle changes.

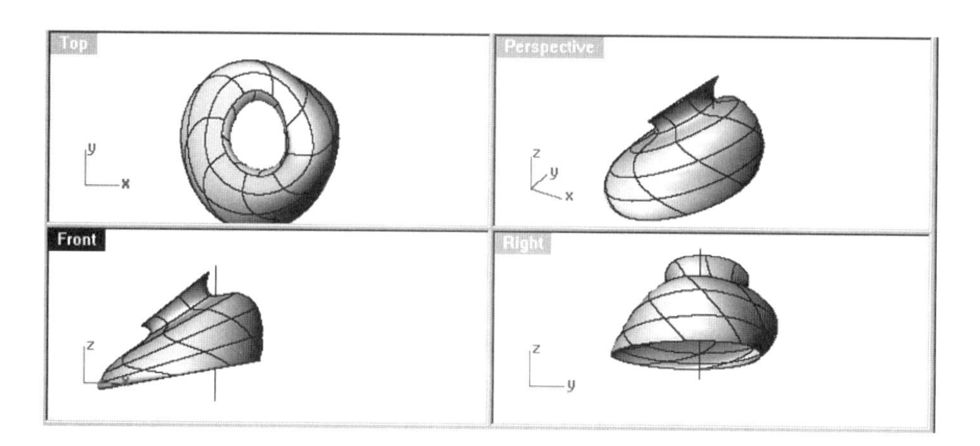

Fig. 6-48. Bent surface.

Flowing Along a Curve

Flowing along a curve changes the shape of the original imaginary backbone, which is a straight line, to the shape of a selected curve. To flow a surface along a curve, perform the following steps.

1. Open the file *TransformSurface3.3dm* from the *Chapter 6* folder on the companion CD-ROM.
2. Select Transform > Flow along Curve, or click on the Flow along Curve button on the Transform toolbar.
3. Select surface A (indicated in figure 6-49) and press the Enter key.
4. Select end point A (indicated in figure 6-49) to specify the original backbone and to indicate the orientation of the backbone.
5. Select end point B (indicated in figure 6-49) to specify the new backbone and to indicate the orientation of the new backbone.

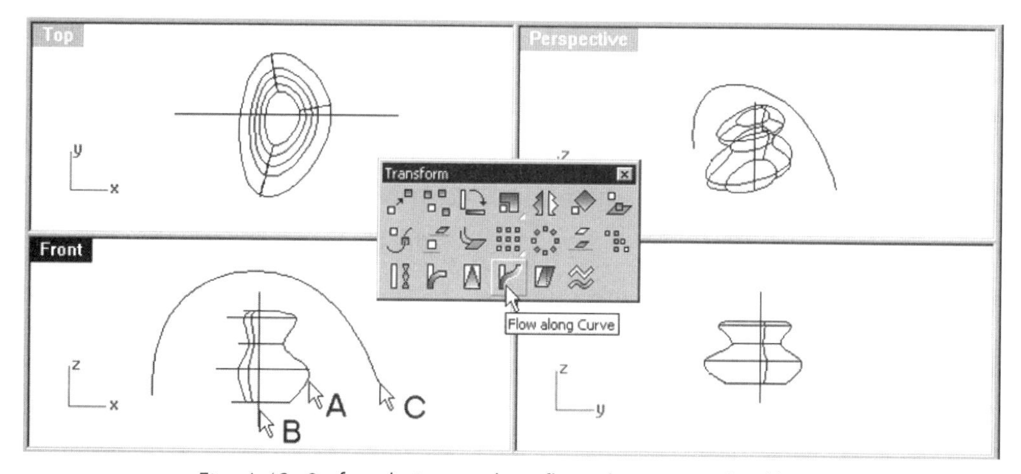

Fig. 6-49. Surface being made to flow along a new backbone.

The surface now flows along the new backbone.

Smoothing

Smoothing removes irregularities from a surface. To smooth a surface, perform the following steps.

1. Select Transform > Smooth, or click on the Smooth button on the Transform toolbar.
2. Select surface A (indicated in figure 6-50) and press the Enter key. (You can display the control points and select them for more subtle changes.)
3. In the Smooth dialog box, check the Smooth X, Smooth Y, Smooth Z, and *World coordinates* options, set the smooth factor to 0.2, and then click on the OK button.

The surface is smoothed.

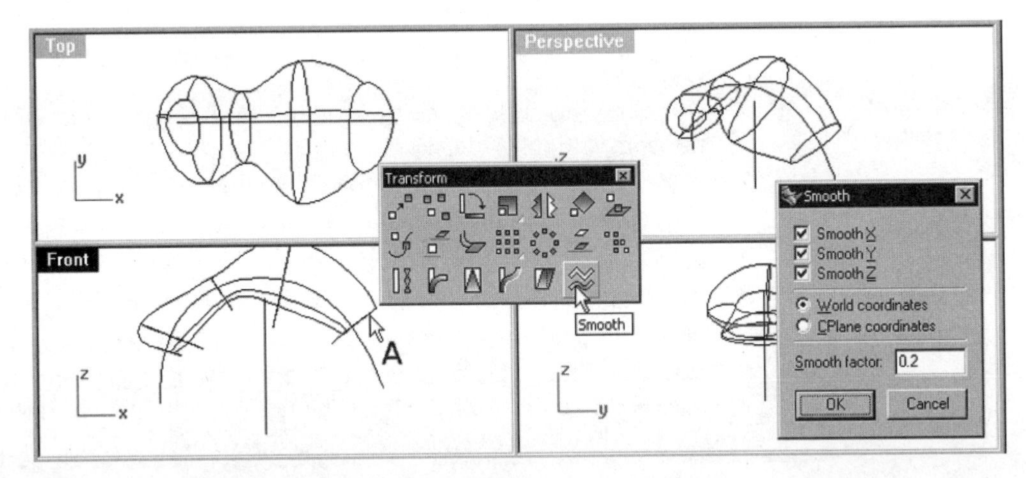

Fig. 6-50. Surface being smoothed.

Surface Analysis Tools

To analyze the surfaces you construct, you use the analysis tools. Basically, most of the tools you use on curves are also applicable to surfaces. (See figure 3-93 and Table 3-6 of Chapter 3.) In particular, most of the tools available from the Analyze, Mass Properties, and Surface Analysis toolbars (shown in figure 6-51) are specific to surface analysis. Table 6-2 outlines the functions of the commands specific to surface analysis and their locations in the main pull-down menu and respective toolbars.

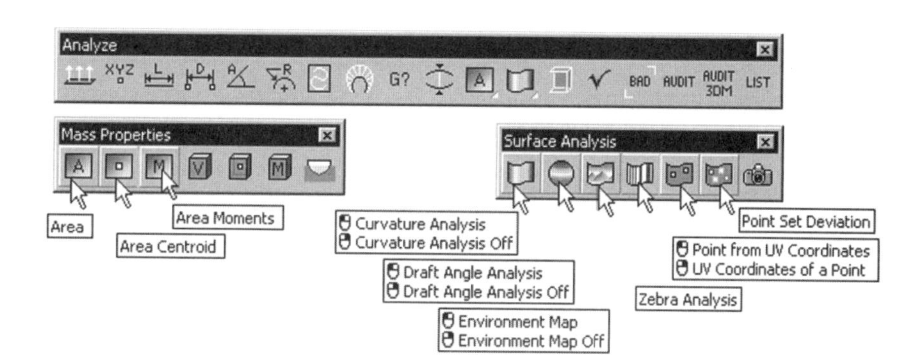

Fig. 6-51. Analyze, Mass Properties, and Surface Analysis toolbars.

Table 6-2: Surface Analysis Commands and Their Functions

Toolbar Button	From Main Pull-down Menu	Function
Area	Analyze > Mass Properties > Area	Evaluating the area of selected surfaces or polysurfaces
Area Centroid	Analyze > Mass Properties > Area Centroid	Constructing a point at the area centroid of selected surfaces or polysurfaces
Area Moments	Analyze > Mass Properties > Area Moments	Evaluating area, area centroid, first area moments, second area moments, product moments, area moments of inertia about World coordinate axes, area radii of gyration about World coordinate axes, area moments of inertia about centroid coordinate axes, and area radii of gyration about centroid coordinate axes
Curvature Analysis	Analyze > Surface > Curvature Analysis	Displaying a color rendering to indicate the curvature values of selected surfaces
Curvature Analysis Off	—	Turning off the rendered curvature analysis display
Draft Angle Analysis	Analyze > Surface > Draft Angle Analysis	Displaying a color rendering to indicate the draft angle values of selected surfaces
Draft Angle Analysis Off	—	Turning off the rendered draft angle analysis display
Environmental Map	Analyze > Surface > Environmental Map	Displaying a color rendering with a bitmap mapped to selected surfaces
Environmental Map Off	—	Turning off the rendered mapped bitmap display

Toolbar Button	From Main Pull-down Menu	Function
Zebra Analysis	Analyze > Surface > Zebra	Displaying the zebra pattern on selected surfaces
Point from UV Coordinates	Analyze > Surface > Point from UV Coordinates	Constructing points on a surface by entering U and V coordinates
UV Coordinates of a Point	Analyze > Surface > UV Coordinates of a Point	Evaluating UV coordinates of points on a surface
Point Set Deviation	Analyze > Surface > Point Set Deviation	Measuring the deviation of points from surfaces and curves

Construction of a Bounding Box

The bounding box of a surface or polysurface consists of a rectangular box enclosing the surface or polysurface. The box indicates an approximate size of the surface in three directions. To construct a bounding box for a surface, perform the following steps.

1. Open the file *AnalyzeSurface1.3dm* from the *Chapter 6* folder on the companion CD-ROM.
2. Select surface A (indicated in figure 6-52) and press the Enter key.
3. Type *W* to use the World coordinate system.

A bounding box is constructed.

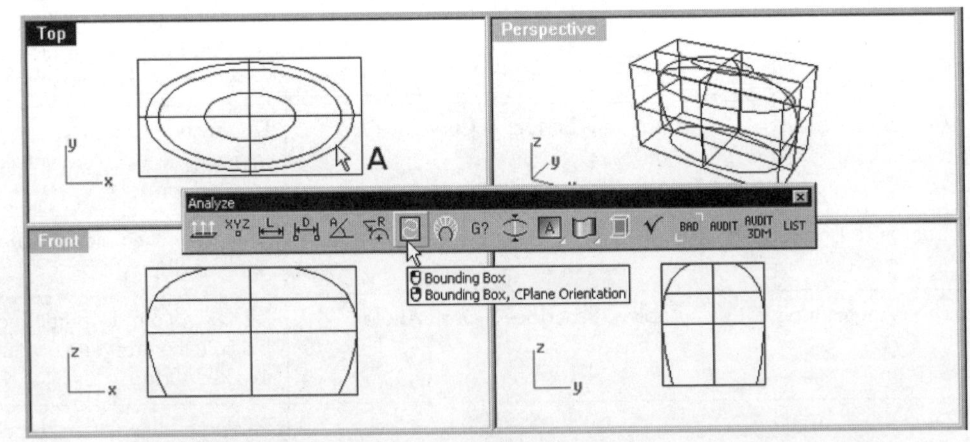

Fig. 6-52. Bounding box constructed.

Area, Area Centroid, and Area Moments Inquiry

To assess the area, area centroid, and area moments of a surface, perform the following steps.

1. Select the bounding box and press the Delete key to delete it.
2. Select Analyze > Mass Properties > Area, or click on the Area button on the Mass Properties toolbar.
3. Select surface A (indicated in figure 6-53) and press the Enter key.

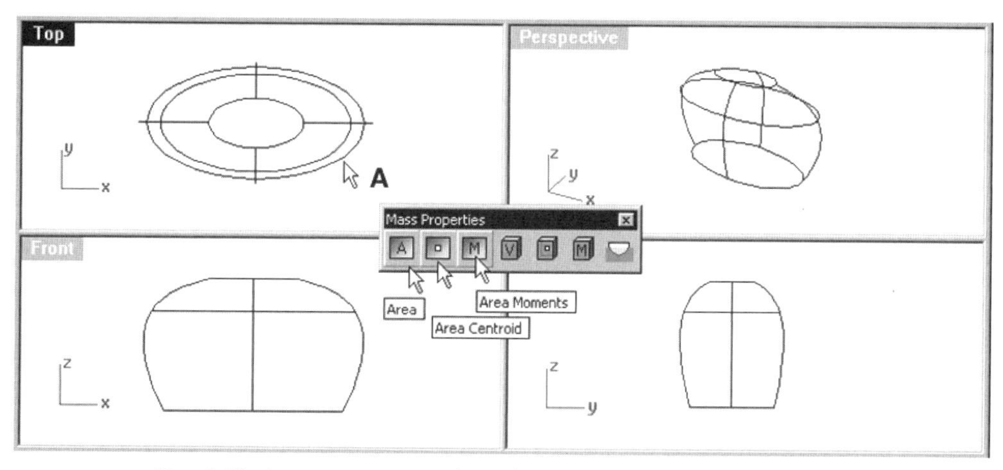

Fig. 6-53. Area, area centroid, and area moments being evaluated.

The surface area is displayed at the command line interface. Continue with the following steps.

4. Select Analyze > Mass Properties > Area Centroid, or click on the Area Centroid button on the Mass Properties toolbar.
5. Select surface A (indicated in figure 6-53) and press the Enter key.

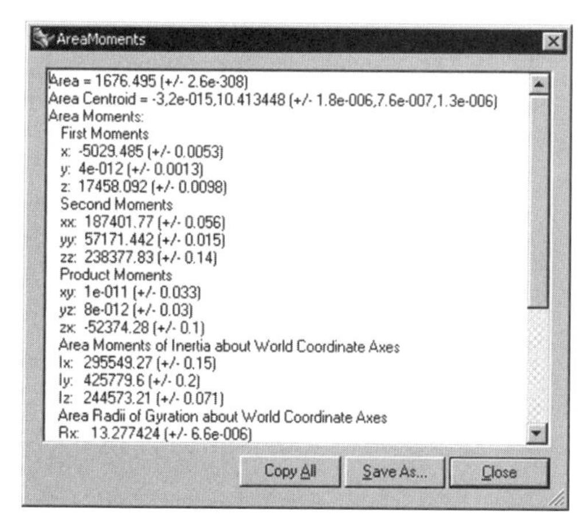

Fig. 6-54. Area Moments dialog box.

The area centroid is displayed and a point is constructed to indicate the location of the centroid. Continue with the following steps.

6. Select Analyze > Mass Properties > Area Moments, or click on the Area Moments button on the Mass Properties toolbar.
7. Select surface A (indicated in figure 6-53) and press the Enter key.

Area moment information is displayed in the Area Moments dialog box, shown in figure 6-54. You may save the information to a file and then click on the Close button to close the dialog box.

Analyzing Surface Profiles

Because a NURBS surface is a smooth surface with no facets, the profile and silhouette of a NURBS surface are displayed using U and V isocurves. To gain a visual picture of the surface, you use curvature rendering, draft angle rendering, environmental map rendering, and zebra stripes rendering and construct curvature arcs.

In regard to individual points on a surface, you construct a point on a surface by specifying the U and V coordinates of the point. Given a point on the surface, you evaluate its U and V coordinates. In addition, you evaluate the deviation of any point from a surface.

Curvature Rendering

To display a color rendering to indicate the curvature values of a surface, perform the following steps. In the rendering, different curvature values are represented by different colors.

1. Maximize the Perspective viewport.
2. Select Analyze > Surface > Curvature Analysis, or click on the *Curvature Analysis/Curvature Analysis Off* button on the Surface Analyze toolbar.
3. Select the surface and press the Enter key.
4. In the Curvature dialog box, select Mean in the Style pull-down list box.

A color rendering showing the curvature values in various colors is displayed, as shown in figure 6-55.

5. Close the Curvature dialog box.

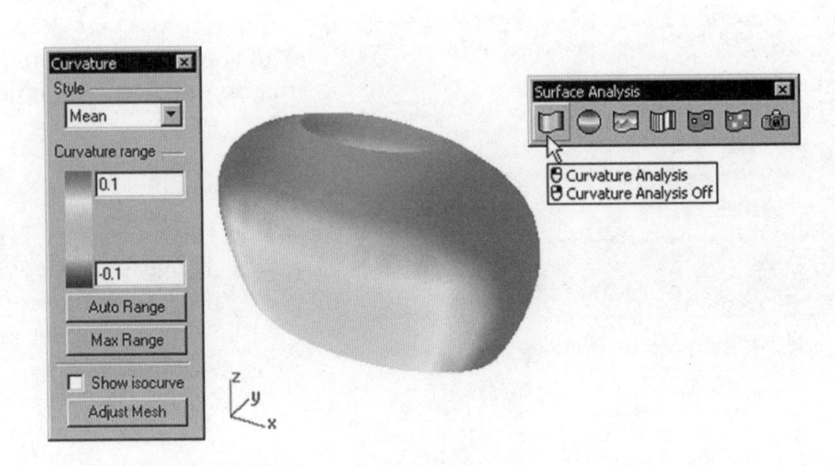

Fig. 6-55. Curvature values displayed in various colors.

Draft Angle Rendering

Draft angles are an essential consideration in mold and die design. To display a color rendering to illustrate the draft angle value of a surface with respect to the construction plane, perform the following steps.

1. Select Analyze > Surface > Draft Angle Analysis, or click on the *Draft Angle Analysis/Draft Angle Analysis Off* button on the Surface Analyze toolbar.
2. Select the surface and press the Enter key.

Draft angle values on the surface are displayed in various colors, as shown in figure 6-56.

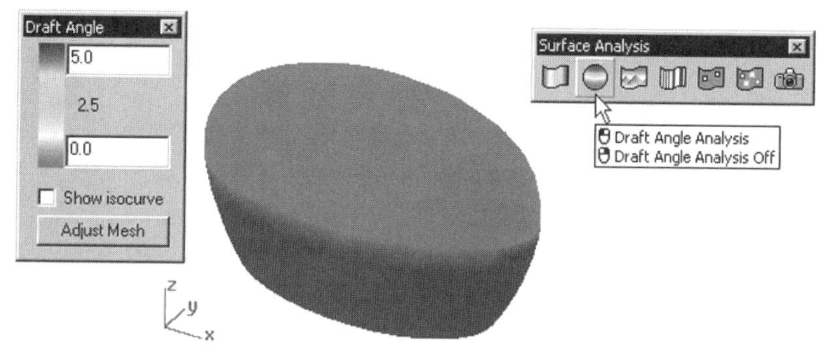

Fig. 6-56. Draft angle values displayed in various colors.

Environment Map Rendering

To simply inspect the smoothness of a surface, you place a bitmap image on the surface. To place a bitmap on a surface and display a rendering of the surface, perform the following steps.

1. Select Analyze > Surface > Environment Map, or click on the *Environment Map/Environment Map Off* button on the Surface Analyze toolbar.
2. Select the surface and press the Enter key.
3. In the Environment Map Options dialog box, select an image.

The selected image is mapped onto the surface, as shown in figure 6-57.

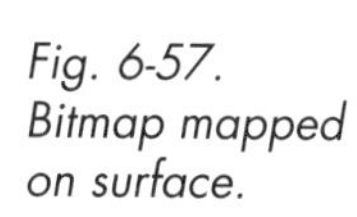

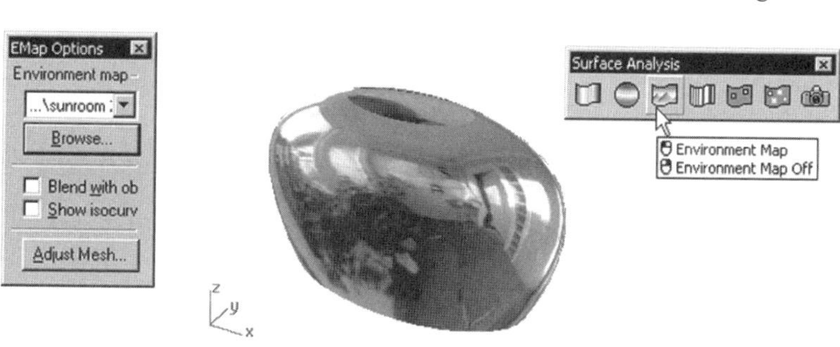

Fig. 6-57. Bitmap mapped on surface.

Zebra Stripes Rendering

To effectively view the smoothness of a surface, you use a set of zebra stripes instead of a bitmap. To use zebra stripes to help visualize the smoothness of a surface, perform the following steps.

1. Select Analyze > Surface > Zebra, or click on the Zebra button on the Surface Analysis toolbar.
2. Select the surface and press the Enter key.
3. In the Zebra Options dialog box, set the stripe direction and stripe size.

Zebra stripes are placed on the surface, as shown in figure 6-58. This is most valuable for checking the continuity between the surfaces of poly-surfaces.

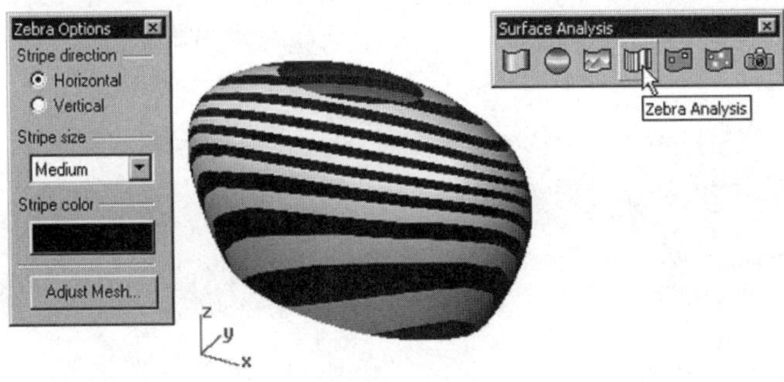

Fig. 6-58. Zebra stripes.

Curvature Arcs

To find out the curvature of a selected point on a surface, you construct two curvature arcs. To construct curvature circles at selected locations of a surface, perform the following steps.

1. Select Analyze > Surface > Curvature Circle.
2. Select the surface.
3. Select a point on the surface and press the Enter key.

The curvature arcs' radii are displayed at the command line interface, as shown in figure 6-59.

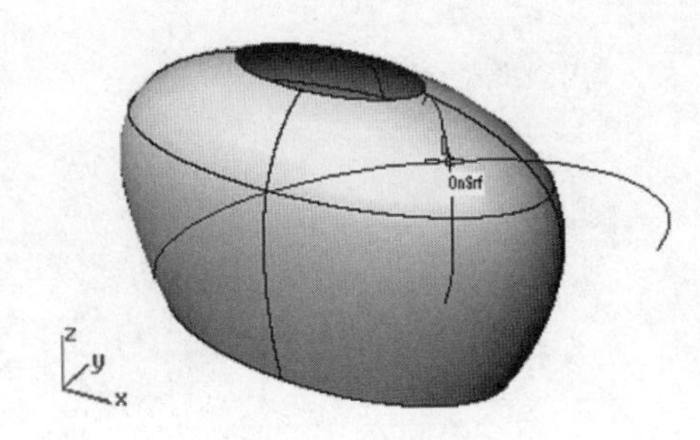

Fig. 6-59. Curvature arcs constructed.

Points on a Surface

A surface has two sets of isocurves in two directions, U and V. Correspondingly, there are two axis directions, U and V. You can represent any point on the surface by a set of U and V coordinates. To construct a point on a surface by specifying the point's U and V coordinate values, perform the following steps.

1. Select Analyze > Surface > Point from UV Coordinates, or click on the *Point from UV Coordinates/UV Coordinates of a Point* button on the Surface Analyze toolbar.
2. Select the surface and press the Enter key.
3. Type *5* to specify the U value.
4. Type *20* to specify the V value.

A point is constructed on the surface, as shown in figure 6-60.

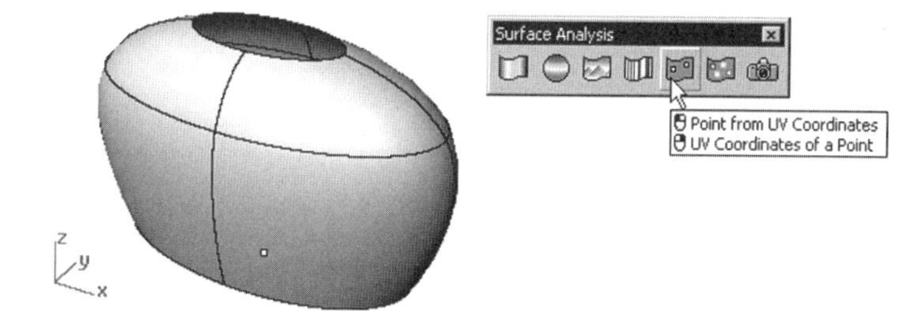

Fig. 6-60. Point constructed on the surface.

Evaluating the U and V Coordinates of a Point

Contrary to constructing a point on a surface by specifying the U and V coordinates, in the following you will evaluate the U and V coordinates of a point on a surface.

1. Select Analyze > Surface > UV Coordinates of a Point, or right-click on the *Point from UV Coordinates/UV Coordinates of a Point* button on the Surface Analyze toolbar.
2. Select the surface.
3. Select point A (indicated in figure 6-61) and press the Enter key.

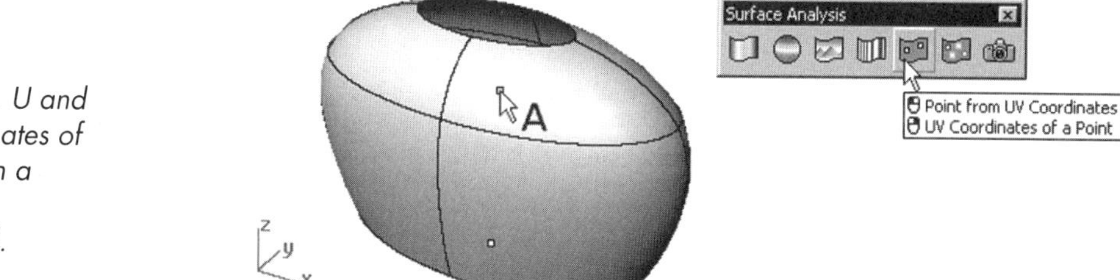

Fig. 6-61. U and V coordinates of a point on a surface evaluated.

The U and V coordinates of the selected point on the surface are displayed at the command line interface, as shown in figure 6-61.

Checking the Deviation of a Point from a Surface

To find out the deviation of a point from a surface, perform the following steps.

1. Open the file *AnalyzeSurface2.3dm* from the *Chapter 6* folder on the companion CD-ROM.
2. Select Analyze > Surface > Point Set Deviation, or click on the Point Set Deviation button on the Surface Analyze toolbar.

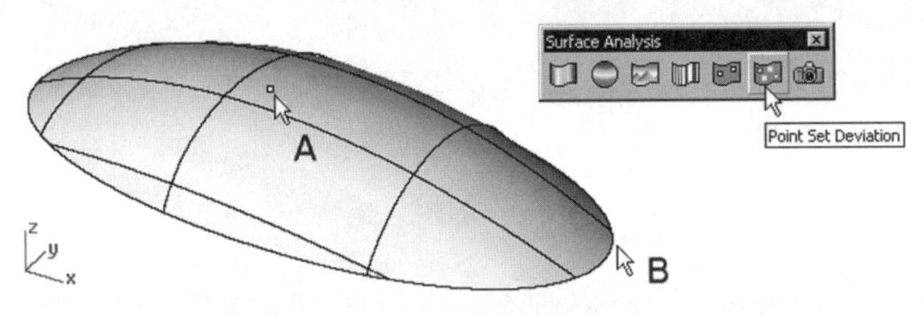

Fig. 6-62. Deviation of a point from a surface being evaluated.

3. Select point A (indicated in figure 6-62) and press the Enter key.
4. Select surface B (indicated in figure 6-62) and press the Enter key.

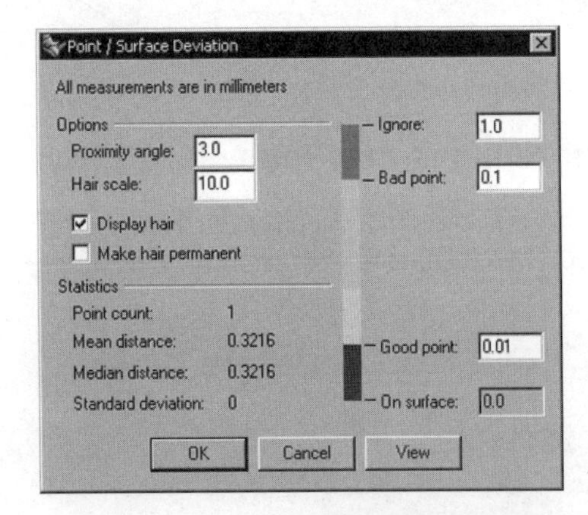

The Point/Surface Deviation dialog box is displayed, as shown in figure 6-63. The information is useful in finding out the deviation of a point or a set of point clouds and the surface constructed from the point objects.

Fig. 6-63. Point/Surface Deviation dialog box.

Direction

To check the normal direction of a surface, perform the following steps. It is very important to determine Booleans when using surfaces and polysurfaces. You will learn about Booleans and polysurfaces in Chapter 7.

1. Select Analyze > Direction, or click on the Direction button on the Analyze toolbar.
2. Select the surface and press the Enter key.

Surface direction arrows are displayed, as shown in figure 6-64.

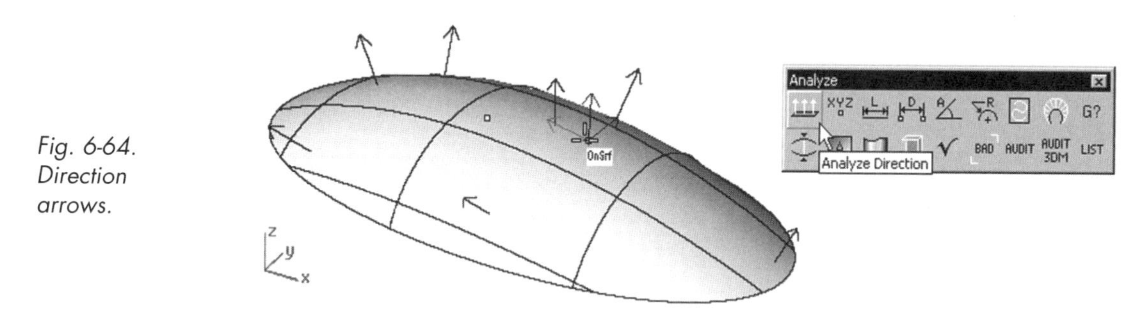

Fig. 6-64. Direction arrows.

Use of Polygon Meshes

Although a polygon mesh is an approximation of a smooth surface, it is still used in many computer applications. For example, STL (a rapid prototyping file format) represents a surface as a set of polygon meshes. When you import surfaces from other computerized applications, you are often importing a set of polygon meshes, which are dealt with via commands on the Mesh toolbar, shown in figure 6-65.

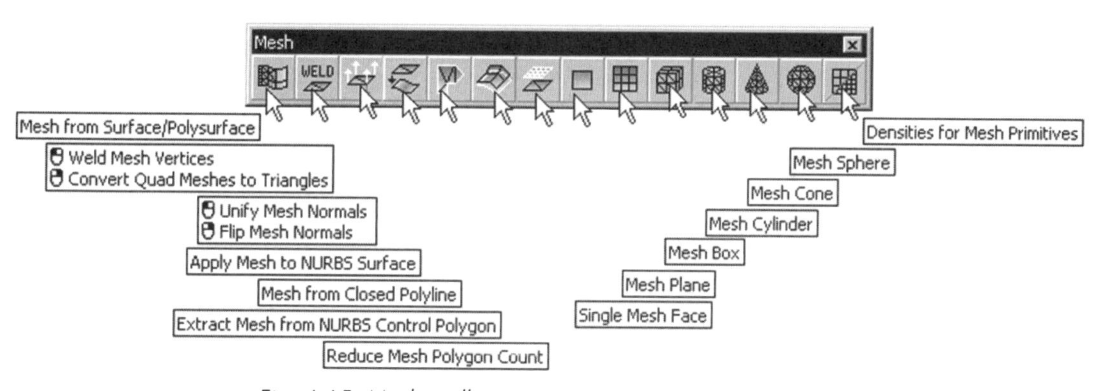

Fig. 6-65. Mesh toolbar.

Table 6-3 outlines the functions of the polygon mesh commands and their locations in the main pull-down menu and respective toolbars.

Table 6-3: Polygon Mesh Commands and Their Functions

Toolbar Button	From Main Pull-down Menu	Function
Mesh from Surface/ Polysurface	Tools > Polygon Mesh > From NURBS Object	Constructing polygon meshes from NURBS surface objects
Weld Mesh Vertices	Tools > Polygon Mesh > Weld	Welding polygon meshes to form a single polygon mesh

Toolbar Button	From Main Pull-down Menu	Function
Convert Quad Meshes to Triangles	—	Converting quadrilateral meshes to triangular meshes
Unify Mesh Normals	Tools > Polygon Mesh > Unify Normals	Unifying the normal direction of a set of joined polygon meshes
Flip Mesh Normals	—	Reversing the direction of normals
Apply Mesh to NURBS Surface	Tools > Polygon Mesh > Apply to Surface	Applying a polygon mesh to a NURBS surface
Mesh from Closed Polyline	Tools > Polygon Mesh > From Closed Polyline	Constructing a polygon mesh from a closed polyline
Extract Mesh from NURBS Control Polygon	Tools > Polygon Mesh > From NURBS Control Polygon	Constructing a polygon mesh from the control points of a NURBS curve or surface
Reduce Mesh Polygon Count	Tools > Polygon Mesh > Reduce	Simplifying a polygon mesh by reducing its number of constituent polygons
Single Mesh Face	Tools > Polygon Mesh Primitives > 3-D Face	Constructing a quadrilateral polygon mesh
Mesh Plane	Tools > Polygon Mesh Primitives > Plane	Constructing a rectangular polygon mesh
Mesh Box	Tools > Polygon Mesh Primitives > Box	Constructing a polygon mesh box
Mesh Cylinder	Tools > Polygon Mesh Primitives > Cylinder	Constructing a polygon mesh cylinder
Mesh Cone	Tools > Polygon Mesh Primitives > Cone	Constructing a polygon mesh cone
Mesh Sphere	Tools > Polygon Mesh Primitives > Sphere	Constructing a polygon mesh sphere
Densities for Mesh Primitives	Tools > Polygon Mesh Primitives > Options	Setting the mesh density of polygon mesh boxes, planes, spheres, cylinders, and cones

In the sections that follow you will learn how to construct and modify polygon meshes.

Mesh Density Setting

Because a polygon mesh is a set of planar polygons that approximates a surface, the number of polygons used in the mesh has a direct impact on the accuracy of representation and the file size. The more numerous the constituent polygons the more accurate the model.

However, file size increases quickly with a decrease in polygon size (smaller size meaning a greater number of polygons in the mesh). Hence, you need to consider using the most appropriate mesh density to provide an optimum balance between the representational accuracy of the model and the size of the file. To work with the Density setting, perform the following steps.

1. Start a new file. Use the metric (Millimeters) template.
2. Select Tools > Polygon Mesh Primitives > Options, or click on the Densities for Mesh Primitives button on the Mesh toolbar.
3. With reference to figure 6-66, set the values in the dialog box and then click on the OK button.

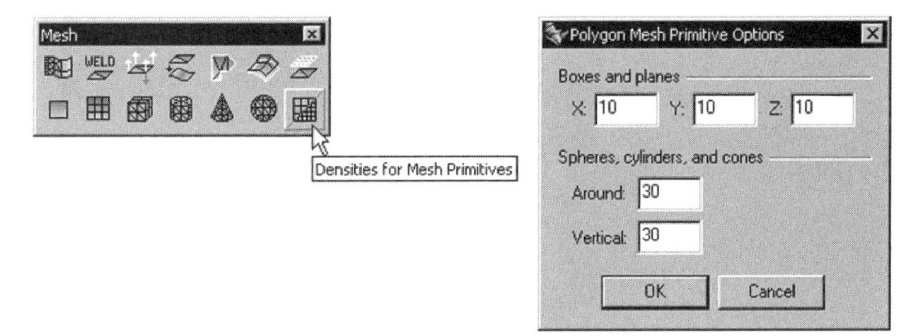

Fig.6-66. Polygon mesh primitives options.

Construction of Mesh Primitives

Mesh primitives are mesh objects of basic geometric shape. There are six types of mesh primitives: cylinder, cone, sphere, box, plane, and 3D face. The sections that follow take you through the construction process for each of these primitives. You construct mesh primitives by selecting a command and specifying the parameters of the primitive.

Mesh Cylinder

To construct a mesh cylinder, perform the following steps.

1. Maximize the Perspective viewport.
2. Select Tools > Polygon Mesh Primitives > Cylinder, or click on the Mesh Cylinder button on the Mesh toolbar.
3. Select two points in the Top viewport to specify the center and radius.
4. Select a point in the Front viewport to indicate the height of the cylinder, as shown in figure 6-67.

A mesh cylinder is constructed.

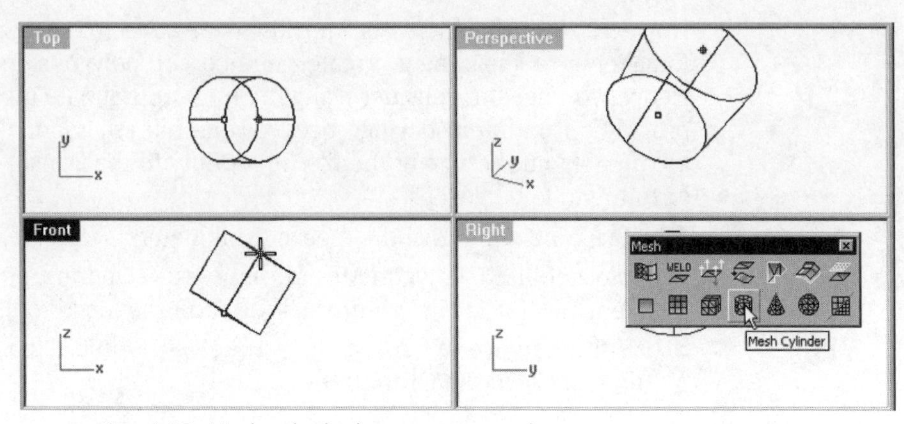

Fig. 6-67. Mesh cylinder being constructed.

Mesh Cone

To construct a mesh cone, perform the following steps.

1. Select Tools > Polygon Mesh Primitives > Cone, or click on the Mesh Cone button on Mesh toolbar.
2. Select two points in the Top viewport to specify the center and radius.
3. Select a point in the Front viewport to indicate the height of the cone, as shown in figure 6-68.

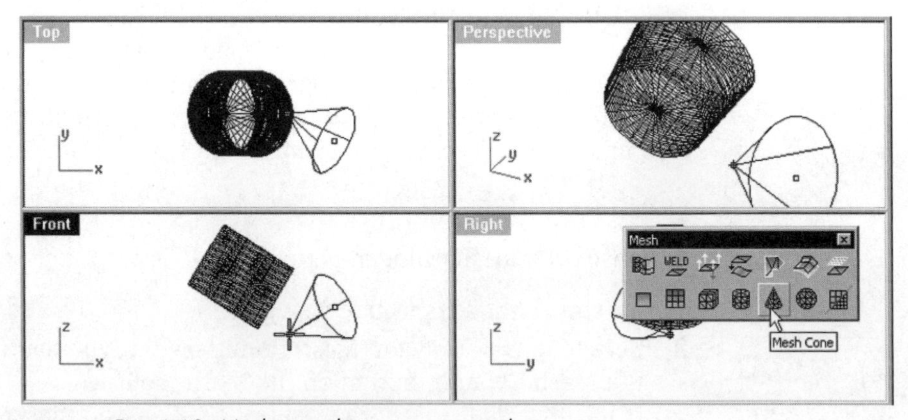

Fig. 6-68. Mesh cone being constructed.

A mesh cone is constructed.

Mesh Sphere

To construct a mesh sphere, perform the following steps.

1 Select Tools > Polygon Mesh Primitives > Sphere, or click on the Mesh Sphere button on the Mesh toolbar.
2 Select two points A and B in the Top viewport to specify the center and radius of the sphere, as shown in figure 6-69.

A mesh sphere is constructed.

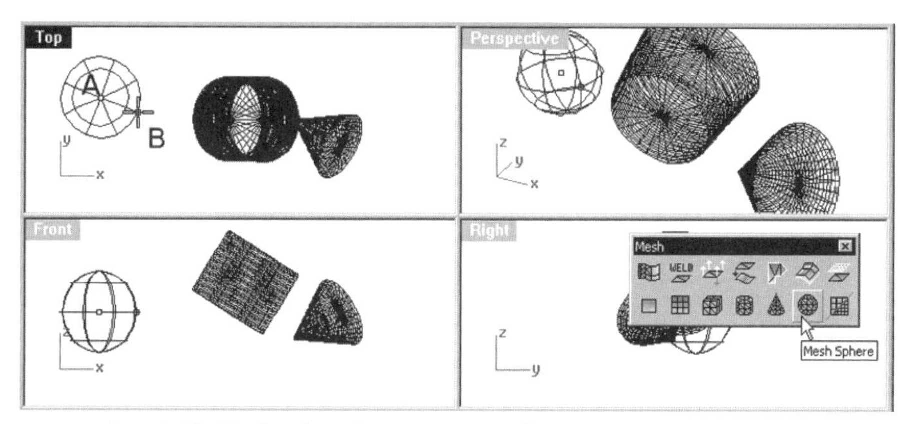

Fig. 6-69. Mesh sphere being constructed.

Mesh Box

To construct a mesh box, perform the following steps.

1 Select Tools > Polygon Mesh Primitives > Box, or click on the Mesh Box button on the Mesh toolbar.
2 Select two points in the Top viewport to specify the diagonal corners of the base of the box.
3 Select a point in the Front viewport to indicate the height of the box, as shown in figure 6-70.

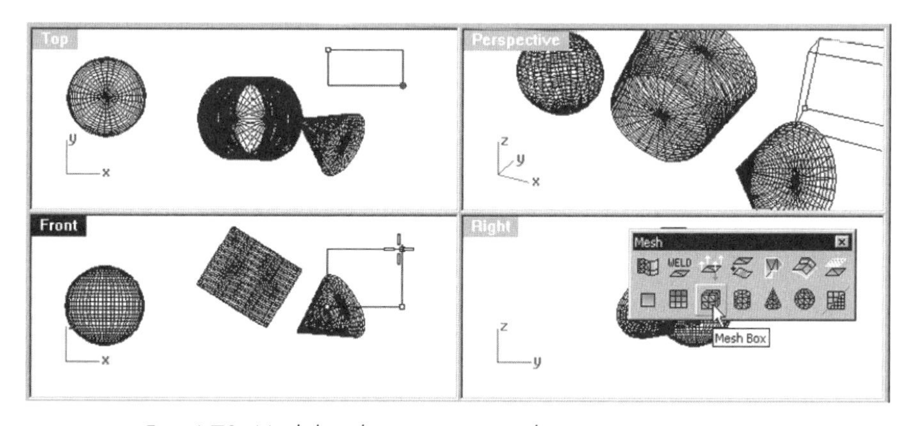

Fig. 6-70. Mesh box being constructed.

A mesh box is constructed.

Mesh Rectangular Plane

To construct a mesh rectangular plane, perform the following steps.

1. Select Tools > Polygon Mesh Primitives > Plane, or click on the Mesh Plane button on the Mesh toolbar.
2. Select two points in the Top viewport to specify the diagonal end points of the rectangular plane, as shown in figure 6-71.

A mesh rectangular plane is constructed.

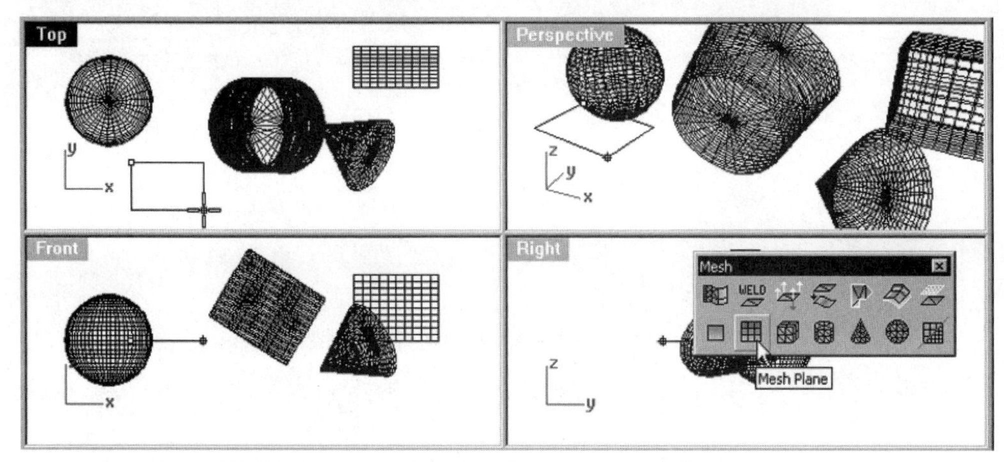

Fig. 6-71. Mesh rectangular plane being constructed.

Mesh 3D Face

To construct a mesh 3D face, perform the following steps.

1. Tools > Polygon Mesh Primitives > 3-D Face, or click on the Single Mesh Face button on the Mesh toolbar.
2. Select three points in the Top viewport, as shown in figure 6-72, and press the Enter key.

A 3D face is constructed. Figure 6-73 shows the rendered Perspective viewport.

3. If you want to save your file, save it as *Mesh1.3dm*.

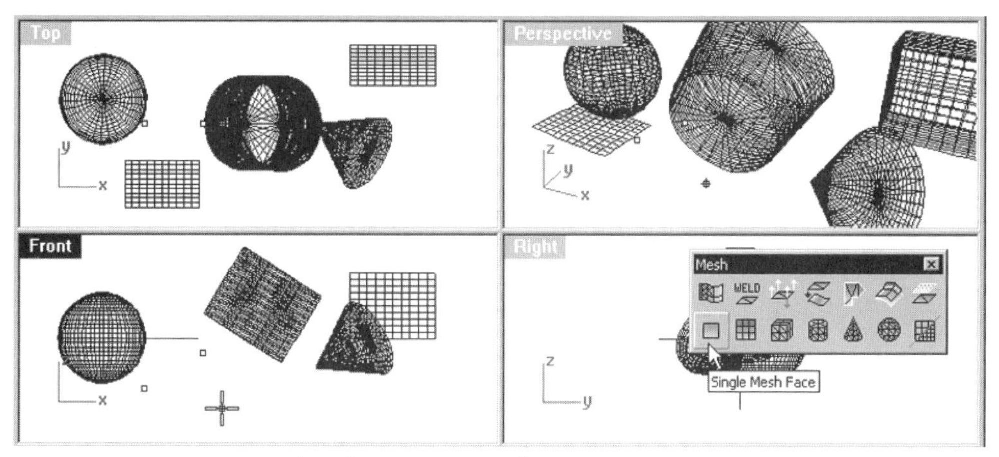

Fig. 6-72. 3D face being constructed.

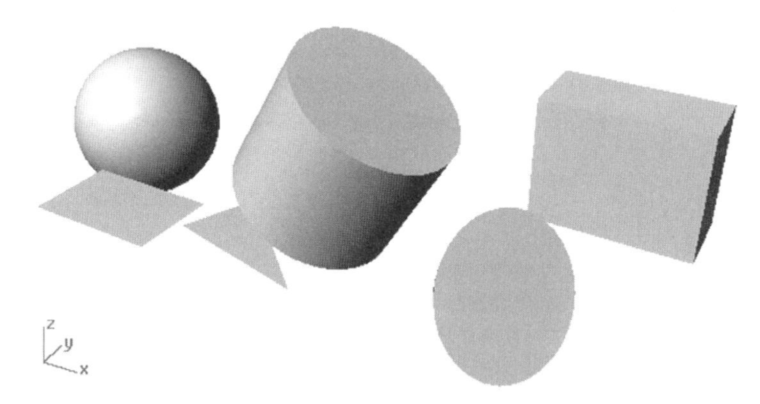

Fig. 6-73. 3D rendered Perspective viewport.

Construction of Derived Polygon Meshes

There are two ways to derive a polygon mesh from existing objects: from closed polylines or from control points of NURBS surfaces. These methods are explored in the sections that follow.

Constructing a Polygon Mesh from a Closed Polyline

You can construct a polygon mesh from a closed polyline, which can be planar or 3D. Perform the following steps.

1. Open the file *Mesh2.3dm* from the *Chapter 6* folder on the companion CD-ROM.
2. Select Tools > Polygon Mesh > From Closed Polyline, or click on the Mesh from Closed Polyline button on the Mesh toolbar.
3. Select polyline A (indicated in figure 6-74).

A polygon mesh is constructed.

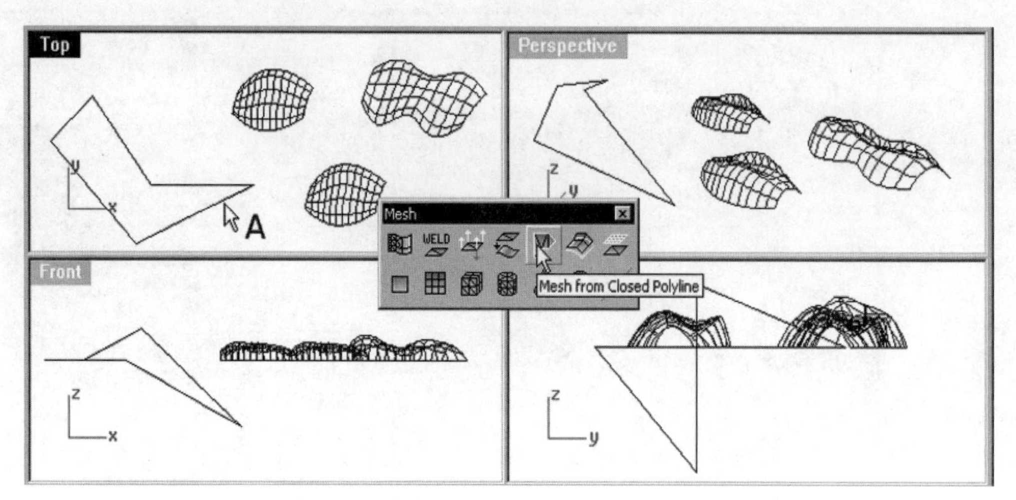

Fig. 6-74. Polygon mesh being constructed from a closed polyline.

Constructing a Polygon Mesh from the Control Points of a NURBS Surface

You can construct a polygon mesh using the control points of a NURBS surface or curves as vertices. Perform the following steps.

1. Maximize the Perspective viewport.
2. Select Tools > Polygon Mesh > From NURBS Control Polygon, or click on the Extract Mesh from NURBS Control Polygon button on the Mesh toolbar.
3. Select surface A (indicated in figure 6-75) and press the Enter key.

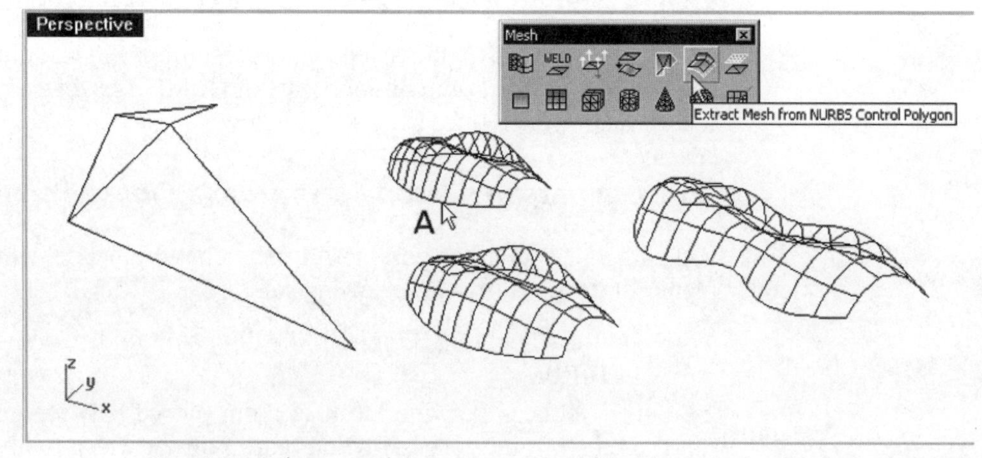

Fig. 6-75. Polygon mesh being constructed from the control vertices of a NURBS surface.

A polygon mesh is constructed.

Constructing a Polygon Mesh from a NURBS Surface

NURBS surfaces and polygon meshes are two different types of objects. A NURBS surface is an exact representation, but the polygon mesh is an approximated representation of the 3D object. If you already have a NURBS surface model, you can construct a polygon mesh from it, as follows.

1. Select Tools > Polygon Mesh > From NURBS Objects, or click on the *Mesh from Surface/Polysurface* button on the Mesh toolbar.
2. Select surface A (indicated in figure 6-76) and press the Enter key.
3. In the Polygon Mesh Options dialog box, set the polygon density by moving the slider bar, and then click on the OK button.

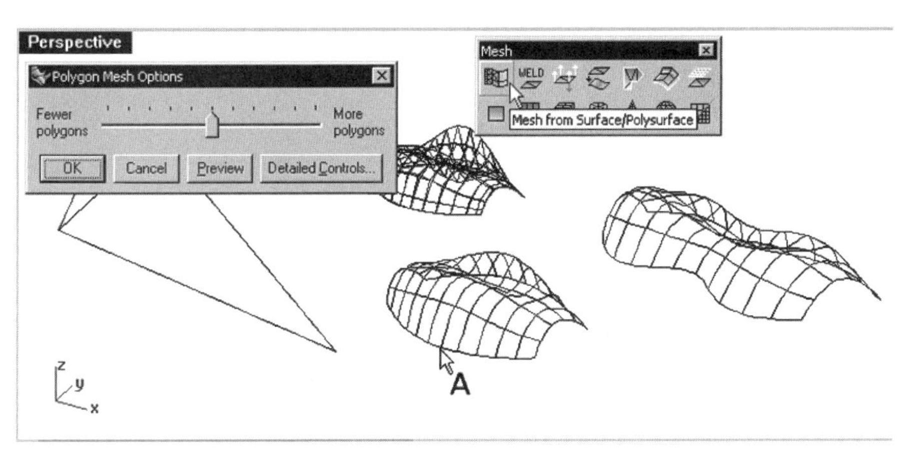

Fig.6-76. Polygon mesh being constructed from a surface profile.

Mapping a Polygon Mesh to a NURBS Surface

In many animation programs that use polygon meshes to represent 3D objects, morphing of two different shapes requires the polygon meshes to have an identical vertex count and mesh structure. To construct two different shapes having identical vertex count and mesh structure, you construct two shapes as NURBS surfaces, construct a polygon mesh from a surface, and map the polygon mesh to the other surface. Perform the following steps.

1. Select Tools > Polygon Mesh > Apply to Surface, or click on the Apply Mesh to NURBS Surface button on the Mesh toolbar.
2. Select mesh A (indicated in figure 6-77).
3. Select surface B (indicated in figure 6-77).

The polygon mesh constructed from a NURBS surface is applied to another surface, as shown in figure 6-78.

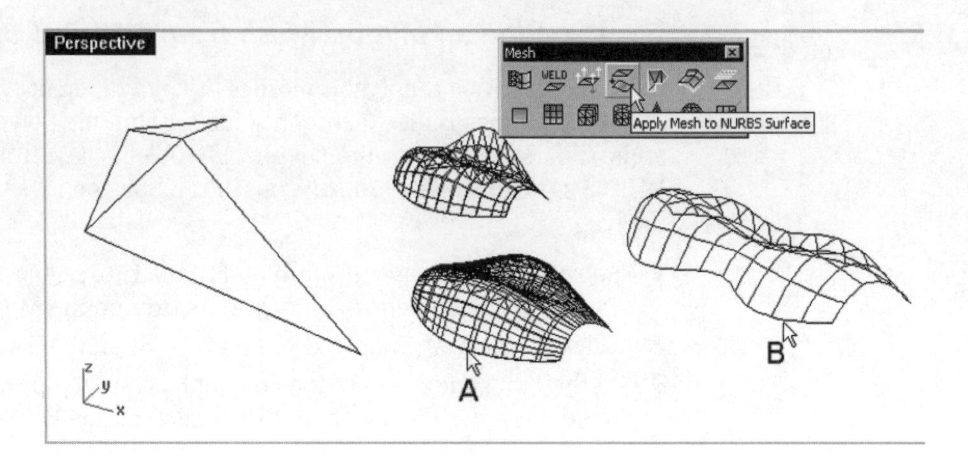

Fig. 6-77. Polygon mesh being applied to a surface.

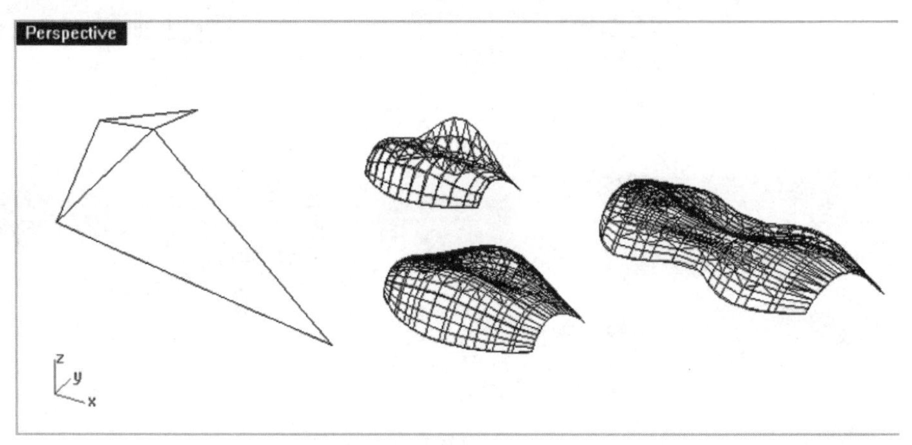

Fig. 6-78. Polygon mesh applied to a surface.

Editing Polygon Meshes

You edit polygon meshes in several ways, including reducing the number of polygons in the mesh, exploding joined polygon meshes into individual polygon meshes, joining contiguous polygon meshes, and welding polygon meshes to form a single polygon mesh object. These operations are explored in the sections that follow.

Exploding Joined Polygon Meshes

Contrary to joining, you can explode joined polygon meshes, as follows.

1. Open the file *Mesh3.3dm* from the *Chapter 6* folder on the companion CD-ROM.
2. Select Edit > Explode, or click on the Explode button on the Main2 toolbar.
3. Select the joined polygon meshes and press the Enter key.

The meshes are exploded into individual polygon meshes, as shown in figure 6-79.

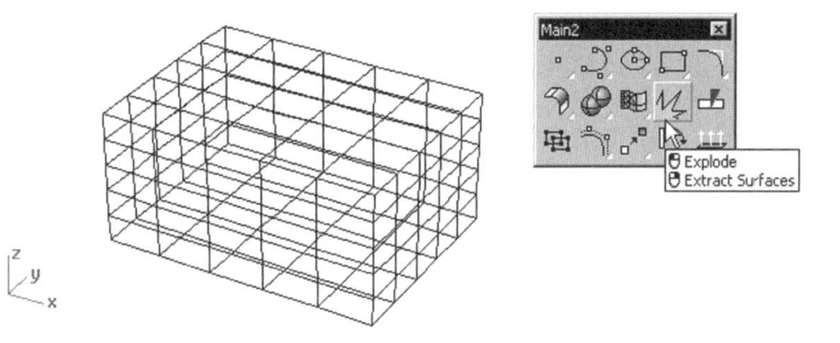

Fig. 6-79. Polygon mesh box being exploded into six meshes.

Joining Polygon Meshes

Like joining curves and NURBS surfaces, you can join two contiguous polygon meshes, as follows.

1. Open the file *Mesh4.3dm* from the *Chapter 6* folder on the companion CD-ROM.
2. Select Edit > Join, or click on the Join button on the Main1 toolbar.
3. Select all the meshes and press the Enter key.

The selected meshes are joined.

Unifying Mesh Normals

When you join polygon meshes, there is a chance their normal direction is not uniform. To unify the normal directions of the joined meshes, perform the following steps.

1. Select Tools > Polygon Mesh > Unify Normals, or click on the Unify Mesh Normals button on the mesh toolbar.
2. Select the joined mesh.

The normal directions of the meshes are unified.

Welding Polygon Meshes

At a glance, welding seems to be analogous to joining. However, these processes are different in that a joined set of polygon meshes still retains the common edges between contiguous meshes and a welded set of polygon meshes has its contiguous duplicated edges removed. Perform the following steps.

1. Select Tools > Polygon Mesh > Weld, or click on the Weld Mesh Vertices button on the Mesh toolbar.
2. Select joined polygon meshes and press the Enter key.
3. Type *270* at the command line interface to specify the angular tolerance between contiguous polygon meshes.

The polygon meshes are welded into a single polygon mesh, as shown in figure 6-80.

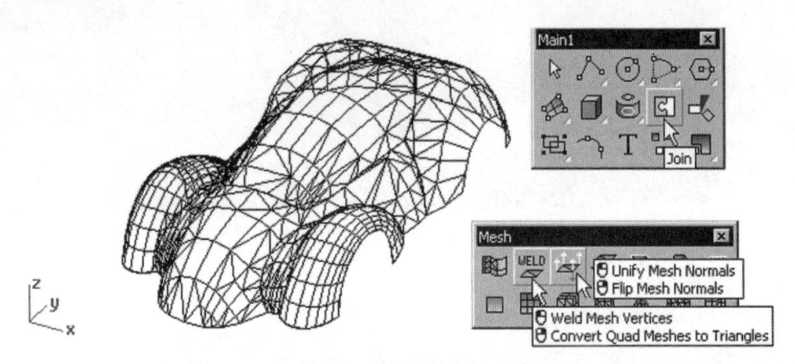

Fig. 6-80. Polygon meshes being welded.

Reduction of Polygon Density

As explained earlier, a high polygon mesh density results in a large file size. If the polygon mesh is overly dense, you can reduce its density. Decreasing the polygon count means simplifying the mesh. To reduce a polygon density, perform the following steps. This process is irreversible, meaning that if you reduce the polygon count and later want to increase the count you have to rebuild the mesh.

1. Select Tools > Polygon Mesh > Reduce, or click on the Reduce Mesh Polygon Count button on the Mesh toolbar.
2. Select the mesh polygon and press the Enter key.
3. In the Reduce Mesh Options dialog box, shown in figure 6-81, reduce the polygon count to 499 and click on the OK button.

The polygon mesh count is reduced, as shown in figure 6-82.

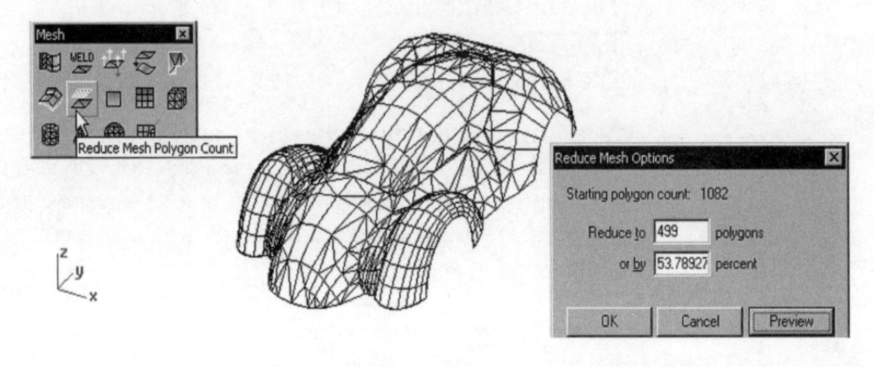

Fig. 6-81. Polygon mesh density being reduced.

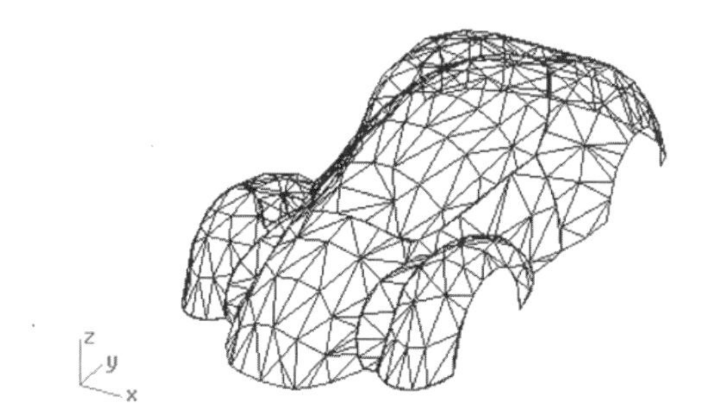

Fig. 6-82. Reduced polygon mesh density.

Transforming and Analyzing Polygon Meshes

The set of tools you use on surfaces also applies to polygon meshes. (See figure 6-51 and Table 6-2.) In addition, you can also use the Section and Contour commands to construct NURBS curves from mesh objects. From the NURBS curves, you construct NURBS surfaces.

Summary

After you construct a NURBS surface, you can modify it in many ways. You can join surfaces to form a polysurface, explode a polysurface into individual surfaces, trim surfaces to remove unwanted portions, and split surfaces. To change a surface profile, you manipulate control points, insert or remove knots, use the Handlebar editor, and use various advanced tools. To manipulate the edges of a surface, you display its edges, split an edge, merge a split edge, join edges, and rebuild edges. To transform a surface, you use translate and deform tools.

In regard to analysis, you can construct a bounding box; evaluate area, area centroid, and area moments; render curvature, draft angle, environment map, and zebra stripes; and construct curvature arcs on a surface and points on a surface. You also learned how to flip the normal direction.

Although NURBS surfaces are the mainstream surface type in design and manufacture, there are still some systems using polygon meshes to represent a free-form body. To cope with these systems, you may need to know how to construct and manipulate polygon meshes.

Review Questions

1. List the methods used to edit and transform surfaces.
2. List the methods you might use to analyze a NURBS surface.
3. Outline the methods of constructing various types of polygon meshes.

CHAPTER 7

Solid Modeling

Introduction

This chapter explores concepts of solid modeling. It also outlines various solid modeling methods, explains Rhino's solid modeling methods, and provides you with practice in creating 3D solids using Rhino.

Objectives

After studying this chapter you should be able to:

- ❐ Describe the key concepts of solid modeling
- ❐ State various types of solid modeling methods
- ❐ Explain Rhino's solid modeling methods and relate them to its surface modeling tools, which you learned about in previous chapters
- ❐ Use Rhino as a design tool to construct solid objects and integrate them with other surface objects

Overview

Among the three types of 3D models, a solid model provides the most comprehensive information about an object. A solid model has associated volume data, and associated face, edge, and vertex information. There are many ways to represent a solid in the computer: pure primitive instancing, generalized sweeping, spatial occupancy enumeration, cellular decomposition, constructive solid geometry, and boundary representation.

Most contemporary solid modeling applications use a hybrid approach. Basically, Rhino represents a solid using a polysurface that encloses a volume (has no gap or opening). In this chapter you will learn solid modeling concepts and methods of representing solids in the computer. You will also construct solid objects using Rhino. In the next chapter you will learn how to produce photorealistic images, as well as 2D engineering drawings.

Solid Modeling Concepts

Unlike a wireframe model (which is a set of unassociated curves that depict the edges of a 3D object) and a surface model (which is a set of unassociated surfaces that delineate the boundary faces of a 3D object), a solid model integrates the mathematical data of a 3D object in the computer. A solid model has associated with it the edge, vertex, face, and volume data of the object it represents. Solid models are used in various downstream computerized operating processes, including finite element analysis, rapid prototyping, and CNC machining.

Because any individual 3D object is unique in shape, it would be impossible to derive a general mathematical formula to represent all types of shapes. Several of the mathematical methods used to derive various types of shapes are explored in the sections that follow.

Primitive Instancing

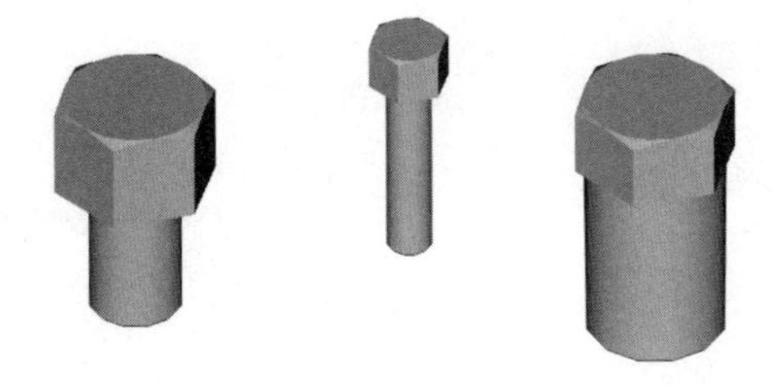

Fig. 7-1. Family of instances derived from a pure primitive.

Primitive instancing predefines a library of primitive solid objects. To make a solid object, you select a primitive from the library and specify the value of the parameters of the primitive. Figure 7-1 shows a family of solid objects derived from a primitive. In this method of constructing a solid model you construct model shapes defined by the primitives contained in the library.

Generalized Sweeping

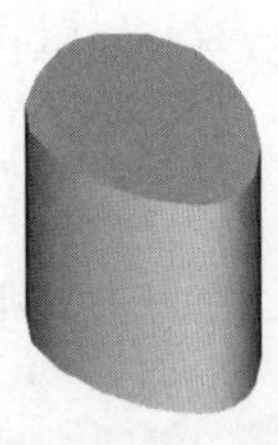

Fig. 7-2. Solid represented by sweeping linearly.

Generalized sweeping creates a solid object by sweeping a 2D or 3D closed-loop curve or lamina in 3D space. You construct a closed-loop curve or lamina and sweep it in a linear direction, about an axis, or along a 3D curve. These methods are depicted in figures 7-2 through 7-4. Generalized sweeping provides a flexible way of constructing solids of various shapes.

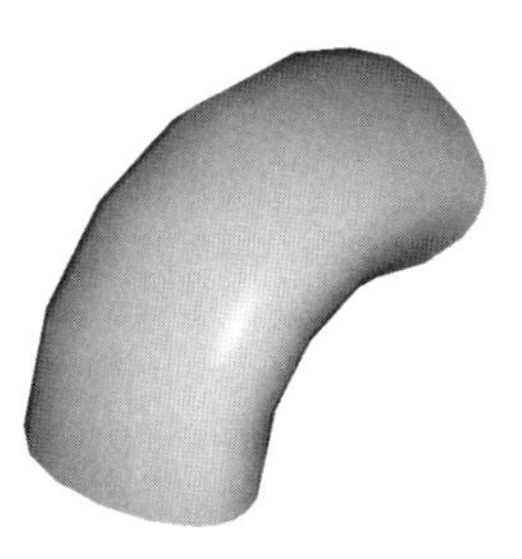

Fig. 7-3. Solid represented by sweeping about an axis.

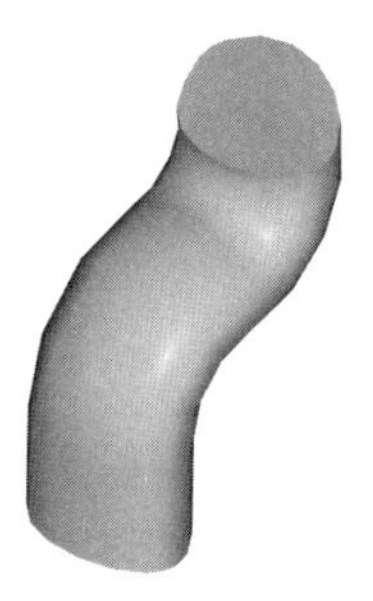

Fig. 7-4. Solid represented by sweeping along a 3D curve.

Spatial Occupancy Enumeration

Spatial occupancy enumeration divides the entire 3D space into a number of identical cubical cells. You construct a solid by listing the cells the object occupies in 3D space. The accuracy of the representation of the solid is a function of the size of the cubical cells.

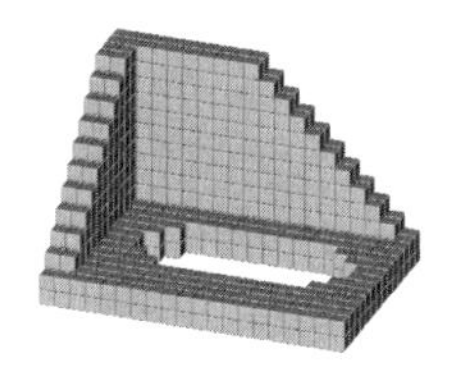

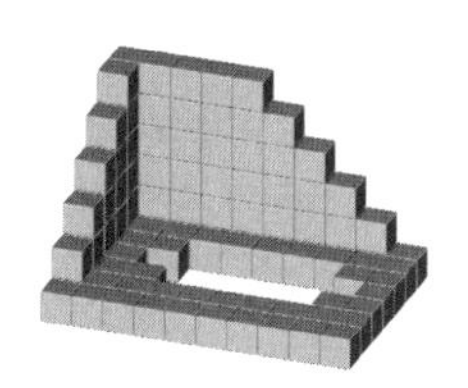

Fig. 7-5. Solids represented as a measure of the cubical cells they occupy in 3D space.

To represent an object more accurately, you use cells of smaller size. As a result of a decrease in cell size, file size will increase tremendously. Figure 7-5 shows a solid object represented at two resolution settings. This method is not widely used in geometry-based modeling systems. However, it is a basis for finite element analysis.

Cellular Decomposition

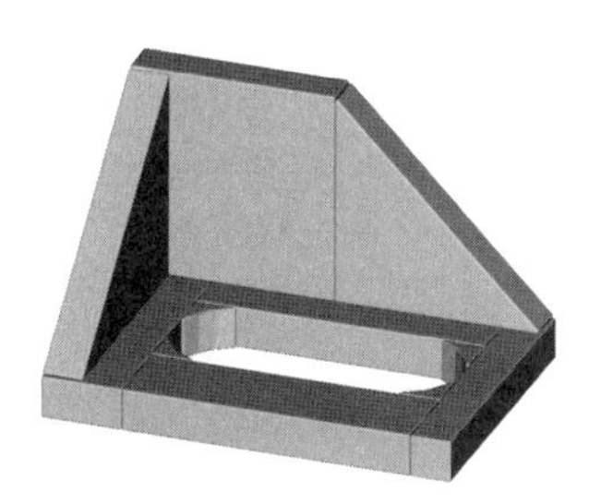

Fig. 7-6. Solid represented by cells of different shapes and sizes.

Cellular decomposition is similar to the spatial occupancy enumeration method in that you construct a solid by listing the cells it occupies in 3D space, an example of which is shown in figure 7-6. However, unlike the spatial occupancy enumeration method the cells are not always identical or cubical in shape. As a result, this method requires less memory to represent objects in a more accurate way. Again, this method is not widely used in geometric modeling.

Boundary Representation

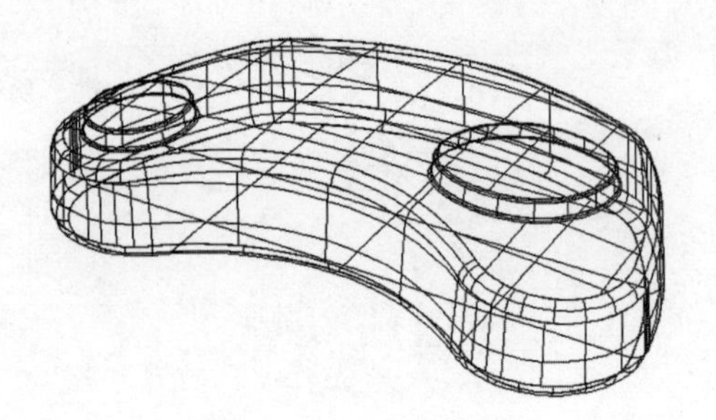
Fig. 7-7. Set of surfaces enclosing a volume.

Boundary representation (B-rep) derives from surface modeling techniques, with the addition of information on the connectivity of surfaces and specification of which side of the surface is solid and which is void. This method defines a solid using a set of surfaces that must not intersect with each other (except at their common vertices or edges) and that must form a closed loop encompassing a volume.

This method of constructing a solid is tedious because you have to construct all surfaces one at a time, as if you were making a surface model. Nevertheless, it enables you to construct a solid object with free-form surfaces, an example of which is shown in figure 7-7.

Constructive Solid Geometry

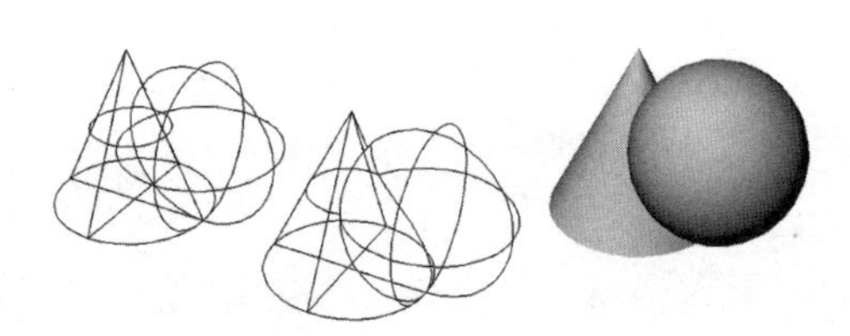
Fig. 7-8. Union of two solids: cone and sphere.

Constructive solid geometry (CSG), also known as set-theoretic modeling, provides a range of primitive solids in a manner similar to the pure primitive instancing method. In addition, it provides an extra facility for you to obtain a solid of complex shape by combining primitive solids via Boolean operations.

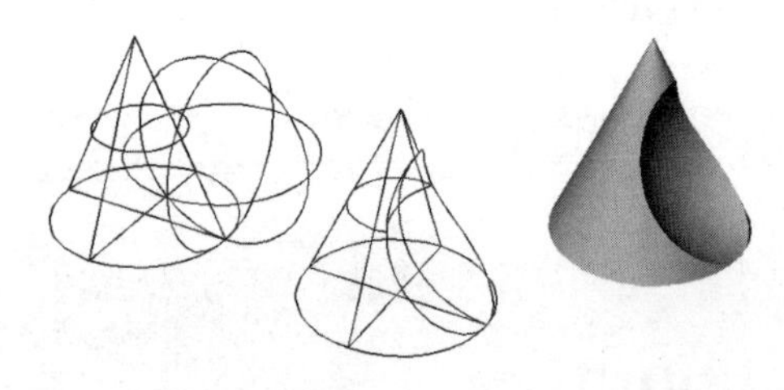
Fig. 7-9. Intersection of two solids: cone and sphere.

To construct a solid of complex shape, you combine solid objects using the union, intersection, and difference operators of set theory. A *union* of two solids, A and B, forms a complex solid representing the volume encompassed by solid A or solid B. An *intersection* of two solids, A and B, forms a complex solid representing the volume encompassed by solid A and solid B. A *difference* of solid A and solid B forms a complex solid representing the volume encompassed by solid A but not solid B. Examples of these states of solids are shown in figures 7-8 through 7-10.

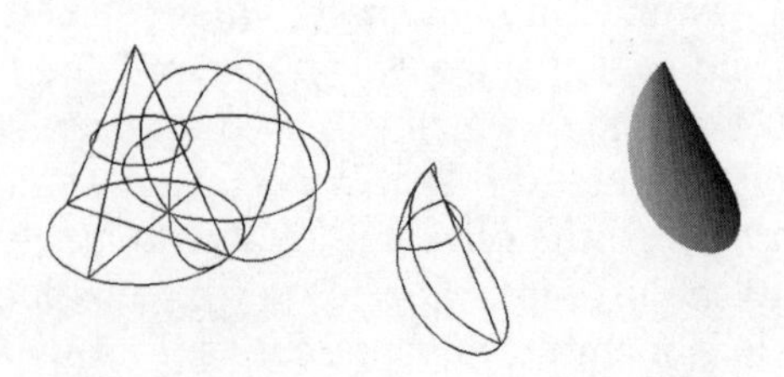
Fig. 7-10. Difference of two solids: cone and sphere.

Hybrid System

Most contemporary computer-aided design applications use a hybrid approach to the representation of solids. In general, they provide the following.

- A set of primitive solids of basic geometric shapes
- A facility for constructing solids by sweeping curves or laminas
- A facility for constructing a solid from a set of surfaces enclosing a volume
- A facility for combining solids using Boolean operations

Rhino Solid Modeling Tools

Rhino uses the B-rep method to represent a solid object. A Rhino solid is a polysurface that encloses a volume without any gap or opening. Individual surfaces of the polysurface must not intersect except at their edges and vertices.

Rhino enables you to construct solid primitives ranging from simple 3D boxes to 3D text objects, to construct solids by extruding planar curves and surfaces, to convert a set of surfaces to a solid by adding planar surfaces to define an enclosed volume, to modify edges of solids by filleting, and to combine solid objects using Boolean operations. You perform these operations from the Solids pull-down menu or from the Solid and Solid Tools toolbars.

Making Primitive Solids

Rhino primitive solids are closed polysurfaces of basic geometric shapes. There are eleven types: box, sphere, ellipsoid, paraboloid, cone, truncated cone, cylinder, tube, pipe, torus, and text. Like primitive surfaces, curves and points are not required. To create a primitive solid, you select a type and specify the parameters. Figure 7-11 shows primitive solids commands on several toolbars.

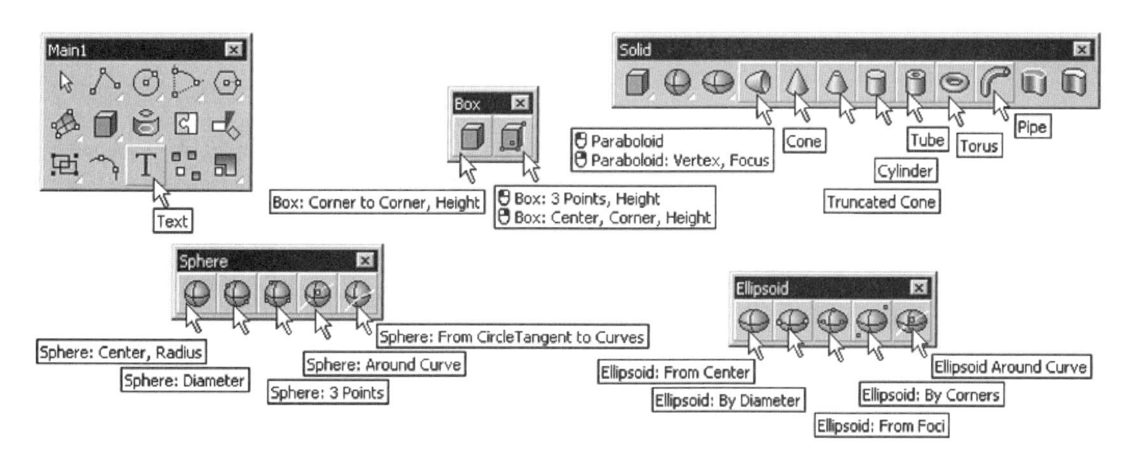

Fig. 7-11. Primitive solids commands on the Solid, Box, Sphere, Ellipsoid, and Main1 toolbars.

Table 7-1 outlines the functions of primitive solids commands and their locations in the main pull-down menu and respective toolbars.

Table 7-1: Primitive Solids Commands and Their Functions

Toolbar Button	From Main Pull-down Menu	Function
Box: Corner to Corner, Height	Solid > Box > Corner to Corner, Height	Constructing a box by specifying two opposite corners of the base and the height
Box: 3 Points, Height; Box: Center, Corner, Height	Solid > Box > 3 Points, Height	Constructing a box by specifying two corners of an edge of the base, width of the base, and the height of the box
Sphere: Center, Radius	Solid > Sphere > Center, Radius	Constructing a sphere by specifying its center and the radius
Sphere: Diameter	Solid > Sphere > Diameter	Constructing a sphere by using two points to specify its diameter
Sphere: 3 Points	Solid > Sphere > 3 Points	Constructing a sphere by specifying three points on the surface of the sphere
Sphere: Around Curve	—	Constructing a sphere by specifying the center point on a curve and the radius of the sphere
Sphere: From Circle Tangent to Curves	—	Constructing a sphere tangent to three curves
Ellipsoid: From Center	Solid > Ellipsoid > From Center	Constructing an ellipsoid by specifying its center, first axis, second axis, and third axis
Ellipsoid: By Diameter	—	Constructing an ellipsoid by specifying the end points of one of the axes and the end points of the remaining axes
Ellipsoid: From Foci	Solid > Ellipsoid > From Foci	Constructing an ellipsoid by specifying the foci and a point on the ellipsoid
Ellipsoid: By Corners	—	Constructing an ellipsoid by specifying the corners of an imaginary bounding box
Ellipsoid Around Curve	—	Constructing an ellipsoid by specifying the center on a curve and the end points of the axes
Paraboloid	Solid > Paraboloid > Focus, Direction	Constructing a paraboloid by specifying the focus, direction, and end point of the paraboloid

Toolbar Button	From Main Pull-down Menu	Function
Paraboloid: Vertex, Focus	Solid > Paraboloid > Vertex, Focus	Constructing a paraboloid by specifying the vertex, direction, and end point of the paraboloid
Cone	Solid > Cone	Constructing a cone by specifying the center, radius of the base, and end point of the cone
Truncated Cone	Solid > Truncated Cone	Constructing a truncated cone by specifying the center, two radii, and end point of the truncated cone
Cylinder	Solid > Cylinder	Constructing a cylinder by specifying its center, radius, and height
Tube	Solid > Tube	Constructing a tube by specifying its center, inner and outer radii, and height
Torus	Solid > Torus	Constructing a torus by specifying the center, radius of the torus, and radius of the cross section
Pipe	Solid > Pipe	Constructing a pipe by specifying a curve, and inner and outer radii at both ends of the pipe
Text	Solid > Text	Constructing text objects

Box

A box is a polysurface with six planar surfaces that are mutually perpendicular to one another. To construct a box, perform the following.

1. Start a new file. Use the metric (Millimeters) template file.
2. Select Solid > Box > Corner to Corner, Height, or click on the *Box: Corner to Corner, Height* button on the Box toolbar.
3. Select locations A, B, and C (indicated in figure 7-12).
4. Select Solid > Box > 3 Points, Height, or click on the *Box: 3 Points, Height/ Box: Center, Corner, Height* button on the Box toolbar.

Two solid boxes are constructed.

5. If you want to save your file, save it as *Boxes.3dm*.

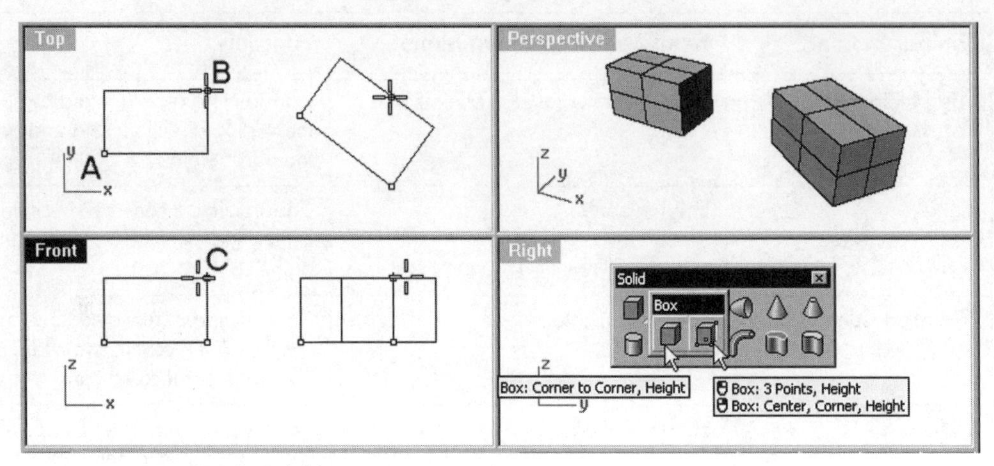

Fig. 7-12. Solid boxes constructed.

Sphere

A sphere is a single closed-loop surface. There are five methods of constructing a sphere, as follows. The first method constructs a sphere by specifying its center and a point on its surface, as follows.

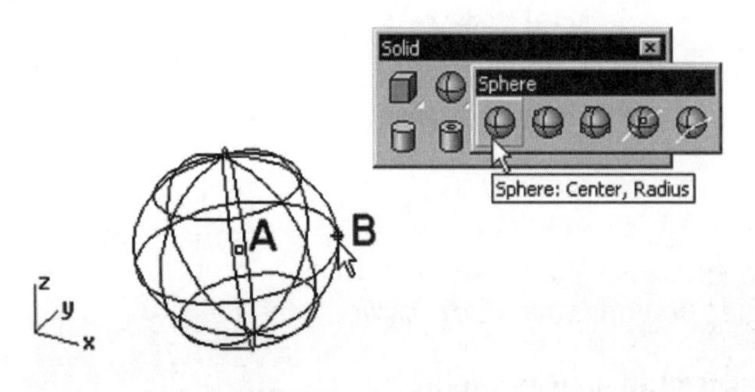

Fig. 7-13. Sphere being constructed by specifying its center and a point on the surface.

1. Start a new file. Use the metric (Millimeters) template file.
2. Maximize the Perspective viewport.
3. Select Solid > Sphere > Center, Radius, or click on the *Sphere: Center, Radius* button on the Sphere toolbar.
4. Select locations A and B (indicated in figure 7-13) to specify the center point and a point on the surface.
5. If you want to save your file, save it as *Sphere1.3dm.*

A sphere is constructed by specifying its center and a point on the surface. A second method of creating a sphere does so by specifying the sphere's diameter, as follows.

1. Start a new file. Use the metric (Millimeters) template file.
2. Maximize the Perspective viewport.

3 Select Solid > Sphere > Diameter, or click on the *Sphere: Diameter* button on the Sphere toolbar.
4 Select locations A and B (indicated in figure 7-14) to specify a diameter for the sphere.
5 If you want to save your file, save it as *Sphere1.3dm*.

A sphere is constructed by specifying its diameter.

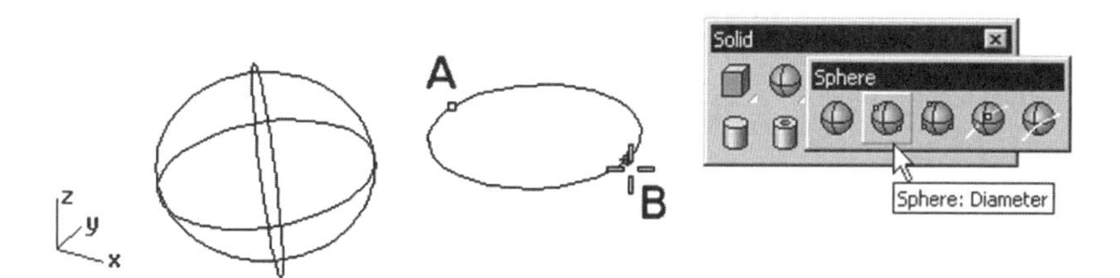

Fig. 7-14. Sphere being constructed by specifying its diameter.

A third method of sphere construction specifies three points on a surface, as follows.

1 Start a new file. Use the metric (Millimeters) template file.
2 Maximize the Perspective viewport.
3 Select Solid > Sphere > 3 Points, or click on the *Sphere: 3 Points* button on the Sphere toolbar.
4 Select locations A, B, and C (indicated in figure 7-15) to specify three points on the surface of the sphere.
5 If you want to save your file, save it as *Sphere1.3dm*.

A sphere is constructed by specifying three points on the surface.

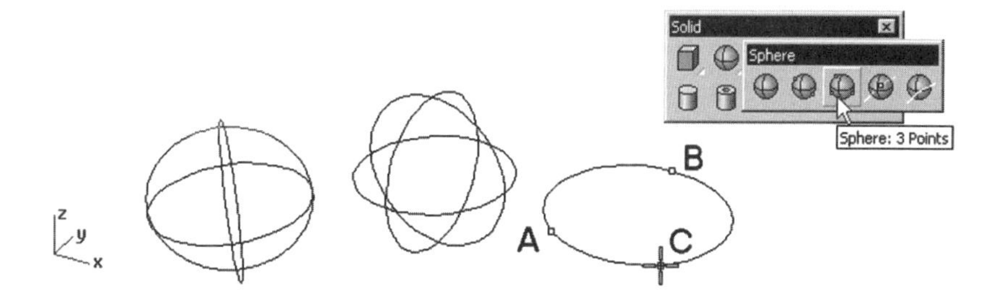

Fig. 7-15. Sphere being constructed by specifying three points on its surface.

A fourth method of constructing a sphere does so by specifying three curves to which the sphere is tangent, as follows.

1 Start a new file. Use the metric (Millimeters) template file.
2 With reference to figure 7-16, construct two curves in the Top viewport and a curve in the Front viewport.

3 Click on the *Sphere: From Circle Tangent to Curves* button on the Sphere toolbar.
4 Select locations A, B, and C (indicated in figure 7-16) on the curves to specify the tangent points.
5 If you want to save your file, save it as *Sphere2.3dm*.

A sphere tangent to three curves is constructed.

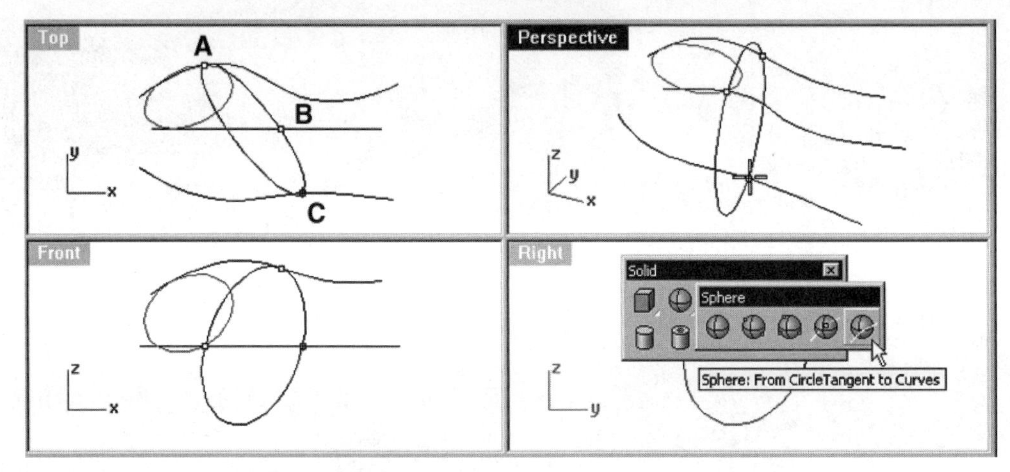

Fig. 7-16. Sphere tangent to three curves being constructed.

A fifth method constructs a sphere on a curve, as follows.

1 Start a new file. Use the metric (Millimeters) template file.
2 With reference to figure 7-17, construct two curves in the Top viewport and a curve in the Front viewport.

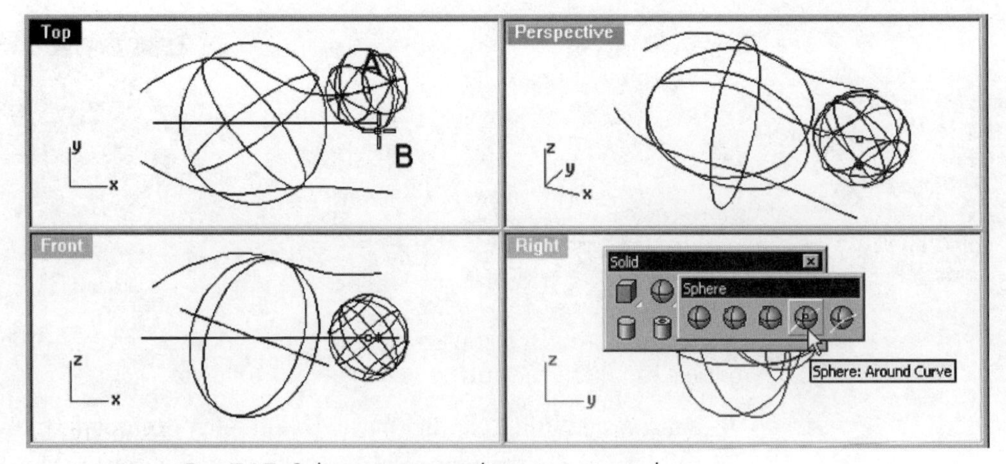

Fig. 7-17. Sphere on a curve being constructed.

3 Click on the *Sphere: Around Curve* button on the Sphere toolbar.
4 Select location A (indicated in figure 7-17) on one of the curves and then select location B (indicated in figure 7-17).
5 If you want to save your file, save it as *Sphere2.3dm*.

A sphere on a curve is constructed, as shown in figure 7-18.

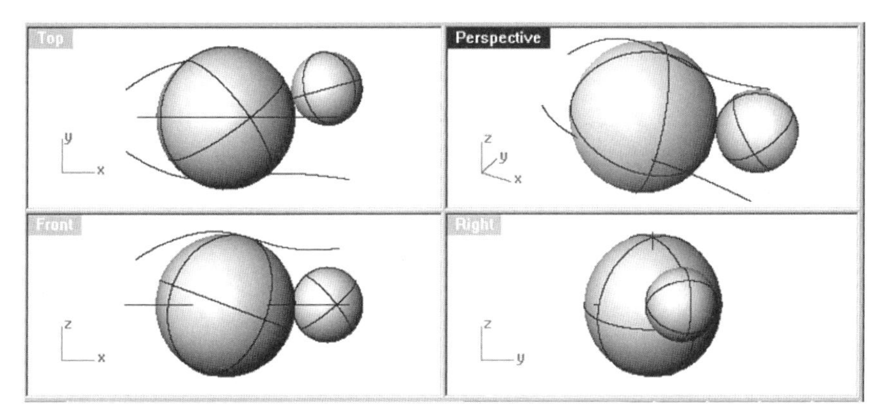

Fig. 7-18. Sphere on a curve.

Ellipsoid

An ellipsoid is also a single closed-loop surface. You can construct an ellipsoid via five methods, as follows. The first method creates an ellipsoid by specifying the ellipsoid's center and axis end points, as follows.

1 Start a new file. Use the metric (Millimeters) template file.
2 Construct a free-form curve.
3 Select Solid > Ellipsoid > From Center, or click on the *Ellipsoid: From Center* button on the Ellipsoid toolbar.
4 Select location A to specify the center and locations B, C, and D (indicated in figure 7-19) to specify the axis end points (indicated in figure 7-19).
5 If you want to save your file, save it as *Ellipsoid.3dm*.

An ellipsoid specified by its center and its axis end points is constructed. A second method creates an ellipsoid by specifying its diameter and two axis end points, as follows.

1 Start a new file. Use the metric (Millimeters) template file.
2 Construct a free-form curve.
3 Click on the *Ellipsoid: By Diameter* button on the Ellipsoid toolbar.
4 Select locations A and B (indicated in figure 7-20) to specify the diameter, and then select locations C and D to specify the axis end points (indicated in figure 7-20).
5 If you want to save your file, save it as *Ellipsoid.3dm*.

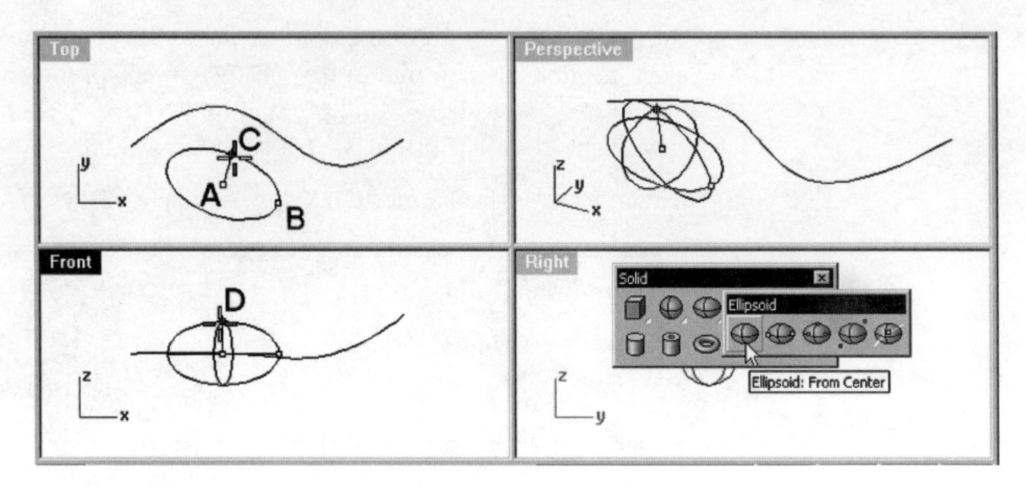

Fig. 7-19. Ellipsoid specified by its center and axis end points.

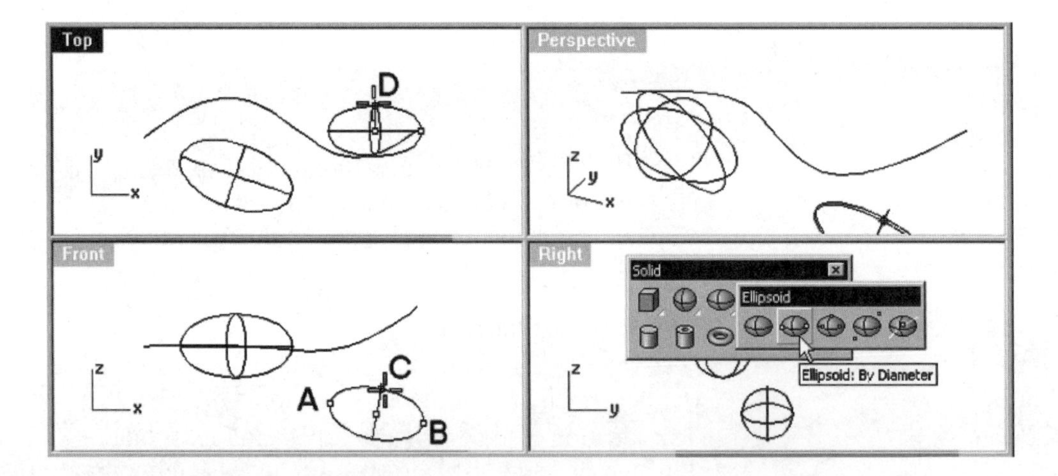

Fig. 7-20. Ellipsoid specified by its diameter being constructed.

An ellipsoid specified by its diameter and two axis end points is constructed. A third method creates an ellipsoid by specifying its foci and a point on the ellipsoid, as follows.

1. Start a new file. Use the metric (Millimeters) template file.
2. Construct a free-form curve.
3. Select Solid > Ellipsoid > From Foci, or click on the *Ellipsoid: From Foci* button on the Ellipsoid toolbar.
4. Select locations A and B (indicated in figure 7-21) to specify the foci, and then select location C to specify a point on the ellipsoid (indicated in figure 7-21).

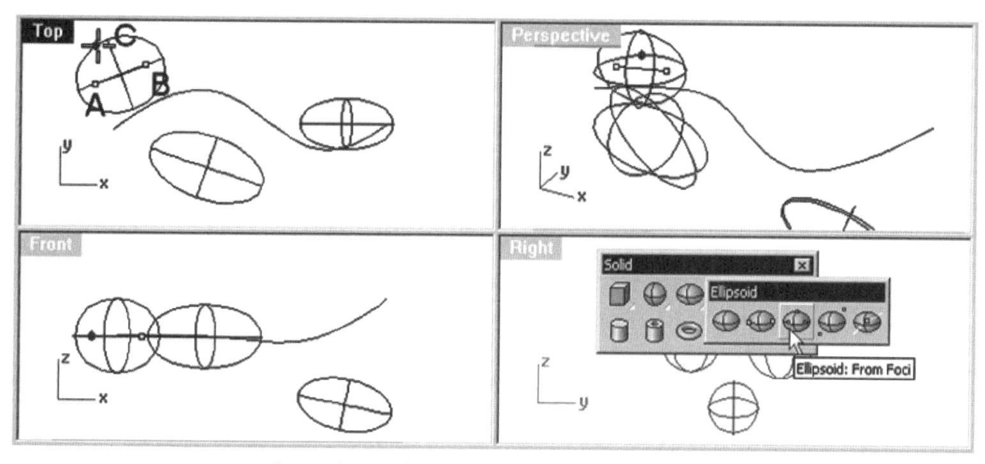

Fig. 7-21. Ellipsoid specified by its foci being constructed.

5 If you want to save your file, save it as *Ellipsoid.3dm*.

An ellipsoid specified by its foci and a point on the ellipsoid is constructed. A fourth method creates an ellipsoid by specifying a bounding box, as follows.

1 Start a new file. Use the metric (Millimeters) template file.
2 Construct a free-form curve.
3 Click on the *Ellipsoid: By Corners* button on the Ellipsoid toolbar.
4 Select locations A, B, and C (indicated in figure 7-22) to specify the corners of the bounding box.
5 If you want to save your file, save it as *Ellipsoid.3dm*.

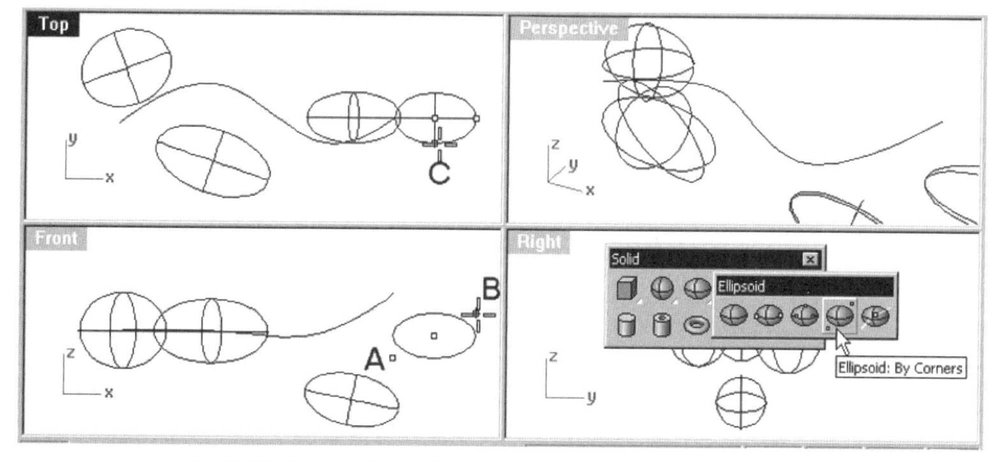

Fig. 7-22. Ellipsoid specified by its bounding box being constructed.

An ellipsoid specified by its bounding box is constructed. A fifth method creates an ellipsoid on a curve, as follows.

1. Start a new file. Use the metric (Millimeters) template file.
2. Construct a free-form curve.
3. Click on the Ellipsoid Around Curve button on the Ellipsoid toolbar.
4. Select one of the curves at location A (indicated in figure 7-23) to specify the center, and then select locations B, C, and D (indicated in figure 7-23) to specify the axis end points.

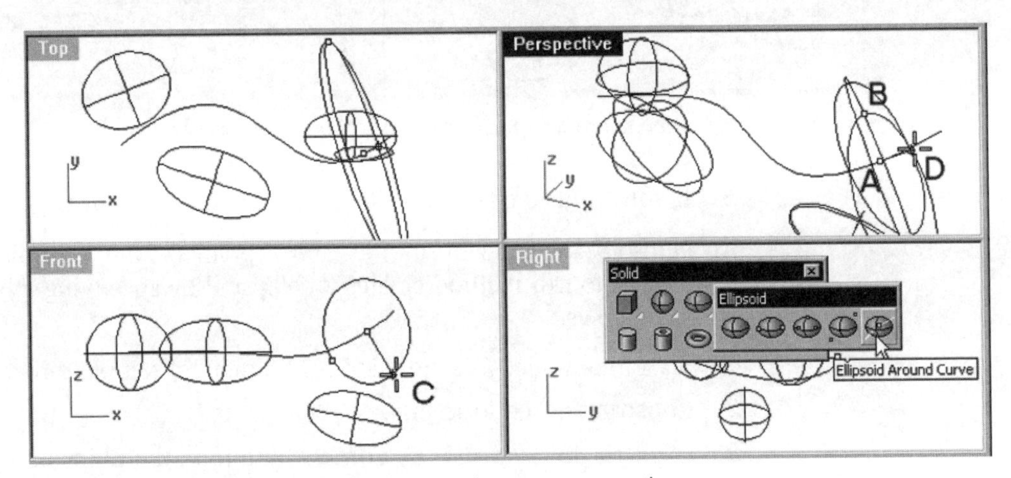

Fig. 7-23. Ellipsoid on a curve being constructed.

5. If you want to save your file, save it as *Ellipsoid.3dm*.

An ellipsoid is constructed on a curve.

Paraboloid

A paraboloid is a single closed-loop surface. To construct a paraboloid, perform the following steps.

1. Start a new file. Use the metric (Millimeters) template file.
2. Maximize the Perspective viewport.
3. Select Solid > Paraboloid > Focus, Direction, or click on the *Paraboloid/Paraboloid: Vertex, Focus* button on the Solid toolbar.
4. Type *C* at the command line area if the prompt there indicates "Cap = No."
5. Select locations A, B, and C (indicated in figure 7-24) to indicate the focus, direction, and end point of the paraboloid.

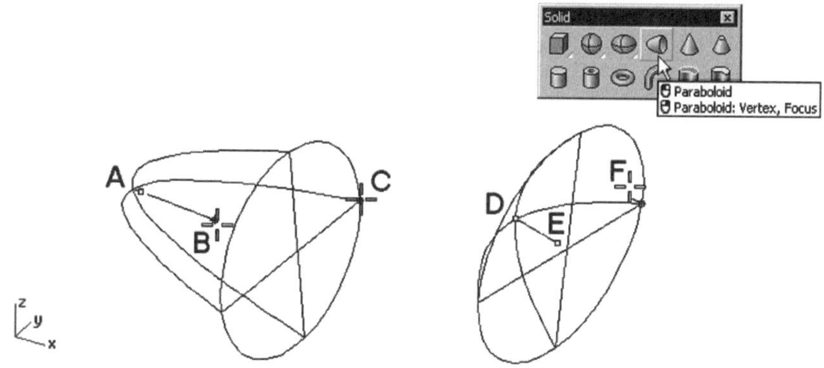

Fig. 7-24. Two paraboloids being constructed.

6 Select Solid > Paraboloid > Vertex, Focus, or right-click on the *Paraboloid/Paraboloid: Vertex, Focus* button on the Solid toolbar.

7 Select locations D, E, and F (indicated in figure 7-24) to specify the vertex, focus, and an end point of the paraboloid.

8 If you want to save your file, save it as *Paraboloid.3dm*.

Two paraboloids are constructed.

Cone

A cone is a polysurface consisting of two joined surfaces. The slant surface is a closed surface and the base is a circular planar surface. To construct a cone, perform the following steps.

1 Start a new file. Use the metric (Millimeters) template file.

2 Select Solid > Cone, or click on the Cone button on the Solid toolbar.

3 Select two points A and B (indicated in figure 7-25) in the Top viewport to indicate the center and radius, and then select location C (indicated in figure 7-25) in the Front viewport to indicate the height.

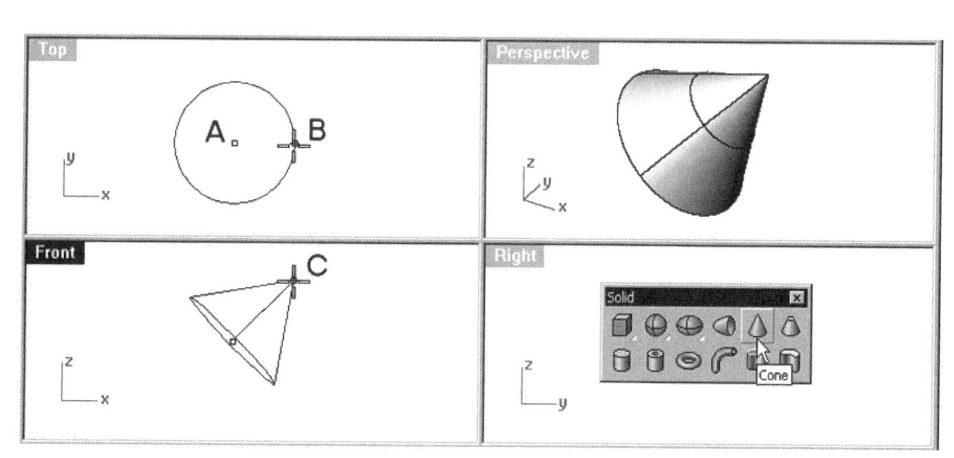

Fig. 7-25. Cone being constructed.

A cone is constructed.

4 If you want to save your file, save it as *Cone.3dm*.

Truncated Cone

A truncated cone is a polysurface consisting of three joined surfaces. The slant surface is a closed surface and the top and bottom surfaces are circular planar surfaces. To construct a truncated cone, perform the following steps.

1 Start a new file. Use the metric (Millimeters) template file.

2 Select Solid > Truncated Cone, or click on the Truncated Cone button on the Solid toolbar.

3 Select three locations A, B, and C (indicated in figure 7-26) in the Top viewport to indicate the center and radii at both ends, and then select location D (indicated in figure 7-26) in the Front viewport to indicate the height.

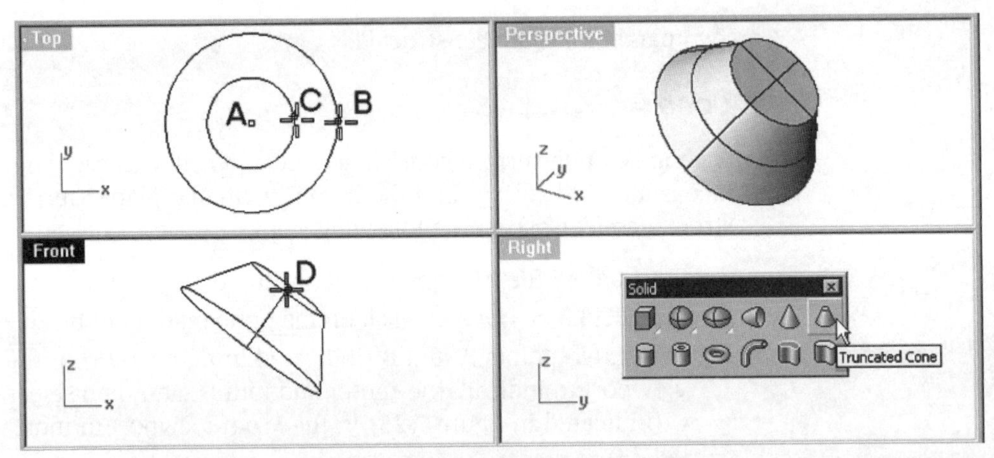

Fig. 7-26. Truncated cone being constructed.

A truncated cone is constructed.

4 If you want to save your file, save it as *TruncatedCone.3dm*.

Cylinder

A cylinder is a polysurface consisting of three surfaces. The body is a cylindrical surface and the top and bottom surfaces are circular planar surfaces. To construct a cylinder, perform the following steps.

1 Start a new file. Use the metric (Millimeters) template file.

2 Select Solid > Cylinder, or click on the Cylinder button on the Solid toolbar.
3 Select locations A and B (indicated in figure 7-27) in the Top viewport to indicate the center and radius.
4 Hold down the Shift key and select location C (indicated in figure 7-27) in the Front viewport to indicate the height. (Holding down the key causes the selected point to be orthogonal to the base of the cylinder. Without holding down the Shift key, you select a point on the Front viewport instead.)

A cylinder is constructed.

5 If you want to save your file, save it as *Cylinder.3dm*.

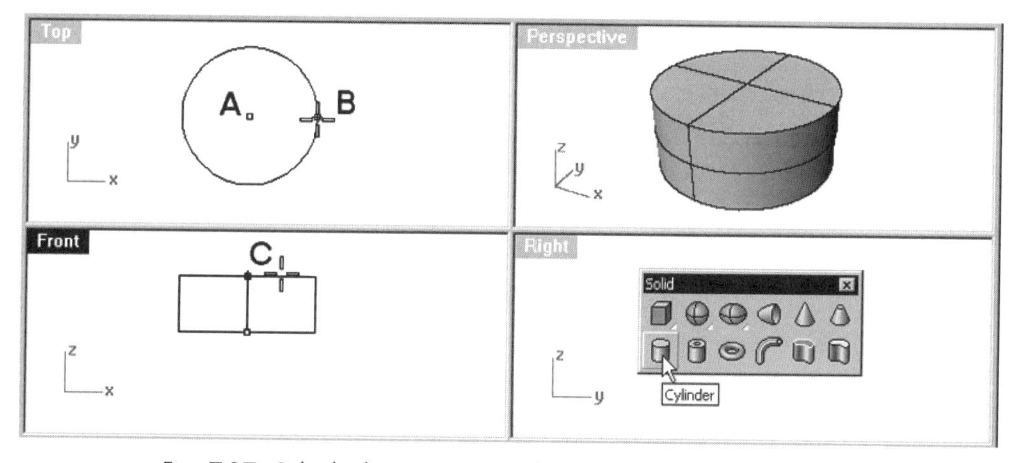

Fig. 7-27. Cylinder being constructed.

Tube

A tube is a polysurface consisting of two cylindrical surfaces and two planar surfaces. To construct a tube, perform the following steps.

1 Start a new file. Use the metric (Millimeters) template file.
2 Select Solid > Tube, or click on the Tube button on the Solid toolbar.
3 Select locations A, B, and C (indicated in figure 7-28) in the Top viewport to indicate the center, inner radius, and outer radius, and then select a point in the Front viewport to indicate the height.

A tube is constructed.

4 If you want to save your file, save it as *Tube.3dm*.

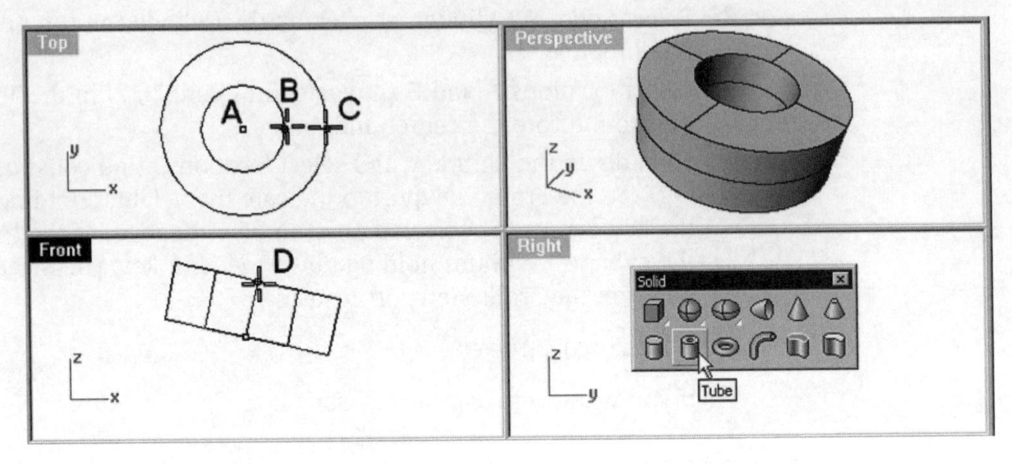

Fig. 7-28. Tube being constructed.

Torus

A torus is single closed-loop surface. To construct a torus, perform the following steps.

1. Start a new file. Use the metric (Millimeters) template file.
2. Select Solid > Torus, or click on the Torus button on the Solid toolbar.
3. Select locations A, B, and C (indicated in figure 7-29) in the Top viewport to indicate the center and radius of the torus and the radius of the tube.

A torus is constructed.

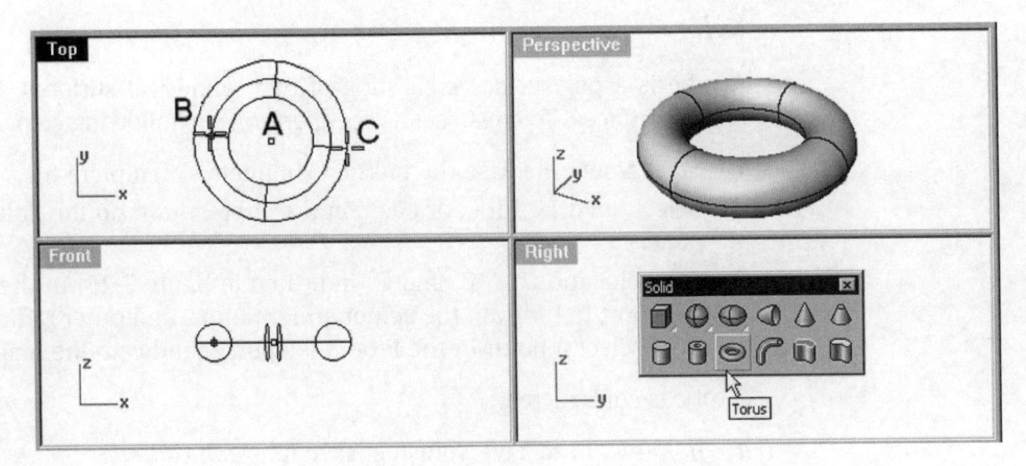

Fig. 7-29. Torus being constructed.

4 If you want to save your file, save it as *Torus.3dm*.

Pipe

A pipe is polysurface consisting of four surfaces: inner pipe, outer pipe, and two planar surfaces. To construct a pipe, perform the following steps.

1. Start a new file. Use the metric (Millimeters) template file.
2. Construct a free-form curve in the Top viewport.
3. Select Solid > Pipe, or click on the Pipe button on the Solid toolbar.
4. Select the free-form curve.
5. Type *C* if the prompt is "Cap = No."
6. Type *T* if the prompt is "Thickness = No."
7. Select locations A and B (indicated in figure 7-30) to specify the starting inner and outer radii of the tube.
8. Press the Enter key if the inner and outer radii of the tube at the other end of the pipe are the same as the start end. Otherwise, select two locations to specify the radii at the end point of the pipe.

A pipe is constructed.

9 If you want to save your file, save it as *Tube.3dm*.

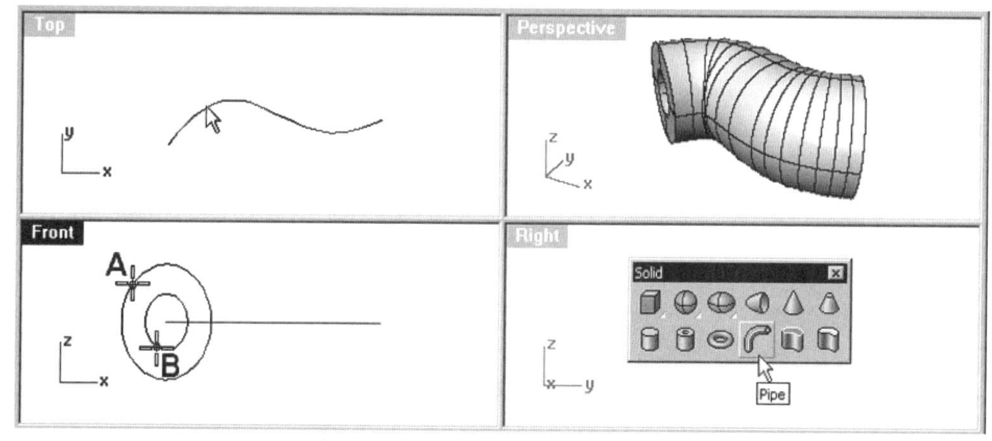

Fig. 7-30. Pipe being constructed.

Text

A text object is a polysurface consisting of three elements: top and bottom planar surfaces that resemble the text, and surfaces joining the top and bottom surfaces. To construct a text object, perform the following steps.

1. Start a new file. Use the metric (Millimeters) template file.

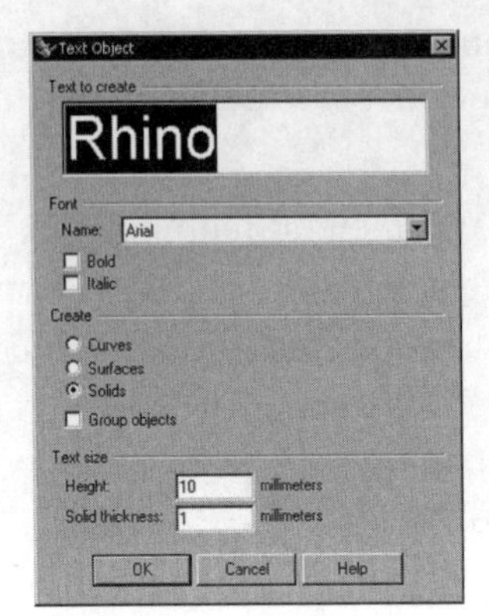

Fig. 7-31. Text Object dialog box.

2. Select Solid > Text, or click on the Text button on the Main1 toolbar.
3. In the Text Object dialog box, shown in figure 7-31, type a text string, select a font and font style, specify text height and thickness, activate the Solids option, and click on the OK button.
4. Select location A (indicated in figure 7-32).

A text object is constructed.

5. If you want to save your file, save it as *Text.3dm*.

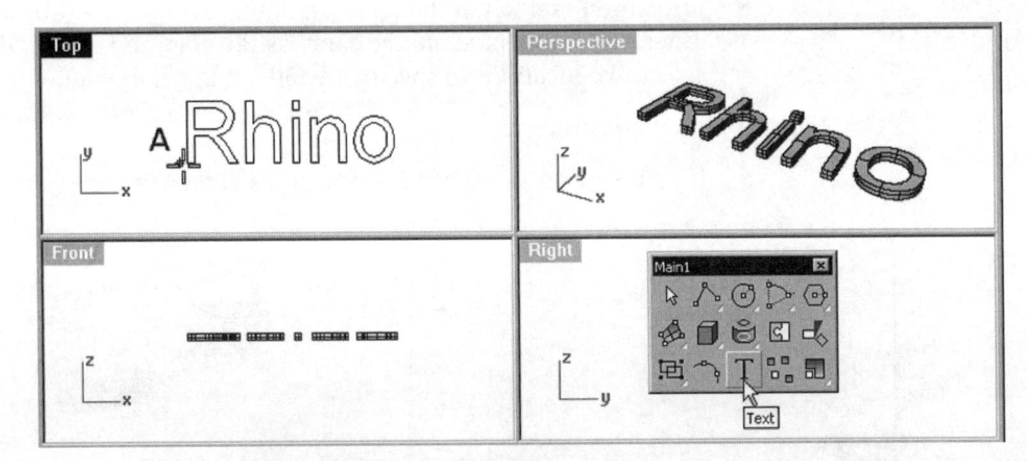

Fig. 7-32. Text object's location being specified.

Extrude Curve and Extrude Surface

As mentioned in Chapter 5, there are four ways to extrude a curve: in a straight line, along a curve, to a point, or in a tapered way. The curve used to construct a surface by extrusion can be a planar curve, nonplanar curve, open curve, or closed curve. The same command can be used to construct a solid if the curve is a closed, planar curve. Apart from extruding a closed, planar curve, you can also extrude a surface to obtain a solid.

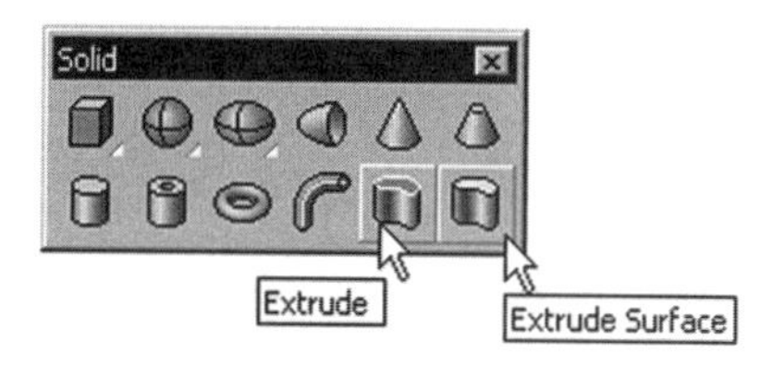

Fig. 7-33. Extrude and Extrude Surface buttons on the Solid toolbar.

To construct an extruded solid, you use the Solids pull-down menu and the Solid toolbar. Figure 7-33 shows the Extrude and Extrude Surface buttons of the Solid toolbar.

Table 7-2 outlines the functions of extrude and fillet commands and their locations in the main pull-down menu and respective toolbars.

Table 7-2: Extrude and Fillet Commands and Their Functions

Toolbar Button	Function
Extrude	Constructing a solid by extruding a planar curve
Extrude Surface	Constructing a solid by extruding a surface

Extrude Planar Curve

By extruding a closed planar curve, you construct a closed polysurface consisting of a surface extruded from the planar curve and two planar surfaces enclosing the extruded surface. If you extrude an open-loop planar curve, you get only an extruded surface. To see the difference between extruding a closed curve and an open planar curve, perform the following steps.

1. Open the file *ExtrudeSolid.3dm* from the *Chapter 7* folder on the companion CD-ROM.
2. Click on the Extrude button on the Solid toolbar.
3. Type *M* if the prompt for "Mode" is not "Straight." Type *S* to change the mode to straight.
4. Select curves A and B (indicated in figure 7-34) and press the Enter key.
5. Select location B (indicated in figure 7-34).

An extruded solid is constructed from the closed planar curve, and a surface is constructed from the open-loop curve. For the closed, planar curve, two planar surfaces (in addition to an extruded surface) are constructed. The surfaces are joined to form a polysurface, and a solid is constructed. To separate them into individual surfaces, you can explode the polysurface.

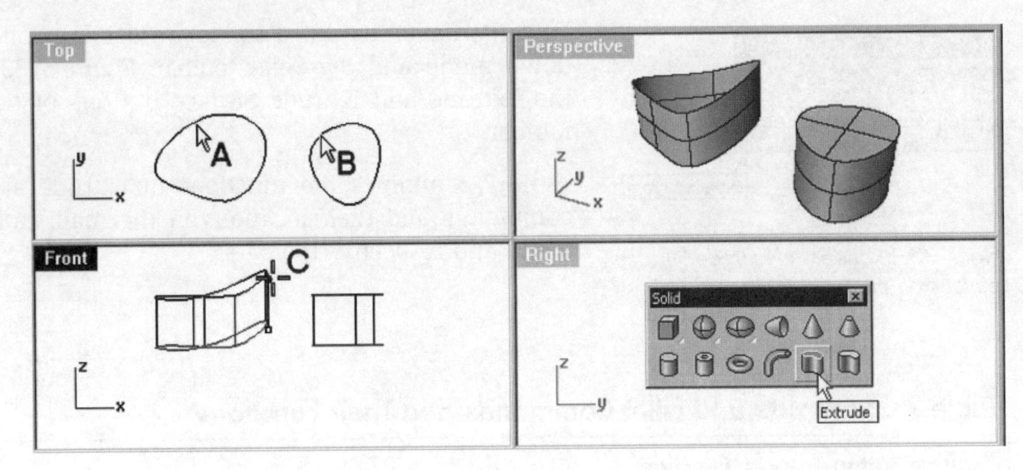

Fig. 7-34. Closed planar curve being extruded.

Extrude Surface

By extruding a surface, you construct a polysurface consisting of two copies of the original surface at a distance apart, and a set of surfaces joining their edges. To construct a solid by extruding a surface, perform the following steps.

1. Open the file *ExtrudeSurface.3dm* from the *Chapter 7* folder on the companion CD-ROM.
2. Click on the Extrude Surface button on the Solid toolbar.
3. Select surface A (indicated in figure 7-35) and press the Enter key.
4. Select location B (indicated in figure 7-35).

An extruded solid is constructed from the surfaces.

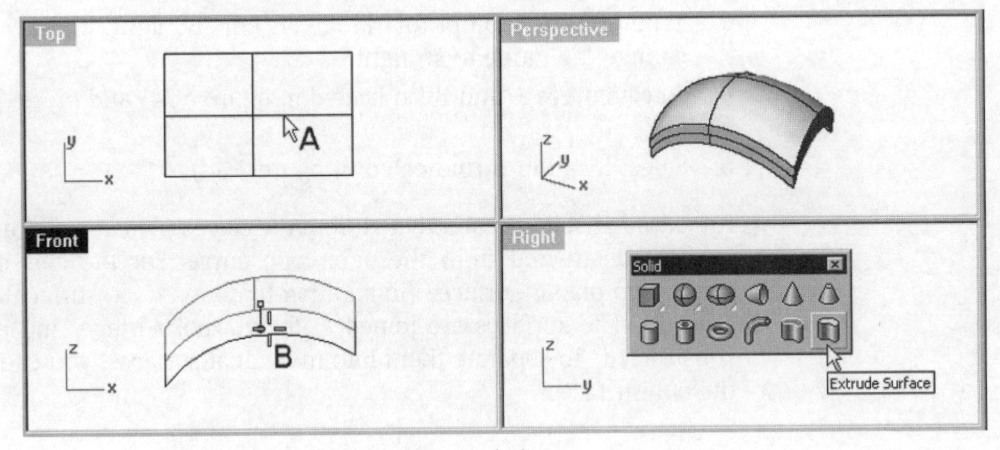

Fig. 7-35. Surfaces being extruded.

Note that extruding a planar surface or a free-form surface obtains a solid.

Filleting

Adding a fillet to edges of a closed polysurface produces filleted surfaces at the selected edges of the polysurface. To construct filleted edges, perform the following steps.

1. Open the file *Fillet.3dm* from the *Chapter 7* folder on the companion CD-ROM.
2. Select Solid > Fillet Edge, or click on the Fillet Edge button on the Solid Tools toolbar.
3. Type *R* at the command line interface and then specify a radius.
4. Select edge A (indicated in figure 7-36) and press the Enter key.

The selected edge is filleted.

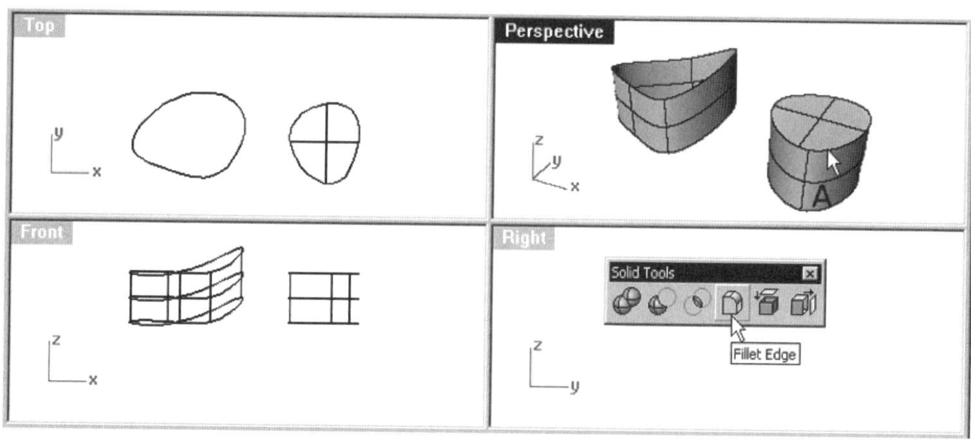

Fig. 7-36. Edges being filleted.

Capping Planar Holes of a Polysurface

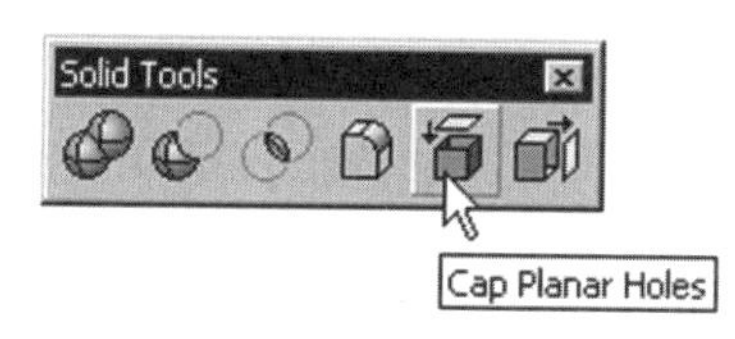

Fig. 7-37. Cap Planar Holes button of the Solid Tools toolbar.

As mentioned, a Rhino solid is a closed polysurface. An existing polysurface that does not form a closed volume because of planar holes can be capped to close the holes. This is done with the Cap Planar Holes button of the Solid Tools toolbar, shown in figure 7-37.

Table 7-3 outlines the function of the Cap command and its location on the main pull-down menu and respective toolbar.

Table 7-3: Cap Command and Its Function

Toolbar Button	From Main Pull-down Menu	Function
Cap Planar Holes	Solid > Cap Planar Holes	Constructing a solid by capping planar holes of a polysurface

To construct a solid by capping planar holes, perform the following.

1. Open the file *Cap.3dm* from the *Chapter 7* folder on the companion CD-ROM.
2. Select Solid > Cap Planar Holes, or click on the Cap Planar Holes button on the Solid Tools toolbar.
3. Select surface A (indicted in figure 7-38) and press the Enter key.

The polysurface is capped. Note that only planar holes can be capped this way. If your polysurface has a lot of openings that are not planar holes, you have to construct surfaces and join them in order to form a solid.

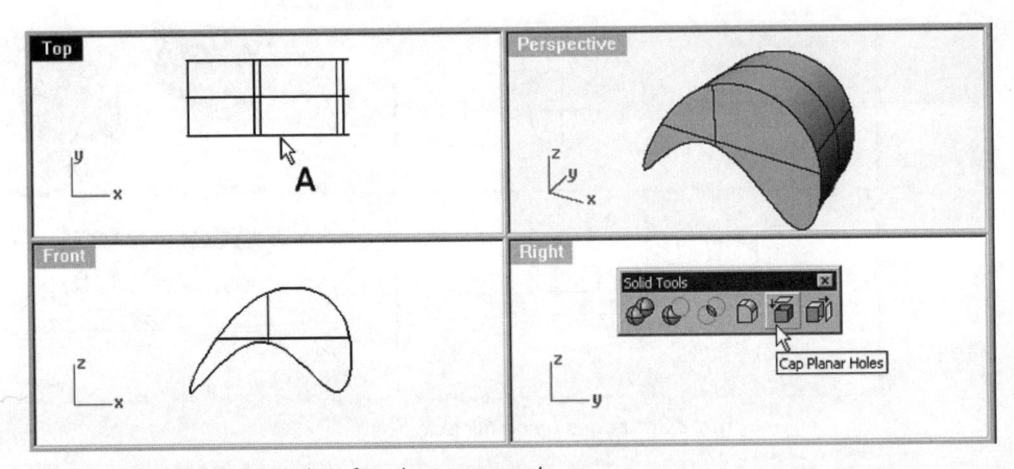

Fig. 7-38. Polysurface being capped.

Combining Solid Objects

There are not many objects that can be represented simply by one of the primitives previously delineated. By combining solids in one of the three ways (union, difference, and intersection), you obtain a more complex object. Booleans generally work better if there are two objects overlapping each other slightly. Figure 7-39 shows Boolean operation commands of the Solid Tools toolbar.

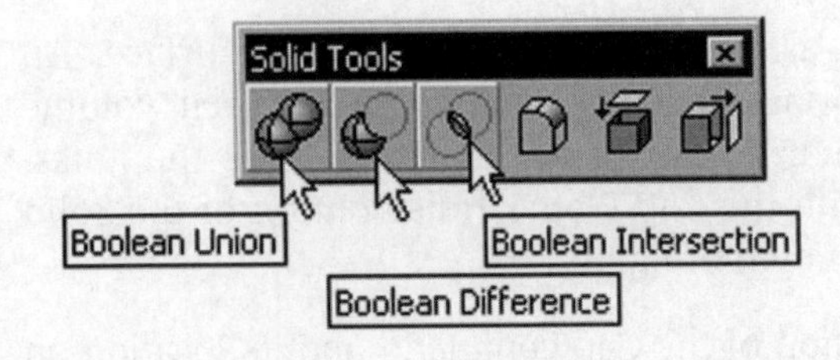

Fig. 7-39. Boolean commands on the Solid Tools toolbar.

Table 7-4 outlines the functions of Boolean operation commands and their locations in the main pull-down menu and respective toolbars.

Table 7-4: Boolean Operation Commands and Their Functions

Toolbar Button	From Main Pull-down Menu	Function
Boolean Union	Solid > Union	Constructing a solid having the volume of all selected solids
Boolean Difference	Solid > Difference	Constructing a solid that has a volume enclosed by the first set of selected solids but not the second set of selected solids
Boolean Intersection	Solid > Intersection	Constructing a solid that has a volume enclosed by the first and second sets of selected solids

Union

A union of a set of solids produces a solid that has the volume of all solids in the set. To construct a solid by uniting two solids, perform the following steps.

1. Open the file *Boolean.3dm* from the *Chapter 7* folder on the companion CD-ROM.
2. Select Solid > Union, or click on the Boolean Union button on the Solid Tools toolbar.
3. Select solids A and B (indicated in figure 7-40) and press the Enter key.

The solids are united.

Difference

A difference of two sets of solids produces a solid that has the volume contained in the first set of solids but not the second set of solids. To construct a solid by subtracting one solid from another solid, perform the following steps.

1. Select Solid > Difference, or click on the Boolean Difference button on the Solid Tools toolbar.
2. Select solid A (indicated in figure 7-41) and press the Enter key. (This is the first set of solids.)
3. Select solid B (indicated in figure 7-41) and press the Enter key. (This is the second set of solids.)

The second set of solids is subtracted from the first set of solids.

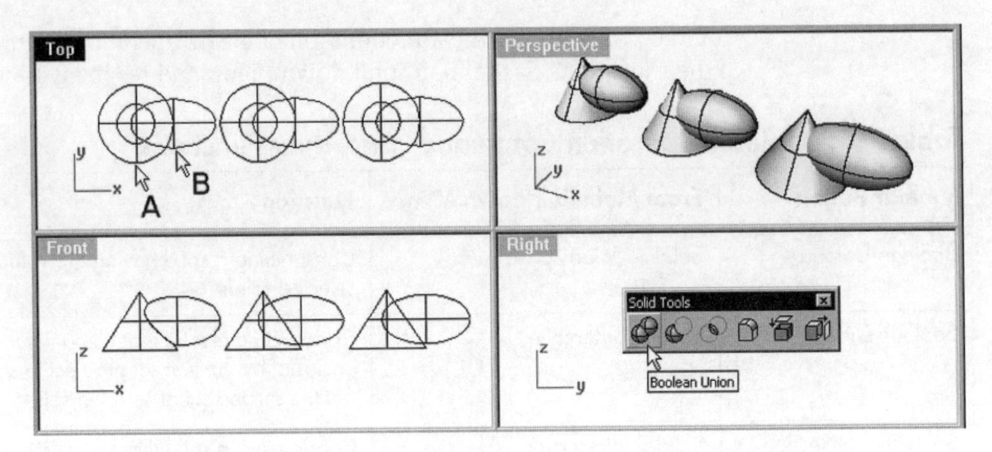

Fig. 7-40. Solids being united.

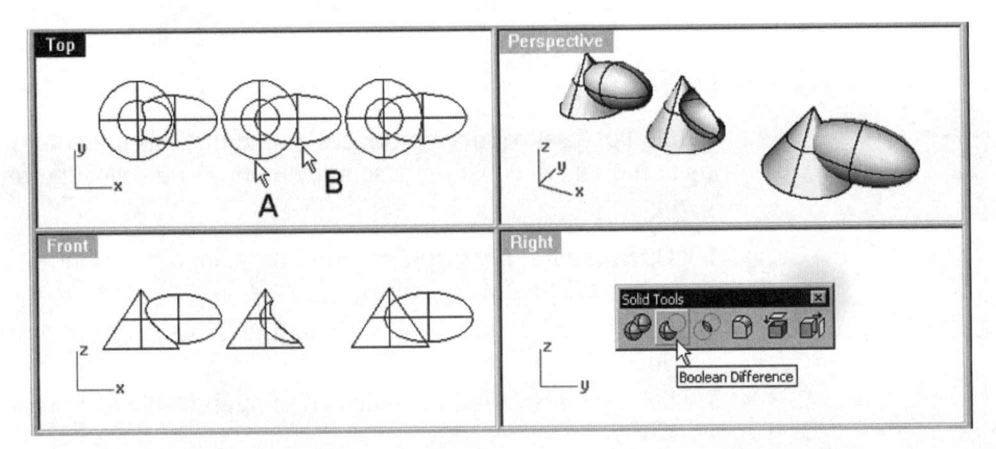

Fig. 7-41. Solid being subtracted,

Intersection

An intersection of two sets of solids produces a solid that has the volume contained in the first set of solids and the second set of solids. To construct a solid that has the volume contained in two solids, perform the following steps.

1. Select Solid > Intersection, or click on the Boolean Intersection button on the Solid Tools toolbar.
2. Select solid A (indicated in figure 7-42) and press the Enter key. (This is the first set of solids.)
3. Select solid B (indicated in figure 7-42) and press the Enter key. (This is the second set of solids.)

A solid of intersection is constructed.

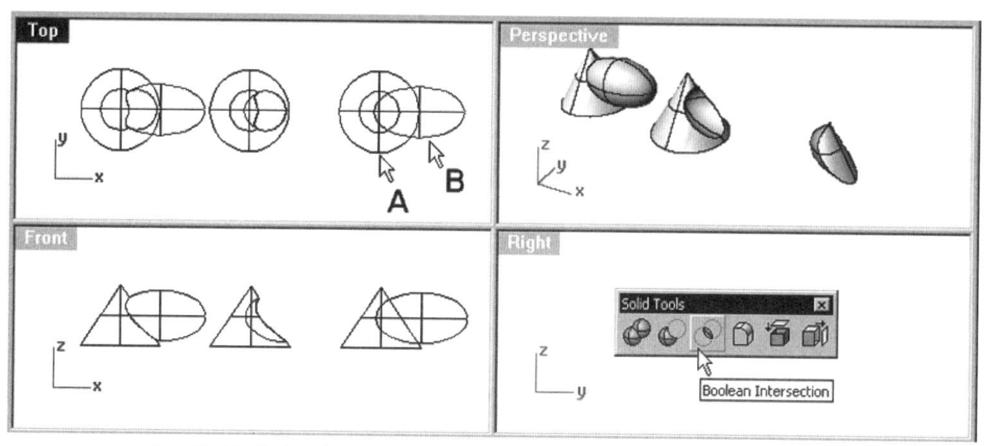

Fig. 7-42. Solids being intersected.

Editing Polysurfaces

There are several ways to edit polysurfaces. By joining surfaces to a closed polysurface, a solid is implied because a Rhino solid is a closed polysurface. To change the view of a solid or polysurface to that of individual surfaces, you explode the object. To separate a surface from a solid or polysurface without exploding it, you extract a surface.

In addition, you trim a solid or polysurface to remove a portion of the surfaces, or split a solid or polysurface to split the object into two objects. Because a solid is a polysurface, extracting a surface from a solid or trimming and splitting a solid creates an open polysurface.

Joining and Exploding

In Chapter 6 you learned how to join surfaces to a polysurface. If the polysurface encloses a volume without any gap or opening, a solid is implied. Contrary to joining, you explode a polysurface into individual surfaces.

Extracting a Surface from a Solid or Polysurface

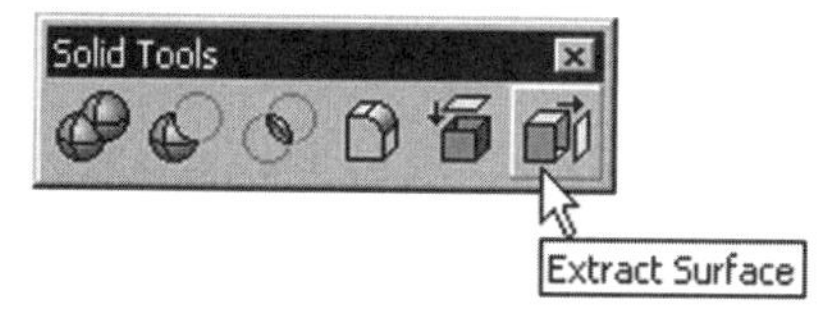

Fig. 7-43. Extract Surface command on the Solid Tools toolbar.

Unlike exploding, which breaks down a joined polysurface into individual surfaces, extracting separates the selected surfaces from the polysurface, leaving the remaining surfaces joined. Figure 7-43 shows the Extract Surface command on the Solid Tools toolbar.

Table 7-5 outlines the functions of the Extract Surface command and its location in the main pull-down menu and respective toolbar.

Table 7-5: Extract Surface Command and Its Function

Toolbar Button	From Main Pull-down Menu	Function
Extract Surface	Solid > Extract Surface	Constructing a surface by duplicating a surface of a polysurface

To extract a surface from a solid, perform the following steps.

1. Open the file *Extract.3dm* from the *Chapter 7* folder on the companion CD-ROM.
2. Select Solid > Extract Surface, or click on the Extract Surface button on the Solid Tools toolbar.
3. Select surface A (indicated in figure 7-44) and press the Enter key.

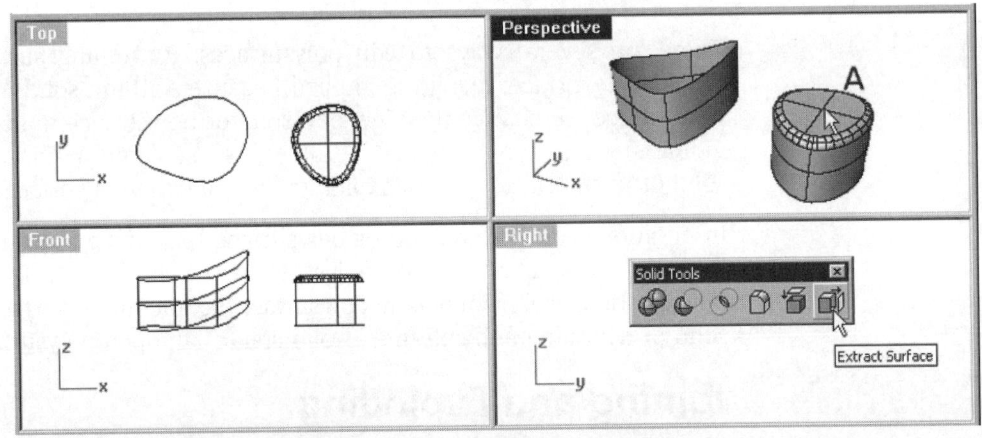

Fig. 7-44. Surface being extracted from a polysurface.

A surface is extracted. To view the extracted surface clearly, you will hide the box. Continue with the following steps.

4. Select Edit > Visibility > Hide, or click on the *Hide/Show* button on the Visibility toolbar.
5. Select the extracted surface, if it is not already selected.

The extracted surface is hidden, as shown in figure 7-45.

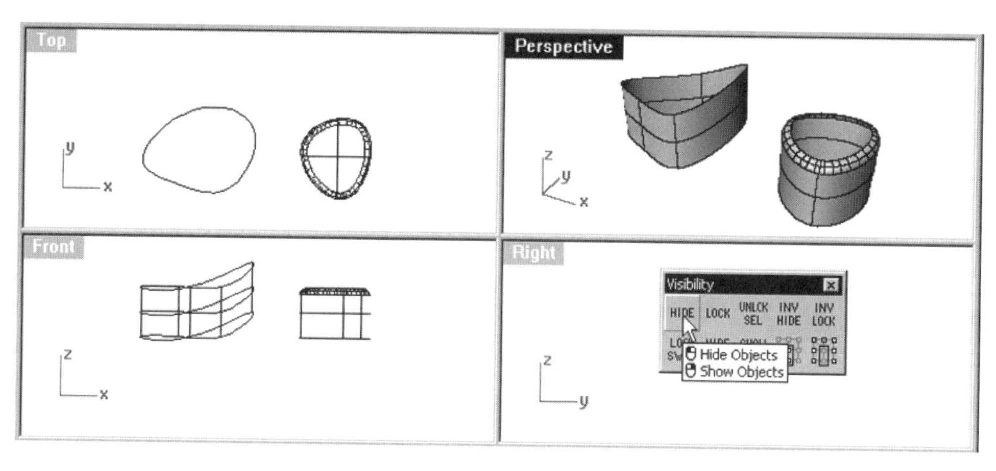

Fig. 7-45. Extracted surface hidden.

Trimming

You can trim a polysurface by using a surface (instead of a curve) as a cutting object, as follows.

1. Open the file *Trim.3dm* from the *Chapter 7* folder on the companion CD-ROM.
2. Select Edit > Trim, or click on the Trim button on the Main1 toolbar.
3. Select surface A (indicated in figure 7-46) and press the Enter key. This is the cutting object.
4. Select polysurface B (indicated in figure 7-46) and press the Enter key.

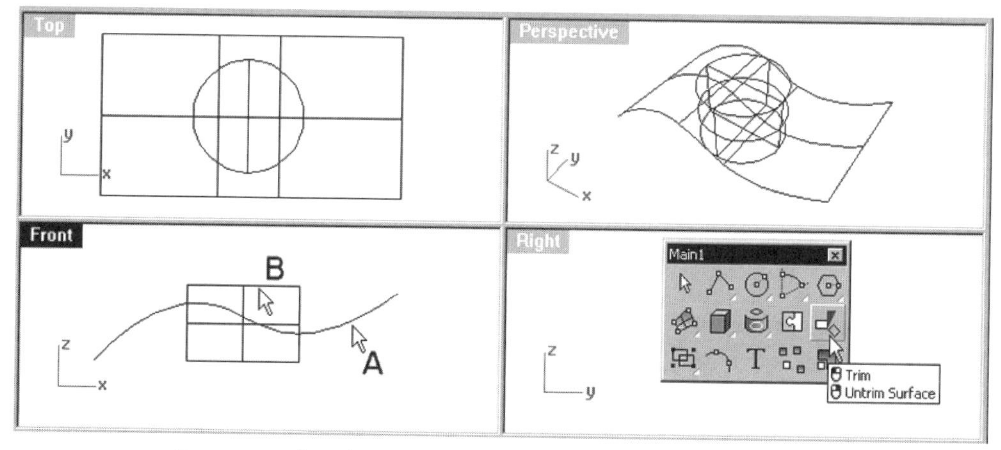

Fig. 7-46. Closed polysurface being trimmed.

The polysurface is trimmed. Note that a trimmed solid will become an open polysurface. To appreciate how the solid, after being trimmed, becomes an open polysurface, continue with the following steps.

5 Select Edit > Visibility > Hide, or click on the *Hide/Show* button of the Visibility toolbar.

Select surface A (indicated in figure 7-47) and press the Enter key.

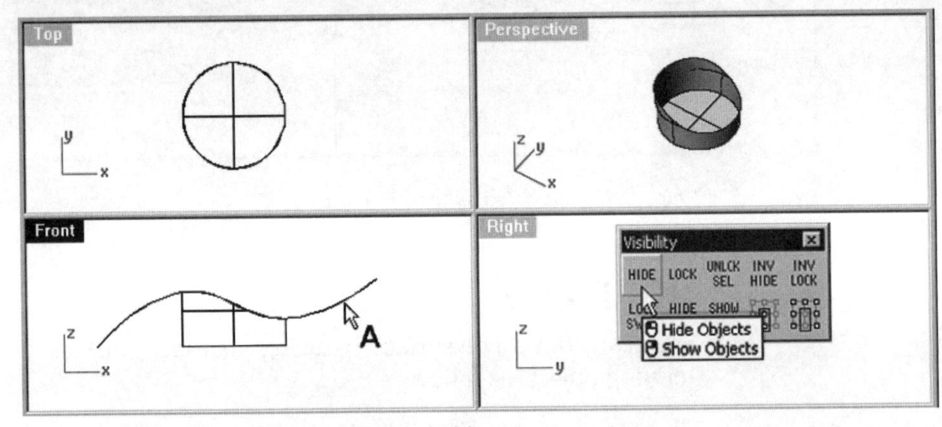

Fig. 7-47. Cutting object being hidden.

To unhide the hidden surface and trim it, continue with the following.

6 Select Edit > Visibility > Show, or right-click on the *Hide/Show* button of the Visibility toolbar.

7 Select Edit > Trim, or click on the Trim button on the Main1 toolbar.

8 Select surface A (indicated in figure 7-48) and press the Enter key. This is the cutting object.

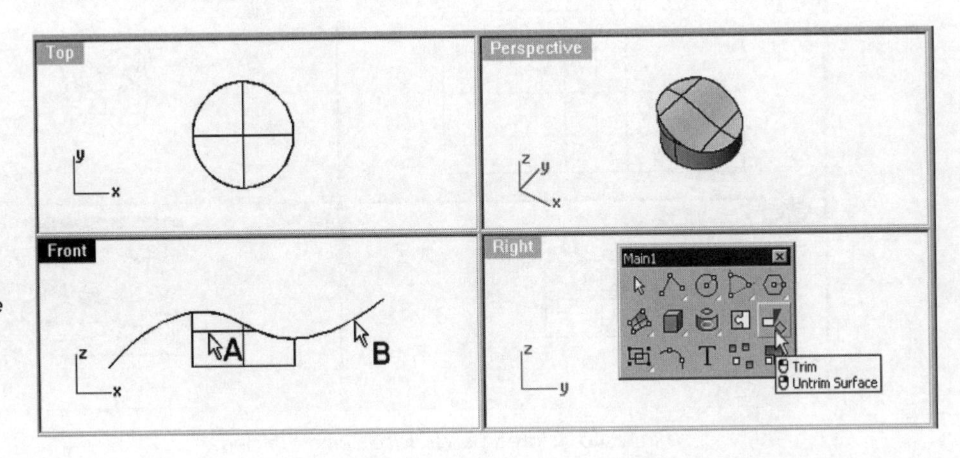

Fig. 7-48. Surface being trimmed by the polysurface.

9 Select polysurface B (indicated in figure 7-48) and press the Enter key.

Joining

Surfaces can be joined to form a polysurface. If the joined surfaces form a closed polysurface, a solid is implied. (You can examine an object's properties via the Detail button.) To join surfaces, perform the following steps.

1 Select Edit > Join, or click on the Join button of the Main1 toolbar.
2 Select surface A and polysurface B (indicated in figure 7-49) and press the Enter key.

A solid is formed.

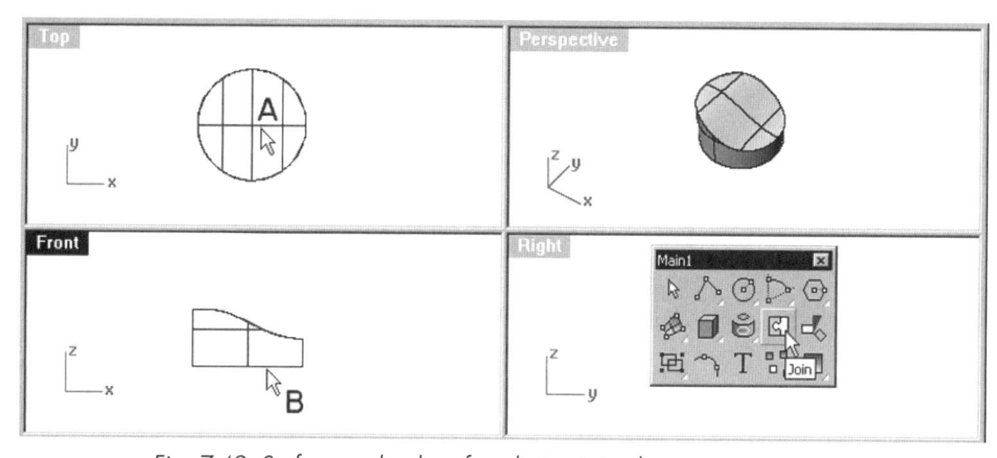

Fig. 7-49. Surface and polysurface being joined.

Splitting

You can split a polysurface into two polysurfaces by using a surface as a cutting object, as follows.

1 Open the file *Split.3dm* from the *Chapter 7* folder on the companion CD-ROM.
2 Select Edit > Split, or click on the Split button on the main toolbar.
3 Select polysurface A (indicated in figure 7-50) and press the Enter key. This is the object to be split.
4 Select surface B (indicated in figure 7-50) and press the Enter key. This is the cutting surface.

The polysurface is split in two.

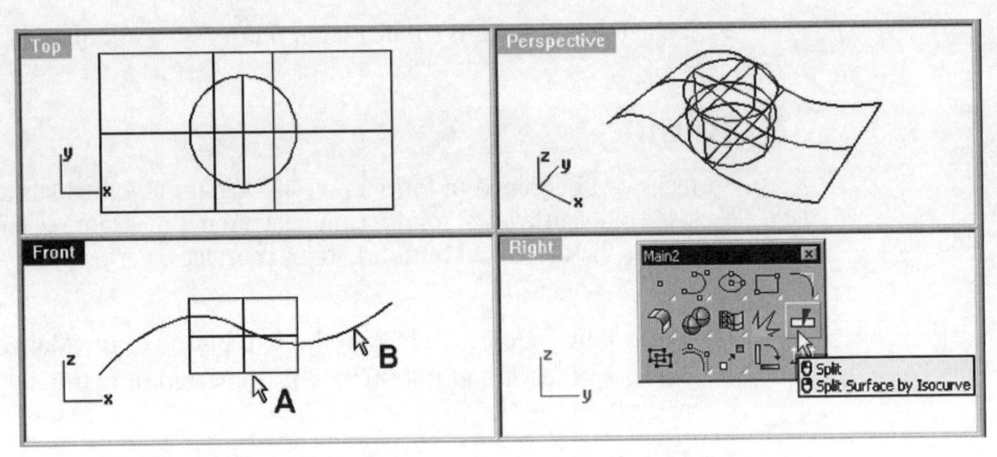

Fig. 7-50. Box being split.

Transforming Polysurfaces

With the exception of the Twist, Bend, Taper, Flow, and Smooth commands (which apply only to individual surfaces), you use the transform commands on polysurfaces to translate and deform them. (See chapters 3 and 6 on using the transform commands.)

Analyzing Solids

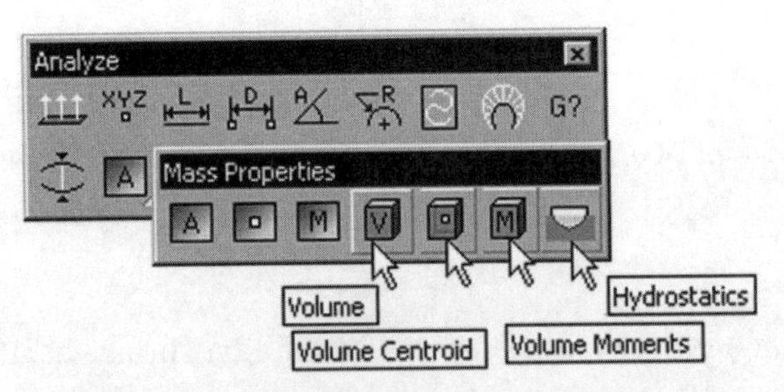

Fig. 7-51. Analysis commands specific to solids.

Most of the analysis tools you learned in Chapter 6 that apply to surfaces and polysurfaces are also applicable to solids. Figure 7-51 shows the analysis commands specific to solids.

Table 7-6 outlines the functions of the analysis commands and their locations on the main pull-down menu and respective toolbars.

Table 7-6: Analysis Commands and Their Functions

Toolbar Button	From Main Pull-down Menu	Function
Volume	Analyze > Mass Properties > Volume	Evaluating the volume of a solid
Volume Centroid	Analyze > Mass Properties > Volume Centroid	Evaluating the volume centroid of a solid

Toolbar Button	From Main Pull-down Menu	Function
Volume Moments	Analyze > Mass Properties > Volume Moments	Evaluating a solid's volume, volume centroid, volume moments (first, second, and product), volume moments of inertia about World coordinate axes, volume radii of gyration about World coordinate axes, volume moments of inertia about centroid coordinate axes, and volume radii of gyration about centroid
Hydrostatics	Analyze > Mass Properties > Hydrostatics	Evaluating the hydrostatics values of surfaces, with the waterline set at the horizontal plane of the World coordinate system

Volume, Volume Centroid, and Volume Moments

To construct a solid box and evaluate its volume, volume centroid, and volume moments, perform the following steps.

1. Open the file *Analyze.3dm* from the *Chapter 7* folder on the companion CD-ROM.
2. Select Analyze > Mass Properties > Volume, or click on the Volume button on the Mass Properties toolbar.
3. Select solid A (indicated in figure 7-52) and press the Enter key.

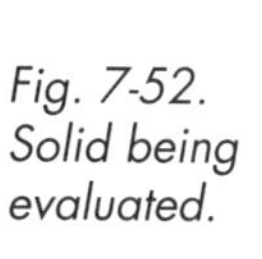

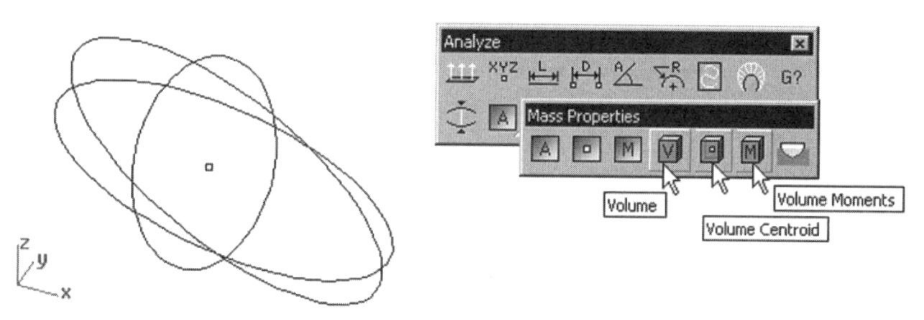

Fig. 7-52. Solid being evaluated.

Volume information is displayed in the command line interface. Continue with the following steps.

4. Select Analyze > Mass Properties > Volume Centroid, or click on the Volume Centroid button on the Mass Properties toolbar.
5. Select solid A (indicated in figure 7-52) and press the Enter key.

Volume centroid information is displayed at the command line interface. Continue with the following steps.

6. Select Analyze > Mass Properties > Volume Moments, or click on the Volume Moments button on the Mass Properties toolbar.

7 Select solid A (indicated in figure 7-52) and press the Enter key.

Volume moment information is displayed in the dialog box shown in figure 7-53.

Hydrostatics

Assuming the waterline is set at the World coordinate plane, you can change the setting by typing *W* at the command line interface and specifying a new waterline. You can evaluate the hydrostatic value of a solid as follows.

1 Select Analyze > Mass Properties > Hydrostatic, or click on the Hydrostatics button on the Mass Properties toolbar.

2 Select the solid and press the Enter key.

Hydrostatics information is displayed in the Hydrostatics dialog box, shown in figure 7-54.

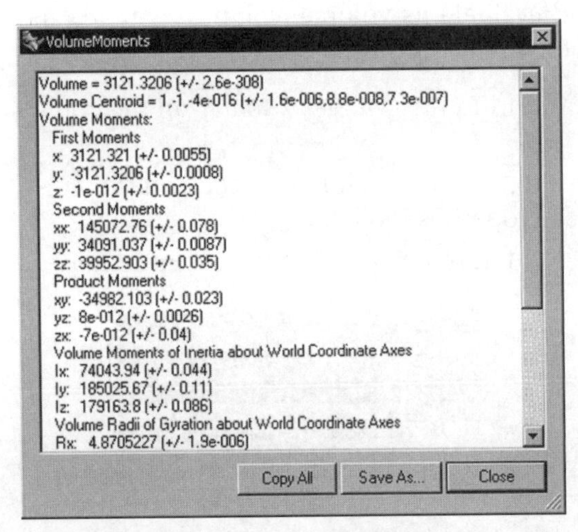

Fig. 7-53. VolumeMoments dialog box.

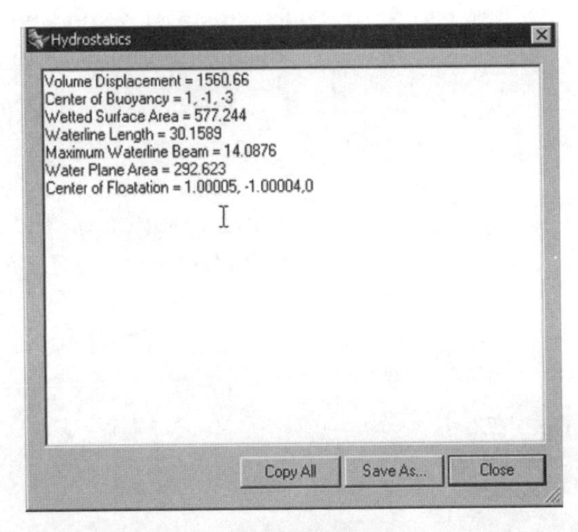

Fig. 7-54. Hydrostatics dialog box.

Summary

A solid model represents integrated data (vertices, edges, surfaces, and volume) of the 3D object the model represents. There are many ways to represent a solid in the computer: primitive instancing, generalized sweeping, spatial occupancy enumeration, cellular decomposition, boundary representation, and constructive solid geometry.

A Rhino solid is a polysurface that encloses a volume with no gap or opening. There are several ways to construct Rhino solids: using primitive polysurfaces, extruding closed-loop curves, extruding surfaces, joining surfaces to form a closed polysurface, and capping planar holes in polysurfaces to form a closed polysurface. To construct solids of more complex shape, you combine solid objects using Boolean operations, add filleted edges, and transform the solid (polysurface) in different ways, which you learned about in chapters 5 and 6.

Because a Rhino solid is a polysurface, you can join surfaces that enclose a volume into a solid. On the other hand, you explode a solid to obtain individual surfaces, and you trim, split, or extract a solid to obtain an open-loop polysurface.

Review Questions

1. Illustrate, with the aid of simple sketches, various ways of representing a solid in the computer.
2. Explain in detail how solids are constructed in Rhino.

CHAPTER 8

Digital Rendering and 2D Drawing Output

Introduction

This chapter explores key concepts involved in rendering. You will be exposed to methods of producing a photorealistic rendered image from 3D models using Rhino's default renderer and Flamingo's Raytrace and Photometric renderers. This chapter also explores the process of constructing 2D engineering drawings.

Objectives

After studying this chapter you should be able to:

- State the key concepts involved in rendering
- Apply materials to objects of a scene
- Add lighting to a scene
- Include environment elements in a scene
- Construct photorealistic rendered images from 3D models
- Use the Rhino renderer, Flamingo Raytrace renderer, and Flamingo Photometric renderer
- Describe the process of constructing 2D engineering drawings

Overview

In addition to exporting to other computerized applications for downstream operations, you can present your 3D models as rendered images for better visualization and as 2D engineering drawings for communication in a conventional engineering form. This chapter explores methods by which 3D objects are "visualized" in the computer, methods of adding realism to a 3D object visually, and methods of representing 3D objects as 2D drawings.

Shaded Display

In reality, there are no curves or lines on a 3D free-form smooth surface. Therefore, you should find only edges on the surface's boundaries. However, boundary edges alone do not provide sufficient information to depict the profile and silhouette of the surface. Hence, a set of isocurves in two orthogonal directions, color shading, or a set of isocurves together with color shading is used to better illustrate a free-form object in the computer display. To select a surface, you select one of the isocurves and the boundary of the surface.

By default, objects are displayed in the viewport in a color assigned to the layer on which the objects reside. In wireframe mode, they are displayed with isocurves. To change the color of an object, you change the layer's color or the object's color. By shading the object, you distinguish it from the viewport background. To shade the object and display isocurves simultaneously, you set the display to shaded mode.

Isocurves and Isocurve Density

In wireframe mode, there are isocurves placed on the surface. For a very simple surface, such as a planar surface, one or two isocurves are adequate to provide enough information on the curvature of the surface. For more complex surfaces, you need more isocurves. Although there are more isocurves to better represent the profile of the surface, selection of individual objects from a bunch of objects with high isocurve density may become difficult. To set the isocurve density of an object, perform the following steps.

1. Open the file *ObjectColor.3dm* from the *Chapter 8* folder on the companion CD-ROM.

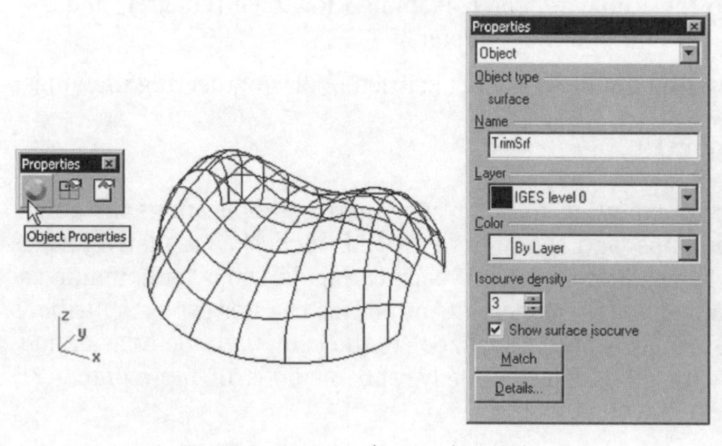

Fig. 8-1. Isocurve density changed.

2. Select Edit > Object Properties, or click on the Object Properties dialog box on the Properties toolbar.
3. Select the polysurface and press the Enter key.
4. In the Properties dialog box, change the isocurve density to 3.

The isocurve density of the surface is changed, as indicated in figure 8-1.

Color Setting

The default color of a Rhino object is determined by the setting *Bylayer*. This means that the color of the object is determined by the color assigned to the layer in which the object resides. You can change an object's color by either changing the layer's color or changing the object's color property.

To change the color assignment of a layer, you use the Layer dialog box by selecting Edit > Layers > Edit Layers. To change an object's color property, you use the Properties dialog box by selecting Edit > Object Properties or by clicking on the Object Properties button on the Properties toolbar and then selecting the object.

Color setting can be done in three ways in the Select Color dialog box, shown in figure 8-2. The first way is to select a color from the preestablished color list. The second way is to use the HSB (hue, saturation, and brightness) color system. The third way is to use the RGB (red, green, blue) system. In a collaborative design environment in which a project is handled by a team, specifying color values (HSB or RGB) can accurately describe the color.

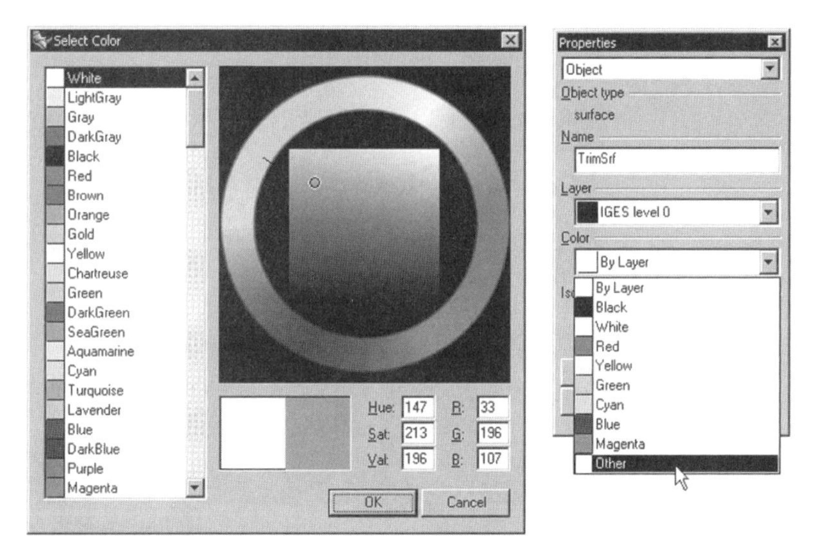

Fig. 8-2. Select Color dialog box.

HSB Color System

In the Select Color dialog box, the HSB settings are labeled Hue, Sat, and Val. The hue represents a color ranging from red through yellow, green, and blue. Saturation describes the intensity of the hue. Brightness concerns the color's value or luminance. To select a color by using the HSB color system, you first select a color from the hue color range; you then set the saturation (intensity) and the brightness (luminance).

RGB Color System

Specifying the RGB value instructs the computer to project a color mix of red, green, and blue to each individual pixel (picture element) of the monitor. To specify color via this method, perform the following steps.

1. In the Properties dialog box, select Other from the Color pull-down box. (If you already closed the Properties dialog box, open it again by selecting Edit > Object Properties, or by clicking on the Object Properties button on the Properties toolbar and then selecting the surface.
2. In the Select Color dialog box, select a color in the circular ring in the color swatch to set the hue. You may regard the hue as the basic color. For example, you click on the green zone to select a green color.
3. In the square box, select a location to set the saturation and value. If you select a location near the left-hand edge of the box, you obtain a fully saturated color. If you select a location near the upper edge of the box, you obtain a higher brightness value.
4. Click on the OK button.

The object's wireframe color is set.

Shaded Display Mode

To shade the display, you use the commands available from the Shade toolbar, shown in figure 8-3.

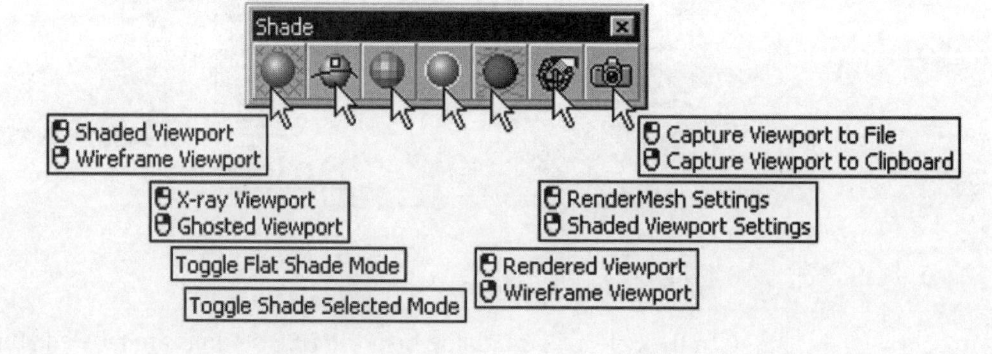

Fig. 8-3. Shade toolbar.

Table 8-1 outlines the functions of the shade commands and their locations in the right-click menu from the viewport's label and respective toolbars.

Table 8-1: Shade Commands and Their Functions

Toolbar Button	Right-click Menu	Function
Shaded Viewport	Shaded Display	Setting the display to shaded mode
Wireframe Viewport	Wireframe Display	Setting the display to wireframe mode
X-ray Viewport	X-ray Display	Setting the display to x-ray mode, displaying all curves and point objects, regardless of being hidden by surfaces in front of them
Ghosted Viewport	Ghosted Display	Setting the display mode to ghosted, in which surfaces are translucent and objects behind are obscured
Toggle Flat Shade Mode	Flat Shade	Toggling the display mode to normal and flat shade without mesh smoothing
Toggle Shade Selected Mode	Shade Selected Objects Only	Toggling the display of selected objects to shaded mode and wireframe mode
Rendered Viewport	—	Emulating the effect of rendering the viewport
Render Mesh Settings	—	Opening the Mesh tab of the Document Properties dialog box
Shaded Viewport Settings	—	Opening the Display tab of the Rhino Options dialog box
Capture Viewport to File	—	Capturing the active viewport to a bitmap file
Capture Viewport to Clipboard	—	Capturing the active viewport to the clipboard as a bitmap

To discover how various shading modes affect the display outcome, perform the following.

1 Select Shaded Viewport, X-ray Viewport, Ghosted Viewport, Toggle Flat Shade Mode, and Rendered Viewport one by one on the Shade toolbar to discover various types of shaded display, as shown in figure 8-4.

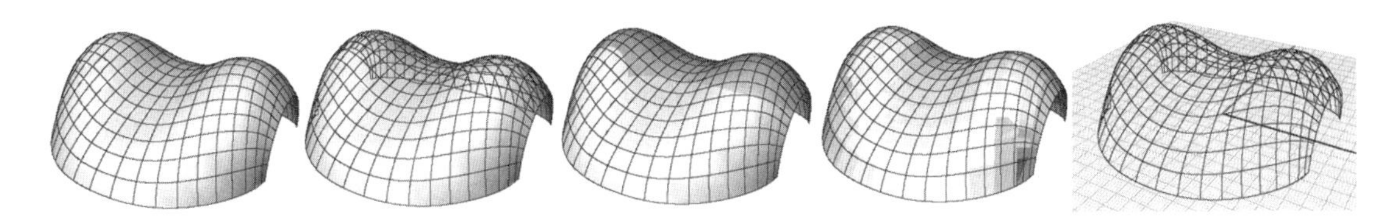

Fig. 8-4. Various shaded display modes.

Digital Rendering

Digital rendering is a method of producing a photorealistic image of an object in a 3D scene in the computer. Unlike shading, which simply applies a shaded color to the surface of an object, rendering takes into account the material properties (color and texture) assigned to the object, the effect of lighting in the scene, and any additional environment factors. To produce a photorealistic image, you need to carry out four basic steps, as follows.

1. Assign material to individual objects.
2. Construct lighting in the scene.
3. Include environment objects.
4. Render the scene.

Rhino, together with Flamingo, incorporates the capacity to produce output from three types of renderers: Rhino, Flamingo Photometric, and Flamingo Raytrace. In the sections that follow you will render images using each of these renderers.

Rhino Renderer

The Rhino renderer is the basic renderer. In producing a rendered image, this renderer takes into account the materials assigned to objects, lights you added in a scene, and simple environment objects (ambient light and background color).

Rhino Renderer Properties

Before you start using the Rhino renderer in creating a rendered image, take some time to study the options that affect the outcome of rendering, as follows.

1. Open the file *Render1.3dm* from the *Chapter 8* folder on the companion CD-ROM.
2. Select Render > Current Renderer > Rhino Render. (If you did not install Flamingo in your computer, this will be the only renderer available. If you installed Flamingo, you will have three renderers: Rhino Renderer, Flamingo Raytrace, and Flamingo Photometric.)
3. Select Render > Properties. This will call out the Rhino Render tab of the Document Properties dialog box, shown in figure 8-5.

The Rhino Render tab includes four option fields, the functions of which are summarized in Table 8-2.

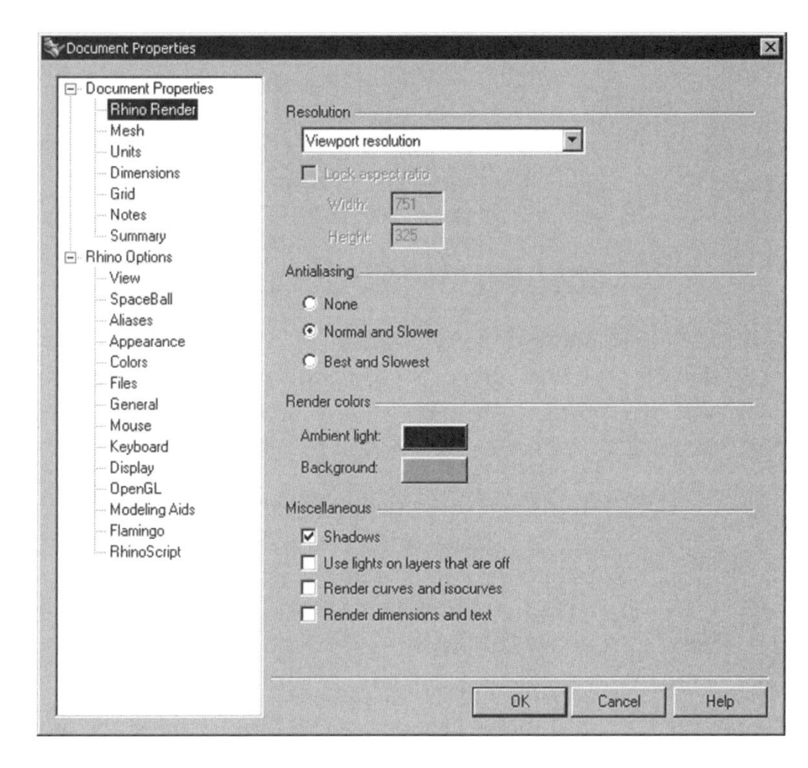

Fig. 8-5. Rhino Render tab of the Document Properties dialog box.

Table 8-2: Rhino Render Tab Option Fields and Their Functions

Field	Function
Resolution	Controls the resolution of the rendered image.
Antialiasing	Controls how jagged edges in an image are treated. Because each discrete pixel in an image has a unique color value, an inclined edge in a rendered image may look jagged. To minimize this jagged effect, adjacent pixels in the inclined edges are averaged. The process is called antialiasing. Antialiasing produces a blurred edge to mask the jagged effect. For example, in a red box on a white background image, pixels are either red or white. Antialiasing produces a set of pinkish pixels between the red and white pixels.
Render Colors	Controls ambient light color and background color. Ambient light is an approximation of all indirect light in the scene. Indirect light is normally referred to as radiosity, which is the effect of reflection from all objects in a scene. To accurately simulate indirect light, you use the Flamingo Photometric renderer.
Misc	Controls the casting of shadows, the use of lights residing on layers that are turned off, the display of curves and isocurves in a rendering, and the display of dimensions and text in a rendering.

4 In the Rhino Render tab, set the resolution to 1024 x 768, check the Best and Slowest checkbox of the Antialiasing field, and check the

Shadows box of the Miscellaneous field. This way, the rendered image's resolution is set, you will obtain the best antialiasing effect, and objects will cast shadows. The higher the antialiasing and the higher the resolution the longer the rendering time. It is sometimes counterproductive.

5 Click on the OK button.

Material Properties

One of the two crucial elements that contribute to a photorealistic rendered image is the digital material you apply to an object. Rhino provides three methods of assigning material properties to an object: via layer, via plug-in, and via the Basic (property assignment) option. In the first method, you assign material properties to a layer and let the object use the layer's material properties. In the second and third method, you assign material properties via plug-in and via the Basic (property assignment) option. Strictly speaking, there are only two categories of material assignment in Rhino: plug-in and basic. That is, the "by layer" method makes use of one of these two categories of material assignment.

Assigning material properties via a plug-in means that you take a material properties source external to the Rhino program and bring it into a library (or create a library). Material properties can then be assigned from that library. You need to use the Flamingo Photometric or Raytrace renderer to be able to use the default material library or to construct a new material library of your own.

Assigning Material Properties by Layer

To set the material properties of a group of objects residing on a layer globally, you assign a material (designated in Rhino as either "plug-in" or "basic") to a layer and let the objects residing on the layer use the layer's material. By default, objects residing on a layer will take on the properties assigned to the layer. To set material properties by layer, perform the following steps.

1 Select Edit > Object Properties, or click on the Object Properties button on the Properties dialog box.
2 Select all objects.
3 In the Properties dialog box, shown in figure 8-6, check the Layer checkbox of the *Assign by* field.
4 Close the Properties dialog box.

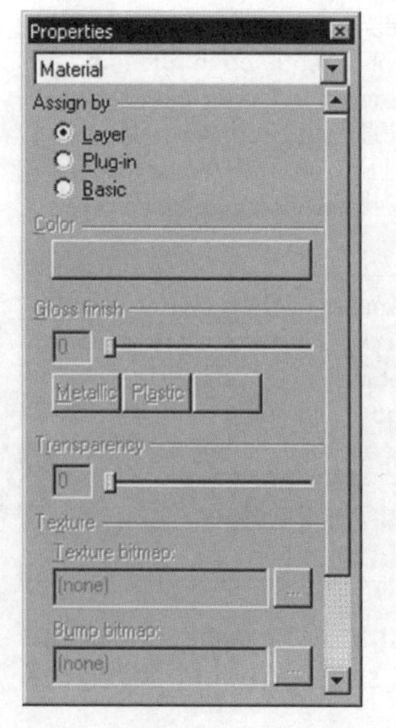

Fig. 8-6. Properties dialog box.

The selected surfaces will now take on the characteristics of the material assigned to the layer on which the surfaces reside. To assign material to the layer, continue with the following steps.

5 Select Edit > Layers > Edit Layers, or click on the Edit Layers button on the Standard toolbar.

6 In the Layers dialog box, shown in figure 8-7, click on the *Material color and name* icon of the *surface(upper)* layer to access the Material dialog box.

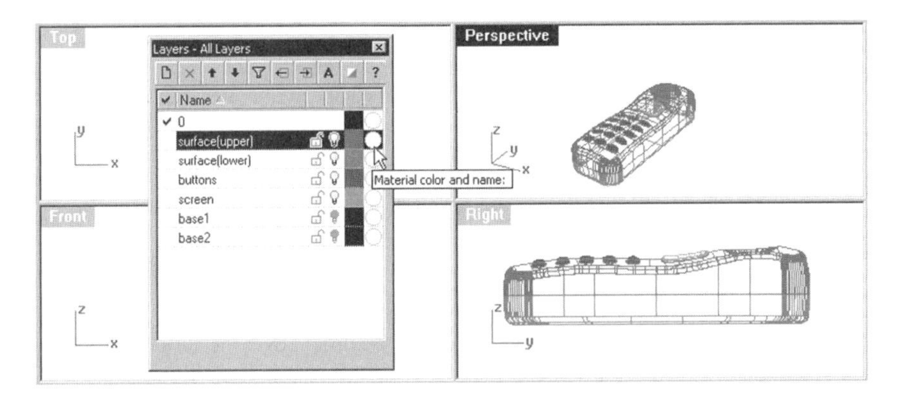

Fig. 8-7. Layers dialog box.

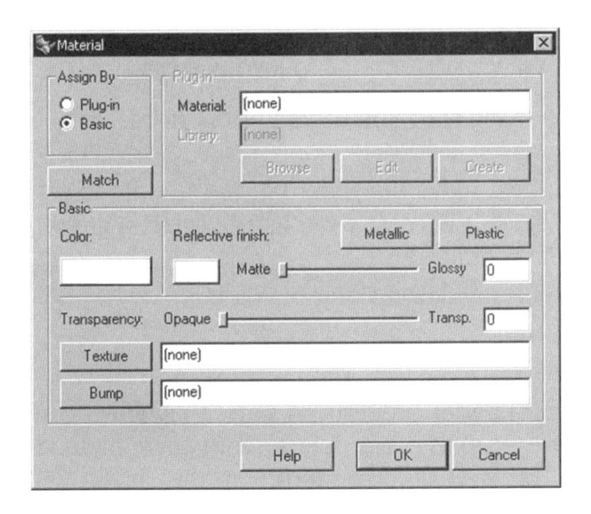

The Plug-in and Basic options for assigning materials are found in the Material dialog box, shown in figure 8-8. In the following section you will explore the basic material options available from the Rhino renderer. You will learn about the plug-in options when you work with the Flamingo renderer later in the chapter.

Fig. 8-8. Material dialog box.

Material Properties Assignment via the Basic Material Option

Rhino's Basic Material option of the Properties dialog box incorporates seven fields: Material Name, Match, Basic Color, Reflective Finish, Transparency, Texture Bitmap, and Bump Bitmap. These fields are outlined in Table 8-3.

Table 8-3: Basic Material Option Fields and Their Functions

Field	Function
Material Name	Assigns a material name to an object for export.
Match	Enables you to assign basic color, reflective finish, transparency, texture, and bump by copying these properties from a selected object.
Basic Color	Represents the main color of the object. You set the basic color by clicking on the color swatch to bring out the Select Color dialog box and selecting a color via the HSB color system or the RGB color system.
Reflective Finish	Establishes how the object reflects light. Here you select a color that the reflective area of the surface will elicit, and the strength of the reflection. By selecting the Metallic option, the reflective color is set to be the same as the basic color. By selecting the Plastic option, the reflective color is set to white. The slider bar lets you set the reflective strength from 0 to 100, giving a completely matt surface through a range of different reflective values to completely glossy.
Transparency	Sets the opacity of the object. You set the object in the range of 100% opaque to 100% transparent.
Texture Bitmap	Fundamentally places an image over the surface with the effect of "painting a picture" on the surface.
Bump Bitmap	Causes the surface to look bumpy, using the color value of the bitmap.

To assign basic material to a layer, perform the following steps.

1. In the Material dialog box, click on the Basic button, if it is not already selected.
2. In the Reflective Finish box, move the slider bar to the right to set the glossy value to 80, select the color swatch, and select a green color from the Select Color dialog box.
3. Click on the OK button on the Select Color dialog box.
4. Click on the Plastic button.
5. Click on the OK button.

The basic material property is now assigned to the layer, *surface(upper)*. All objects with their material assigned to layer and residing on this layer will inherit the material property. To appreciate the effect of the material properties assigned to the layer, continue with the following steps to render the scene.

6. Select Render > Render, or click on the Render button on the main toolbar. The result is shown in figure 8-9. Set the color for the other layers per Table 8-4, which follows.

Fig. 8-9. Result of step 6.

Table 8-4: Layer Rendering Settings

Layer	Color	Reflective Finish	Transparency
surface(lower)	Green	Plastic: 80 Glossy	0
button	Cyan	Metal: 80 Glossy	0
screen	Light Gray	Plastic: 80 Glossy	90

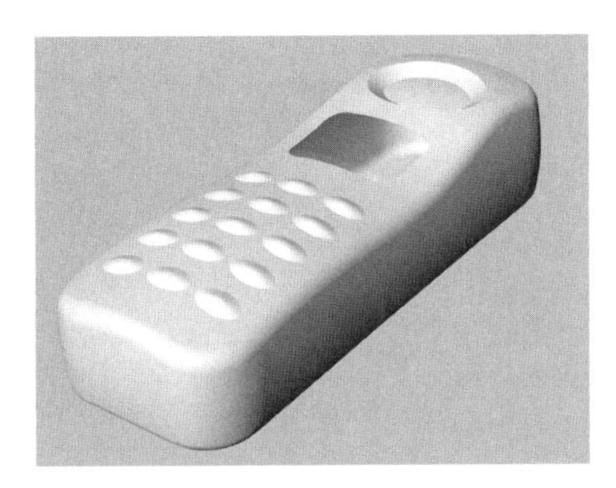

7 Render the Perspective viewport. The result is shown in figure 8-10.

Fig. 8-10. Rendered image.

Assigning Material Properties to Individual Objects

To set the material properties of selected objects individually, you assign a material to them via the Properties dialog box. To apply a basic material to selected surfaces, perform the following steps.

1. Maximize the Perspective viewport.
2. Select Edit > Object Properties, or click on the Object Properties button on the Properties dialog box.
3. Select surfaces A and B, indicated in figure 8-11. (Hold down the Shift or Ctrl key while selecting enables you to make multiple object selection.)
4. In the Properties dialog box, select Material from the pull-down list box and click on the Basic checkbox.

You will find that the Properties dialog box's fileds, although different in arrangement, are basically the same as those of the Material dialog box. Continue with the following steps.

5. Change the color to Red.
6. Click on the Metallic button.
7. Set glossy finish to 100.

Establish a texture bitmap by continuing with the following steps.

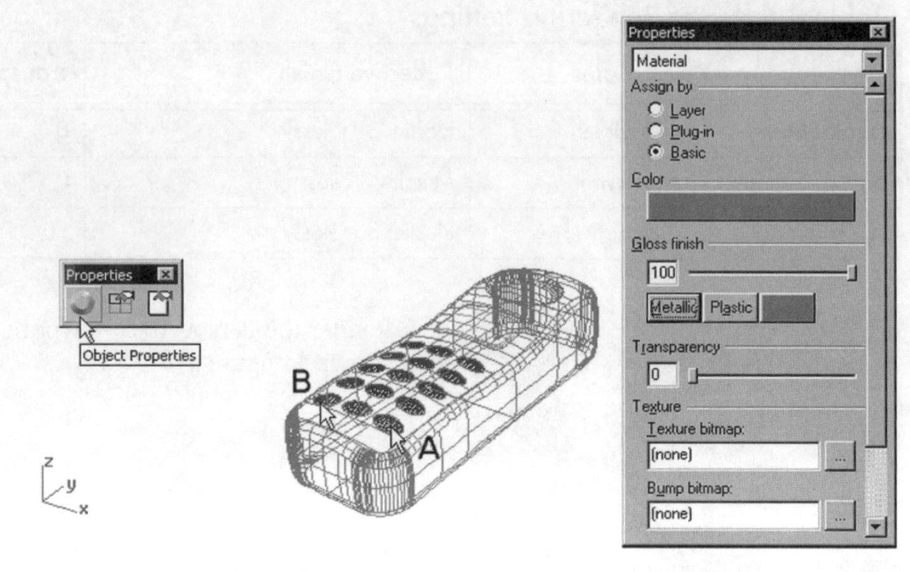

Fig. 8-11. Material tab of the Properties dialog box.

8 Turn on layer *base1*.

9 Select Edit > Object Properties, or click on the Object Properties button on the Properties dialog box.

10 Select surface A (indicated in figure 8-12).

11 In the Properties dialog box, click on the Texture Bitmap button and select the image *Render1.jpg* from the *Chapter 8* folder on the companion CD-ROM.

12 Render the Perspective viewport, as shown in figure 8-13.

Establish a bump bitmap by continuing with the following steps.

13 Turn off layer *base1* and turn on layer *base2*.

14 Select Edit > Object Properties, or click on the Object Properties button on the Properties dialog box.

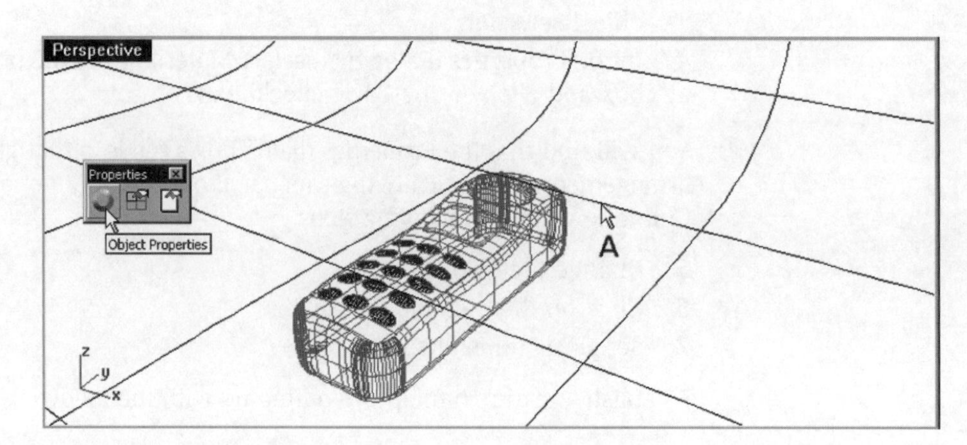

Fig. 8-12. Texture image applied.

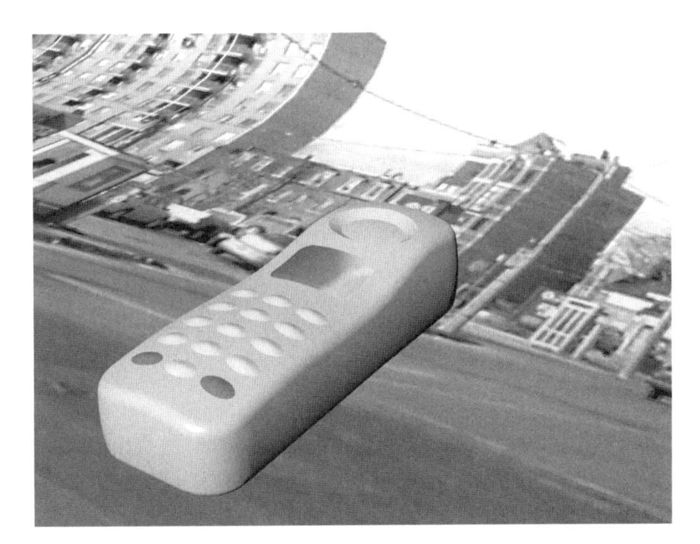

Fig. 8-13. Rendered image.

15 Select surface A (indicated in figure 8-14). Click on the Bump bitmap button and select the image *Render1.jpg* from the *Chapter 8* folder on the companion CD-ROM.

16 Render the Perspective viewport, as shown in figure 8-15.

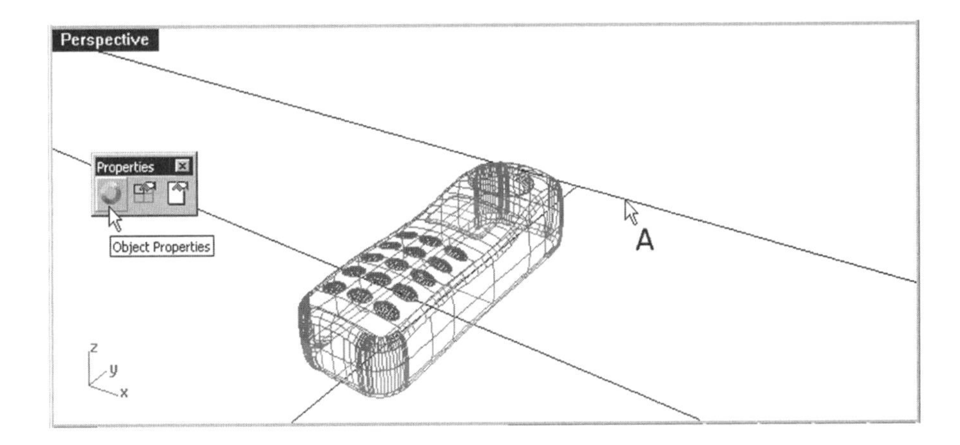

Fig. 8-14. Bump bitmap applied.

Fig. 8-15. Rendered image.

Rhino Lighting

In addition to material assignment, lighting is also a crucial element that contributes to the final digital rendered image. The purpose of lighting is to illuminate the 3D scene and to achieve specific lighting effects. In a Rhino file, there is a default light. If you imagine the screen is a camera viewer through which you look into the 3D space in which you construct the model, the default light is located over your left shoulder. The intensity of the light is always good enough to illuminate the entire 3D space, no matter how large it is. That is why you have a well-illuminated scene in the rendering processes described previously, even though you have not added any light.

If you are satisfied with the default lighting effect, you do not need to add any light. Note that once a light of any kind is added the default light is turned off automatically. It will turn on again if you delete all lights you add.

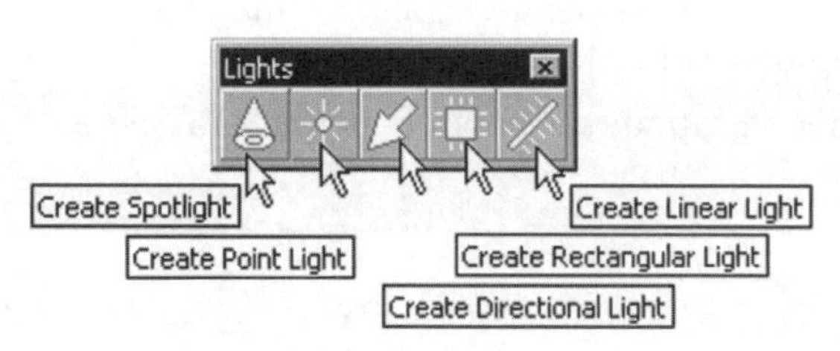

Fig. 8-16. Lights toolbar.

To better control the lighting in a scene, you add virtual lights and apply sunlight and environmental lighting effects. There are five categories of virtual lighting: spotlights, point light, directional light, rectangular light, and linear light. Using the Rhino renderer, you can construct spotlight, point light, and directional light. Using the Flamingo renderer, you can construct all five types of lights. To construct a light in the scene, you use the commands available from the Lights toolbar, shown in figure 8-16.

Spot light, point light, and directional light are examined in the following sections. Rectangular light and linear light are discussed under the section on Flamingo's renderers later in the chapter.

Spotlights

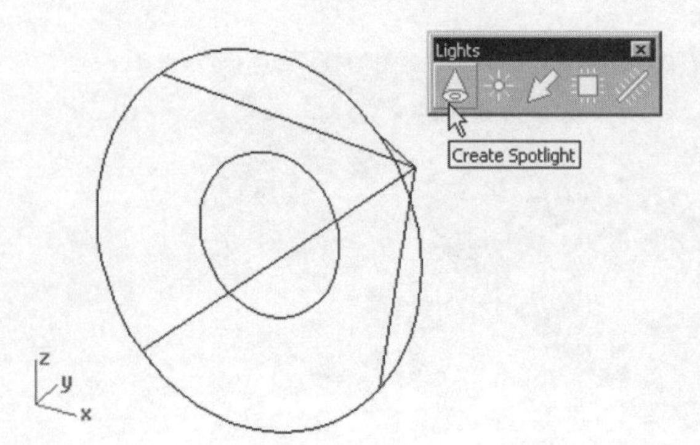

A spotlight is characterized by a light source emitting a conical beam of light toward a target. It illuminates a conical volume in the scene. In working with spotlights in Rhino, you specify the light source, the target, and the diameter of the base of the cone. Figure 8-17 shows a spotlight constructed in a scene.

Fig. 8-17. Spotlight in a scene.

In the scene, a spotlight is represented by two cones with a common vertex. They represent the direction of the light, not the range of the light. The inner cone represents the area of full brightness of the light, and brightness will decrease from the inner cone toward the outer cone. To adjust the size and location of the spotlight, you can turn on the control points and drag them to new locations. A spotlight that has narrower cones produces more detail than a spotlight with wider cones.

Point Light

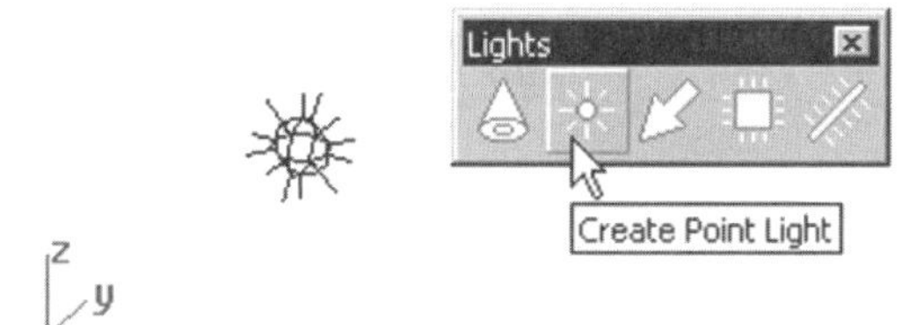

Fig. 8-18. Point light in a scene.

A point light is analogous to an ordinary electric light bulb that emits light in all directions. Point light illuminates an entire scene. In working with point light in Rhino, you specify a location of the light source. Figure 8-18 shows a point light in a scene.The point light in a scene is a symbol representing the source of a point light emitting light in all directions.

Directional Light

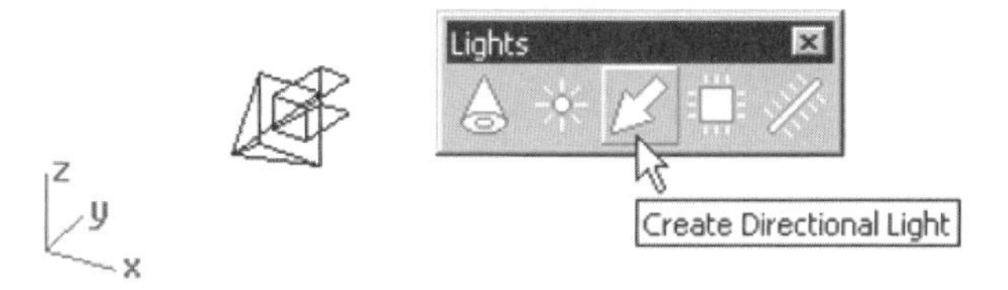

Fig. 8-19. Directional light in a scene.

Directional lighting derives from a light source at a distance, such as the sun. This lighting, generally represented as a beam or parallel beams of light, illuminates an entire scene. You specify a direction vector for the directional light. Figure 8-19 shows a directional light in a scene.

A directional light in a scene is also a symbol. It depicts the direction. It does not matter where it is placed.

Working with Lighting Effects

In this section you will learn how to incorporate spotlight, point light, and directional light in a scene.

1. Open the file *Render2.3dm* from the *Chapter 8* folder on the companion CD-ROM.
2. Set the current layer to *spotlight*. You will construct the light-source objects in this layer.

3 Select Render > Create Spotlight, or click on the Create Spotlight button on the Lights toolbar.

4 Type *V* at the command line area to use the Vertical option.

5 With reference to figure 8-20, select location A in the Top viewport to specify the base of the cone, select point B in the Top viewport to specify the radius of the cone, and select point C in the Right viewport to specify the end of the cone.

A spotlight vertical to the Top viewport is constructed. Note that once a light of any type is constructed the default light will be turned off automatically.

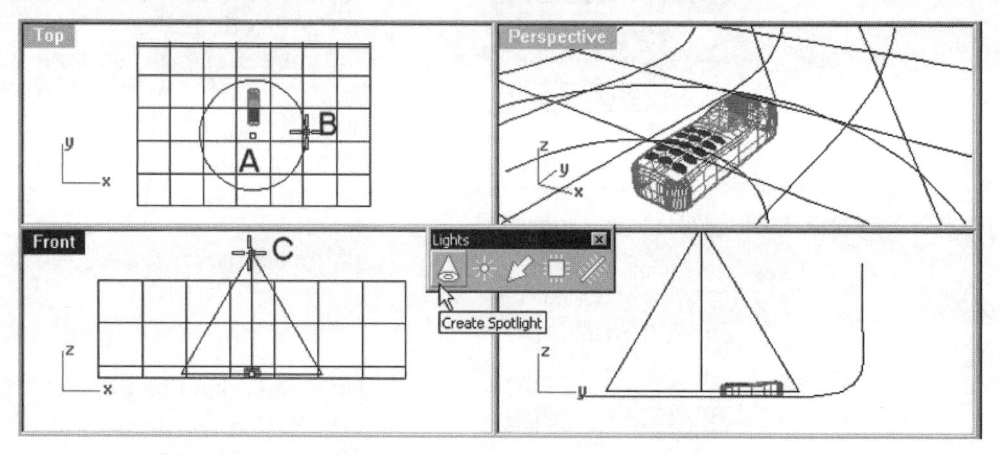

Fig. 8-20. Spotlight constructed.

6 Maximize the Perspective viewport.

7 Select Render > Render, or click on the *Render/Render Properties* button on the Standard toolbar.

Fig. 8-21. Scene rendered.

Because the position of your light may not be the same as that shown in figure 8-21, the rendered image will be different. Adjust the properties of the light by continuing with the following steps.

8 Select Edit > Object Properties, or click on the Object Properties button on the Properties toolbar.

9 Select spotlight A.

10 Select Light from the pull-down list box of the Properties dialog box, shown in figure 8-22.

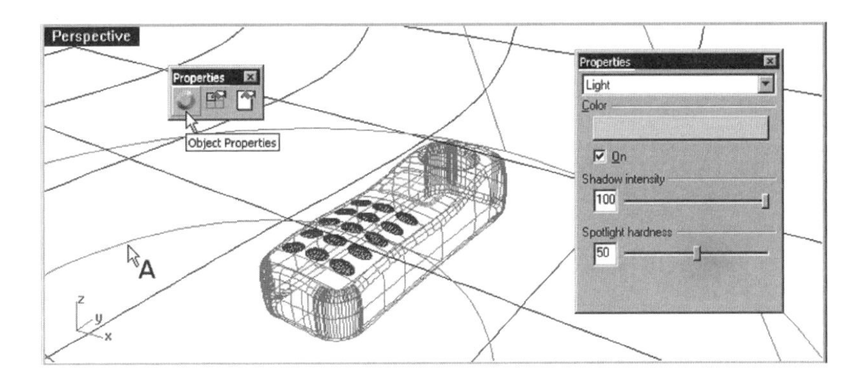

Fig. 8-22. Spotlight's properties being modified.

You can modify three parameters of a spotlight: color, shadow darkness, and lighting hardness. Color refers to the color of the light. Shadow Darkness sets the darkness of the shadow for spotlights. Spotlight Hardness defines the hardness of the spotlight edge.

11 Select the color swatch, set the color to cyan, and click on the OK button to exit.

The light's color is changed. To move the light, continue with the following steps.

12 Set the display to a four-viewport configuration by double clicking on the Perspective viewport's label.

13 Select Edit > Control Points > Control Points On, or click on the *Control Points On/Points Off* button on the Point Editing toolbar.

14 Select the spotlight and press the Enter key.

15 Select control point A in the Front viewport, hold down the mouse button, drag the control point to position B, and release the mouse button. (See figure 8-23.)

16 Select control point C in the Top viewport, hold down the mouse button, drag the control point to location D, and release the mouse button. (See figure 8-23.)

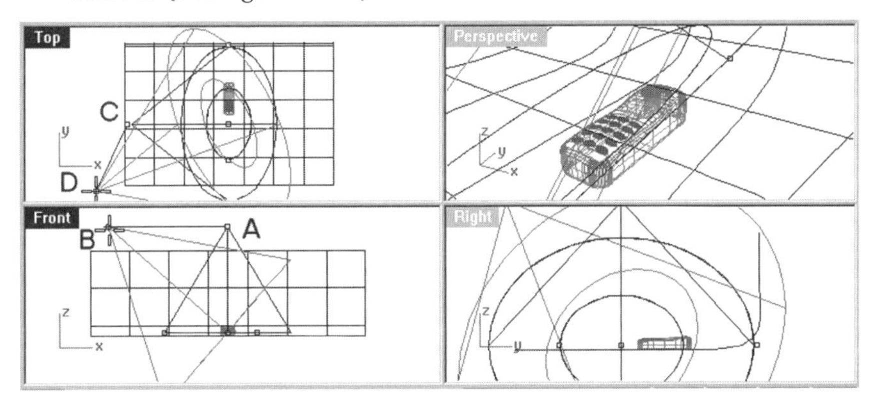

Fig. 8-23. Light's location changed.

17 Click on the Perspective viewport.

18 Select Render > Render, or click on the *Render/Render Properties* button on the Standard toolbar. The result is shown in figure 8-24.

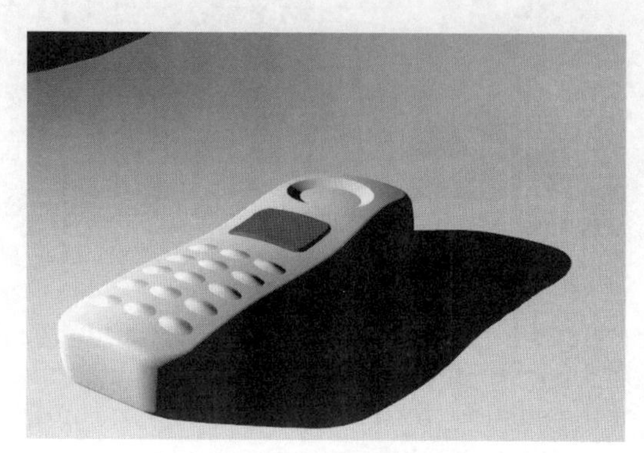

Fig. 8-24. Scene rendered.

Turn off the spotlight and add a point light to the scene by continuing with the following steps.

19 Select Edit > Object Properties, or click on the Object Properties button on the Properties toolbar.

20 Select the spotlight.

21 In the Light tab of the Properties dialog box, deselect the On option and close the dialog box.

22 Set the current layer to *pointlight*.

23 Select Render > Create Point Light, or click on the Create Point Light button on the Lights toolbar.

24 With reference to figure 8-25, select a point in the Front viewport.

A point light is constructed.

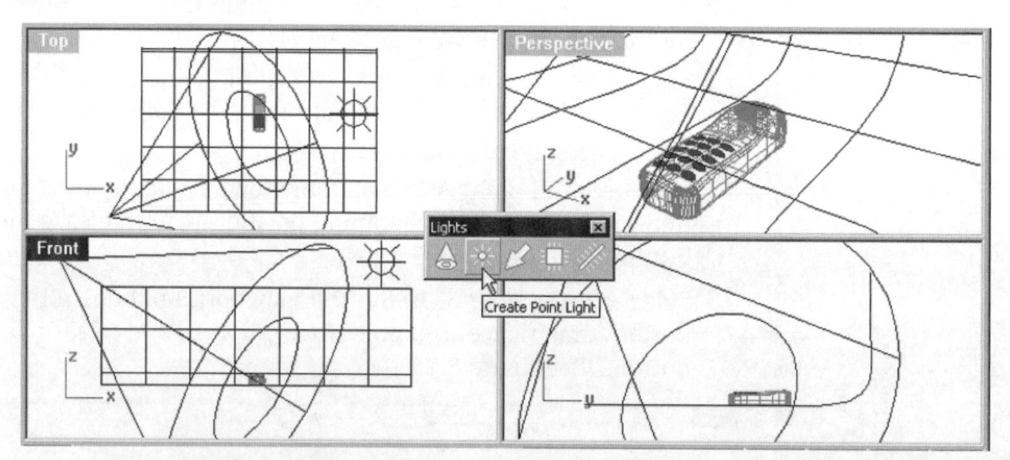

Fig. 8-25. Point light being constructed.

25 Select Edit > Object Properties, or click on the Object Properties button on the Properties toolbar.

26 Select the point light and press the Enter key.

27 Select the Light tab from the Properties dialog box.

28 Adjust the light's intensity to 60, as shown in figure 8-26.

29 Click on the Perspective viewport.

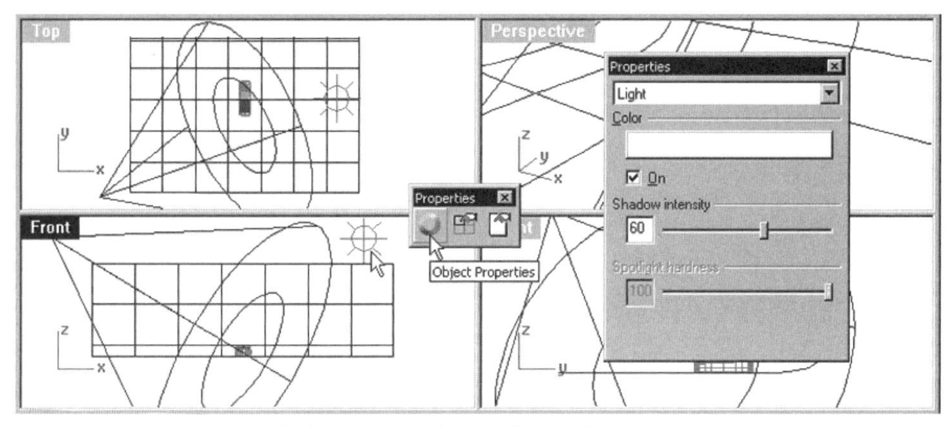

Fig. 8-26. Point light's intensity being changed.

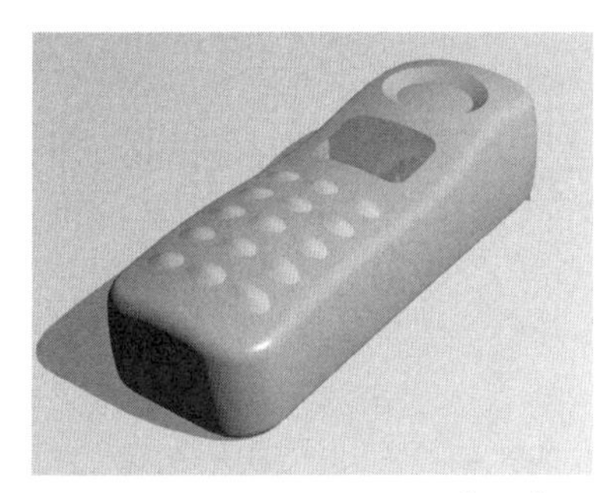

Fig. 8-27. Viewport rendered.

30 Select Render > Render, or click on the *Render/ Render Properties* button on the Standard toolbar. The result is shown in figure 8-27.

Turn off the point light and add a directional light by continuing with the following steps.

31 Select Edit > Object Properties, or click on the Object Properties button on the Properties toolbar.

32 Select the point light and press the Enter key.

33 Deselect the On option to turn off the light.

34 Select Render > Create Directional Light, or click on the Create Directional Light button on the Lights toolbar.

35 With reference to figure 8-28, select locations A and B in the Right viewport to specify the start and end of the light direction vector.

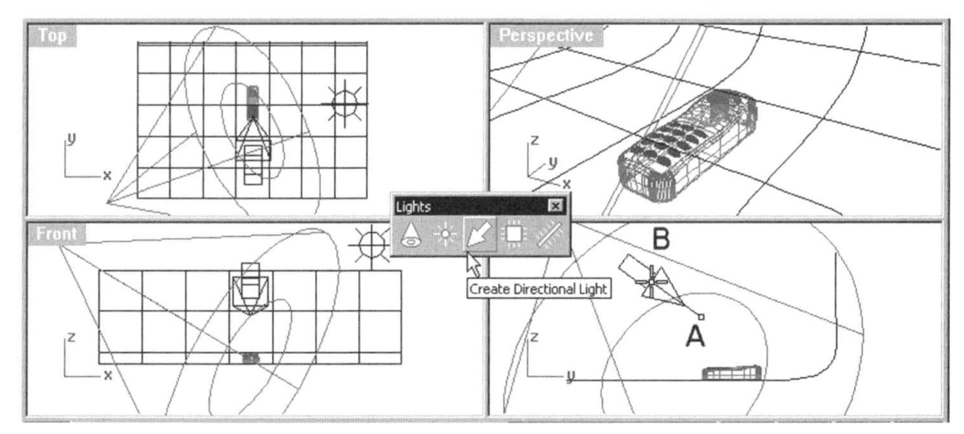

Fig. 8-28. Directional light being constructed.

36 Render the scene, as shown in figure 8-29.

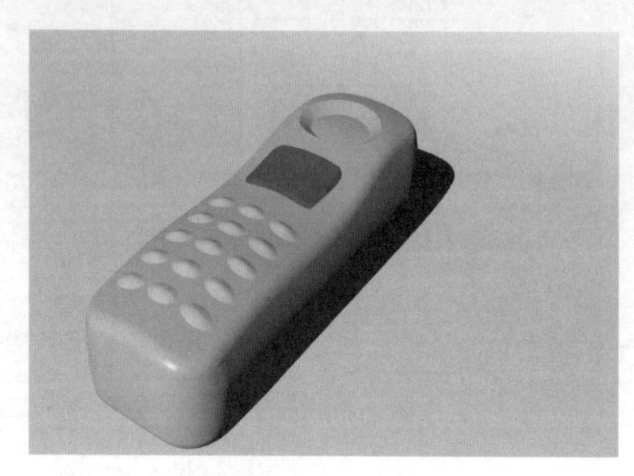

Fig. 8-29. Viewport rendered with directional light.

Environment Objects

Environment objects are objects you include in the scene to enhance its appearance. Using Rhino's basic renderer, you can adjust the ambient light and the background color.

Ambient Light

In reality, the combined effect of all lightings in a scene causes objects not directly under any light source to be illuminated. This is done by reflection of lights in the entire scene. The effect is commonly known as radiosity. The effect of radiosity depends on the strength of the light sources and how the lights are reflected by the objects in the scene. For example, an object placed inside a white box will be better illuminated than an object placed inside a box of dark color. The reason is that more light is reflected by the white box. Ambient light provides a means of approximating the radiosity effect. Using Rhino's basic renderer, you set the ambient light's color.

Background Color

To simulate a real environment, an image placed in the background may save a lot of effort in object construction and material assignment. Using Rhino's basic renderer, background image is not available. However, you can still change the background's color. To overcome the constraint resulting from the absence of background images, you can construct a backdrop in the form of a flat, cylindrical, or spherical surface; place the backdrop in the background; and apply to the backdrop surface a material incorporating an image map.

Setting Ambient Color and Background Color

To establish ambient lighting and a background color, perform the following steps.

1. Open the file *Render3.3dm* from the *Chapter 8* folder on the companion CD-ROM.
2. Select Render > Properties.
3. In the Rhino Render tab of the Document Properties dialog box, select the Ambient Light color swatch.
4. In the Select Color dialog box, select green, and then click on the OK button.

Fig. 8-30. Ambient light changed and scene rendered.

5. Close the Document Properties dialog box.
6. Render the scene, as shown in figure 8-30.
7. Select Render > Properties.
8. In the Rhino Render tab of the Document Properties dialog box, select the Background color swatch.
9. In the Select Color dialog box, select red, and then click on the OK button.
10. Render the scene, as shown in figure 8-31.

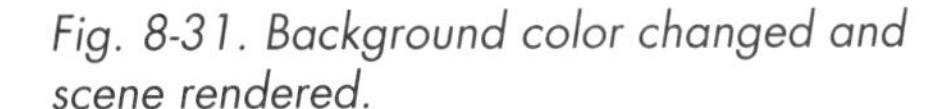

Fig. 8-31. Background color changed and scene rendered.

Flamingo Renderers

Flamingo is an add-on application that runs under Rhino's modeling environment. If you have installed Flamingo in your computer, you have two more types of rendering effect available to you via the Raytrace renderer and the Photometric renderer. (A Flamingo Eval version is available with the Rhino 3.0 CD or the Rhino 3.0 Eval CD. The Flamingo Eval version places horizontal lines across the finished rendered image.)

Flamingo Raytrace Renderer

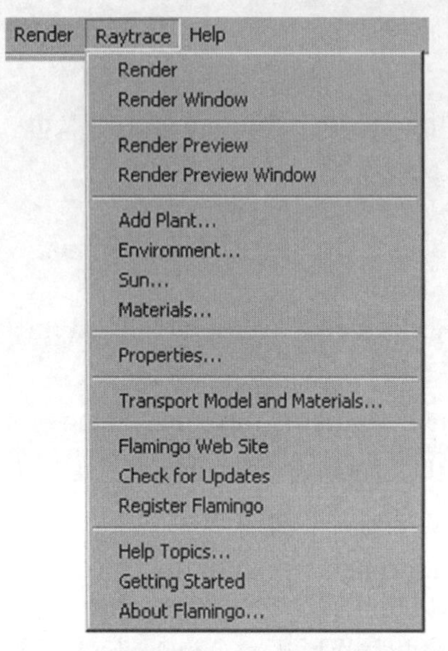

Fig. 8-32. Raytrace menu.

Raytracing is a technique that incorporates the effect of the path of light rays from all light sources in a scene to the viewer's eye. Raytracing takes into account the intensity of the light source (or sources) and the transparency and reflectivity of objects between the light source and the viewer's eye. The Flamingo Raytrace renderer uses raytracing techniques to compute the combined effect of all direct lighting. It is best suited for simulation of a studio environment via various special lighting effects. To use the Raytrace renderer, perform the following.

1 Select Render > Current Renderer > Flamingo Raytrace.

You will find the Raytrace menu, shown in figure 8-32, to the right of the Render menu.

Among the many options available from the Raytrace menu, five of them are significant: Add Plant, Environment, Sun, Materials, and Transport Model and Materials. These options are examined later in the chapter.

Flamingo Document Properties

You should spend some time examining the options available before using the Raytrace renderer.

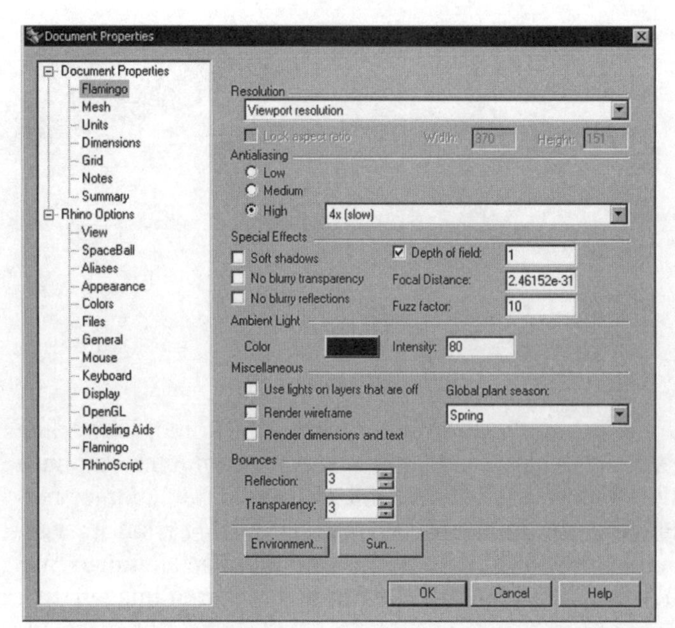

1 Select Raytrace > Properties.

The Flamingo tab of the Document Properties dialog box is displayed, as shown in figure 8-33. This tab has a number of fields, including Resolution, Antialiasing, Special Effects, Ambient Light, Miscellaneous, Bounces, Environment, and Sun. Table 8-5 outlines the fields of the Flamingo tab.

Fig. 8-33. Flamingo tab of the Document Properties dialog box.

Table 8-5: Flamingo Tab Fields and Their Functions

Field	Function
Resolution	Sets the resolution of the rendered image.
Antialiasing	Removes the jagged edges of the rendered image.
Soft shadows	Softens the edges of shadows in the rendered image.
No blurry transparency	Keeps the transparency sharp without any blur. Normally, blurriness is introduced to transparent material to simulate a real-life situation.
No blurry reflections	Keeps the reflection sharp without any blur that is otherwise introduced by default.
Depth of field	Sets the depth of field, which is the distance between the near-field boundary and the far-field boundary.
Focal Distance	Sets the focal length of the imaginary camera.
Fuzz factor	Sets the size of an acceptable circle of confusion.
Ambient Light Color	Sets the color of ambient light.
Ambient Light Intensity	Sets the intensity of the ambient light.
Use Lights on layers that are off	Enables the use of lights that reside on turned-off layers.
Render wireframe	Renders the wireframe.
Render dimensions and text	Renders dimensions and text.
Global plant season	Sets the season for any plants inserted in the scene.
Bounces Reflection	Determines the number of levels of reflections. The number refers to the level of reflection that will be raytraced.
Bounces Transparency	Determines the number of levels of transparency. The number refers to the number of transparent surfaces that can be seen through.
Environment	Lets you insert environment objects in the scene.
Sun	Lets you turn on the sun light source and set its parameters.

Flamingo Photometric Renderer

Radiosity refers to the indirect lighting in a scene. Unlike rendered ambient light, which approximates indirect lighting, radiosity is a more accurate representation of the light emanating from all directions. You use radiosity for scenes in which the indirect lighting effect is significant. The Flamingo Photometric renderer adds to the rendered image subtle radiosity lighting effects produced by the combination of all indirect lighting sources. The Photometric renderer is suitable for interiors for which you want to obtain a more realistic visualization of how the real lights will light the scene. To use the Photometric renderer, perform the following.

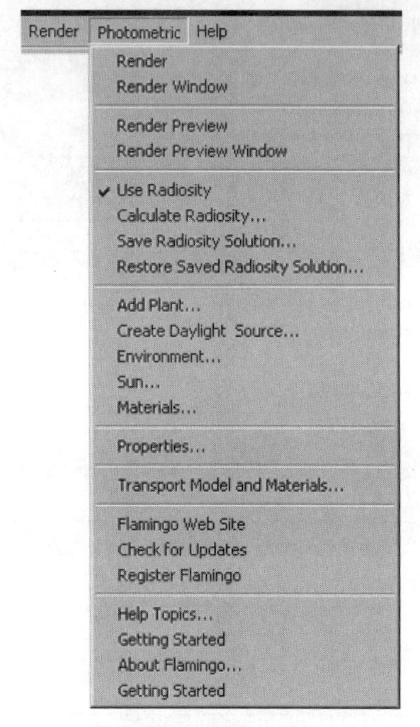

Fig. 8-34. Photometric menu.

1 Select Render > Current Renderer > Flamingo Photometric.

You will find the Photometric menu, shown in figure 8-34, to the right of the Render menu. In addition to the options of the Raytrace menu, there are several options in the Photometric menu, as follows: Use Radiosity, Calculate Radiosity, Save Radiosity Solution, and Restore Saved Radiosity Solution. These options affect the radiosity of a scene.

Flamingo Document Properties and Photometric Properties

You should take some time to explore the settings involved in photometric rendering. Perform the following.

1 Select Photometric > Properties.

The Flamingo tab of the Document Properties dialog box for the Photometric renderer, shown in figure 8-35, is basically the same as that for the Raytrace renderer, except that the ambient lighting color is not modifiable. The reason is that ambient light, which is an approximation of radiosity, is replaced by radiosity calculation in photometric rendering. In addition to the Flamingo tab, there is an additional tab relevant to photometric rendering, the Photometric tab.

2 Select the Photometric tab.

The Photometric tab is displayed, as shown in figure 8-36. Table 8-6 outlines the fields of the Photometric tab.

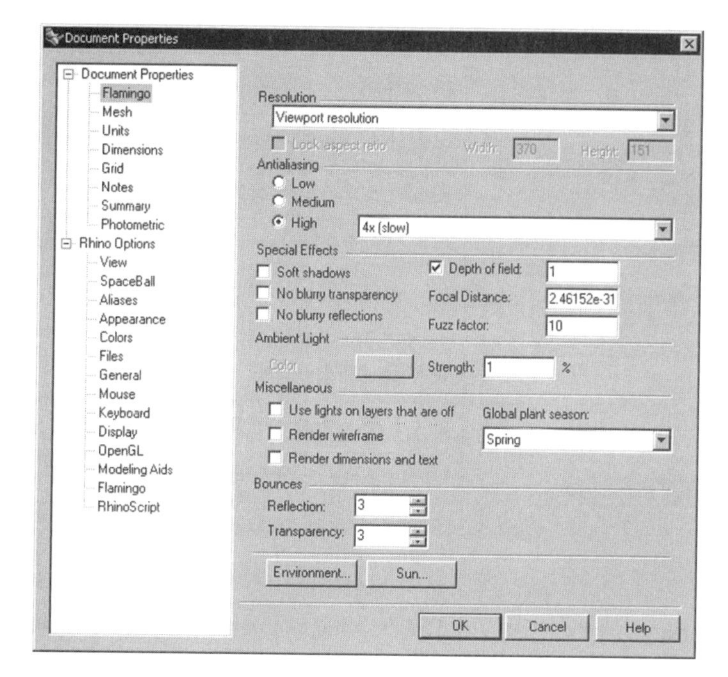

Fig. 8-35. Flamingo tab of the Photometric renderer.

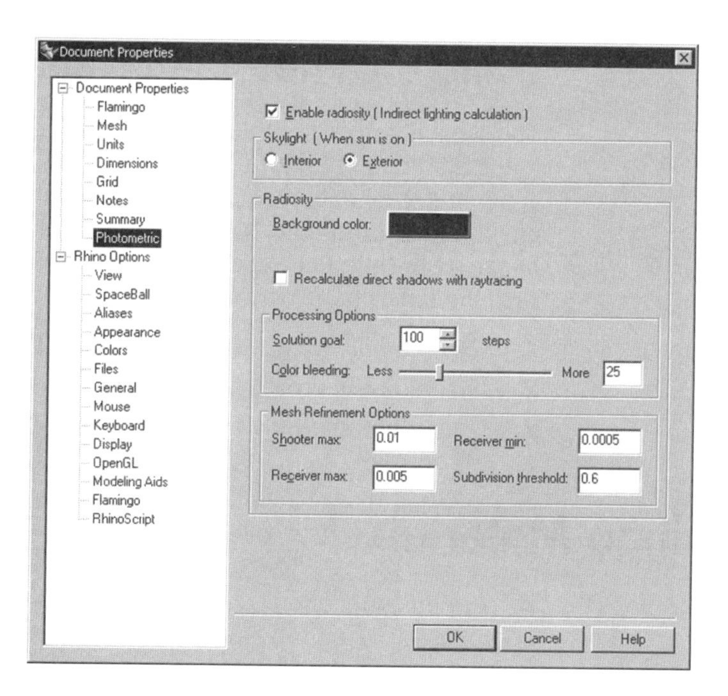

Fig. 8-36. Photometric tab for Photometric renderer.

Table 8-6: Photometric Tab Fields and Their Functions

Field	Function
Enable radiosity	Calculates and renders a radiosity solution.
Skylight Interior	Estimates the ambient light at all points for rendering interior scenes.
Skylight Exterior	Estimates the ambient light at all points for rendering outdoor scenes.
Background color	Sets the background's color.
Recalculate direct shadows with raytracing	During the radiosity pre-process, replaces shadows casted from primary lights with raytraced shadows.
Solution goal	Limits the number of steps for radiosity calculation.
Color bleeding	Controls the color saturation of reflected light.
Shooter max.	Sets the size of radiosity mesh faces reflecting lights.
Receiver max.	Sets the maximum size of radiosity mesh faces receiving lights.
Receiver min.	Sets the minimum size of radiosity mesh faces receiving lights.
Subdivision threshold	Controls the sensitivity of the adaptive subdivision algorithm.

Using the Photometric Renderer

The steps involved in producing a photorealistic image via the Flamingo renderers are virtually the same as those for the Rhino renderer. You apply materials to objects in the scene, construct lights in the scene, add environment objects to the scene, and use the renderer to produce the image. Using the Photometric renderer, you can calculate radiosity and save the radiosity by performing the following steps.

1. Select Render > Current Renderer > Flamingo Photometric.
2. Select Photometric > Use Radiosity.
3. Select Photometric > Calculate Radiosity.
4. Select Photometric > Save Radiosity Solution.

Flamingo Materials

As previously mentioned, there are two major types of materials: basic and plug-in. Using the Flamingo Raytrace or Photometric renderer, you can use plug-in materials. In addition, you can apply two types of transparency, cast shading during raytracing, adjust mapping and tiling of bitmaps applied on an object, add decals, and include waves.

Material Properties Assignment via Plug-in

The "plug-in" material library is available when you use the Flamingo Raytrace renderer or the Flamingo Photometric renderer. Using plug-in material is simple. You select a material from the plug-in material library, set the parameters of the material if necessary, and apply the material to the object. Like basic material, you can apply plug-in material to individual objects or to all objects of a layer by setting the material property of the objects to *bylayer.*

Material Library

To open the Material Library dialog box (via which you select a plug-in material, edit plug-in material, or construct new plug-in material), perform the following steps.

1 Open the file *FlamingoMaterial1.3dm* from the *Chapter 8* folder on the companion CD-ROM.

If you assign plug-in material to individual objects, follow steps 2 through 4 and skip steps 5 through 9. If you assign plug-in to a layer, skip steps 2 through 4 and steps 8 and 9, and follow steps 5 through 7. If you are using the Raytrace or Photometric renderer and want to set up a material, perform step 8.

2 Select Edit > Object Properties, or click on the Object Properties button on the Properties dialog box.
3 Select polysurface A.
4 In the Properties dialog box, shown in figure 8-37, select Material from the pull-down list and then click on the Browse button.

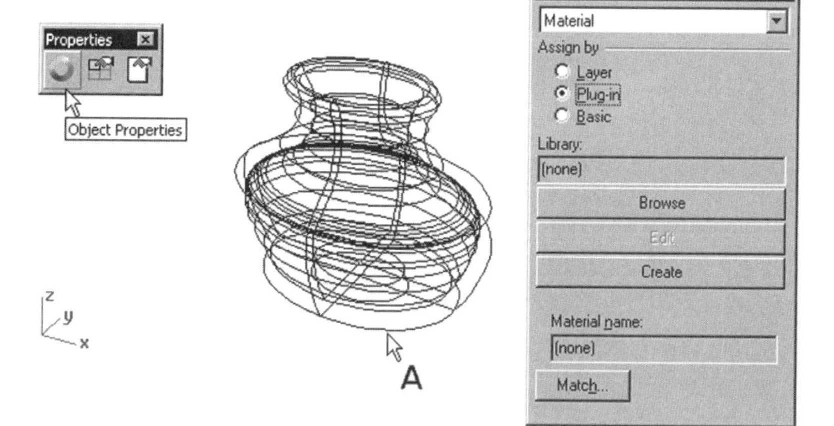

Fig. 8-37. Accessing the Material library from the Properties dialog box.

5 Select Edit > Layers > Edit Layers, or click on the Edit Layers button on the Standard toolbar.

6 In the Layers dialog box, select a layer and then click on the *Material color and name* icon.

7 In the Material dialog box, shown in figure 8-38, click on the Browse button.

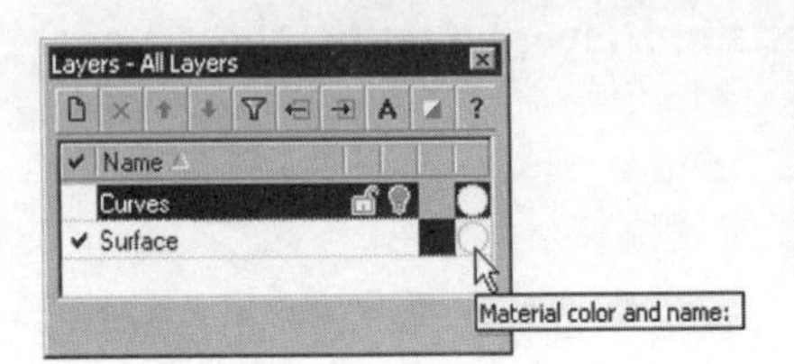

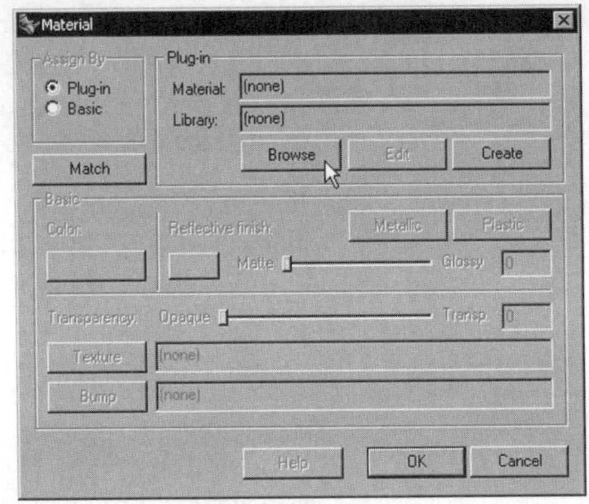

Fig. 8-38. Accessing the Material library from the Material dialog box.

8 To use the Raytrace renderer, select Raytrace > Materials. To use the Photometric renderer, select Photometric > Materials.

Performing either steps 1 through 3 or 4 through 6 will access the Material Library dialog box, shown in figure 8-39. The Material Library dialog box contains a number of panes. The left-hand pane displays various types of materials arranged in folders and subfolders. The central pane displays the materials available from each subfolder. After you select a material from the central pane, a preview showing what the material will look like when mapped to a sphere or a cube is displayed in the preview panes on the right-hand side of the dialog box. If you find a material suitable for use, click on the OK button. The selected material will apply to selected layers or selected objects.

9 In the Material Library dialog box, expand the Plastic folder and the Smooth subfolder.

10 Select Orange from the Name list.

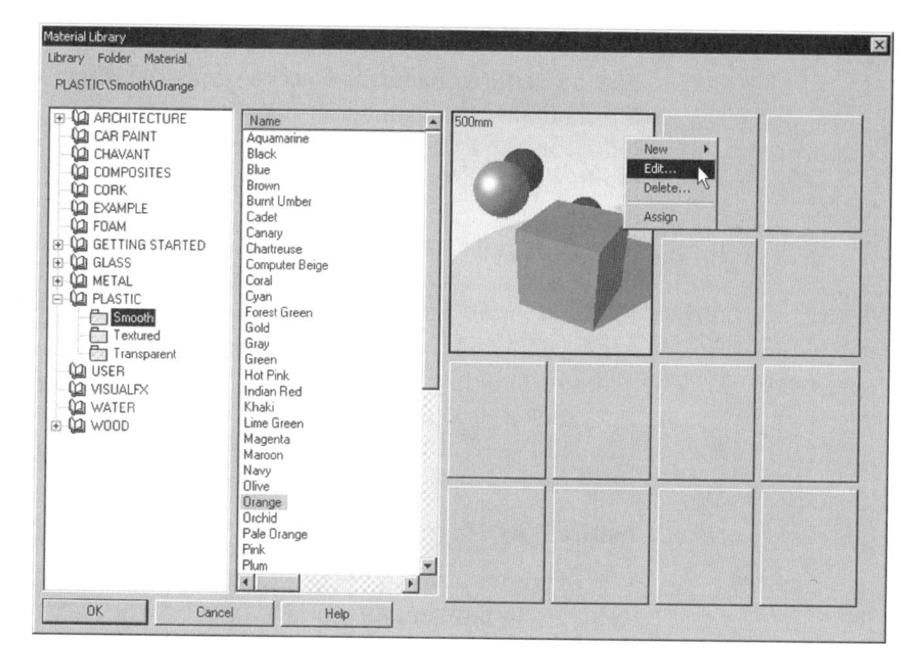

Fig. 8-39. Material Library dialog box.

Material Editor

To edit an existing plug-in material, you use the Material editor. The Material Editor dialog box, shown in figure 8-40, has a procedures pane at the left, four tabs at the center, and a number of preview panes at the right. To examine this dialog, perform the following.

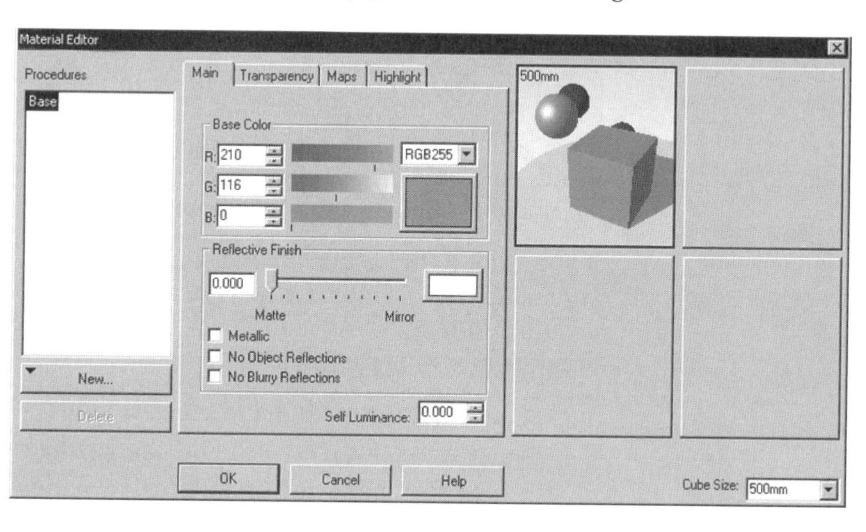

Fig. 8-40. Material Editor dialog box.

1. Click on the preview pane.
2. Right-click and select Edit.

Procedures Pane

The Procedures pane displays the material components that make up the material. If the material consists of only one material component, this pane shows the name of the component. If the material consists of a number of combined material components, this pane will show a hierarchy of the material components. Because the combination of two or more material components involves mathematical procedures, the material is called procedural material. In essence, a procedural material is a combination of two or more material components. Each of these components can also be a procedural material. The plain stainless steel shown here is a material that consists of a single material component. To modify a material component, you use the Main, Transparency, Maps, and Highlight tabs of the Procedures pane. These tabs are examined in the sections that follow.

Main Tab

The Main tab lets you set the base color, reflective finish, and self-luminance. You may set base color via the HSB or RGB color systems. Reflective finish concerns how the material will reflect lights and objects in the environment. Here you use the slider bar to set how much the material reflects from 0 (matte finish) to 1 (mirror finish). The default reflective finish color is white. To achieve special effects, you select a reflective finish color by clicking on the color swatch. If you check the Metallic option, the reflective finish color setting will be the same as that of the base color.

By default, a reflective material will reflect light as well as the objects around it. If you do not want reflection of objects in the environment, check the No Object Reflections box. By default, a small amount of noise is introduced to blur the reflection in order to simulate a real-world environment. To achieve a sharp reflection without any blur, check the No Blurry Reflections box.

It you want to make the material appear to glow, set a value in the Self Luminance box. It must be noted that a self-luminous material will not illuminate the environment. To set the material via the Main tab, perform the following steps.

1. Click on the lower left-hand preview pane.
2. Change the reflective finish and change the reflective color.
3. Click on the upper right-hand preview pane.
4. Check the Metallic box.
5. Click on the lower right-hand preview pane.
6. Check the No Object Reflections box, shown in figure 8-41.

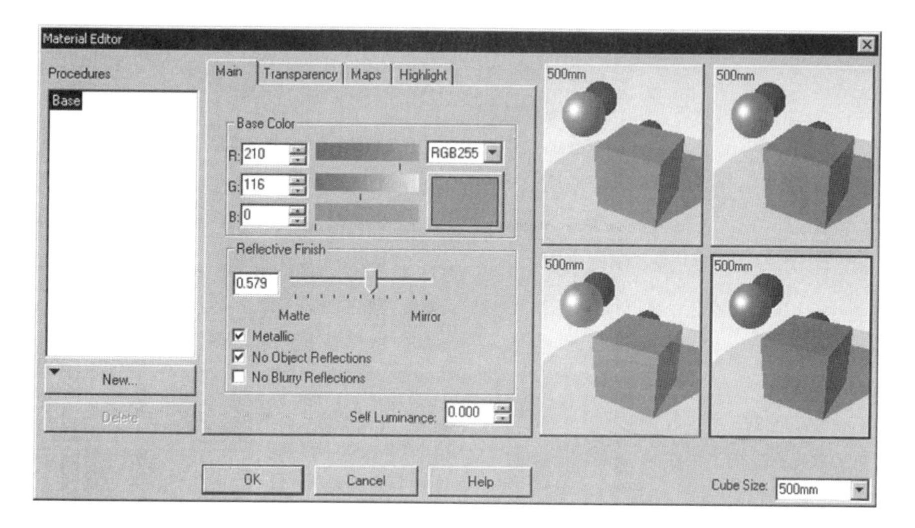

Fig. 8-41. Different versions of basic color and reflective finish.

You now have four different versions of the material. If you want to save one of these materials, perform the following steps.

7 Click on a preview pane you want to save and then click on the OK button.

8 In the Overwrite Warning dialog box, click on the Save As button and specify a name for the new material.

Transparency Tab

The Transparency tab establishes the transparency of material. Here you set the transparency and transparency finish of the material. A transparent material allows light to pass through it and makes the objects behind it visible. By moving the slider bar, you set the transparency of a material, from completely opaque to completely transparent.

When light passes through a transparent material, it is refracted and some of it is absorbed. The index of refraction is a measure of how much refraction occurs when looking through the transparent material. Attenuation is a measure of the amount of light being absorbed as it passes through the transparent material.

The transparent finish ranges from clear to frosted. If a transparent material is completely frosted, it allows light to pass through but the objects behind it are not visible. If a material is partly transparent, there is a small amount of default noise introduced to the transparency to make the material look more natural. If you do not want to have this default noise, check the No Blurry Transparency option. To set up a transparent material, perform the following steps.

1 Select the Transparency tab of the Material Editor dialog box.

2 Click on the upper left-hand preview pane.
3 Set the transparency to 0.5, index of refraction to 1.2, and attenuation to 0.3.
4 Set the transparency finish to 0.2. (See figure 8-42.)
5 Click on the OK button.

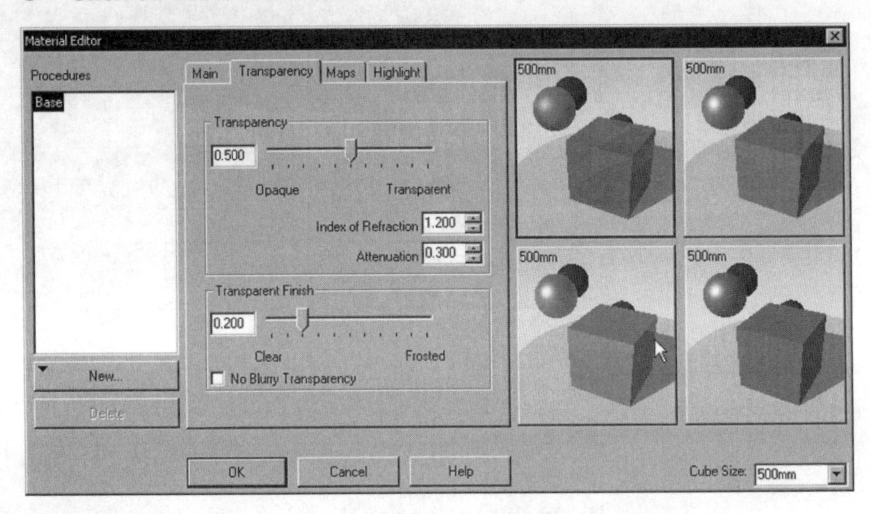

Fig. 8-42. Refractive finishes modified.

6 In the Overwrite Warning dialog box, click on the Save As button.
7 Specify the material name *Orange1* in the Save Material As dialog box and then click on the OK button.

To render the Perspective viewport, continue with the following steps.

8 Select Render > Current Renderer > Flamingo Raytrace.
9 Select Raytrace > Render.

Fig. 8-43. Rendered image.

Figure 8-43 shows the rendered image.

Maps Tab

The Maps tab contains two fields: Image Mapping and Procedural Bumps. They enable you to paste an image bitmap or a procedural bitmap onto the material. An image bitmap is an ordinary bitmap produced from a digital photograph, scanned picture, or picture created via a paint program. A procedural bitmap is a set of predefined mathematical formulae through which you construct a bitmap pattern by specifying a number of parameters.

Image Mapping: Perform the following steps to construct a material via image bitmapping and procedural bitmapping.

1. Open the file *FlamingoMaterial2.3dm* from the *Chapter 8* folder on the companion CD-ROM.
2. Select Edit > Layers > Edit Layers, or click on the Edit Layers button on the Standard toolbar.
3. In the Layers dialog box, select the Layers option and click on the *Material color and name* icon.
4. In the Material dialog box, click on the Browse button.
5. Click on the preview pane, right-click, and select New > Default Gray.
6. Select the Maps tab of the Material Editor dialog box, shown in figure 8-44.

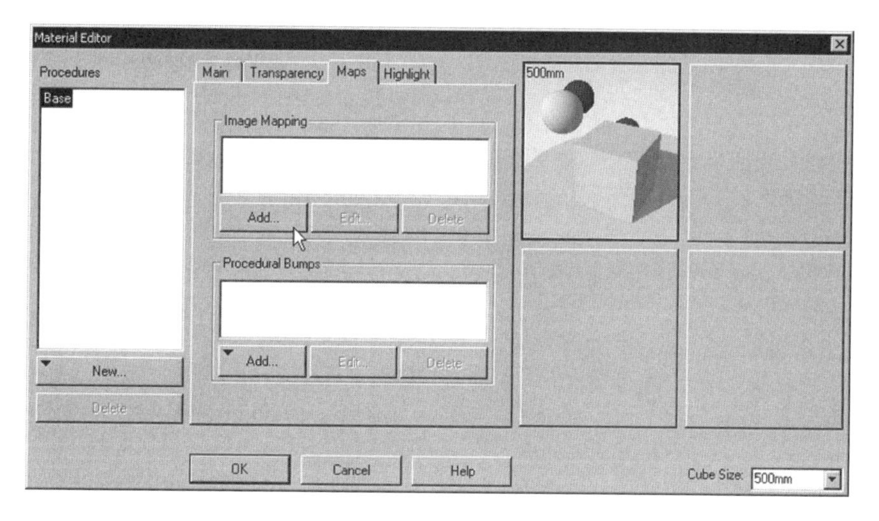

Fig. 8-44. Maps tab of the Material Editor dialog box.

7. In the Image Mapping field, click on the Add button.
8. Select bitmap *FlamingoMaterial2* from the *Chapter 8* folder on the companion CD-ROM.

Main Tab of the Image Mapping Dialog Box: After you select a bitmap, the Image Mapping dialog box appears, as shown in figure 8-45. The Image Mapping dialog box has four tabs: Main, Map, Orientation, and Advanced.

The Main tab sets the tile size and the strength of the bitmap. It also enables you to mirror-image tiles. By setting the tile size, you adjust the size of the image bitmap to suit the object on which you map the bitmap.

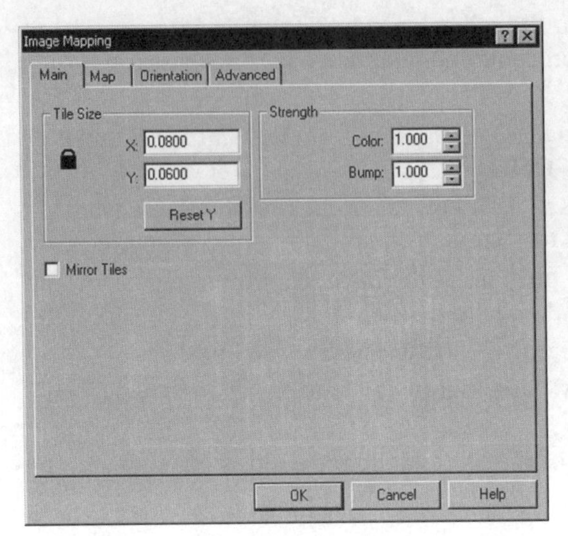

Fig. 8-45. Main tab of the Image Mapping dialog box.

To determine how much color from the material under the bitmap is going to show through the bitmap, you set the color strength, which ranges from 0 to 1. If you set the color strength to 1, the underlying material will be covered entirely by the bitmap. If you set color strength to less than 1, some color attributes of the underlying material will display through the bitmap.

To use the color value of the bitmap to give the illusion of a bumpy appearance, you set the bump strength of the bitmap, which ranges from –1 to 1. If the bump strength is 0, there will be no bumpy effect. If you set the bump strength to a positive value, a darker color of the bitmap will appear to rise above the surface in rendering. If you set the bump strength to a negative value, the bump direction will be reversed (i.e., indented rather than raised). To work with the bump strength setting, perform the following steps.

1. Set the tile size to 0.08.
2. Click on the Reset Y button.
3. Set the color strength to 1.
4. Set the bump strength to 1.

Map Tab of the Image Mapping Dialog Box: The Map tab lets you reselect a bitmap and apply a masking effect to the bitmap. In case you change your mind, you can click on the Browse button and reselect a bitmap. Masking refers to restriction of the use of a portion of the bitmap. Basically, masking causes a portion of the bitmap to be masked (or ignored), thus causing the underlying material to show. Perform the following.

1. Select the Map tab.

There are two ways to mask a bitmap. One is to mask selected colors of the bitmap (via the Color option), and the other is to mask the bitmap in accordance with the value of the alpha channel stored in the bitmap (via the Alpha Channel option).

To use the Color option, you select a masking color by selecting the color swatch and then selecting a color from the Select Color dialog box. Alternatively, you can select the Color Dropper and select a color from the bitmap preview. Sensitivity causes a range of color to be included as a masking color. Blur causes a partial masking. By clicking on the Reverse button, the effect of masking is reversed, causing pixels that are masked to become unmasked (and vice versa if you click on the button again). If you click on the Transparent button, the area of the surface masked by the bitmap will become transparent.

If you have a bitmap image that incorporates an alpha channel, in addition to the RGB channels you can use the alpha channel as a masking color. Thus, alpha masking provides the Reverse and Transparent options only. Continue with the following.

2 Select Color in the Masking field, as shown in figure 8-46, and view the result.

Orientaion Tab of the Image Mapping Dialog Box: The Orientation tab lets you orient or reposition the bitmap image with respect to the material. Here you select a reference plane, rotate the image, and specify an offset value to orient the image's origin. To examine the Orientation tab, perform the following steps.

1 Select the Orientation tab.

2 Accept the default, shown in figure 8-47.

Advanced Tab of the Image Mapping Dialog Box: The advanced tab lets you set the components altered by the image and set the sampling noise. Components affected can be the base color, the transparent color, the mirror color, transparent intensity, and mirror intensity. To introduce random noise on the bitmap, you select Low, Medium, or High from the Sampling Noise field. To examine the Advanced tab, perform the following steps.

1 Select the Advanced tab, shown in figure 8-48.

2 Click on the OK button to return to the Maps tab of the Material Editor dialog box.

This will bring you back to the Material Editor dialog box.

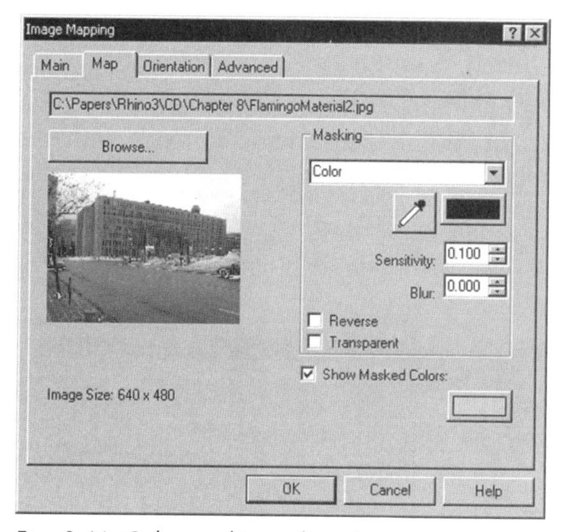

Fig. 8-46. Color masking selected.

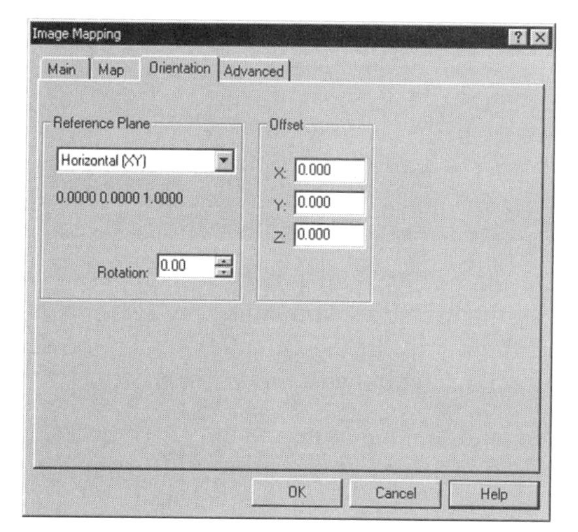

Fig. 8-47. Orientation tab.

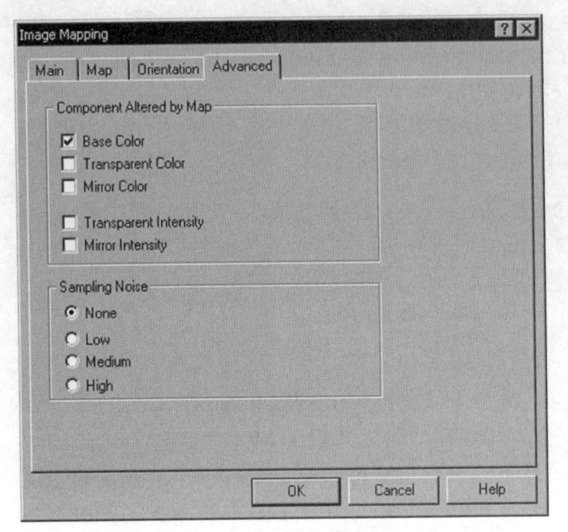

Fig. 8-48. Advanced tab of the Image Mapping dialog box.

Procedural Bumps: The second field of the Map tab of the Material editor is the Procedural Bumps field. If you want to add a bumpy pattern to the material, you can use procedural bumps instead of using an image bitmap. The main advantage of using a procedural bump is that less memory space is required and computation is faster than using an image bump.

In the Procedural Bumps field of the Maps tab of the Material Editor dialog box, you can use a procedural bump to provide the effect of bumping instead of using an image bitmap. Examples of procedural bumps are Sandpaper, Rubble, and Pyramid. To work with the procedural bump functionality, perform the following steps.

1. Click on the Add button and then select Sandpaper, as shown in figure 8-49.

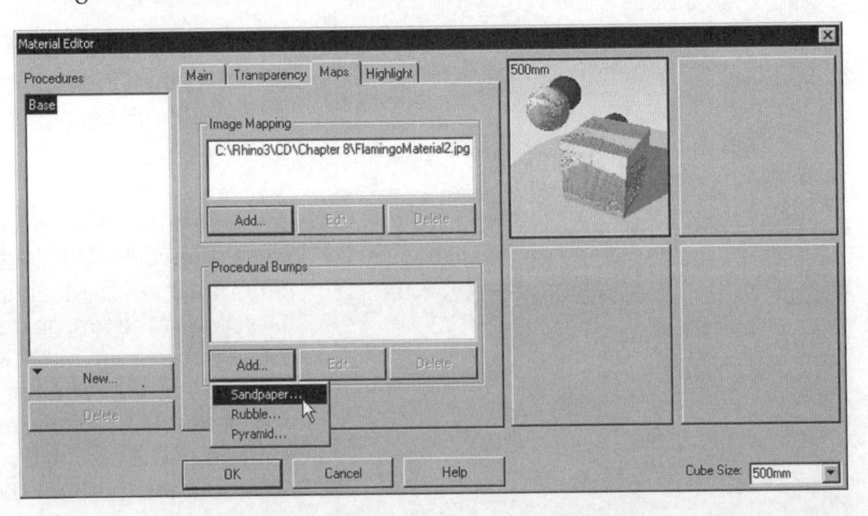

Fig. 8-49. Selecting a procedural bump map.

The Edit Sandpaper Bumpmap dialog box, shown in figure 8-50, has two tabs: Main and Orientation. In the Main tab, you adjust the scale of the bump map and the height of the bump. In the Orientation tab, you set the orientation of the bump map. Continue with the following steps.

2. Select the tabs one by one and make any necessary adjustments.
3. Click on the OK button on the Edit Sandpaper Bumpmap dialog box.
4. Click on the OK button on the Material Editor dialog box.

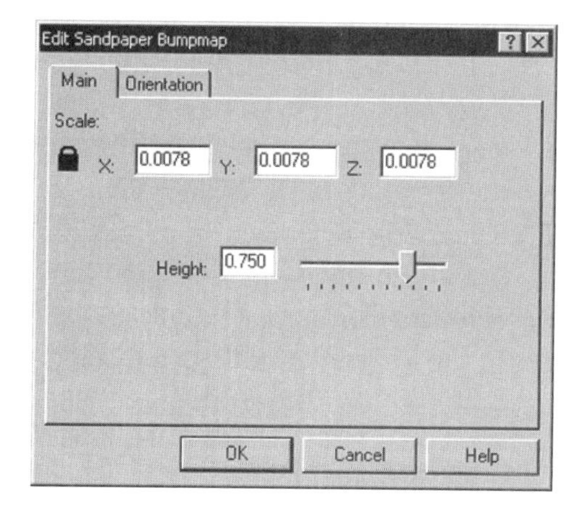

Fig. 8-50. Edit Sandpaper Bumpmap dialog box.

5. In the Save Material As dialog box, specify the material name *mystyle*.
6. Click on the OK button on the Material Editor dialog box.
7. Render the scene, the result of which is shown in figure 8-51.

Fig. 8-51. Rendered image.

Highlight Tab

The Highlight tab lets you set the glossiness of the material. Here you set the sharpness, intensity, and color of the highlight. If you want to have highlight on the material but do not want it to be reflective, you should set the highlight of the material by checking the Specify Highlight box on the Highlight tab of the Material Editor dialog box.

Setting a lower sharpness gives a broader highlight, and setting a higher sharpness value gives a narrower highlight. The intensity value determines the strength of the highlight. The color specifies the highlight color. To work with the various highlight settings, perform the following steps.

1. Open the file *FlamingoMaterial3.3dm* from the *Chapter 8* folder on the companion CD-ROM.
2. Select Edit > Object Properties, or click on the Object Properties button on the Properties dialog box.
3. Select ball A (indicated in figure 8-52).
4. In the Properties dialog box, select Plug-in and then click on the Edit button.

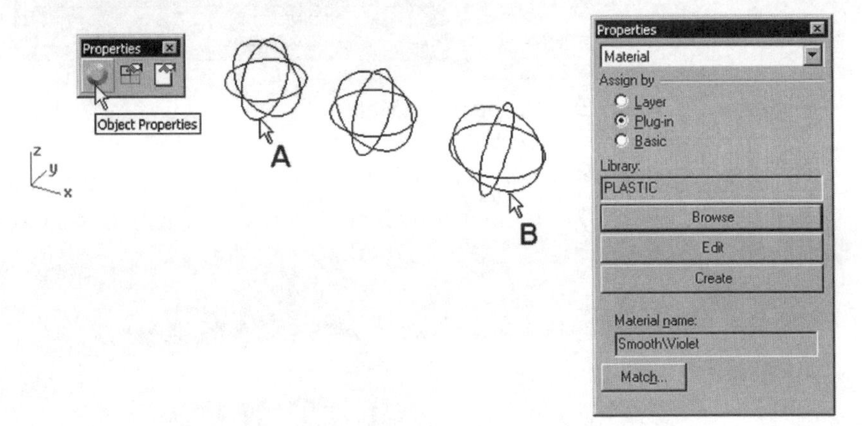

Fig. 8-52. Ball selected.

5. Select the Highlight tab of the Material Editor dialog box, shown in figure 8-53.
6. Set sharpness to 28 and intensity to 2.
7. Click on the OK button.
8. Save the material as *Violet1*.

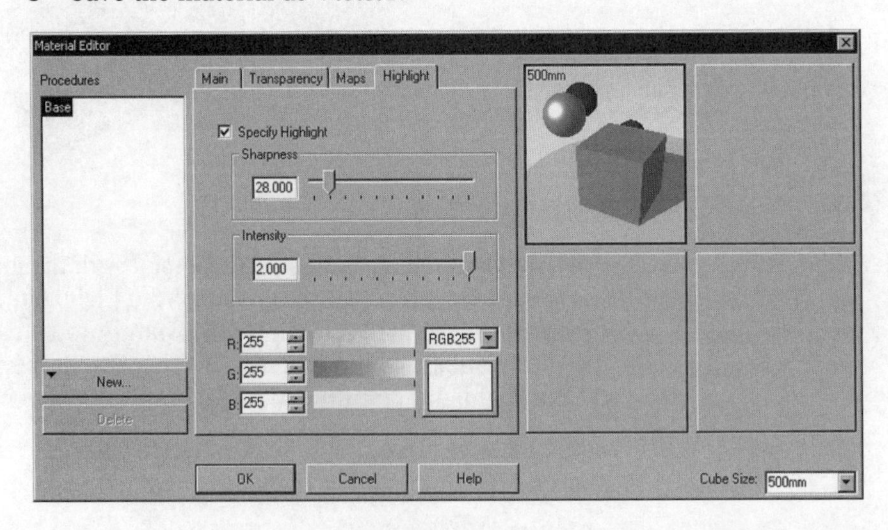

Fig. 8-53. Highlight tab.

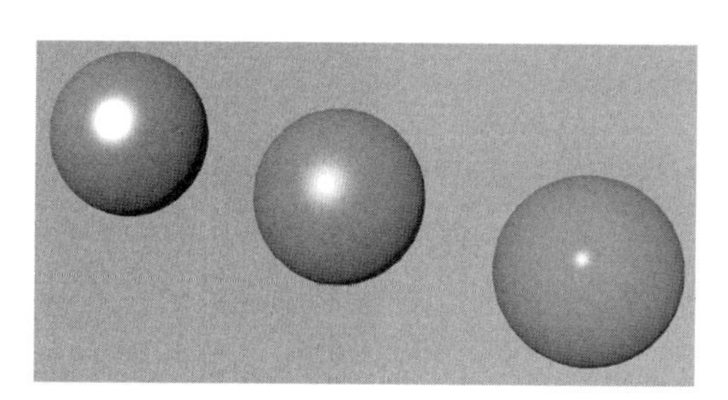

9 Repeat steps 2 through 8 to change the sharpness value of ball B (indicated in figure 8-52) to 1, and then save the material as *Violet2*.

10 Render the scene, as shown in figure 8-54.

Fig. 8-54. Different highlight settings.

Procedural Materials

As mentioned previously, the Procedures pane of the Material editor controls the combination of two or more material components and how the components are combined. Because each individual procedural material has its unique set of material components and its way of combining the components, delineating all of them is not possible. However, the following steps show you how to use a procedural material.

1 Select Raytrace > Material.

2 In the left-hand pane of the Material Library dialog box, select Wood > Solid.

3 In the central pane, select Maple Natural, Polished.

4 Click on the preview pane, right-click, and select New > Use Current Material as Template. The result is shown in figure 8-55.

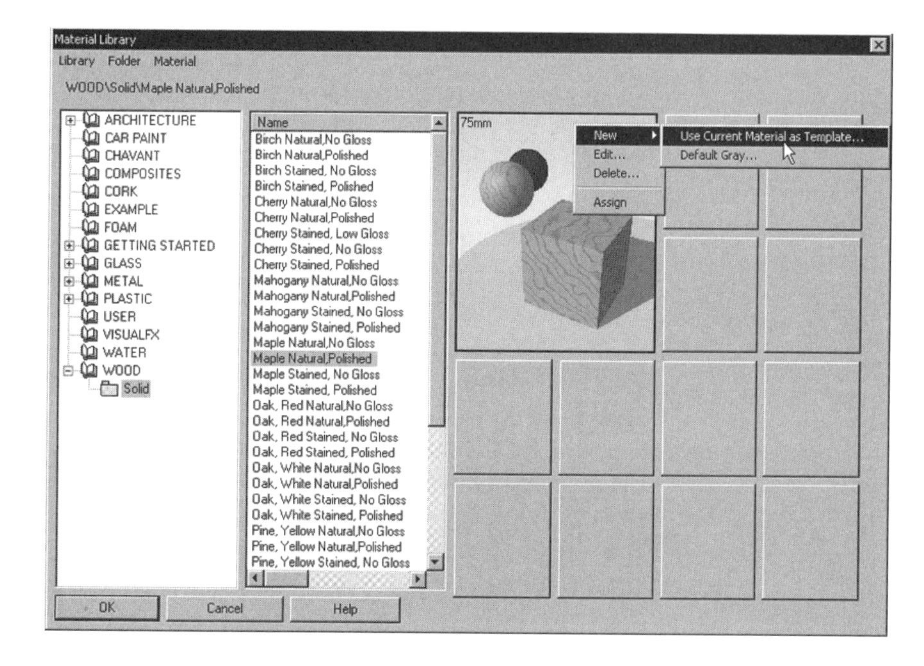

Fig. 8-55. New material being activated.

Initially, the central portion of the Material Editor dialog box has only one tab, the Blend tab, as shown in figure 8-56.

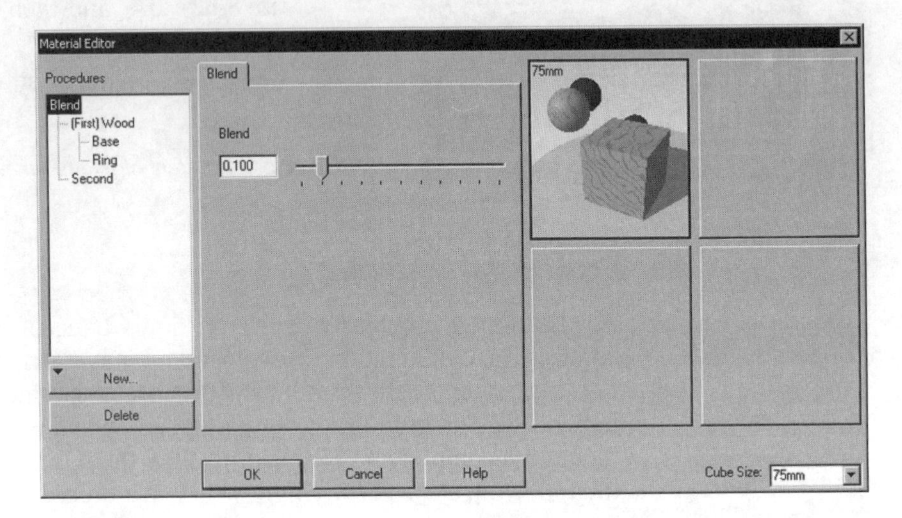

Fig. 8-56. Material Editor for polished natural maple solid wood.

On the Procedures pane, you will find a hierarchy indicating that this material has two components, First and Second. The first component is also a procedural material consisting of two material components, Base and Ring. If you do not find the hierarchy in this pane, double click on Blend. Continue with the following steps.

5 Move the slider bar of the Blend tab to adjust how the first material is blended with the second material.

6 Click on *(First) Wood*. You will find two tabs in the central portion of the dialog box: Wood and Orientation. This is a procedure for adjusting how its components, base and ring, are combined.

7 Click on Base. You will find the four tabs (discussed previously). You may adjust this component to set a material of the base component.

8 Do the same for the Ring component.

9 When finished, click on OK and specify a material name (or click on Cancel to exit).

Flamingo Transparency, Shadow Casting, and Mapping

Using Flamingo, you can set two types of transparency, establish casting shading while raytracing, and position the bitmap in a number of ways. You do this from the Flamingo option of the Properties dialog box, shown in figure 8-57.

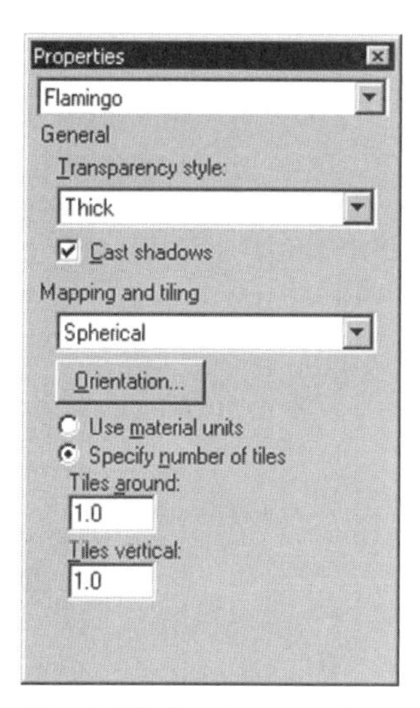

Fig. 8-57. Flamingo option of the Properties dialog box.

Transparency

There are two types of transparency: thin and thick. To appreciate their difference, perform the following steps.

1. Open the file *FlamingoMaterial4.3dm* from the *Chapter 8* folder on the companion CD-ROM.
2. Set the material residing on layer *Ellipsoid* to the plug-in material settings Plastic, Smooth, and Blue.
3. Set the material residing on layer *Box* to the plug-in material settings Plastic, Transparent, and Aquamarine.
4. Select Edit > Object Properties, or click on the Object Properties button on the Properties toolbar.
5. Select object A (indicated in figure 8-58).
6. In the Properties dialog box, select Flamingo from the pull-down list.
7. Select Thin.
8. Render the scene, as shown in figure 8-59.

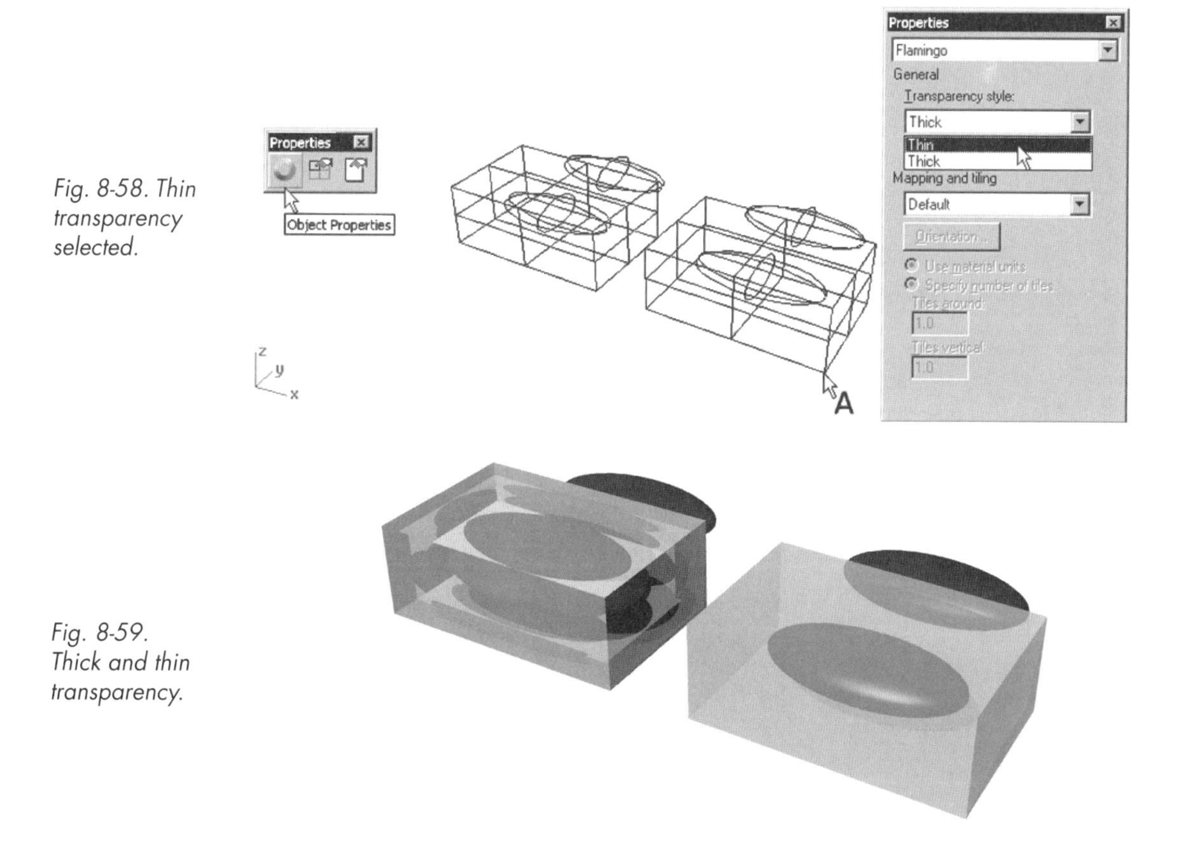

Fig. 8-58. Thin transparency selected.

Fig. 8-59. Thick and thin transparency.

Fig. 8-60. Rendered scene.

Shadow Casting

To determine whether an individual object will cast a shadow or not while raytracing, you check or uncheck the Cast Shadows checkbox. To work with the shadow-casting feature, perform the following steps.

1 Open the file *FlamingoMaterial5.3dm* from the *Chapter 8* folder on the companion CD-ROM.
2 Render the scene, as shown in figure 8-60.

Both cones cast shadows on the box.

3 Select Edit > Object Properties, or click on the Object Properties button on the Properties toolbar.
4 Select object A.
5 Deselect the *Cast shadows* checkbox, shown in figure 8-61.
6 Render the scene again, as shown in figure 8-62.

One of the cones has no shadow at all.

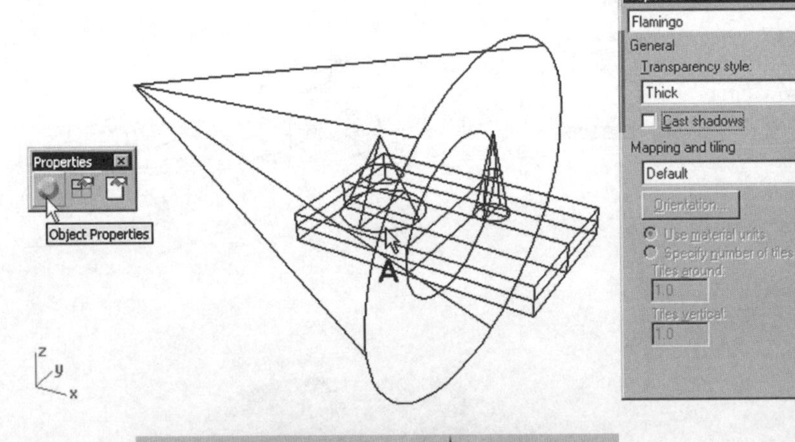

Fig. 8-61. Object's property being changed.

Fig. 8-62. Final rendered scene.

Mapping and Tiling

To position a bitmap on an object to achieve a desired outcome, you use the options provided on the Mapping and Tiling area of the Properties dialog box. There are four types of bitmap mapped to a surface: planar, cube, cylindrical, and spherical.

The planar method projects the bitmap perpendicularly onto the object from a selected plane. The cube method places the bitmap on the walls of a cube and projects the bitmaps orthographically onto the object in six directions. You can adjust the origin and orientation of the cube. The cylindrical method places the bitmap on an imaginary cylinder and then projects the bitmap from the cylinder onto the object.

You can specify the center, rotation, and axis of the imaginary cylinder, and you can adjust these parameters, as well as the number of tiles of bitmap to be placed. The spherical method places the map on the wall of an imaginary sphere and projects the bitmap from the sphere onto the object. The center and rotation of the sphere are adjustable, and you can specify the number of tiles placed on the imaginary sphere in two directions. To work with the mapping and tiling functionality, perform the following steps.

1. Open the file *FlamingoMaterial6.3dm* from the *Chapter 8* folder on the companion CD-ROM.
2. Select Edit > Object Properties, or click on the Object Properties button on the Properties toolbar.
3. Select object A (indicated in figure 8-63).
4. In the Properties dialog box, select Flamingo and then Planar.

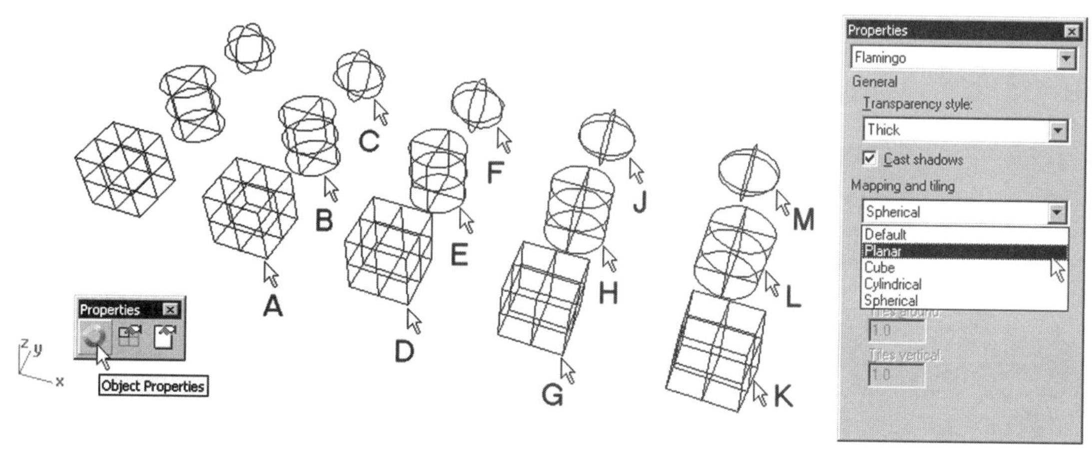

Fig. 8-63. Mapping and tiling methods being changed.

5. With reference to figure 8-63, repeat steps 3 and 4 for objects B and C, setting their mapping and tiling to Planar.
6. Repeat steps 3 and 4 for objects D, E, and F, setting their mapping and tiling to Cube.
7. Repeat steps 3 and 4 for objects G, H, and J, setting their mapping and tiling to Cylindrical.
8. Repeat steps 3 and 4 for objects K, L, and M, setting their mapping and tiling to Spherical.
9. Render the scene, as shown in figure 8-64.

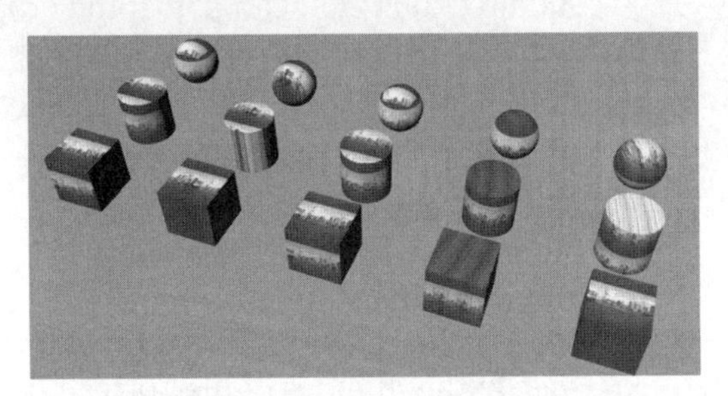

Fig. 8-64. Rendered image.

Decals

A decal is a special type of bitmap that covers only a part of a surface. Typically, you use decals to place a label, add a sign, or place a specific bitmap on a designated area of a surface. You apply a decal via the Decal option of the Properties dialog box. Here you add a decal, edit the decal's placement location, modify its properties, delete a decal you already applied, and move a decal up or down in the decal hierarchy. To manipulate decals, you use the Decal option of the Properties dialog box.

Adding a Decal

To add a decal to a surface, perform the following steps.

1. Open the file *FlamingoMaterial7.3dm* from the *Chapter 8* folder on the companion CD-ROM.
2. Select surface A (indicated in figure 8-65).
3. Select Edit > Object Properties, or click on the Object Properties button on the Properties toolbar.
4. In the Properties dialog box, select Decal from the pull-down list box and then click on the Add button.
5. In the Open Bitmap dialog box, select the bitmap file *Decal* from the *Chapter 8* folder on the companion CD-ROM.

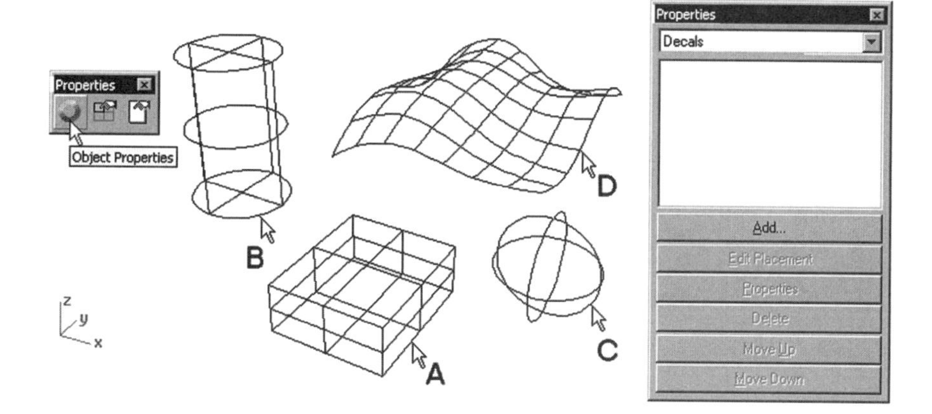

Fig. 8-65. Decal being placed on an object.

Placement

After you select a bitmap, you need to specify how to map the bitmap onto the surface. You do this via one of four method types: planar, cylindrical, spherical, and UV. The planar, cylindrical, and spherical methods work in much the same way as those associated with mapping a bitmap. However, the UV method is somewhat different. It maps the bitmap onto the entire selected surface. While mapping, the image is stretched or distorted in accordance with the U and V isocurves of the surface. To place a decal, perform the following steps.

1. In the Decal Mapping Style dialog box, select Planar and then click on the OK button.
2. Select end points A, B, and C (indicated in figure 8-66) to position the decal.
3. Press the Enter key when finished.

The decal is placed.

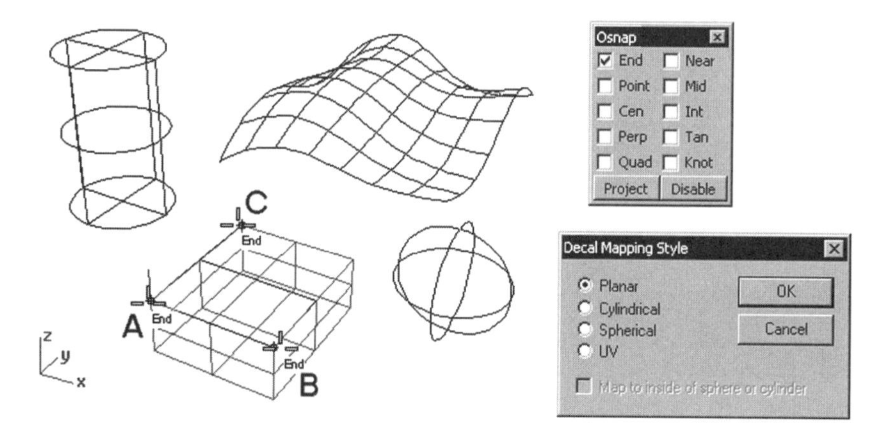

Fig. 8-66. Decal being positioned.

Fig. 8-67. Edit Decal dialog box.

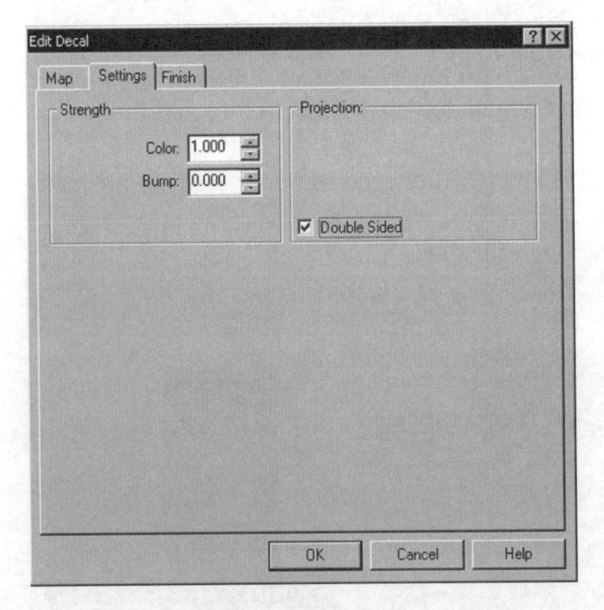

Fig. 8-68. Settings tab of the Edit Decal dialog box.

Properties

After specifying a location and orientation for a decal, you need to specify its properties via the Edit Decal dialog box, shown in figure 8-67.

The Edit Decal dialog box has three tabs: Map, Settings, and Finish. In the Settings tab, shown in figure 8-68, you can reselect a bitmap if you change your mind. Here you can also apply a masking effect to the decal bitmap in exactly the same way you mask an ordinary bitmap. You use the Color option to mask selected colors of the bitmap, and use the Alpha Channel option to mask a portion of the bitmap in accordance with the alpha channel's color. You can also reverse the effect of masking, causing the masked portion to be unmasked and the unmasked portion to be masked. You can also set the underlying material to be transparent in the masked areas of the bitmap.

The Settings tab controls the color and bump strengths of the decal and decides whether the reverse side of the surface gets the decal mapping. Like an ordinary bitmap, the color strength sets the strength of the decal in respect to the underlying material. If it is set to a value of 1, the color of the final rendered object will come from the decal. If it is less than 1, the underlying object will show through the decal. The bump strength also ranges from –1 to 1. A value of 0 has no bumpy effect, 0 to 1 causes the dark color to rise from the surface, and –1 to 0 causes the dark color to recess into the surface. By checking the Double Sided option, the decal is mapped to both sides of a surface.

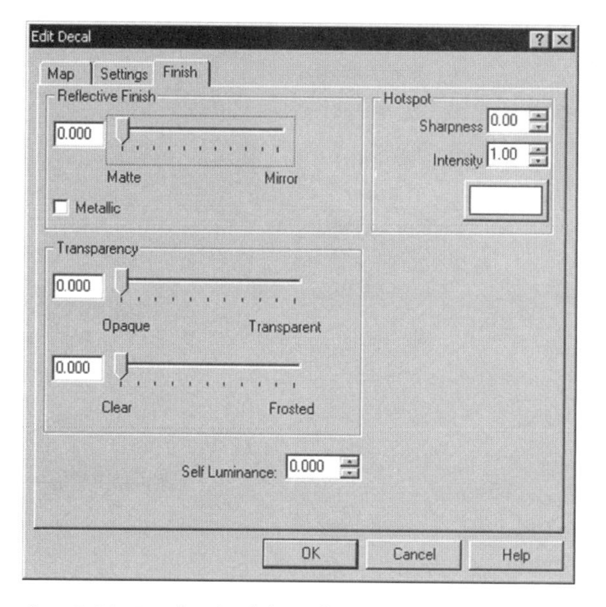

Fig. 8-69. Finish tab of the Edit Decal dialog box.

The Finish tab of the Edit Decal dialog box, shown in figure 8-69, establishes reflective finish, transparency, hotspot, and self-luminance parameters. In the Reflective Finish field, you set the reflectivity of the decal, from matte to mirror finish. To obtain a metallic finish, you select the Metallic option. In the Transparency field, you set the decal to be completely opaque to completely transparent. A transparent decal will cause the underlying material to be transparent. For a transparent decal, you can determine the state of transparency, clear or frosted. The Hotspot field controls the highlight. It sets the sharpness and intensity of the highlight of the decal. To examine the decal-editing functionality, perform the following.

1. In the Edit Decal dialog box, click on the OK button.

A decal is mapped on a planar face.

Moving Up and Down

If you apply more than one decal to a material, the second decal will be placed beneath the first one, the third beneath the second, the fourth beneath the third, and so on. To reorder the sequence, you move a decal to an upper position or move it down to a lower position.

Planar, Cylindrical, Spherical, and UV Mapping

As mentioned previously, there are four decal-mapping method types: planar, cylindrical, spherical, and UV. To map decals using the cylindrical, spherical, and UV methods, perform the following steps.

1. Open the Properties dialog box by selecting Edit > Object Properties, or click on the Object Properties button on the Properties toolbar.
2. Select cylindrical surface B (indicated in figure 8-65).
3. In the Properties dialog box, click on the Add button.
4. In the Open Bitmap dialog box, select the bitmap file *Decal* from the *Chapter 8* folder on the companion CD-ROM.

5. In the Decal Mapping Style dialog box, select Cylindrical and then click on the OK button.
6. Select center A and location B (indicated in figure 8-70) to position the decal.

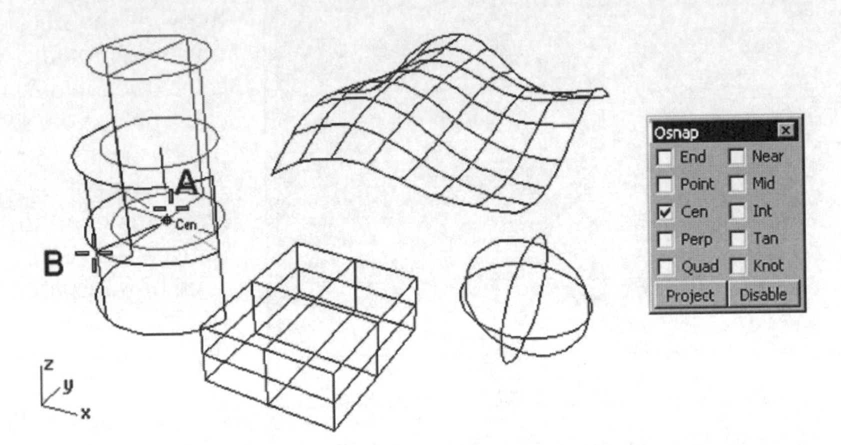

Fig. 8-70. Position selected.

7. Double click on the Perspective viewport to change to a four-viewport display.
8. Double click on the Front viewport to maximize it.
9. Check the Ortho option on the status bar.
10. Select a control point of the decal and drag it to position A (indicated in figure 8-71).
11. Press the Enter key.
12. Click on the OK button in the Edit Decal dialog box.

A decal is mapped in a cylindrical way. Continue with the following steps.

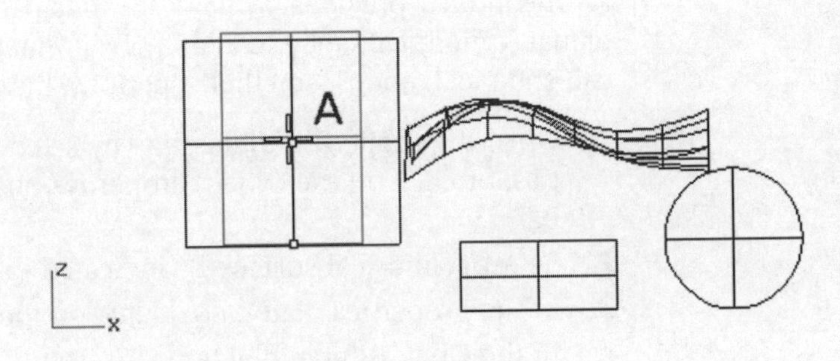

Fig. 8-71. Decal being moved.

13. Maximize the Perspective viewport.

14 Select spherical surface C (indicated in figure 8-65).

15 In the Properties dialog box, click on the Add button.

16 In the Open Bitmap dialog box, select the bitmap file *Decal* from the *Chapter 8* folder on the companion CD-ROM.

17 In the Decal Mapping Style dialog box, select Spherical and then click on the OK button.

18 Select center A and location B (indicated in figure 8-72).

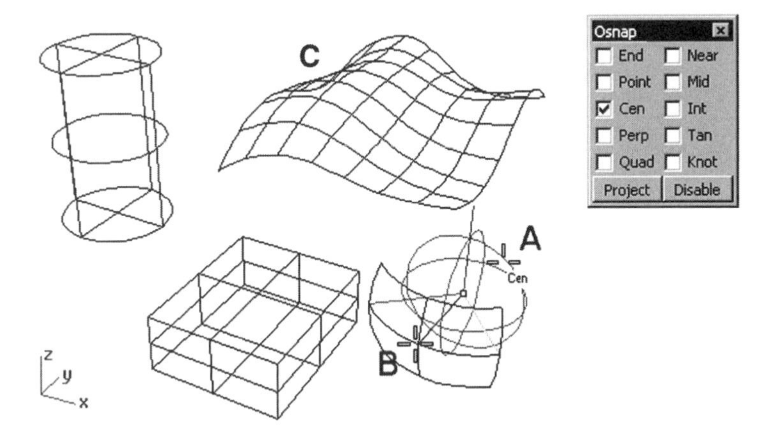

Fig. 8-72. Spherical mapping of a decal.

19 Press the Enter key.

20 Click on the OK button in the Edit Decal dialog box.

A decal is mapped spherically. Continue with the following steps.

21 Select surface C (indicated in figure 8-72).

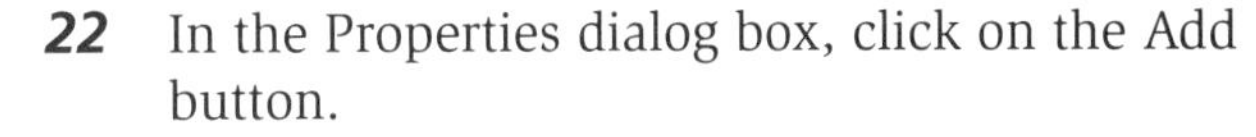

22 In the Properties dialog box, click on the Add button.

23 In the Open Bitmap dialog box, select the bitmap file *Decal* from the *Chapter 8* folder on the companion CD-ROM.

24 In the Decal Mapping Style dialog box, select UV and then click on the OK button.

25 Click on the OK button in the Edit Decal dialog box.

A decal is mapped in accordance with the UV of a surface.

Fig. 8-73. Decals mapped.

26 Render the scene, as shown in figure 8-73.

Waves

Waves and ripples can be added to the rendered image of an object without really having to model them using the curve and surface modeling tools. Waves and ripples are special types of bump effects on a surface. To add waves and ripples, you use the Wave option of the Properties dialog box, as follows.

1. Open the file *FlamingoMaterial8.3dm* from the *Chapter 8* folder on the companion CD-ROM.
2. Select Edit > Object Properties, or click on the Object Properties button on the Properties toolbar.
3. Select polysurface A (indicated in figure 8-74).
4. In the Properties dialog box, select Waves and then click on the Add button.
5. Select location B and press the Enter key.

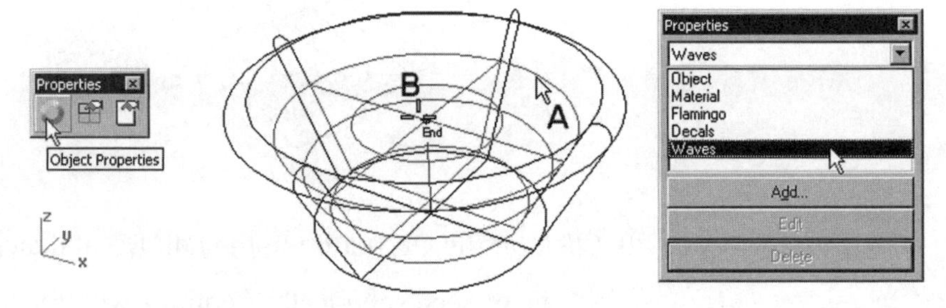

Fig. 8-74. Wave being applied.

After you select a location, you set the parameters of the wave in the Edit Waves dialog box, shown in figure 8-75.

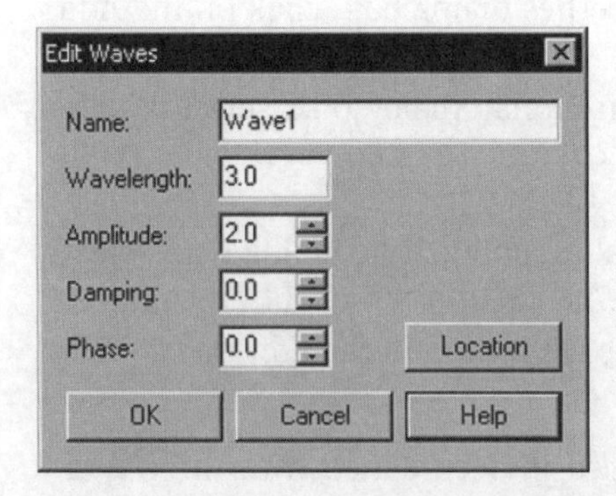

Fig. 8-75. Edit Waves dialog box.

The Edit Waves dialog box has a number of settings: Name, Wavelength, Amplitude, Damping, Phase, and Location. The Name box lets you specify the name of the wave. Wavelength sets the distance between adjacent waves. Amplitude sets the height of the wave. Damping causes the wave height to decrease from the wave origin. Phase causes the wave to start at a different point in the cycle. Location lets you redefine the origin of the wave. Continue with the following.

6. In the Edit Waves dialog box, set Wavelength to 3 and Amplitude to 2, and then click on the OK button.

The wave effect is applied.

7. Render the scene, as shown in figure 8-76.

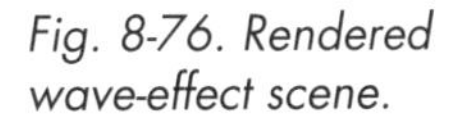
Fig. 8-76. Rendered wave-effect scene.

Flamingo Lights and Sunlight

Using Flamingo renderers, you can add Rhino lights as well as rectangular and linear lights. You can also include daylight effects. The sections that follow explore lighting effects achieved via Flamingo.

Flamingo Lights

The Raytrace and Photometric Flamingo renderers provide two types of light in addition to those provided by the Rhino renderer. Combining those of the Rhino renderer and the Flamingo renderers, you have five types of lights from which to choose: spotlight, point light, directional light, rectangular light, and linear light. Rectangular light and linear light are examined in the sections that follow.

Rectangular Light

Rectangular lighting simulates the effect of a fluorescent light box. You specify the location, width, and length of the rectangular light. In a scene, a rectangular light is a rectangular symbol depicting a rectangular light box. Like other types of lights, a rectangular light is not shown in the rendered image. To add a rectangular light to a scene, perform the following steps.

1. Open the file *FlamingoLight1.3dm* from the *Chapter 8* folder on the companion CD-ROM.
2. Select Render > Create Rectangular Light, or click on the Create Rectangular Light button on the Lights toolbar.
3. Select locations A, B, and C (indicated in figure 8-77).

A rectangular light is constructed.

4. Render the scene, as shown in figure 8-78.

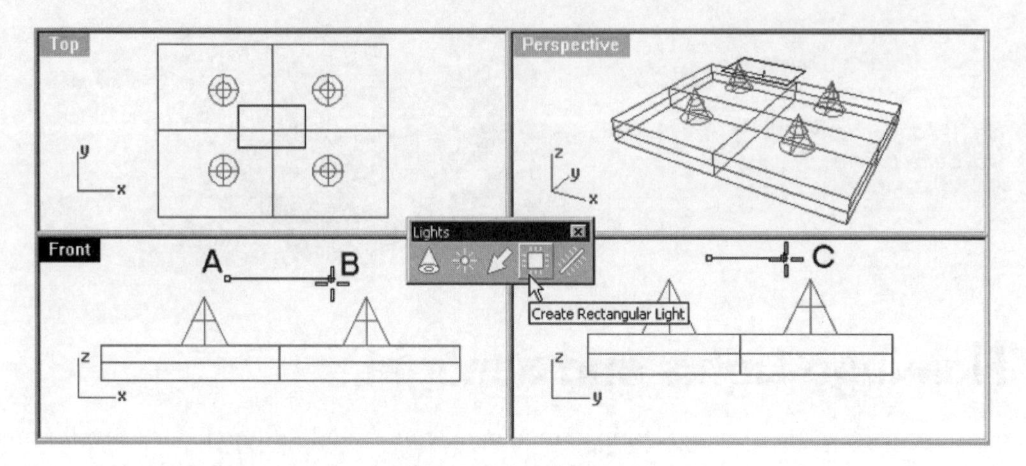

Fig. 8-77. Rectangular light being constructed.

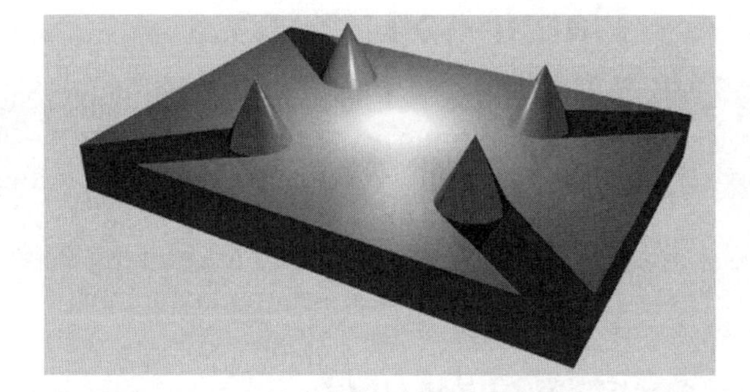

Fig. 8-78. Rendered rectangular-light scene.

Linear Light

Linear light simulates the effect of a fluorescent tube being added to a scene. You specify the start point and the end point of the linear light. In a scene, a linear light is a symbol in the shape of a cylinder. It is not displayed in the rendered image. To turn off the rectangular light and add a linear light to a scene, perform the following steps.

1. Select Edit > Object Properties, or click on the Object Properties button on the Properties toolbar.
2. Select the rectangular light.
3. In the Properties dialog box, select Light and then deselect the On box.

The rectangular light is turned off.

4. Select Render > Create Linear Light, or click on the Create Linear Light button on the Lights toolbar.

5 Select locations A and B (indicated in figure 8-79).

A linear light is constructed.

6 Render the scene, as shown in figure 8-80.

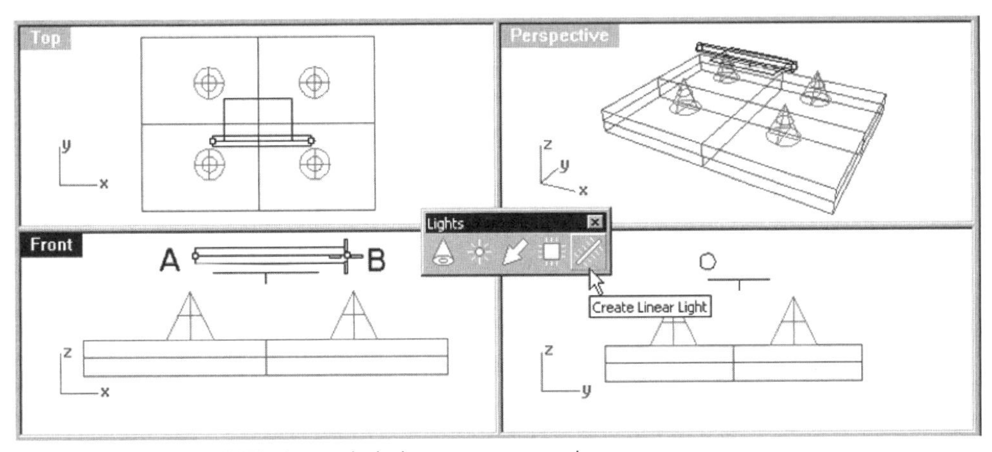

Fig. 8-79. Linear light being constructed.

Fig. 8-80. Rendered liner-light scene.

Daylighting

Daylighting concerns illumination of the scene by introducing direct sunlight and indirect sunlight via the sky, the ground and other objects. Sun and sky settings can be accessed by selecting Raytrace > Sun (using the Raytrace renderer) or Photometric > Sun (using the Photometric renderer). Alternatively, you can click on the Sun button on the Flamingo tab of the Document Properties dialog box.

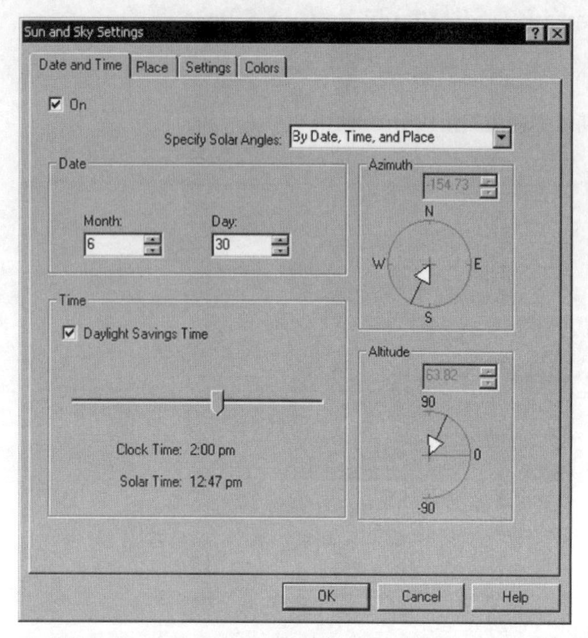

Fig. 8-81. Sun and Sky Settings dialog box.

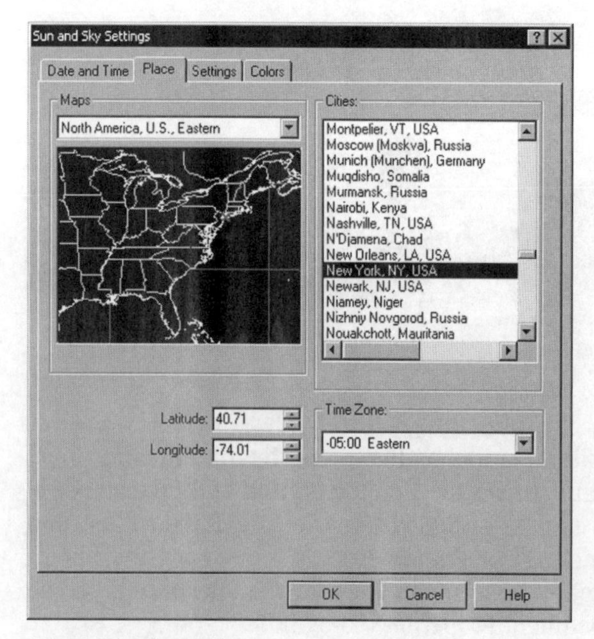

Fig. 8-82. Place tab.

1. Open the file *House1.3dm* from the *Chapter 8* folder on the companion CD-ROM.
2. Select Raytrace > Sun (for the Raytrace renderer) or Photometric > Sun (for the Photometric renderer).

This accesses the Sun and Sky Settings dialog box, shown in figure 8-81. The Sun and Sky Settings dialog box has four tabs: Date and Time, Place, Settings, and Colors. The Date and Time tab lets you turn on the sun effect and decide how the sunlight's direction is determined, either by specifying date and time in this tab (and specifying placement in the Place tab) or by selecting Directly from the Specify Solar Angles pull-down list box and then specifying the azimuth and altitude. Continue with the following steps.

3. Specify month, day, and time in the Date and Time tab.
4. Select the Place tab, shown in figure 8-82.

If you decided to specify the sunlight's direction by date, time, and place in the Date and Time tab, you specify the place in the Place tab, as follows.

5. In the Place tab, select a place.
6. Select the Settings tab, shown in figure 8-83.

The Settings tab allows you to specify cloudiness and intensity of sun and sky, and to establish the north direction in a Rhino file. Cloudiness ranges from 0 (clear) to 1 (completely overcast). Sun intensity refers to the intensity of direct sunlight, and "sky" refers to the intensity of indirect sunlight. By default, the Y direction of a Rhino file is north. You may change the north direction here. Continue with the following steps.

7 Make any necessary changes to the options of the Settings tab.
8 Select the Colors tab, shown in figure 8-84.

The Colors tab enables you to determine the color of the direct light and indirect light. Continue with the following.

9 Click on the OK button.

The sun effect is turned on and its parameters specified.

10 Select Photometric > Render.
11 Render the scene, as shown in figure 8-85.

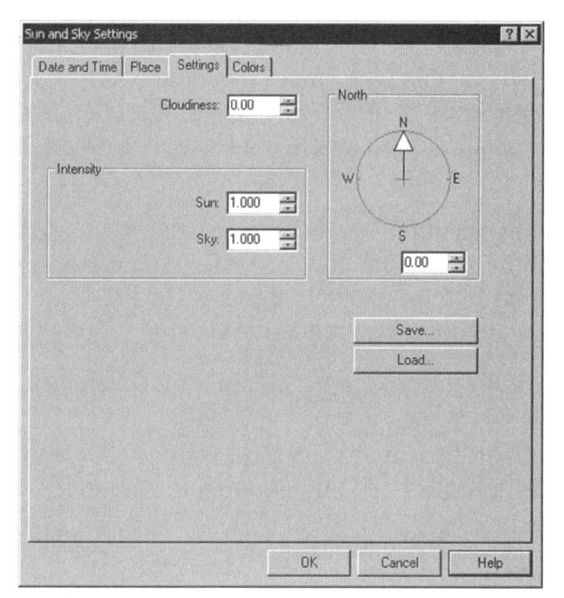

Fig. 8-83. Settings tab.

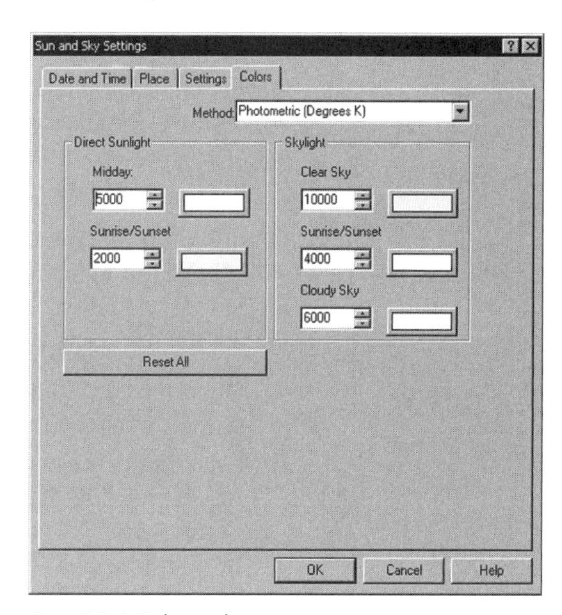

Fig. 8-84.Colors tab.

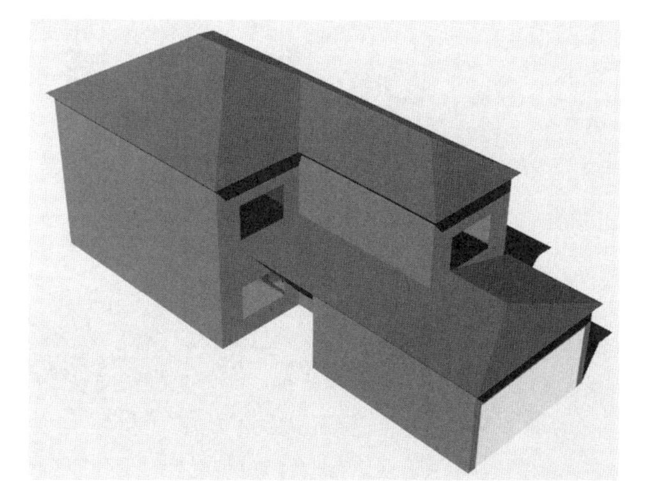

Fig. 8-85. Rendered day-lighting scene.

Plantlife

Plant imagery is incorporated in Rhino as procedural objects you insert from a Plant library. In a Rhino file, an inserted plant is shown as a symbol representing the respective plant. However, in a rendered image the plant is represented realistically, in accordance with parameters you specify. You can add plants to your scene from the Plant library, accessed by selecting Raytrace > Add Plant, Photometric > Add Plant, or by clicking on the *Plants/Edit Plants* button on the Flamingo toolbar.

The Plant Library dialog box has three major areas that display folders of plants on the left-hand side, a list of plants from selected plant folders at the center, and a set of preview panes on the right. Along with each preview pane, there is a pull-down list enabling you to specify the season in which the plant is grown: early spring, spring, late spring, summer, late summer, autumn, or winter. In addition to specifying its season, you can modify a plant's structure and size.

In addition to setting the plant season here, you can also set the season globally in the Flamingo tab of the Document Properties dialog box. In the following you will add a number of plants to a scene. Note that adding plants to a scene dramatically increases rendering time.

1. Open the file *House2.3dm* from the *Chapter 8* folder on the companion CD-ROM.
2. Select Raytrace > Add Plant, Photometric > Add Plant, or click on the *Plants/Edit Plants* button on the Flamingo toolbar, shown in figure 8-86.
3. In the Plant Library dialog box, shown in figure 8-87, select a plant.

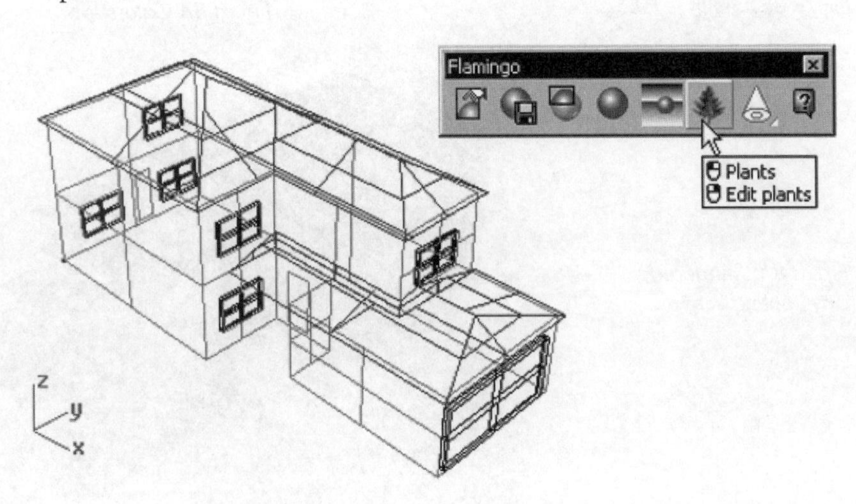

Fig. 8-86. Adding a plant.

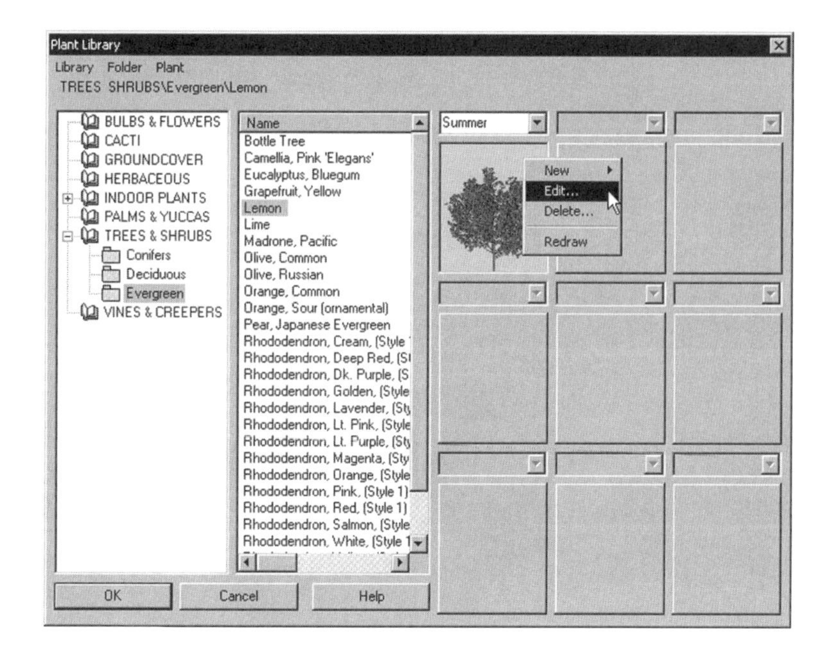

Fig. 8-87. Plant Library dialog box.

4 In the Plant preview pane, right-click and then select Edit.

5 In the Plant Editor dialog box, shown in figure 8-88, set the parameters for the plant and then click on the OK button.

6 Return to the Plant Library dialog box and click on the OK button.

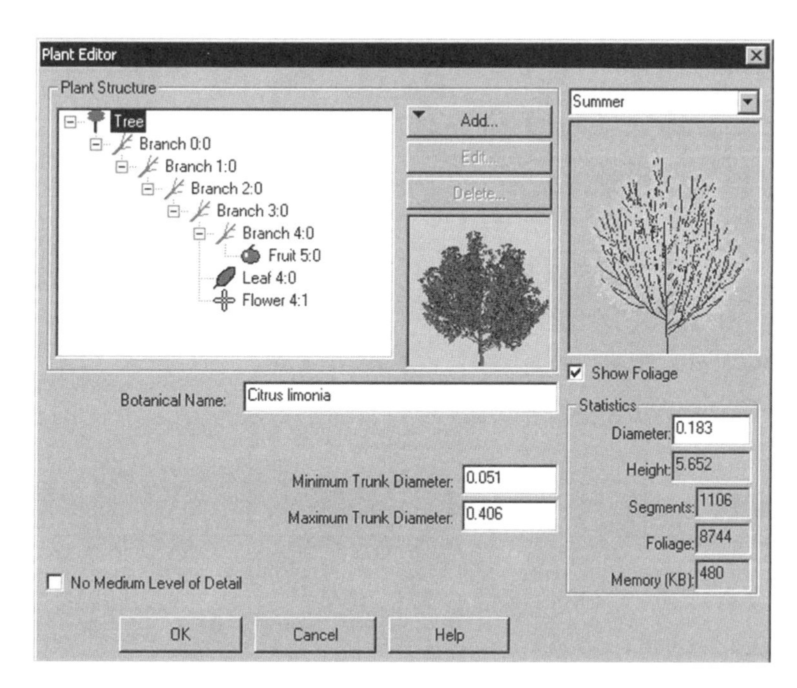

Fig. 8-88. Plant Editor dialog box.

7 Select a location in the scene.
8 Repeat steps 2 through 7 to add more plants to the scene.

Select Photometric > Render. The result is shown in figure 8-89.

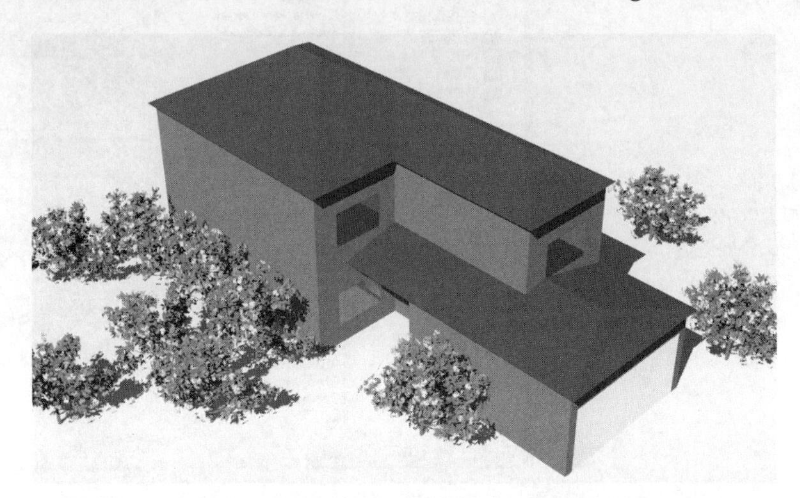

Fig. 8-89. Plants added and scene rendered.

Flamingo Environment Objects

Environment objects are objects included in a rendered scene that specify the nature of the entire environment. There are six environment-object options: Background, Background Image, Clouds, Haze, Ground Plane, and Alpha Channel. To work with environment objects, perform the following steps.

1. Open the file *House3.3dm* from the *Chapter 8* folder on the companion CD-ROM.
2. Turn on the sun effect.
3. Select Raytrace > Properties (for the Raytrace renderer) or Photometric > Properties (for the Photometric renderer).

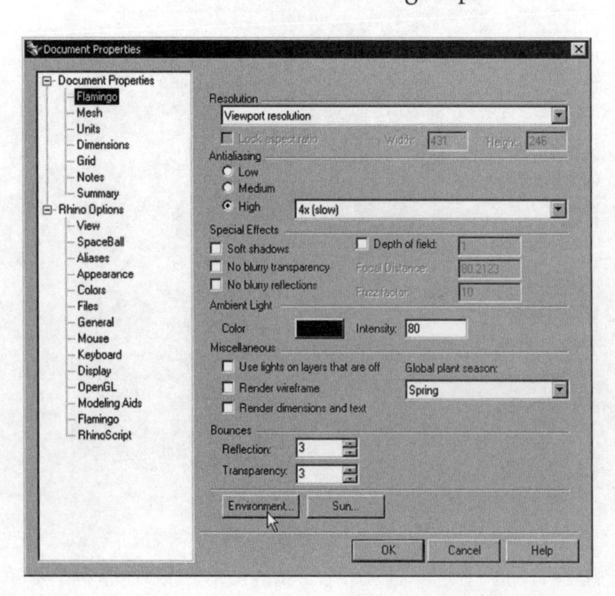

Fig. 8-90. Environment dialog box.

4 In the Flamingo tab of the Document Properties dialog box, click on the Environment button. This accesses the Environment dialog box, shown in figure 8-90.

Initially, the Environment dialog box has only one tab, Main. This tab has two fields: Background Color and Advanced. In the Background field, you can specify a sky effect, solid color, and a two-color or three-color gradient. In the Advanced field, you can specify a background image, clouds, haze, ground plane, and alpha channel. Selecting any of these items will access a tab in the dialog box that allows you to modify a selection's parameters. Continue with the following steps.

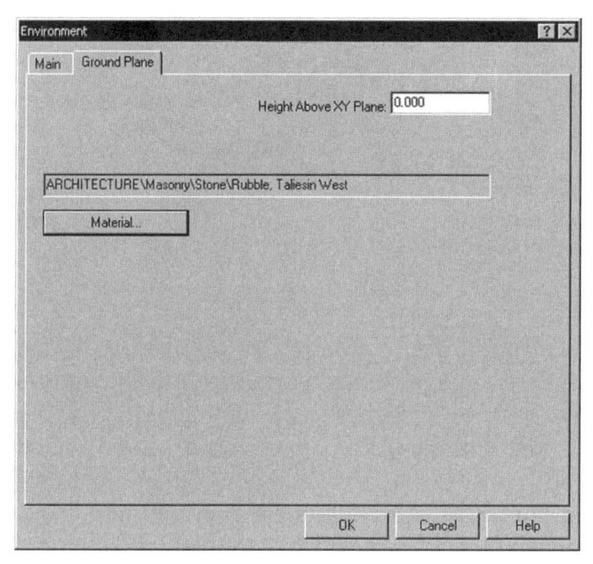

Fig. 8-91. Ground Plane tab.

5 Check the Automatic Sky box. Click on the OK button in the Environment dialog box. Click on the OK button in the Document Properties dialog box.

6 Check the Ground Plane box.

7 On the Ground Plane tab, shown in figure 8-91, select Material.

8 In the Material Library dialog box, select Architecture > Masonry > Stone > Rubble, Taliesin West, and then click on the OK button.

9 Click on the OK button on the Environment dialog box and on the Document Properties dialog box.

10 Render the scene, as shown in figure 8-92.

Fig. 8-92. Rendered environment-object scene.

Instead of accepting the default sky effect, you can check the Background Image box of the Main tab and select an image. You can also add clouds and haze, and create an alpha channel. Figure 8-93 shows the rendered scene with haze added.

Fig. 8-93. Haze added.

Save Small

A NURBS surface is an accurate representation of a smooth surface. However, producing a rendered image of such surfaces typically requires the existence of a set of polygon meshes representing the surfaces. When you invoke the Render command to construct a rendered image, the computer automatically constructs a set of polygon meshes and uses the meshes for rendering purposes. Normally, these polygon meshes are preserved when you save your file. The next time you render the object (if it is not modified), the computer will use the preserved meshes (saved in the file) for rendering. As a result, rendering time is reduced because the program does not have to construct the polygon meshes.

The inclusion of polygon mesh data, however, makes file sizes larger. To minimize the storage requirement for polygon mesh data, you can use the Save Small command, which saves NURBS surfaces but not the polygon meshes required for construction of a rendered image. However, as a result, the next time you open the file rendering time will be longer because the computer will have to reconstruct the set of polygon meshes. To save at a smaller file size (which does not preserve polygon meshes), perform the following.

1 Select File > Save.

2D Drawings

2D orthographic engineering drawings are the conventional means of communication among engineering personnel. Although the advent of computer-aided design applications replaced some of the uses of 2D drawings, there remain many situations or purposes for which 2D drawings and/or 2D drawing output is useful or necessary.

One important function of 2D drawings is to specify precisely the dimensions of the objects they represent, along with annotations that convey other information about the object or objects represented. In Rhino, you use the Dimension toolbar, shown in figure 8-94, to create 2D drawings and incorporate dimensions and annotations.

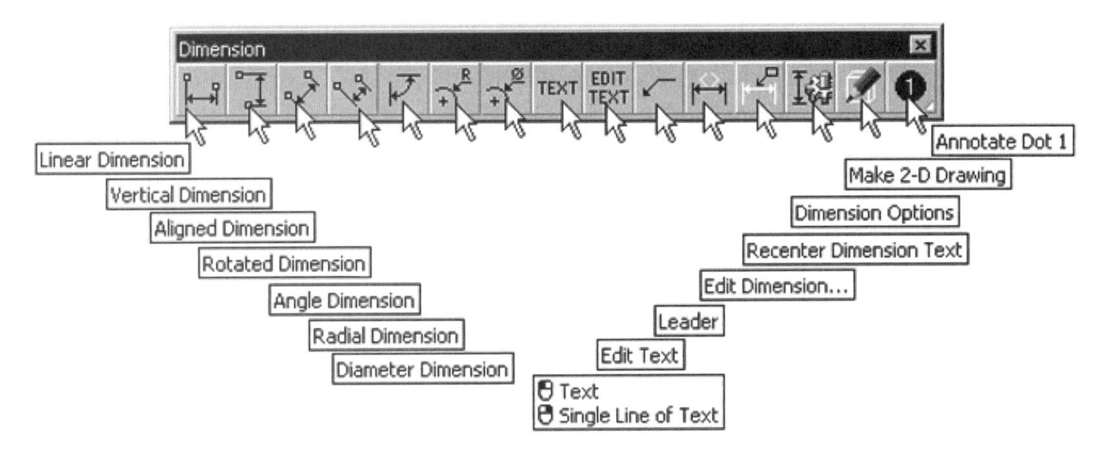

Fig. 8-94. Dimension toolbar.

Orthographic Drawings

The process of outputting a 3D model as an engineering (2D) drawing is fairly simple using Rhino. Basically, you select a command and let the computer do all of the 2D drawing construction work, as follows.

1. Open the file *2Ddrawing.3dm* from the *Chapter 8* folder on the companion CD-ROM.
2. Select Dimension > Make 2-D Drawing, or click on the Make 2-D button on the Dimension toolbar.
3. Select the polysurface and press the Enter key. (See figure 8-95.)

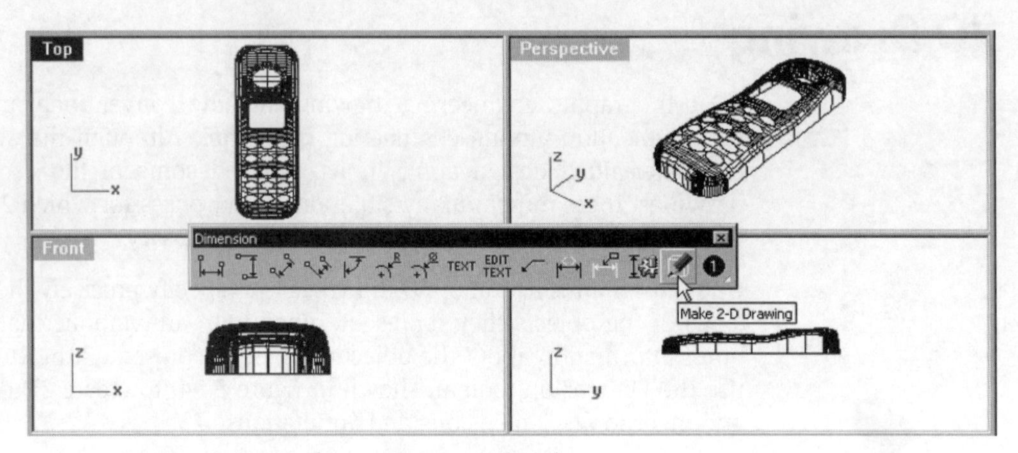

Fig. 8-95. 2D drawing being produced.

4 In the 2-D Drawing Options dialog box, shown in figure 8-96, select 4 View (USA) and then click on the OK button.

5 Set the current layer to *Make2dvisiblelines* and turn off the layer *Polysurface.*

6 Maximize the Top viewport.

A 2D drawing is constructed, as shown in figure 8-97.

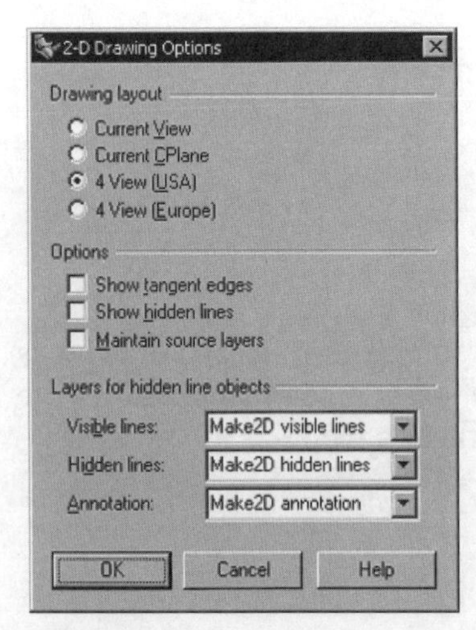

Fig. 8-96. 2-D Drawing Options dialog box.

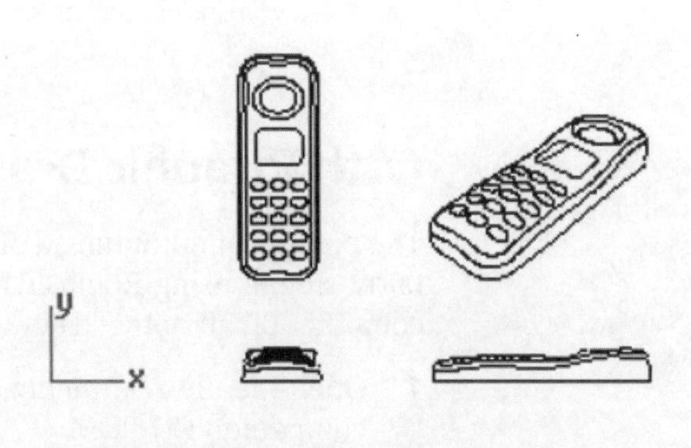

Fig. 8-97. Front, side, top, and isometric views constructed.

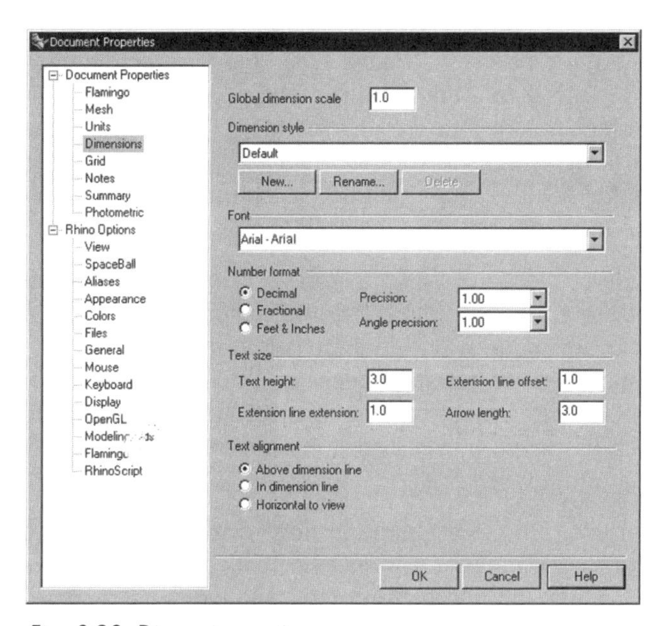

Fig. 8-98. Dimension options.

Dimensioning and Annotation

2D drawings typically indicate the dimensions of the 3D object or objects they represent, and typically incorporate informative annotations. To add dimensions and annotations to a 2D Rhino drawing, perform the following steps.

1. Select Dimension > Dimension Properties, or click on the Dimension Options button on the Dimension toolbar.
2. In the Dimensions tab of the Document Properties dialog box, shown in figure 8-98, set the text height to 3 units and the arrow length to 3 units, and then click on the OK button.
3. With reference to figure 8-99, add dimensions and annotations.

The 2D drawing is complete.

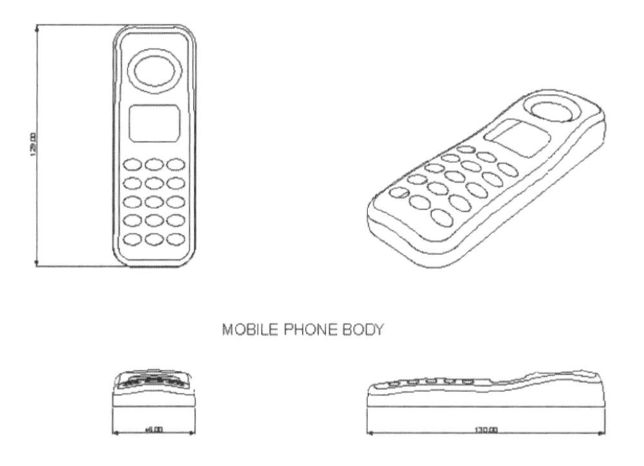

Fig. 8-99. Dimensions and annotations added.

Summary

A smooth surface does not have any thickness or any profile curves on it. In essence, you see only its boundaries and its silhouette. To enhance visualization, isocurves are placed on it. To contrast an object from its surrounding background, you shade the viewport. Shading is a means of enhancing the visual representation of a 3D object by applying a color shade to the surface of the object.

To produce a photorealistic image from objects you construct in Rhino, you use one of the three renderers: Rhino's basic renderer, Flamingo's Raytrace renderer, or Flamingo's Photometric renderer. There are four major steps in producing a photorealistic rendered image from the objects you construct in Rhino. You apply material properties to the objects, construct lights in the scene, include environment objects, and use a renderer to produce an image.

Using the basic renderer, you add basic materials to objects, establish lighting effects, and set ambient and background color. Using the Flamingo renderers, you add plug-in material (in addition to basic material), specify additional types of lighting effects, and set various types of environment objects. The Raytrace renderer is most suitable for studio applications in which lights are traced from their source to the viewer's eye as seen from one of the viewports.

In additional to raytracing, the Photometric renderer takes into account the radiosity effect of all light sources in a scene. This produces a more natural image and is most suitable for producing images that replicate exterior (outdoor) environments. The Photometric renderer also provides a radiosity solution, which refers to calculation of information attached to an image for the purpose of adjusting lighting effects automatically. The radiosity feature also works in paint programs such as Photoshop.

To add realism to an object, you apply material properties. Strictly speaking, there are two option categories of material properties: Plug-in and Basic. A huge repertoire of material is available in the plug-in material library. In this process, you choose a material, establish the parameters of application of the material, and apply the material to the object. The other category of material is Basic. In this process, you specify the basic color, reflective finish, transparency, texture, and bumpiness.

To make a scene more authentic, you can add lighting effects. Lighting options include spotlight, point light, and directional light. Lighting effects are applied via a choice of three renderers: Rhino, Flamingo Raytrace, and Flamingo Photometric. The Rhino renderer is the basic renderer. The Flamingo Raytrace renderer traces beams of light from light sources, and the Flamingo Photometric renderer accurately computes the combined effect of all indirect light sources.

Polygon meshes are constructed and incorporated in a file whenever a rendered image is constructed. To reduce file sizes, you can "save small" by not saving the polygon meshes. However, this requires that meshes be reconstructed if you subsequently want to render images saved this way.

2D orthographic engineering drawings are a standard means of communication among engineers. In Rhino, you create 2D drawings by selecting an object and letting the computer generate the drawing views. A Rhino 2D drawing can incorporate information on the dimensions of the object or objects the drawing represents, as well as other annotation (textual information).

Review Questions

1. Explain the methods by which an object is shaded in the viewport.
2. State the methods by which material properties are applied to an object.
3. Describe the procedures of setting up spotlight, point light, and directional light.
4. Explain the concepts of raytrace and radiosity.
5. Explain how 2D engineering drawings are produced from computer models of 3D objects.

CHAPTER 9

Digital Modeling Projects

Introduction

This chapter consolidates what you learned in previous chapters. In this chapter you will have an opportunity to apply your knowledge of creating digital models.

Objectives

After studying this chapter you should be able to:

- ❒ Use Rhino as a tool in constructing 3D curves and 3D surfaces models
- ❒ Use Flamingo as a tool in producing photorealistic images from 3D digital models

Overview

The prime goal of surface modeling is to construct a set of surfaces and put them together to represent an object depicting a design. Creating surfaces inevitably requires curves and points. In comparison to surface construction, curve construction is a tedious job.

To enhance your skill for curve and surface construction, you will work on a number of digital modeling projects. Because the location and shape of the curves have a direct impact on the shape of surfaces constructed from them, the particulars of most curves in these projects are given to you. While working on these projects, you should try to relate the 3D curves to the 3D surfaces. It is hoped that you reverse the process, seeing the curves when a surface is given.

NURBS Surface Modeling

The ultimate product of using Rhino as a design tool is a computer model (surface model or solid model) that represents a 3D free-form object. Because Rhino solids are closed polysurfaces, both surface and solid modeling enter into the 3D NURBS surface creation process. Hence, you need a very thorough understanding of the various methods of constructing points, NURBS curves, and NURBS surfaces.

Design development using the computer is a two-stage iterative process. With an idea or concept in mind, you start deconstructing the 3D object into discrete surface elements by identifying and matching various portions of the object with various types of surfaces. Unless the individual surfaces are primitive surfaces, you need to think about the shapes and locations of curves and points required to build the surface, as well as the surface construction commands to be applied on the curves and points. This is a top-down thinking process.

After this thinking process, you create the curves and points and apply appropriate surface construction commands. By putting the surfaces together properly in 3D space, you obtain the 3D object. This is the bottom-up construction process. If the surfaces you construct do not conform to your concept, you think about the curves and points again. With practice you gain the experience that makes this process more efficient.

Logically, you construct points and curves, and from these construct surfaces and solids. In detailing your design, you add surface features. To construct surface features that conform in shape to existing surfaces, you create points and curves from existing objects.

Joypad Project

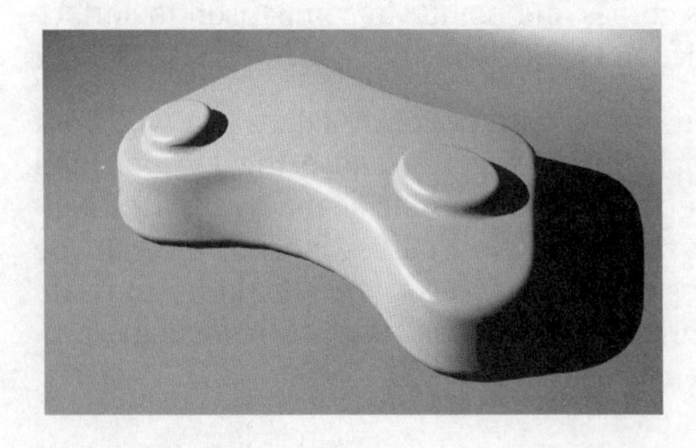

Fig. 9-1. Rendered image of the joypad.

Figure 9-1 shows the rendered image of a joypad model, and figure 9-2 shows the surface model from which the rendered image is produced. Before you start creating the surface model, first think for a while about what surfaces are there that make up the surface model. Following this thinking process, you should try to identify what wireframes are needed in constructing the surfaces. After this critical top-down thinking process, you can start the bottom-up process of creating the wireframes and then the surfaces.

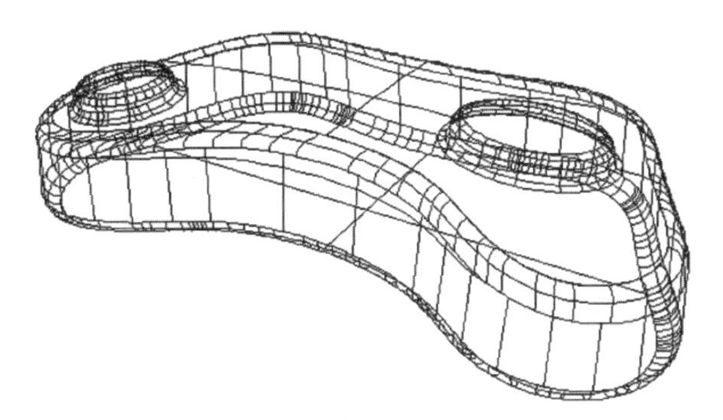

Fig. 9-2. Surface model.

Depending on the design intent and minute details of the model, you can construct the models in many different ways. To help you realize the process of creating a surface model developed from curves to surfaces, you will work through the steps that follow. However, you should try your own way of creating the model after following the steps.

Main Body

In essence, the main body of the joypad consists of a planar surface (bottom), an extruded surface (side), and a loft surface (top). Apart from these main features, you will construct a blended surface between the top face and the side face, and a filleted surface between the side face and the bottom face.

Curves for Side and Bottom Faces

The same set of curves will be used to construct the bottom and side faces. It is a set of joined arcs and lines. Because a line is degree 1 curve and an arc is a degree 2 curve, you will rebuild the curve after joining. To help visualize the model, perform the following steps to set the grip extents.

1. Start a new file. Use the metric (Millimeters) template file.
2. Select File > Properties.
3. In the Document Properties dialog box, select the Grid tab.
4. In the Grid tab, set the grid extents to 250 mm and then click on the OK button.

Continue with the following steps to construct two circles.

5. Maximize the Top viewport.
6. Select Curve > Circle > Center, Radius, or click on the *Circle: Center, Radius* button on the Circle toolbar.
7. Type *50,50* at the command line area to specify the center of the circle.
8. Type *30* to specify the radius.
9. Repeat steps 5 through 8 to construct another circle of 30 units radius at 70,100.

10 Select Zoom Extents from the Standard toolbar. Two circles are constructed, as shown in figure 9-3.

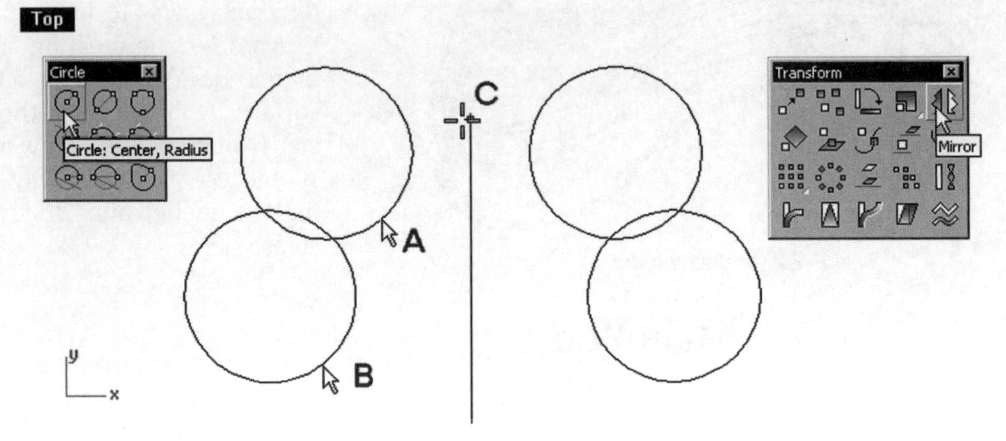

Fig. 9-3. Two circles constructed and being mirrored.

Continue with the following steps to mirror the circles.

11 Select Transform > Mirror, or click on the Mirror button on the Transform toolbar.
12 Select circles A and B (indicated in figure 9-3) and press the Enter key.
13 Type *120,0* at the command line area to specify the start of the mirror plane.
14 Select the Ortho option on the status bar.
15 Select location C (indicated in figure 9-3). Because the Ortho option is selected, the mirror line defined this way will always be vertical.
16 Deselect the Ortho option on the status bar.
17 Select Zoom Extents from the Standard toolbar.

Continue with the following steps to construct two tangent circles and two tangent lines.

18 Construct a tangent circle. Select Curve > Circle > Tangent, Tangent, Radius, or click on the *Circle: Tangent, Tangent, Radius* button on the Circle toolbar.
19 Select locations A and B (indicated in figure 9-4).
20 Type *150* at the command line area to specify the radius.
21 Repeat steps 18 through 20 to construct one more tangent circle tangent to locations C and D (indicated in figure 9-4) of radius 80 mm.

22 Construct a tangent line. Select Curve > Line > Tangent to 2 Curves, or click on the *Line: Tangent to Two Curves* button on the Lines toolbar.

23 Select locations E and F (indicated in figure 9-4).

24 Repeat steps 22 and 23 to construct one more tangent line tangent to locations G and H (indicated in figure 9-4).

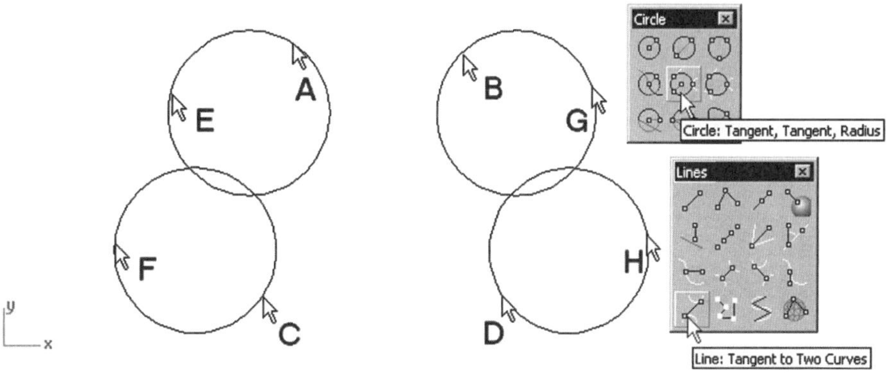

Fig. 9-4. Tangent circles and tangent lines being constructed

Continue with the following steps to trim the circles.

25 Select Edit > Trim, or click on the Trim button on the Main1 toolbar.

26 Select circles A, B, C, and D (indicated in figure 9-5) and press the Enter key.

27 Select locations E and F (indicated in figure 9-5) and press the Enter key.

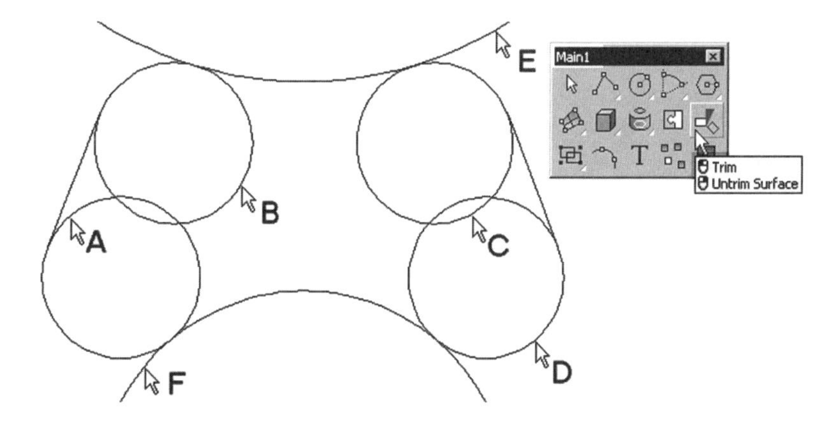

Fig. 9-5. Circles being trimmed.

28 Select Edit > Trim, or click on the Trim button on the Main1 toolbar.

29 Select curves A, B, C, and D (indicated in figure 9-6) and press the Enter key.

30 Select locations E, F, G, and H (indicated in figure 9-6) and press the Enter key.

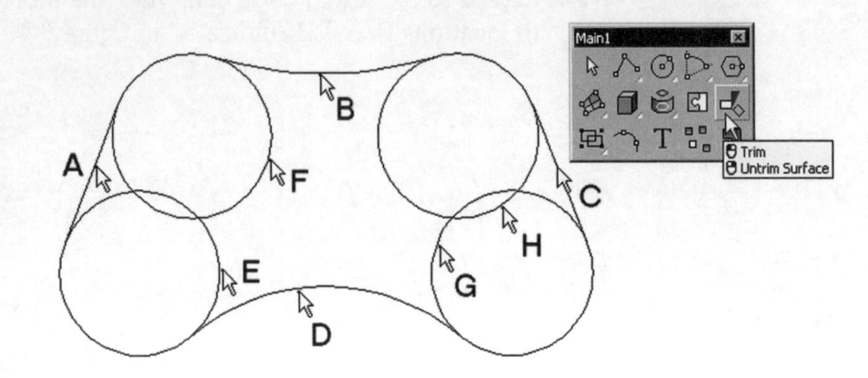

Fig. 9-6. Circles being further trimmed.

Because lines are degree 1 curves and circles are degree 2 curves, continue with the following steps to join these elements into a single curve and rebuild the joined curve.

31 Select Edit > Join, or click on the Join button on the Main1 toolbar.

32 Select curves A, B, C, D, E, F, G, and H (indicated in figure 9-7).

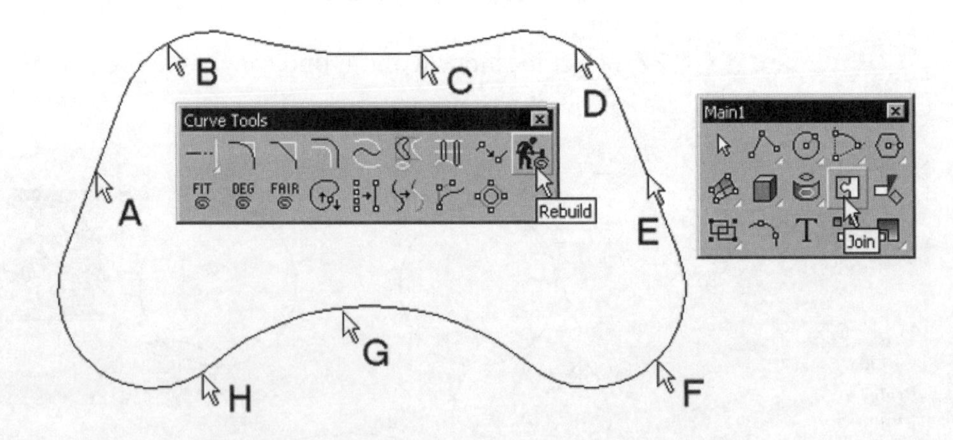

Fig. 9-7. Curves being joined and rebuilt.

33 In the course of selecting the curves, if a Join dialog box such as that shown in figure 9-8 pops up, click on the Yes button.

34 Select Edit > Rebuild, or click on the Rebuild button on the Curve Tools toolbar.

35 Select joined curve A (indicated in figure 9-7).

36 In the Rebuild Curve dialog box, shown in figure 9-9, set the point count to 24 and degree to 3, and then click on the OK button.

The curves for the bottom and top faces are complete.

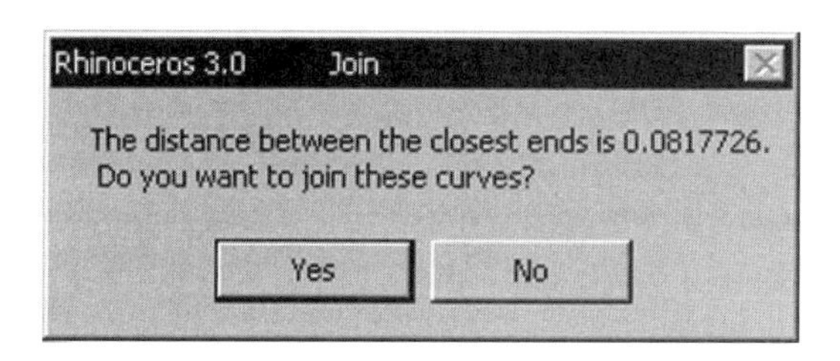

Fig. 9-8. Join dialog box.

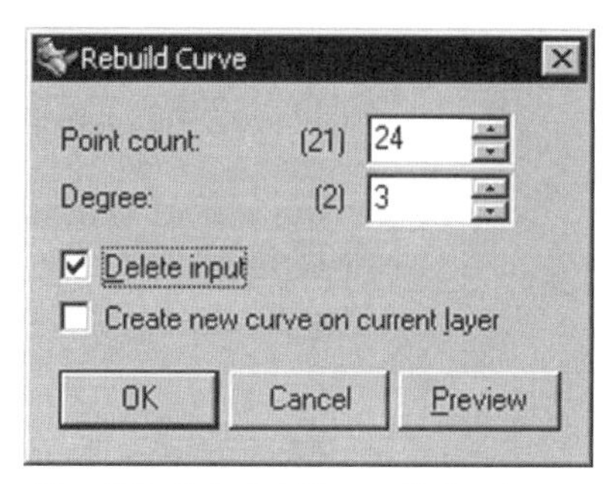

Fig. 9-9. Rebuild Curve dialog box.

Curves for Top Face

The top face is a lofted surface from which you will construct a free-form curve to be copied. Perform the following steps to construct a curve in the Front viewport.

1 Maximize the Front viewport.

2 Select Curve > Free-Form > Interpolate Points, or click on the *Curve: Interpolate Points/Curve: Interpolate on Surface* button on the Curve toolbar.

3 Type *0,20* at the command line area to specify the first point.

4 Type *120,30* at the command line area to specify the second point.

5 Type *240,20* at the command line area to specify the third point and press the Enter key.

A free-form curve is constructed, as shown in figure 9-10.

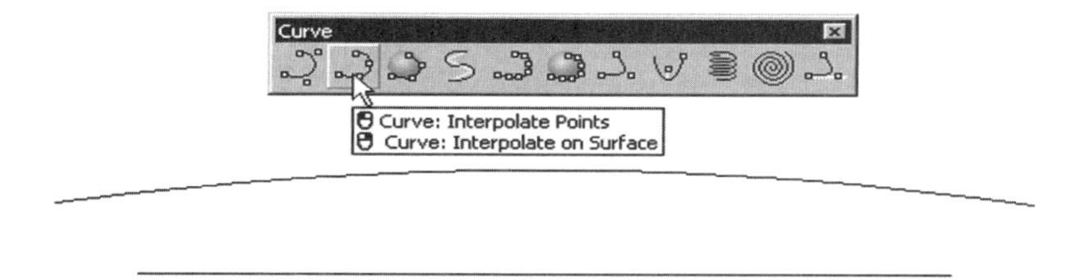

Fig. 9-10. Free-form curve constructed.

Continue with the following steps to copy the free-form curve twice.

6 Maximize the Perspective viewport.
7 Select Transform > Copy, or click on the Copy button on the Transform toolbar.
8 Select curve A (indicated in figure 9-11) and press the Enter key.
9 Select any point on the screen to specify a point to copy from.
10 Type *r0,80,5* at the command line area to specify a relative distance for the copy.
11 Type *r0,160* at the command line area to specify the distance of the second copy and press the Enter key.

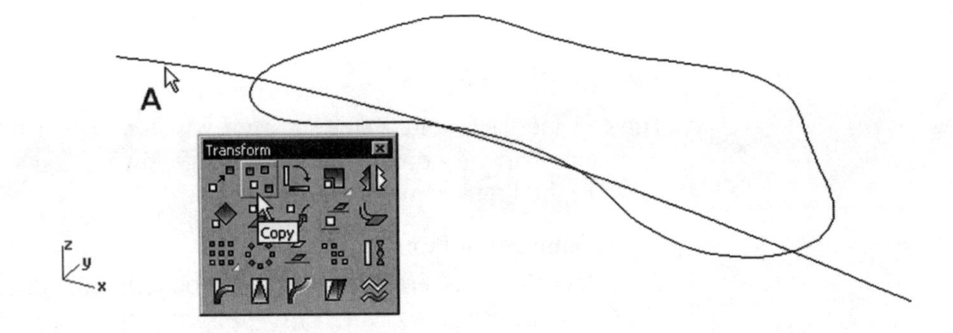

Fig. 9-11. Free-form curve being copied.

Top, Side, and Bottom Faces

As previously mentioned, the bottom face is a planar surface, the side face is an extruded surface, and the top face is a lofted surface. Perform the following steps to construct a planar surface, an extruded surface, and a lofted surface.

1 Set the current layer to *Layer 01*.
2 Select Surface > Planar Curves, or click on the Surface from Planar Curves button on the Surface toolbar.
3 Select curve A (indicated in figure 9-12) and press the Enter key.
4 Select Surface > Extrude > Straight, or click on the Extrude Straight button on the Extrude toolbar.
5 Select curve A (indicated in figure 9-12).
6 Because the location you select has two edges and a surface (the curve, the edge curve of the planar surface, and the planar surface), select Curve in the pop-up menu.
7 Press the Enter key.

8 Type *B* at the command line interface if the screen reads "Both-Sides = Yes."
9 Type *M* at the command line interface to change the mode of extrusion.
10 Type *T* at the command line interface to set the extrusion mode to Tapered.
11 Type *R* at the command line interface to set the draft angle.
12 Type *–5* at the command line interface to specify a taper angle of –5 degrees.
13 Type *50* at the command line interface to specify the extrusion height.
14 Select Surface > Loft, or click on the Loft button on the Surface toolbar.
15 Select curves B, C, and D (indicated in figure 9-12) and press the Enter key.
16 In the Loft Options dialog box, click on the OK button.

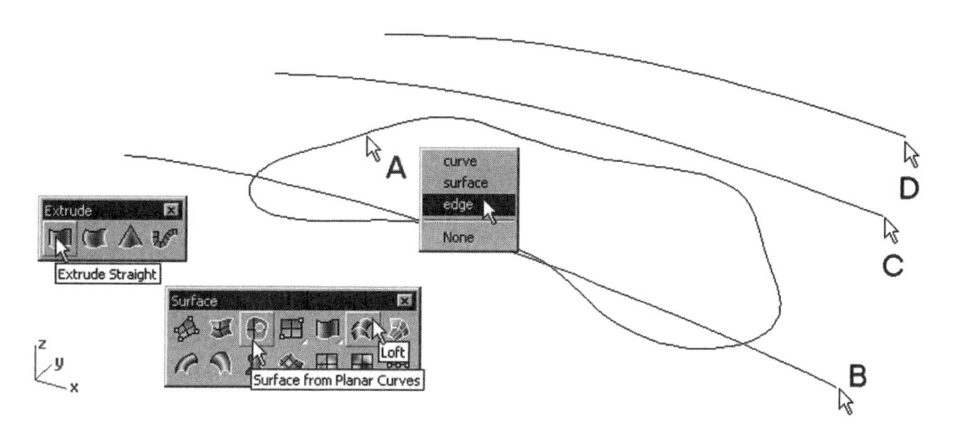

Fig. 9-12. Planar surface, extruded surface, and lofted surface being constructed.

Blend and Filleted Surfaces

There are two ways to provide a smooth transition between two adjacent surfaces. One way is to construct a blended surface, and the other way is to construct a filleted surface. To construct a filleted surface, the adjacent surfaces need to touch or intersect each other. To construct a blended surface, there must be a gap between the adjacent surfaces.

In the following you will construct a blended surface between the top and the side. Therefore, you will have to trim the top and side faces in order to provide a gap. To trim the surfaces, you will construct offset surfaces from the top and side faces and use these offset surfaces as cutting objects for trimming. Perform the following steps to construct two offset surfaces.

1. Turn off the default layer.
2. Select Surface > Offset Surface, or click on the Offset Surface button on the Surface Tools toolbar.
3. Select surface A (indicated in figure 9-13).
4. Type *F* at the command line interface if the offset direction is not the same as that indicated in figure 9-13.
5. Type *5* at the command line interface to specify the offset distance.

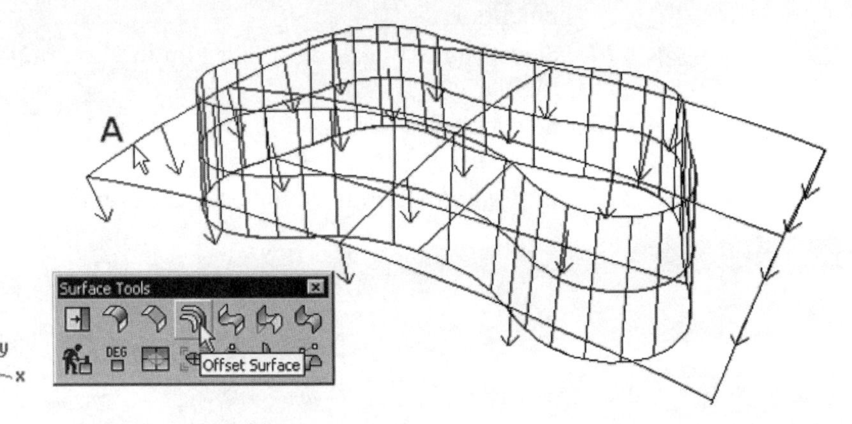

Fig. 9-13. An offset surface being constructed from the extruded surface.

6. Select Surface > Offset Surface, or click on the Offset Surface button on the Surface Tools toolbar.
7. Select surface A (indicated in figure 9-14).
8. Type *F* at the command line interface if the offset direction is not the same as that indicated in figure 9-14.
9. Type *5* at the command line interface to specify the offset distance.

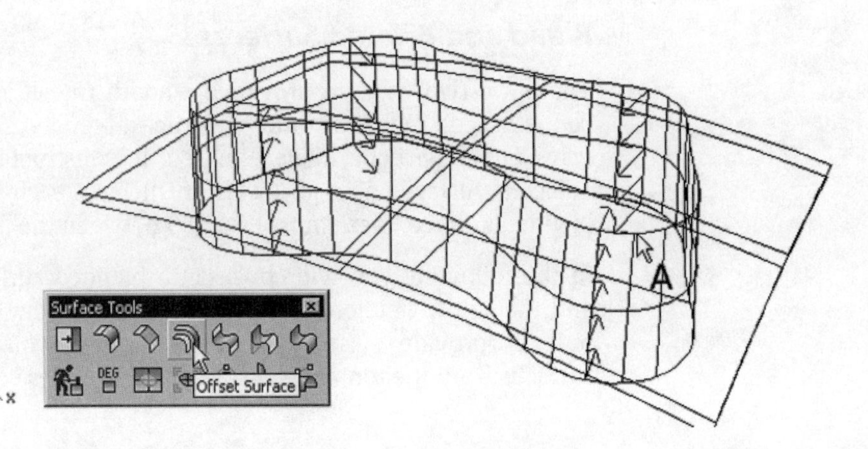

Fig. 9-14. An offset surface being constructed from the lofted surface.

Continue with the following steps to trim the extruded surface and the lofted surface using the offset surfaces as cutting objects and to delete the offset surfaces.

10 Select Edit > Trim, or click on the *Trim/Untrim Surface* button on the Main1 toolbar.

11 Select surface A (indicated in figure 9-15), the surface offset from the lofted surface, and press the Enter key.

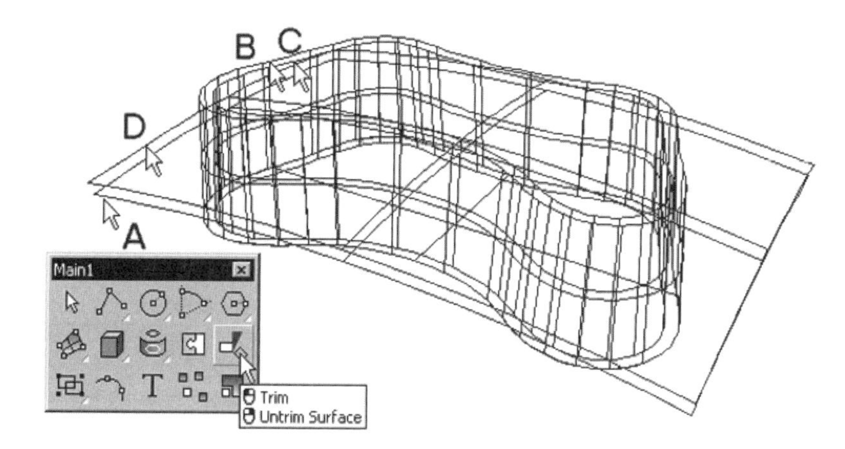

Fig. 9-15. Extruded surface and lofted surface being trimmed by the offset surfaces.

12 Select surface B (indicated in figure 9-15), the extruded surface, and press the Enter key.

13 Select Edit > Trim, or click on the *Trim/Untrim Surface* button on the Main1 toolbar.

14 Select surface C (indicated in figure 9-15), the surface offset from the extruded surface, and press the Enter key.

15 Select surface D (indicated in figure 9-15), the lofted surface, and press the Enter key.

16 Select Edit > Delete.

17 Select surfaces A and B (indicated in figure 9-16) and press the Enter key.

Continue with the following steps to construct a filleted surface and a blended surface.

18 Select Surface > Fillet, or click on the Fillet Surface button on the Surface Tools toolbar.

19 Type *R* at the command line interface to use the Radius option.

20 Type *5* at the command line interface to specify a filleted radius of 5 units.

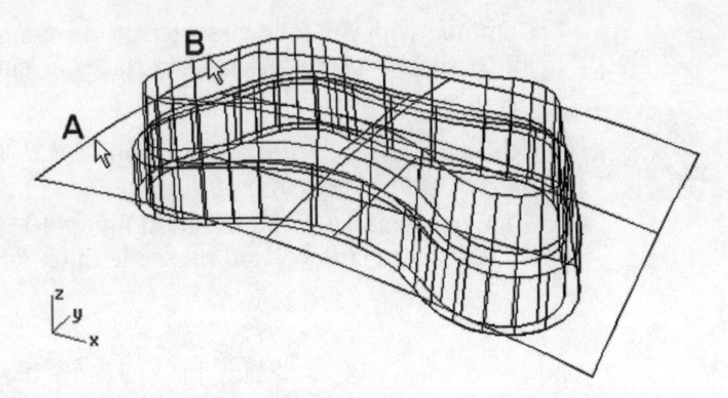

Fig. 9-16. Offset surfaces being deleted.

21 Select A (the upper part of the extruded surface) and B (the planar surface), indicated in figure 9-17.

22 Select Surface > Blend Surface, or click on the Blend Surface button on the Surface Tools toolbar.

23 Select C (the upper edge of the extruded surface) and D (the edge of the lofted surface), indicated in figure 9-17.

24 In the Adjust Blend Bulge dialog box, click on the OK button.

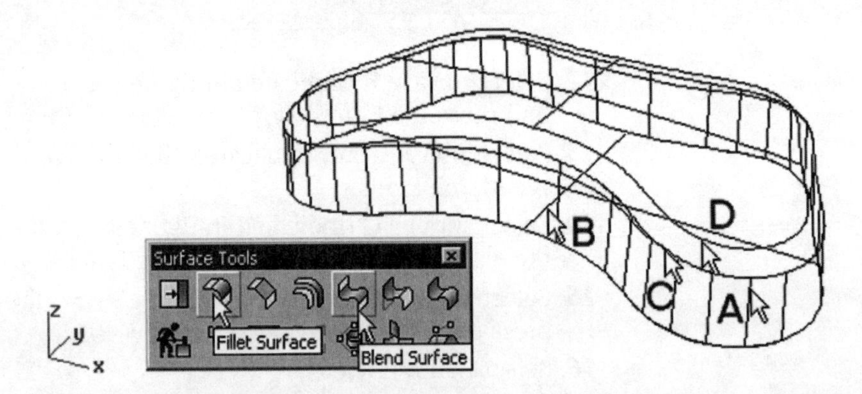

Fig. 9-17. Filleted surface and blended surface being constructed.

Raised Features on the Top Face

To construct smooth and natural raised features, in the following you will construct offset surfaces from the top face. To define the contour of the raised features, you will construct two curves and then construct extruded surfaces from the curves. After constructing these surfaces, you will construct filleted surfaces as transition faces. To construct a circle and an ellipse for creating two extruded surfaces, perform the following steps.

1 Set the current layer to the default layer.

2 Click on the *Circle: Deformable* button on the Circle toolbar.
3 Type *50,50* at the command line interface to specify the center.
4 Type *18* at the command line interface to specify the radius.
5 Select Curve > Ellipse > From Center, or click on the *Ellipse: From Center* button on the Ellipse toolbar.
6 Type *170,75* at the command line area to specify the center.
7 Type *r25 < 10* at the command line area to specify the location of the end of the first axis. (r25 < 10 is a polar coordinate. It means a distance of 25 units within 10 degrees counterclockwise of the last point.)
8 Type *r20 < 110* at the command line area to specify the location of the end of the second axis.

A circle and an ellipse are constructed, as shown in figure 9-18.

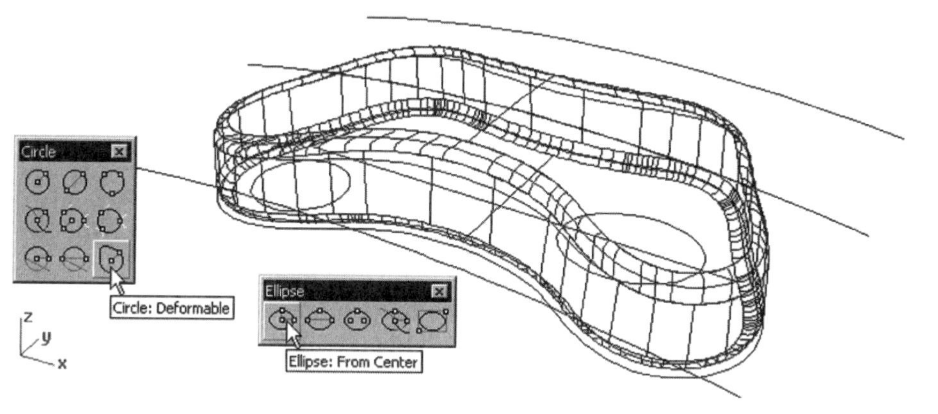

Fig. 9-18. Circle and ellipse constructed.

Because an ellipse is a degree 2 curve and has four kink points, extruding it may result in a polysurface of four surfaces instead of a single surface. Continue with the following steps to edit the ellipse and change its polynomial degree to 3 and to make it periodic. A periodic curve is a closed curve with no kink point.

9 Select Edit > Rebuild, or click on the Rebuild button on the Curve Tools toolbar.
10 Select the ellipse and press the Enter key.
11 In the Rebuild Curve dialog box, set the point count to 12 and degree to 3, and then click on the OK button.

Continue with the following steps to construct two extruded surfaces.

12 Set the current layer to *Layer 01*.
13 Select Surface > Extrude > Tapered.

14 Select the circle and the ellipse and then press the Enter key.
15 Type *R* at the command line interface to set the draft angle.
16 Type *–5* at the command line interface to specify a taper angle of –5 degrees.
17 Type *50* at the command line interface to specify the extrusion height.

Two extruded surfaces are constructed, as shown in figure 9-19.

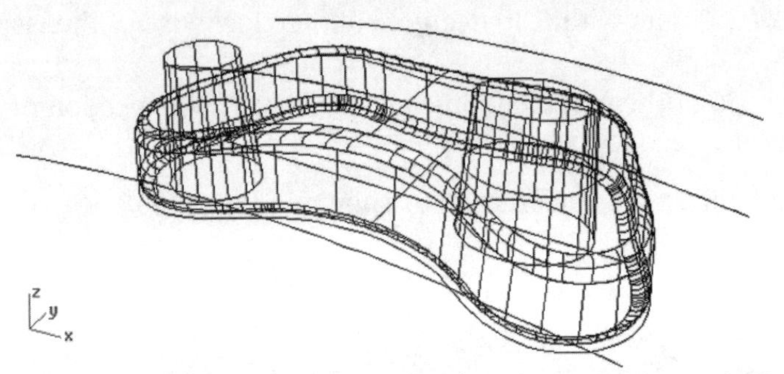

Fig. 9-19. Extruded surfaces constructed.

Continue with the following steps to construct two filleted surfaces.

18 Turn off the default layer.
19 Select Edit > Visibility > Hide, or click on the Hide Objects button on the Visibility toolbar.
20 Select all surfaces except those shown in figure 9-20.
21 Select Surface > Fillet, or click on the Fillet Surface button on the Surface Tools toolbar.
22 Type *R* at the command line interface to use the Radius option.
23 Type *3* at the command line interface to specify a filleted radius of 3 units.

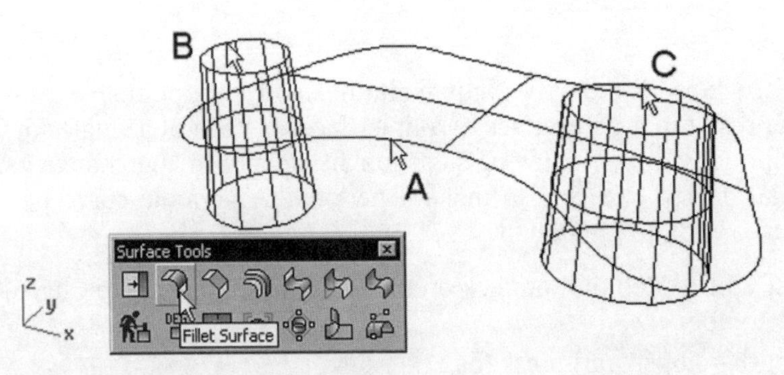

Fig. 9-20. Surfaces hidden and filleted surfaces being constructed.

24 Select A and B (indicated in figure 9-20).
25 Select Surface > Fillet, or click on the Fillet Surface button on the Surface Tools toolbar.
26 Select A and C (indicated in figure 9-20).

Continue with the following steps to complete the raised features.

27 Construct an offset surface (offset distance of 7 units) with reference to figure 9-21.

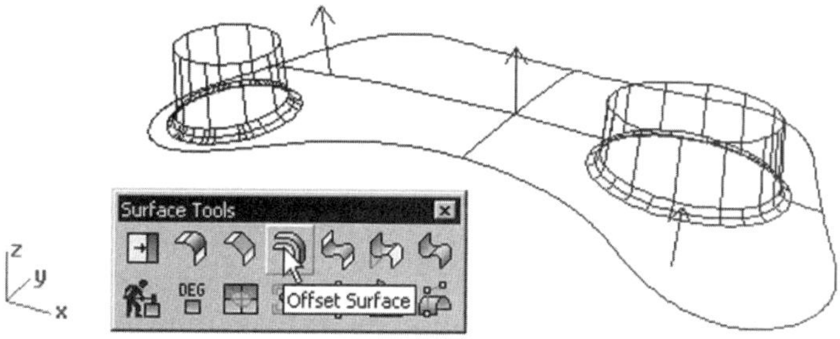

Fig. 9-21. Offset surfaces being constructed.

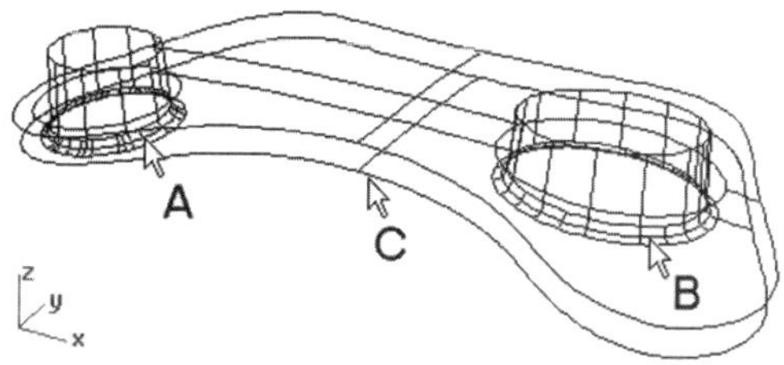

Fig. 9-22. Surfaces being hidden.

28 Hide surfaces A, B, and C (indicated in figure 9-22).

29 Select Edit > Properties, or click on the Object Properties button on the Properties toolbar.

30 Select surface B (indicated in figure 9-23).

31 In the Properties dialog box, select Object and then set *Isocurve density* to 6.

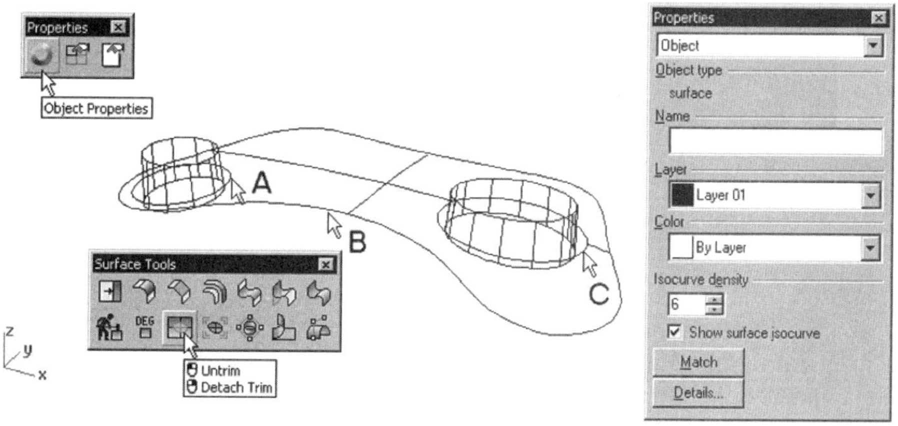

Fig. 9-23. Surface's isocurve density changed and boundary edge being untrimmed.

32 Select Surface > Edit Tools > Untrim, or click on the *Untrim/ Detach Trim* button on the Surface Tools toolbar.

33 Select boundary edge A (indicated in figure 9-23).

34 Repeat steps 32 and 33 twice to untrim boundary edges B and C (indicated in figure 9-23).

35 Select Transform > Copy, or click on the Copy button on the Transform toolbar.

36 Select surface A and press the Enter key.

37 Type *i* at the command line interface to use the Inplace option. (Inplace means copying the object at its original location.)

38 Set the display to a four-viewport configuration.

39 Select Surface > Fillet, or click on the Fillet Surface button on the Surface Tools toolbar.

40 Select location B and edge C (indicated in figure 9-24).

41 Select Surface > Fillet, or click on the Fillet Surface button on the Surface Tools toolbar.

42 Select location D and edge E (indicated in figure 9-24).

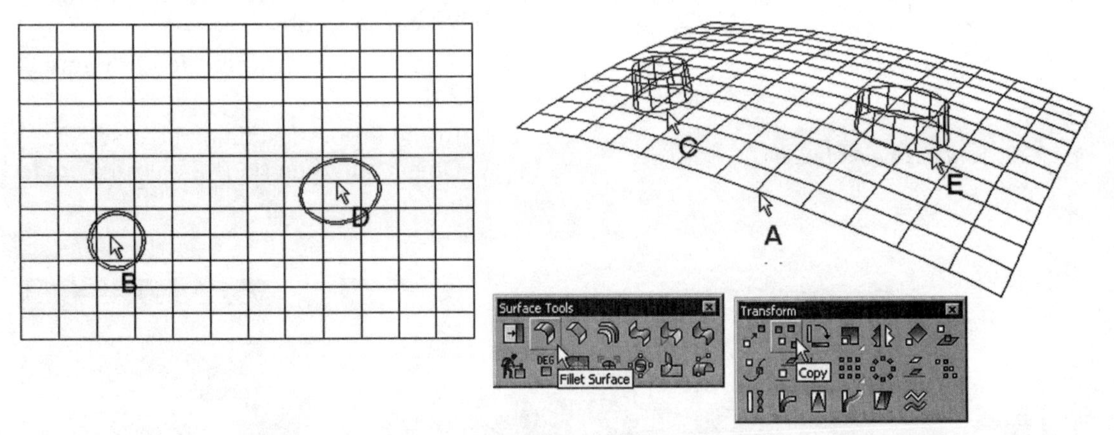

Fig. 9-24. Surface copied in place and filleted surfaces being constructed.

Continue with the following steps to unhide hidden surfaces and shrink the trimmed surfaces.

43 Select Edit > Visibility > Show, or right-click on the *Hide Objects/ Show Objects* button on the Visibility toolbar.

44 Select Surface > Surface Edit Tools > Shrink Trimmed Surface, or click on the Shrink Trimmed Surface button on the Surface Tools toolbar.

45 Select all surfaces and press the Enter key.

The model is complete, as shown in figure 9-25.

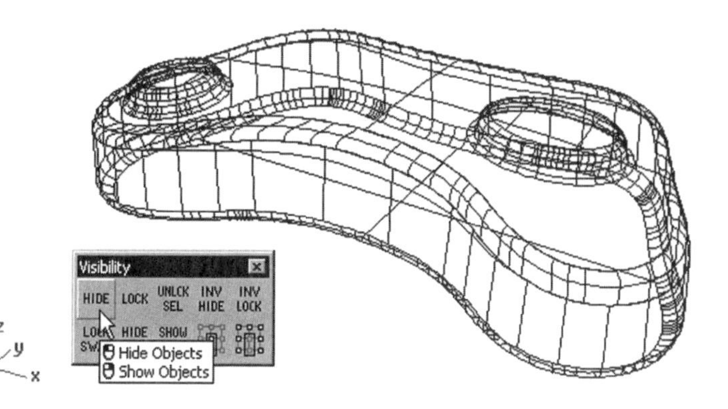

Fig. 9-25. Completed model.

Closed Polysurfaces

The set of surfaces representing the joypad model you constructed encloses a volume. Therefore, you can join the individual surfaces to form a solid. However, there may be some minute gaps between contiguous surfaces that prevent the surfaces from joining properly. To rectify any such problem, you use the edge tools to join the naked edges of those problematic contiguous surfaces, as follows.

1. Select Analyze > Edge Tools > Join 2 Naked Edges, or click on the Join 2 Naked Edges button on the Edge Tools dialog box.
2. Select edge A (indicated in figure 9-26). Because there are two edges, you will select one of the edges from the pop-up dialog box.
3. If the gaps between the edges of contiguous surfaces are too large, an Edge Joining dialog box will pop up. Click on the OK button. Note that the joining tolerance of your model may not be the same as that shown in figure 9-26.
4. Repeat steps 1 through 3 for all edges.

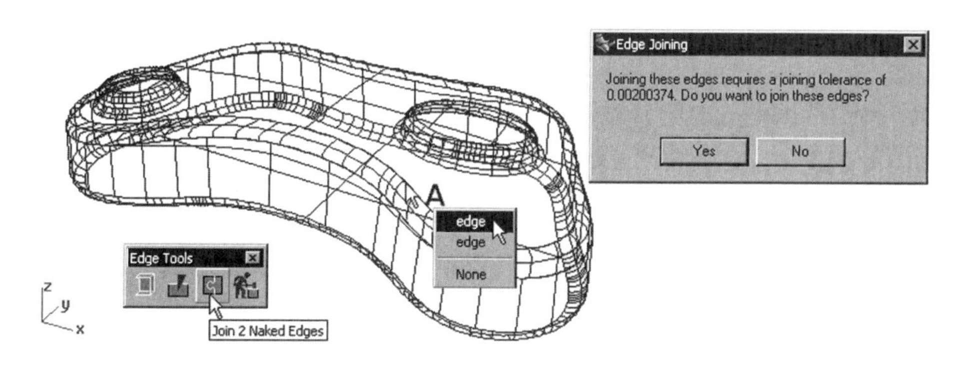

Fig. 9-26. Edges of contiguous surfaces being joined.

Rendering

In the following you will construct a background surface, apply material properties to the surfaces, add lightings to the scene, and produce a rendered image.

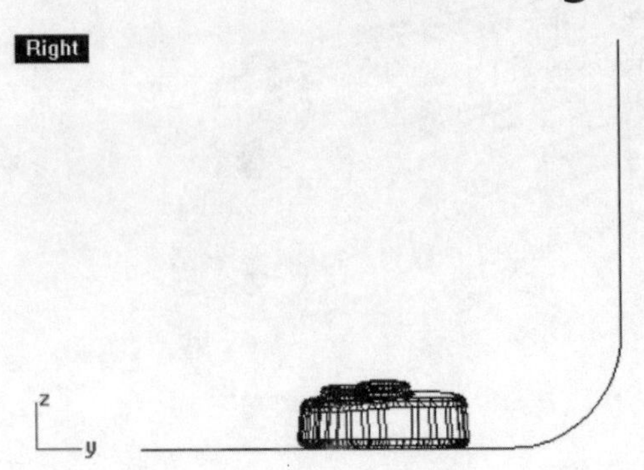

Fig. 9-27. Lines and arc constructed, joined, and rebuilt.

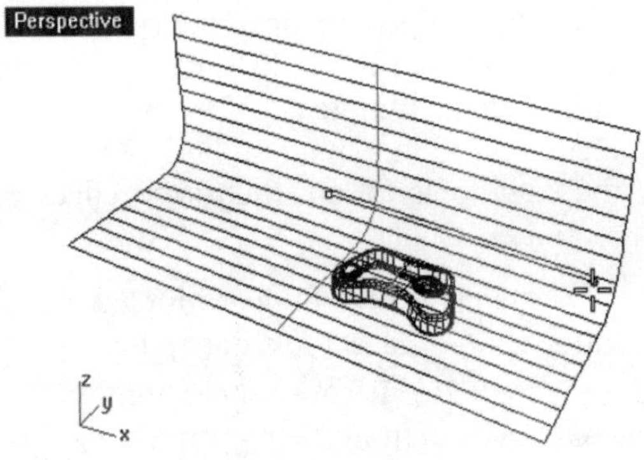

Fig. 9-28. Joined curve extruded.

1. Maximize the Right viewport and construct two lines and an arc. The exact dimensions of the lines and arc are unimportant. You only need to keep the horizontal line touching the base of the joypad model, as shown in figure 9-27.
2. Join the lines and arc to form a single curve and rebuild the joined curve with point count 20 and degree 3.
3. Maximize the Perspective viewport.
4. Extrude the joined curve in accordance with figure 9-28.
5. With reference to figure 9-29, construct a spotlight.
6. Add material properties to the background surface and the joypad.
7. Render the scene.
8. Save your file as *JoyPad.3dm*.

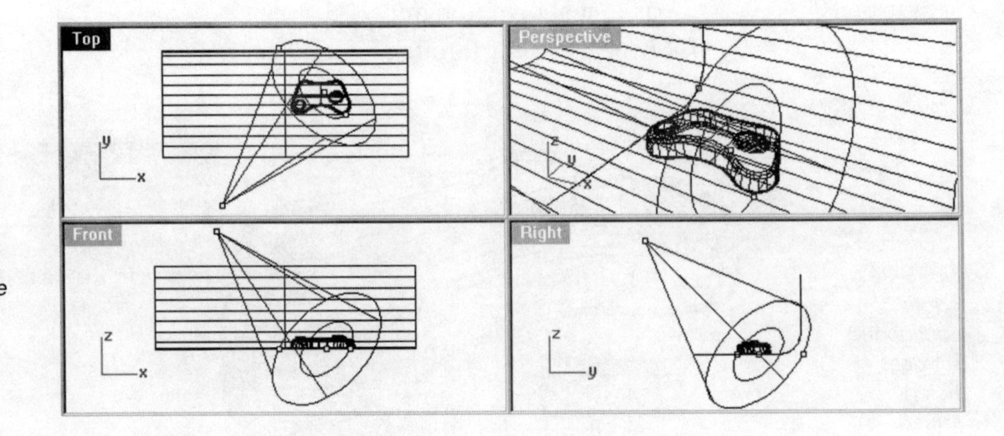

Fig. 9-29. Spotlight added to the scene.

Electric Cable Project

Figure 9-30 shows the rendered image of a piece of electric cable, and figure 9-31 shows the surface model of the component. In this project, you will construct helix curves and circles around curves and construct sweep surfaces from the curves.

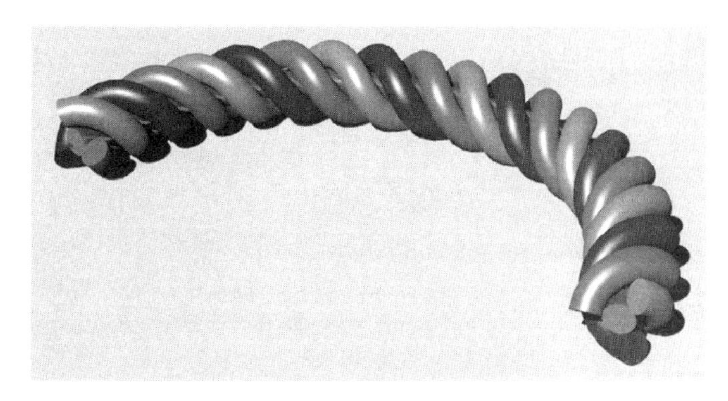

Fig. 9-30. Rendered image of the electric cable.

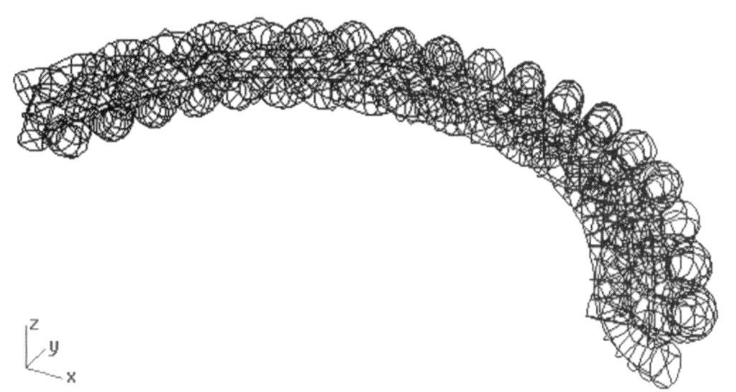

Fig. 9-31. Electric cable model.

Points and Curves

From the rendered image, you see that the electric cable is a set of four wires: a central core and three additional wires wound on the core. In essence, these surfaces are sweep surfaces with both ends capped. To make these surfaces, you will first construct a free-form curve to define the path for making the central core. Then you will construct three helixes around the free-form curve to define the path for the wires winding on the core.

To properly phase the helix on the core, you will construct a circle around the path of the core, set the construction plane to the circle, and construct an array of points to define the starting points of the helix curves. After making the helix curves, you will make several circles around the curves for use as cross sections of the sweep surfaces. Perform the following steps to construct a free-form curve and a circle around the curve.

1. Start a new file. Use the metric (Millimeters) template file.
2. Maximize the Perspective viewport.
3. With reference to figure 9-32, construct a free-form curve that passes through the following points: –30,–15; –20,5; 20,5; and 30,-15.

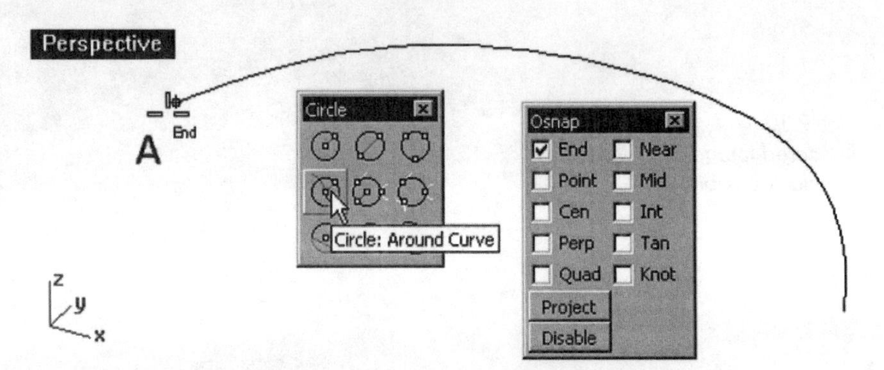

Fig. 9-32. Free-form curve constructed and a circle around the curve being constructed.

In the Osnap dialog box, check the End box and clear all other boxes.

4. Click on the *Circle: Around Curve* button on the Circle toolbar.
5. Select curve A (indicated in figure 9-32) near its end point.
6. Type *3* at the command line interface to specify the radius.

Continue with the following steps to construct a point object.

7. In the Osnap dialog box, check the Quad box and clear all other boxes.
8. Select Curve > Point Object > Single Point, or click on the Single Point button on the Point toolbar.
9. Select the Quad point A (indicated in figure 9-33).

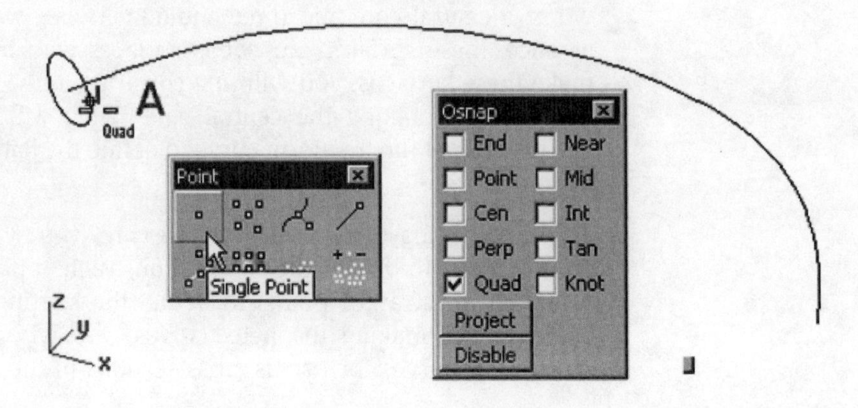

Fig. 9-33. Point object being constructed at the quadrant position of the circle.

Continue with the following steps to set the construction plane and then an array of point objects.

10 Select View > Set CPlane > To Object, or click on the Set CPlane to Object button on the Set CPlane toolbar.

11 Select circle A (indicated in figure 9-34).

12 In the Osnap dialog box, check the End box and clear all other boxes.

13 Select Transform > Array > Polar, or click on the Polar Array button on the Array toolbar.

14 Select point B (indicated in figure 9-34) and press the Enter key.

15 Select end point C (indicated in figure 9-34) as the center of the polar array.

16 Type *3* at the command line interface to set the number of arrayed objects.

17 Type *360* at the command line interface to specify the angle to be filled.

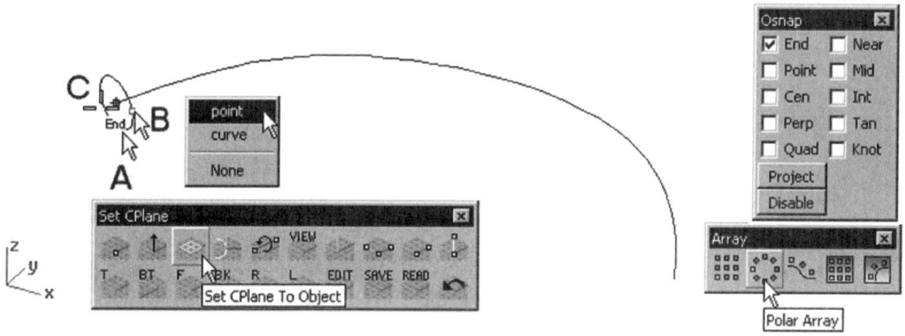

Fig. 9-34. Construction plane set and point object being arrayed.

Continue with the following steps to construct three helix curves.

18 In the Osnap dialog box, check the Point box and clear all other boxes.

19 Select Curve > Helix, or click on the *Helix/Vertical Helix* button on the Curve toolbar.

20 Type *A* at the command line interface to construct a helix around a curve.

21 Type *T* at the command line interface to set the number of turns.

22 Type *6* at the command line interface to specify 6 turns.

23 Select curve A (indicated in figure 9-35).

24 Select point B (indicated in figure 9-35).

25 Repeat steps 19 through 24 twice to construct two more helix curves around curves A at points C and D (indicated in figure 9-35).

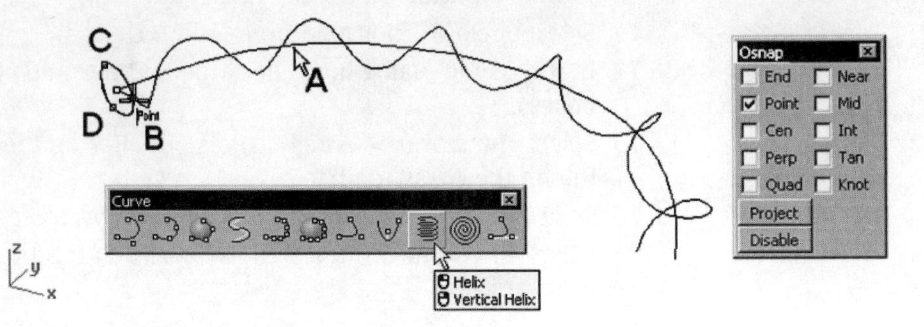

Fig. 9-35. Helix curves being constructed.

Continue with the following steps to construct four around-curve circles.

26 In the Osnap dialog box, check the End box and clear all other boxes.

27 Click on the *Circle: Around Curve* button on the Circle toolbar.

28 Select curve A (indicated in figure 9-36) near its end point.

29 Type *1.5* at the command line interface to specify the radius.

30 Repeat steps 28 through 30 three more times to construct three more around-curve circles at B, C, and D (indicated in figure 9-36).

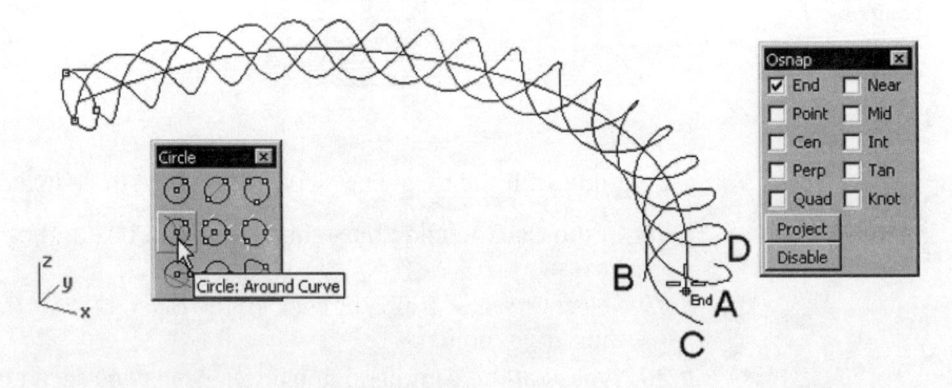

Fig. 9-36. Circles being constructed.

Surfaces and Solids

After making the curves, surface construction is simple. You need only select the appropriate surface construction commands. Perform the following steps to construct four sweep surfaces.

1 Set the current layer to *Layer 01*.

2 Select Surface > Sweep1 Rail, or click on the Sweep 1 Rail button on the Surface toolbar.
3 Select curve A (indicated in figure 9-37) as the path.
4 Select curve B (indicated in figure 9-37) as the cross section and press the Enter key.
5 In the Sweep 1 Rail Options dialog box, click on the OK button.
6 Repeat steps 2 through 5 three more times to construct three more sweep surfaces from curves C and D, curves E and F, and curves G and H.

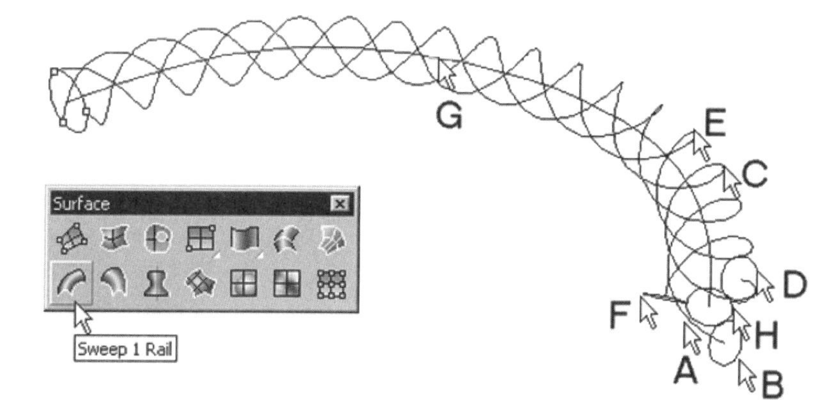

Fig. 9-37. Sweep surfaces being constructed.

Continue with the following steps to cap the surfaces.

7 Select Solid > Cap Planar Holes, or click on the Cap Planar Holes button on the Solid Tools toolbar.
8 Select surface A (indicated in figure 9-38).
9 Repeat steps 7 and 8 three more times to cap surfaces B, C, and D (indicated in figure 9-38).

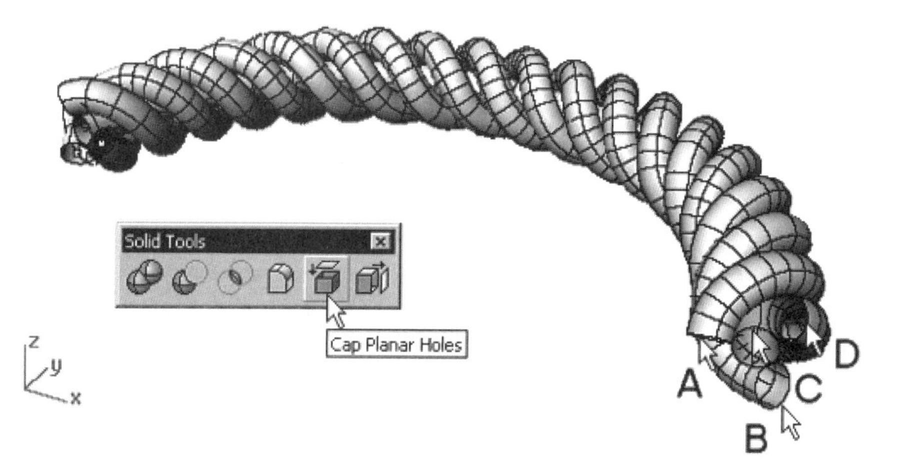

Fig. 9-38. Planar holes being capped.

Material and Rendering

The model is complete. Add material properties and produce a rendering from the model, as follows.

1. Turn off the default layer.
2. Add material properties to individual polysurfaces, setting them to different colors.
3. Add a ground plane at a distance of –4 units from the XY plane.
4. Render the scene.
5. Save your file as *ElectricCable.3dm.*

The model is complete.

Toothbrush Project

Figure 9-39 shows the rendered image of a toothbrush, and figure 9-40 shows the surface model. In the following you will construct the model and produce a rendered image from the model.

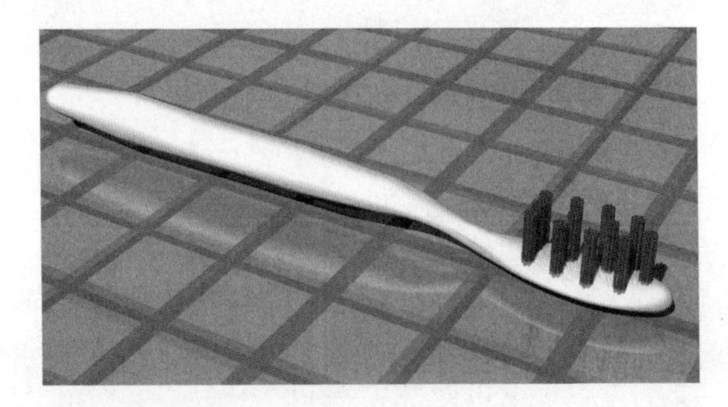

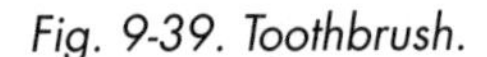

Fig. 9-39. Toothbrush.

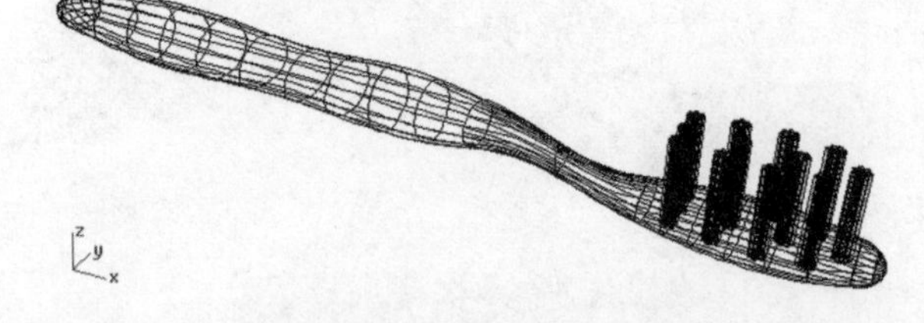

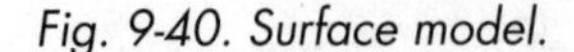

Fig. 9-40. Surface model.

Main Body of the Toothbrush

The main body of the toothbrush consists of four surfaces: two surfaces from curve networks, a patch surface, and a blended surface. In making the surfaces from a curve network, you will construct curves in the Top and Front viewports, and then construct cross-section profiles from the curves. Toward an appreciation of two different ways of tackling the problem, one of the surfaces from curve network will have openings at both ends and the other will have a smooth, closed end and an open end. To close one of the ends of the former surface, you will construct a path surface.

To finish the main body, you will move the former surface-from-curve network and construct a blended surface to fill the gap between the two surfaces of the curve network. Construct two free-form curves in the Top viewport and two free-form curves in the Front viewport, as follows.

1. Start a new file. Use the metric (Millimeters) template file.
2. Set the grid extents to 200 mm.
3. Maximize the Top viewport.
4. Construct a free-form curve passing through points A, B, and C (indicated in figure 9-41).
5. Zoom the view to its extents.
6. Check the Ortho button on the status bar.
7. Mirror the free-form curve with the start of the mirror plane at 0,0; then select point D as the end point of the mirror plane.

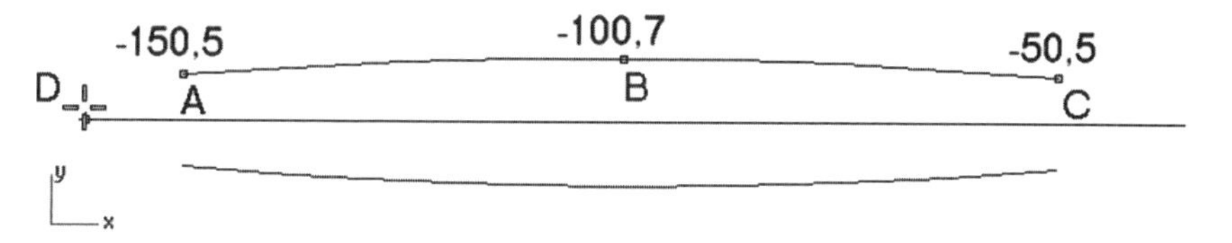

Fig. 9-41. Free-form curve constructed in the Top viewport and being mirrored.

8. Deselect the Ortho button on the status bar.
9. Maximize the Front viewport.
10. Construct two free-form curves in accordance with figure 9-42.

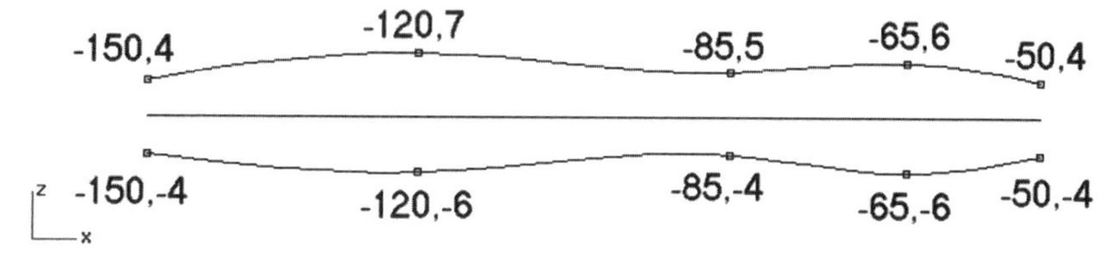

Fig. 9-42. Two free-form curves constructed in the Front viewport.

11. Set the display to a four-viewport configuration.
12. Select Curve > Cross-Section Profiles, or click on the Curve from Cross Section Profiles button on the Curve Tools toolbar.
13. Select curves A, B, C, and D (indicated in figure 9-43) and press the Enter key.

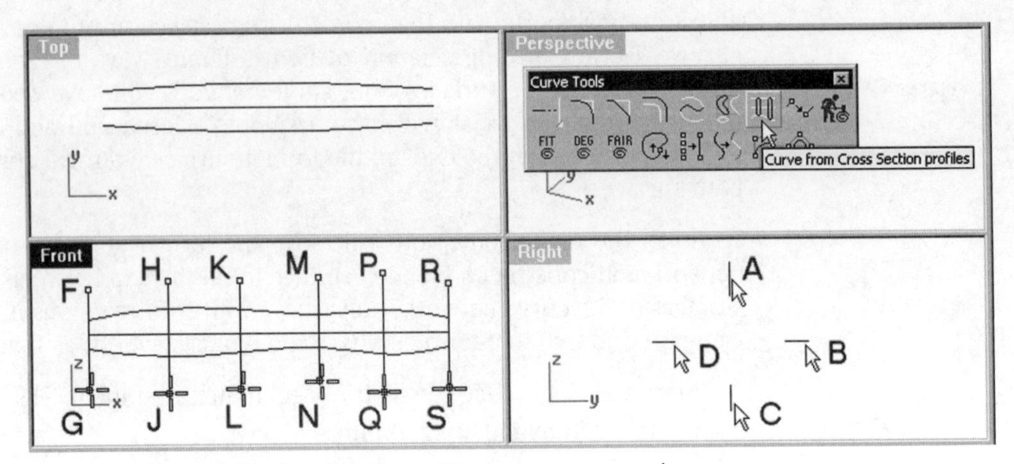

Fig. 9-43. Cross-section profiles being constructed.

14 Select F and G, H and J, K and L, M and N, P and Q, and R and S to construct a series of cross-section profiles.

15 Press the Enter key.

Continue with the following steps to construct a surface-from-curves network.

16 Set the current layer to *Layer 01*.

17 Select Surface > Curve Network, or click on the Surface from Network of Curves button on the Surface toolbar.

18 Select all curves and press the Enter key.

19 Click on the OK button in the Surface From Curve Network dialog box.

A surface-from-curves network is constructed, as shown in figure 9-44.

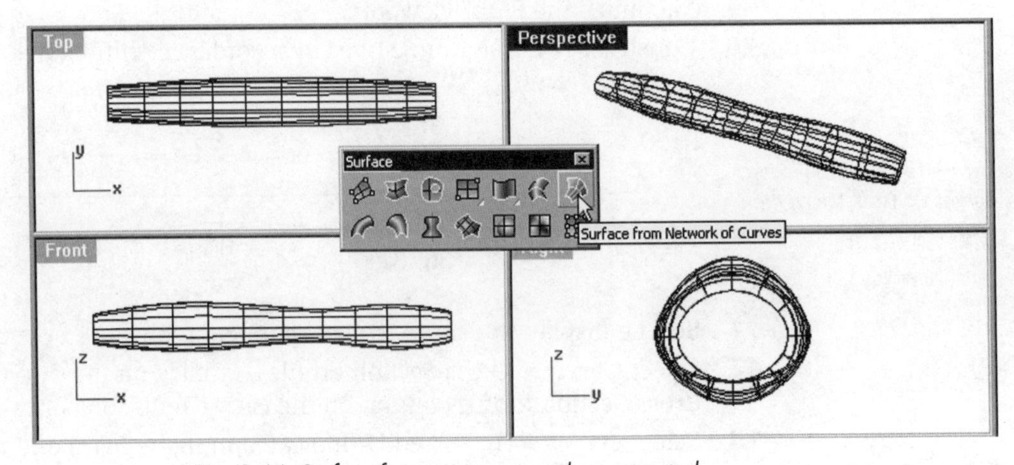

Fig. 9-44. Surface-from-curves network constructed.

The surface-from-curves network has two openings. To close one of the ends, continue with the following steps to construct a patch surface.

20 Maximize the Front viewport.
21 Set the current layer to the default layer.
22 Construct a point object at A (indicated in figure 9-45).
23 Set the current layer to *Layer 01*.
24 Select Surface > Patch, or click on the Patch button on the Surface toolbar.
25 Select point A and edge B (indicated in figure 9-45).
26 In the Patch Surface Options dialog box, click on the OK button.

A patch surface is constructed.

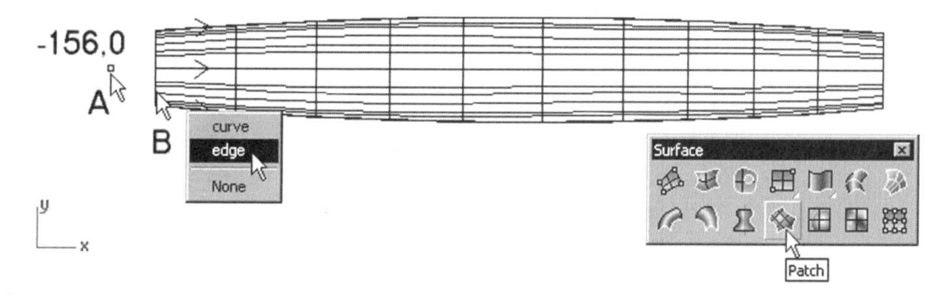

Fig. 9-45. Patch surface being constructed.

27 Select Analyze > Edge Tools > Join 2 Naked Edges, or click on the Join 2 Naked Edges button from the Edge Tools toolbar.
28 Select edge A (indicated in figure 9-46) twice.
29 Click on the OK button in the Edge Joining dialog box.

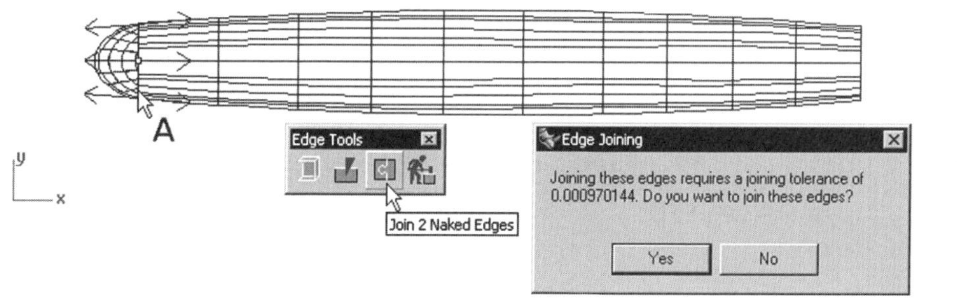

Fig. 9-46. Edges being joined.

Continue with the following steps to construct another surface-from-network curve.

30 Set the current layer to the default layer.

31 Maximize the Top viewpoint.

32 Construct a free-form curve passing through points A, B, C, D, E, F, and G in accordance with figure 9-47.

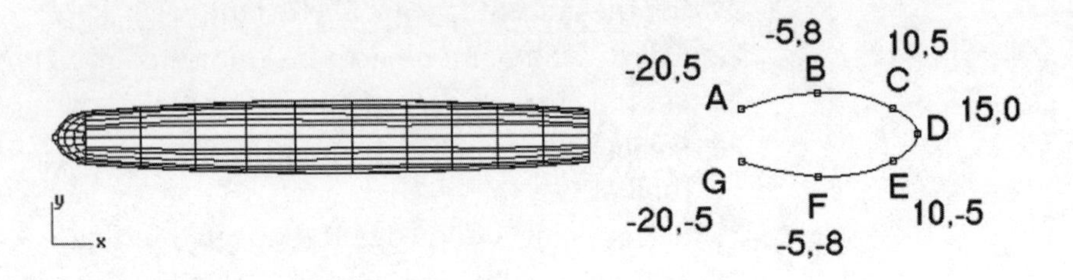

Fig. 9-47. Curve and line constructed in the Top viewport.

33 Maximize the Front viewport.

34 Construct two lines AB and DE, and an arc BCD, in accordance with figure 9-48.

35 Join the lines and arc into a single curve.

36 Rebuild the joined curve with 20 control point count and degree 3.

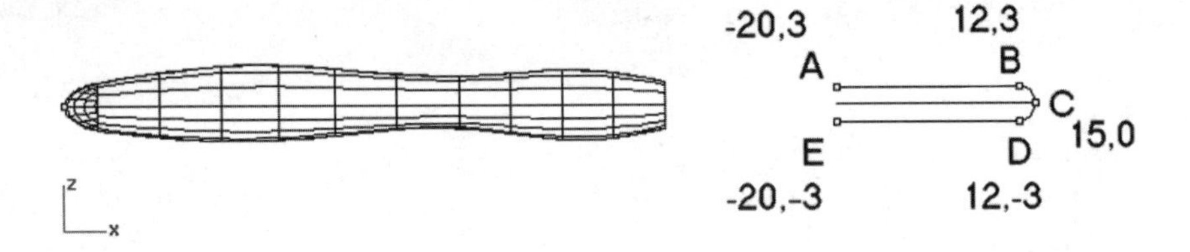

Fig. 9-48. Lines and arc constructed in the Front viewport.

37 Maximize the Perspective viewport.

38 Construct a line AB in accordance with figure 9-49.

39 Split curves C and D in two, using line AB as the cutting object, as shown in figure 9-49.

40 Delete line AB after splitting curves C and D.

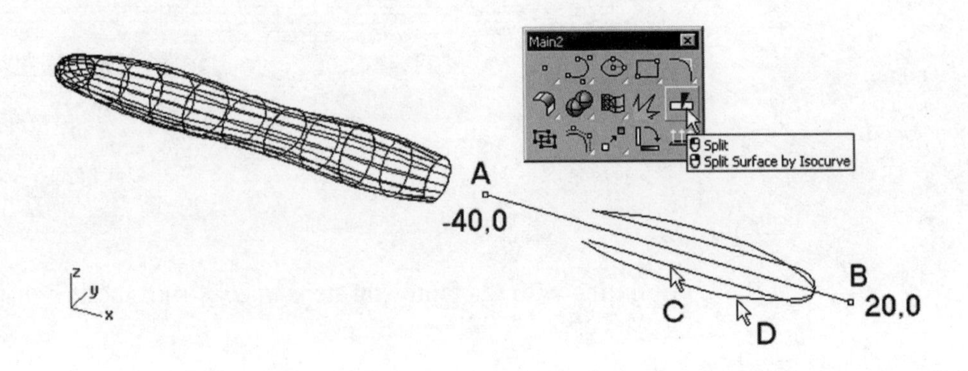

Fig. 9-49. Curves being split.

41 With reference to figure 9-50, construct a series of cross-section profiles.

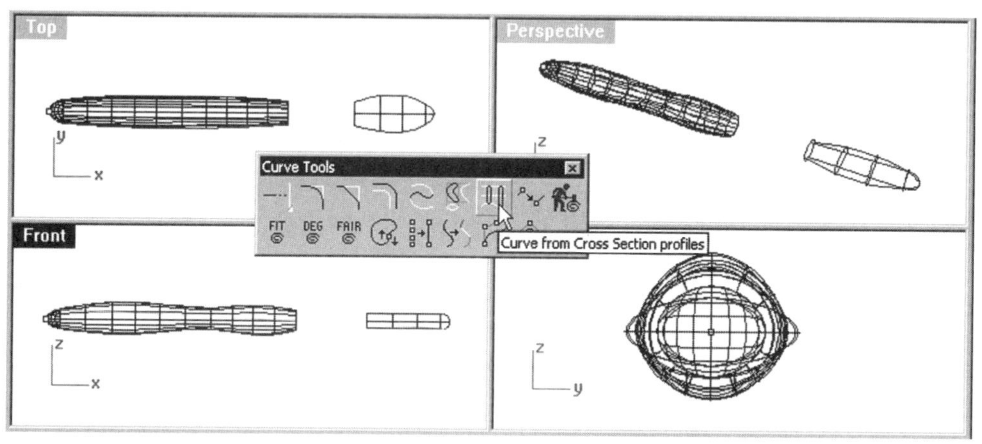

Fig. 9-50. Cross-section profiles being constructed.

42 Turn off the default layer and set the current layer to *Layer 01*.

43 Construct a surface-from-curves network, as shown in figure 9-51.

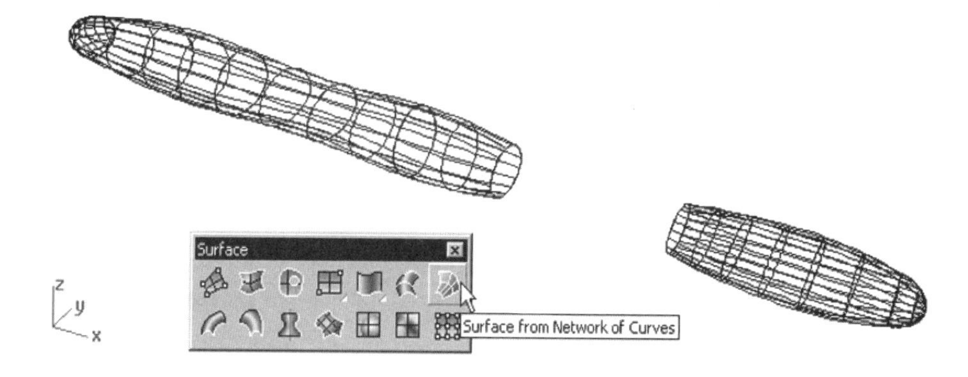

Fig. 9-51. Surface-from-curves network.

Continue the following steps to move a polysurface, construct a blended surface, and join the surfaces.

44 Set the display to a four-viewport configuration.

45 Select Transform > Move, or click on the Move button on the Transform toolbar.

46 Select polysurface A (indicated in figure 9-52) and press the Enter key.

47 Select location B and then type *r5 < 90* at the command line interface.

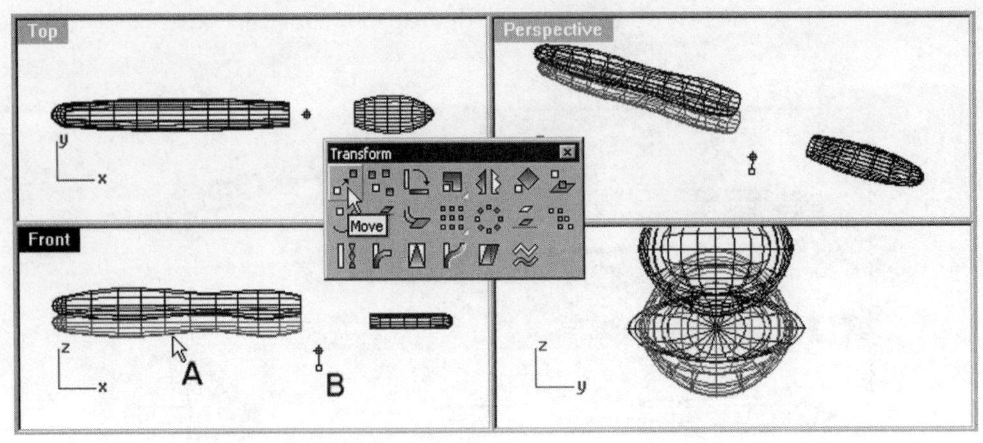

Fig. 9-52. Polysurface being moved.

The selected polysurface is moved.

48 Maximize the Perspective viewport.

49 Select Surface > Blend Surface, or click on the Blend Surface button on the Surface Tools toolbar.

50 Select edges A and B (indicated in figure 9-53).

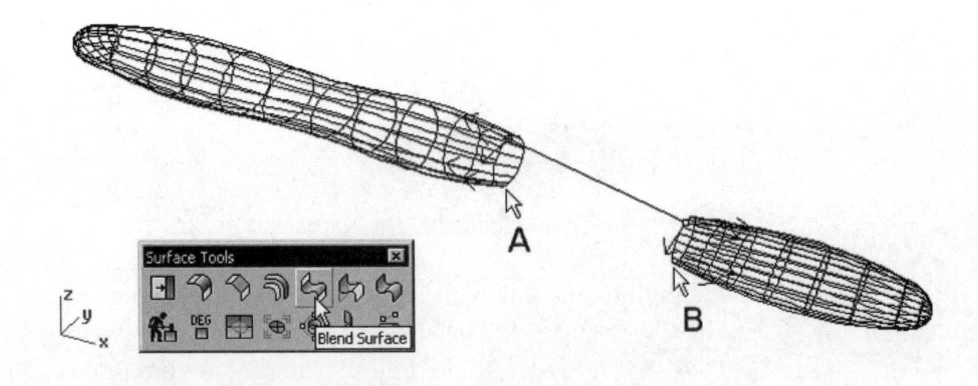

Fig. 9-53. Blended surface being constructed.

51 Press the Enter key.

52 Click on the OK button on the Adjust Blend Nudge dialog box.

53 Join the surfaces along edges A and B (indicated in figure 9-54).

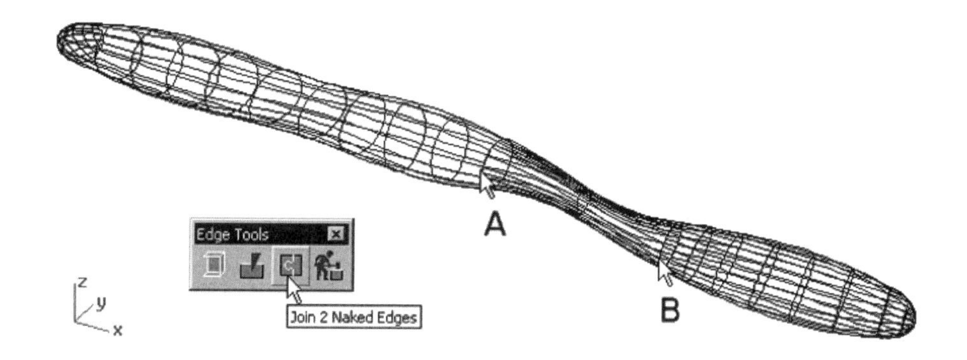

Fig. 9-54. Edges being joined.

Hairs of the Toothbrush

To model the hairs of the toothbrush, you will construct bunches of cylinders. To save time in repeating the bunches of cylinders, you will group them to facilitate selection, array them along a curve, array them in a rectangular way, and mirror them. Perform the following steps to create a curve for use as a path for arraying.

1 Maximize the Top viewport and zoom the display.
2 Set the default layer as the current layer.
3 Construct a free-form curve passing through points A, B, C, and D (indicated in figure 9-55).

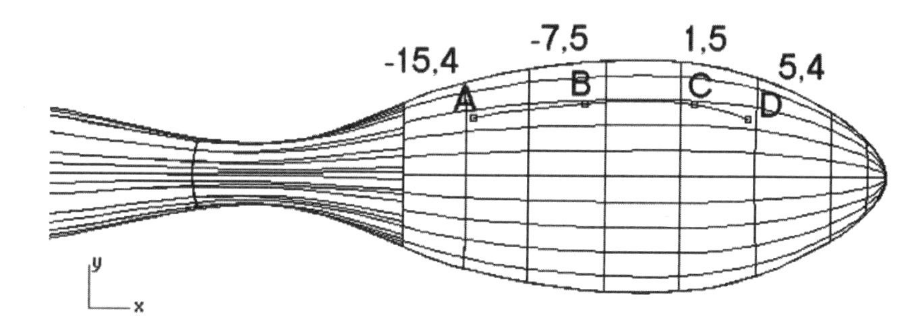

Fig. 9-55. Curve constructed.

Continue with the following steps to construct a cylinder.

4 Set the current layer to *Layer 02*.
5 Set the display to a four-viewport configuration.
6 Select Solid > Cylinder, or click on the Cylinder button on the Solid toolbar.
7 Type *–15,4* at the command line interface to specify the center.
8 Type *0.4* at the command line interface to specify the radius.
9 Move the curve to the Front viewport.

10 Type *r14 < 90* at the command line interface to specify the end point of the cylinder.

A cylinder is constructed, as shown in figure 9-56.

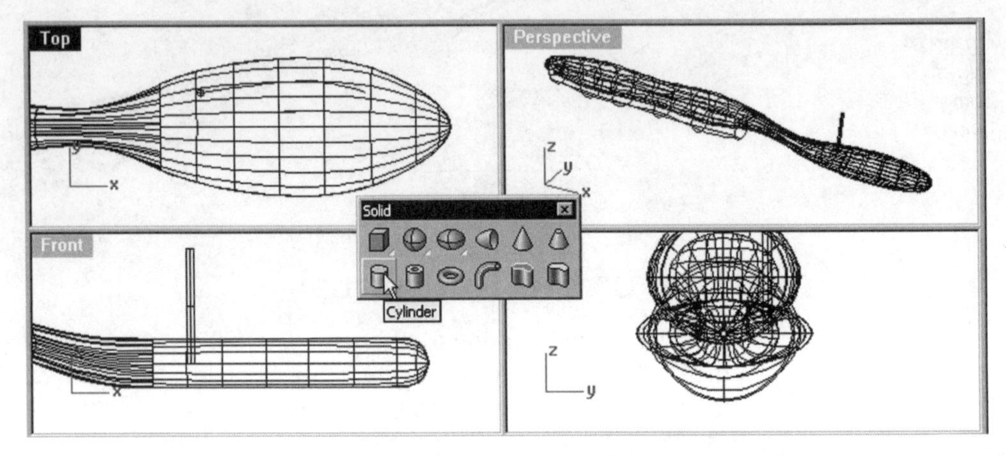

Fig. 9-56. Cylinder constructed.

Continue with the following steps to copy the cylinder you just constructed.

11 Maximize the Top viewport.

12 Select Transform > Copy, or click on the Copy button on the Transform toolbar.

13 Select cylinder A (indicated in figure 9-57) and press the Enter key.

14 Select any point on the screen.

15 Type *r1 < 0* at the command line interface and press the Enter key.

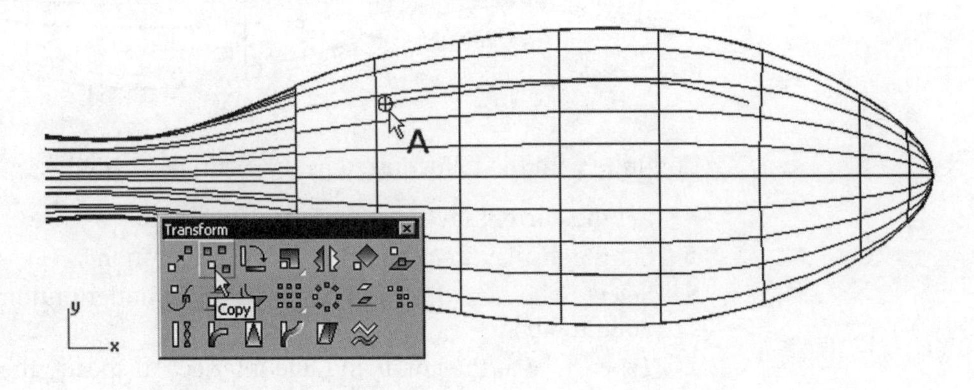

Fig. 9-57. Cylinder being moved.

Continue with the following steps to make a polar array of the cylinder.

16 Select Transform > Array > Polar, or click on the Polar Array button on the Transform toolbar.

17 Select cylinder A (indicated in figure 9-58) and press the Enter key.

18 Check the Cen button on the Osnap dialog box.

19 Select center point B (indicated in figure 9-58).

20 Type *7* at the command line interface to specify the number of items.

21 Type *360* at the command line interface to specify the angle to be filled.

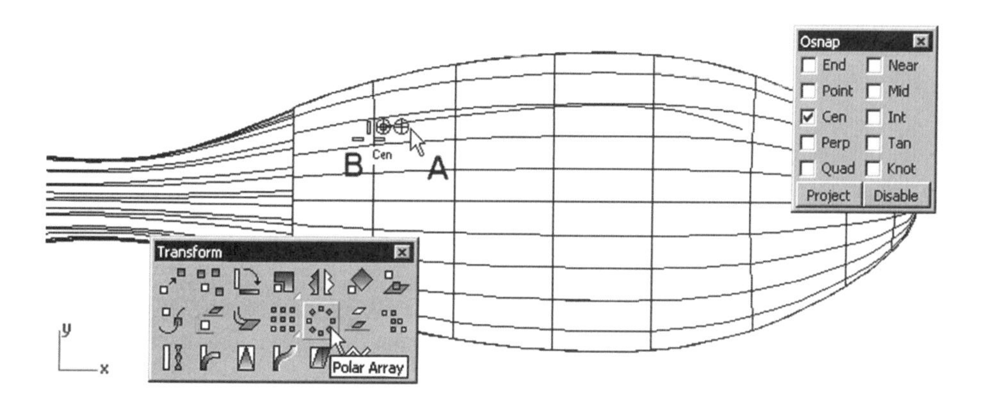

Fig. 9-58. Polar array being constructed.

Continue with the following steps to group the cylinders and construct an array along a curve.

22 Select Edit > Group > Group, or click on the Group button on the Grouping toolbar.

23 Select the seven cylinders and press the Enter key.

24 Select Transform > Array > Along Curve, or click on the Array along Curve button on the Array toolbar.

25 Select group A (indicated in figure 9-59) and press the Enter key.

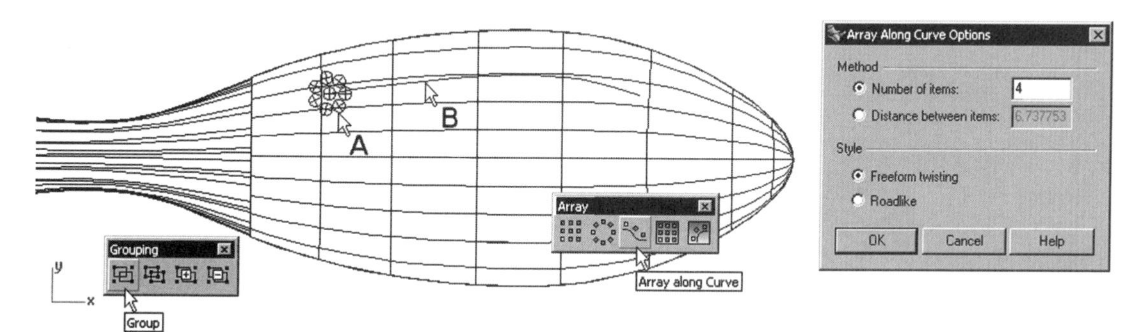

Fig. 9-59. Group of cylinders being arrayed along a curve.

26 Select curve B (indicated in figure 9-59).

27 In the Array Along Curve Options dialog box, set the number of items to 4 and then click on the OK button.

Continue with the following steps to copy the group of cylinders and make a rectangular array of the copied cylinders.

28 Select Transform > Copy, or click on the Copy button on the Transform toolbar.

29 Select group A (indicated in figure 9-60) and press the Enter key.

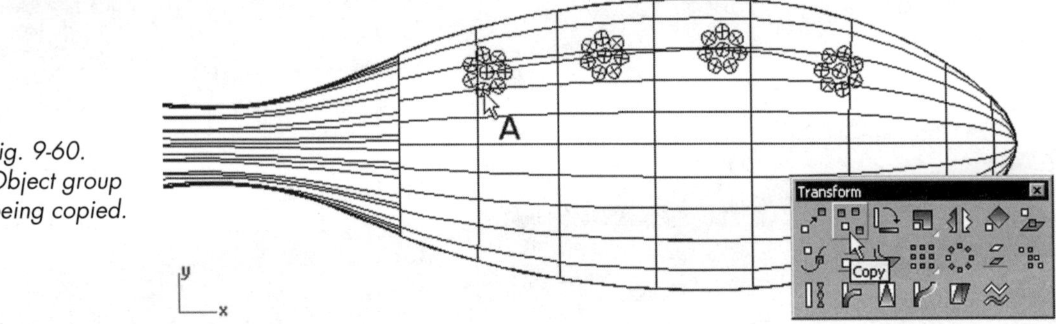

Fig. 9-60. Object group being copied.

30 Select any point on the screen.

31 Type *r4 < 270* at the command line interface and press the Enter key.

32 Select Transform > Array > Rectangular, or click on the Rectangular Array button on the Array toolbar.

33 Select group A (indicated in figure 9-61) and press the Enter key.

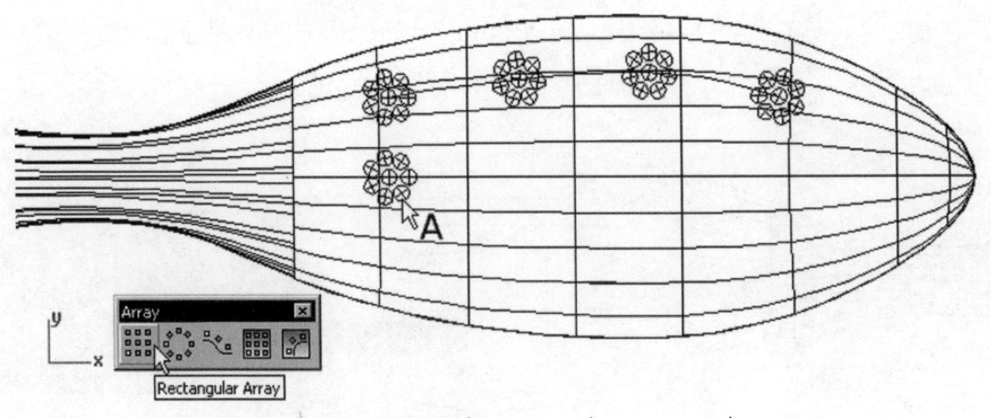

Fig. 9-61. Object group being arrayed.

34 Type *4* at the command line interface to specify the number of items along the X axis.

35 Type *1* at the command line interface to specify the number of items along the Y axis.

36 Type *1* at the command line interface to specify the number of items along the Z axis.

37 Type *8* at the command line interface to specify the distance between items along the X axis.

Continue with the following steps to complete the model.

38 Select Transform > Mirror, or click on the Mirror button on the Transform toolbar.

39 Select groups A, B, C, and D (indicated in figure 9-62) and press the Enter key.

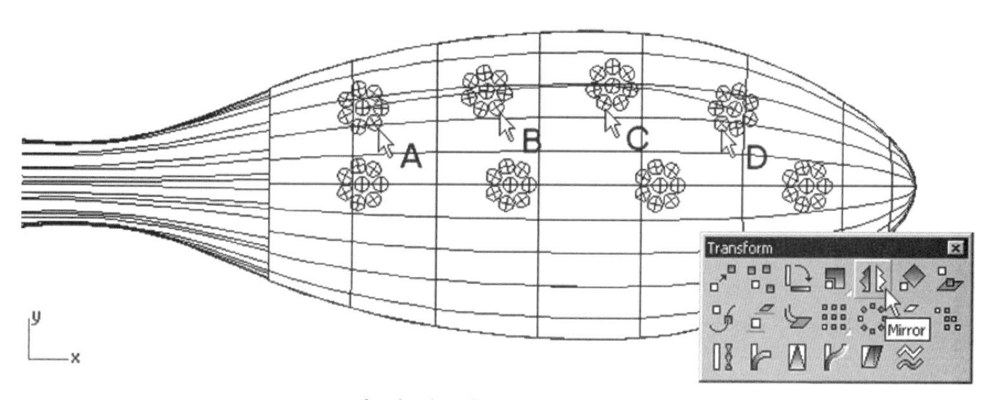

Fig. 9-62. Groups of cylinders being mirrored.

40 Type *0,0* at the command line interface to specify the start of the mirror plane.

41 Type *20,0* at the command line interface to specify the end of the mirror plane.

42 Turn off the default layer.

The model is complete.

Material Assignment and Rendering

To assign and render material, perform the following steps.

1 Apply plug-in material CarPaint Yellow to *Layer 01*.

2 Apply plug-in material Plastic Transparent Cyan to *Layer 02*.

3. Add a ground plane at –5 units from the XY plane with plug-in material Ceramic Tile Mosaic Square Gray High Gloss.
4. Use the Photometric renderer to render the scene.
5. Save your file as *ToothBrush.3d*.

Medal Project

Figure 9-63 shows the rendered image of a medal, and figure 9-64 shows the surface model. This is a simple model consisting of a surface-from-height field of bitmap images and a cylinder.

Fig. 9-63. Rendered image of the medal.

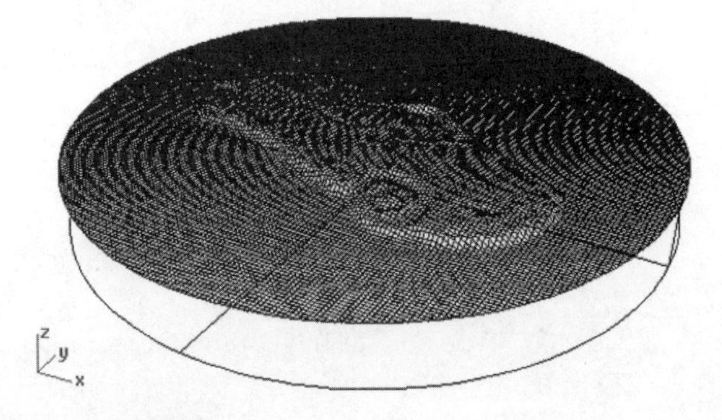

Fig. 9-64. Surface model of the medal.

Surface Model

The surface model is a simple cylindrical object with the top face derived from a bitmap image. To construct an extruded surface and a surface-from-height field, perform the following steps.

1. Start a new file. Use the metric (Millimeters) template file.
2. Select Curve > Circle > Center, Radius, or click on the *Circle: Center, Radius* button on the Circle toolbar.

3 Select point A (indicated in figure 9-65) in the Front viewport to specify the center.

4 Select point B (indicated in figure 9-65) in the Top viewport to specify the radius.

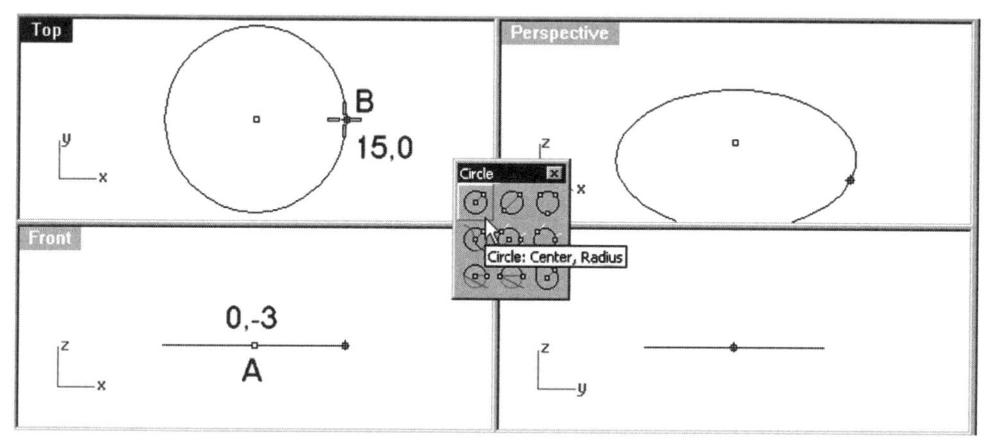

Fig. 9-65. Cylinder being constructed.

5 Select Surface > Extrude > Straight, or click on the Extrude Straight button on the Extrude toolbar.

6 Select the circle and press the Enter key.

7 Place the cursor in the Front viewport, as shown in figure 9-66.

8 Specify an extrusion distance of 6 units in the command line interface.

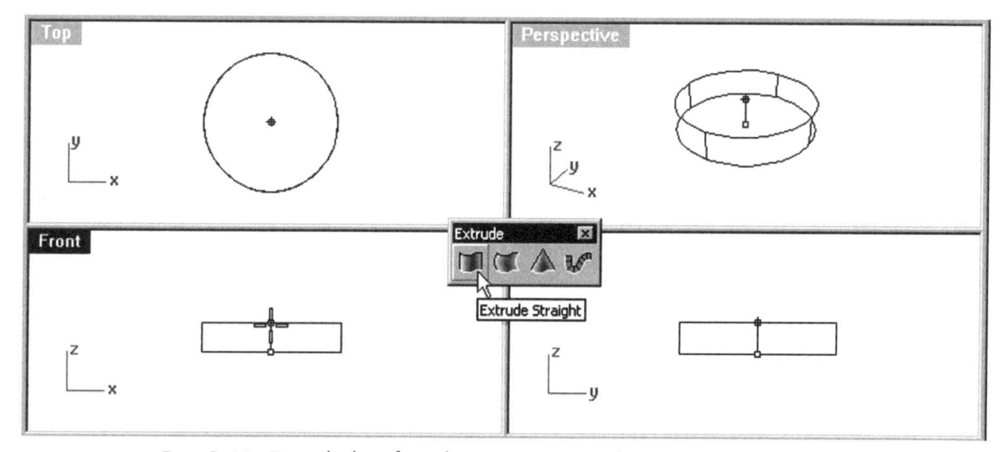

Fig. 9-66. Extruded surface being constructed.

9 Select Surface > Heightfield from Image, or click on the Heightfield from Image button on the Surface toolbar.

10 In the Open dialog box, select the image file *Car.tga* from the *Chapter 9* folder on the companion CD-ROM.

11 Select locations A and B (indicated in figure 9-67).

12 In the Heightfield dialog box, set the number of sample points to 200 times 200, set the height to 0.5, and then click on the OK button.

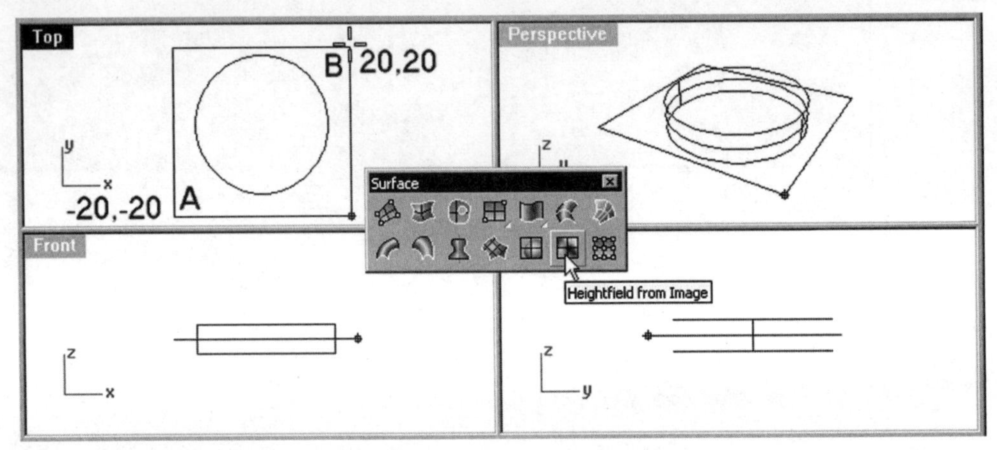

Fig. 9-67. Surface-from-bitmap height field being constructed.

Continue with the following steps to trim the surface and the cylinder.

13 Select Edit > Trim, or click on the *Trim/Untrim Surface* button on the Main1 toolbar.

14 Select A and B (indicated in figure 9-68) and press the Enter key. These are the cutting objects

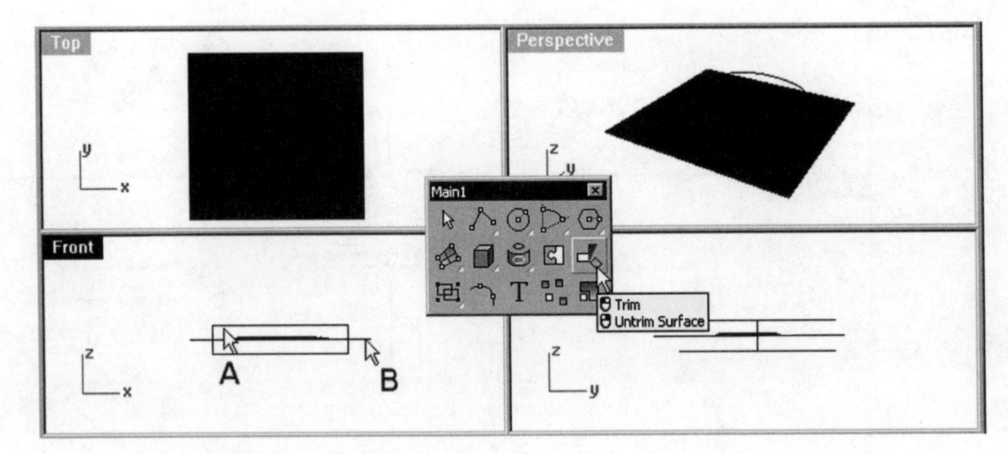

Fig. 9-68. Surfaces being trimmed.

15 Select A and B (indicated in figure 9-68) and press the Enter key. These are the portions to be trimmed.

16 Press the Enter key.

Continue with the following steps to join the naked edges and cap the planar hole.

17 Select Analyze > Edge Tools > Join 2 Naked Edges, or click on the Join 2 Naked Edges button on the Edge Tools toolbar.

18 Select edge A (indicated in figure 9-69) twice and then click on the OK button in the Edge Joining dialog box.

19 Select Solid > Cap Planar Holes, or click on the Cap Planar Holes button on the Solid Tools toolbar.

20 Select polysurface A (indicated in figure 9-69) and press the Enter key.

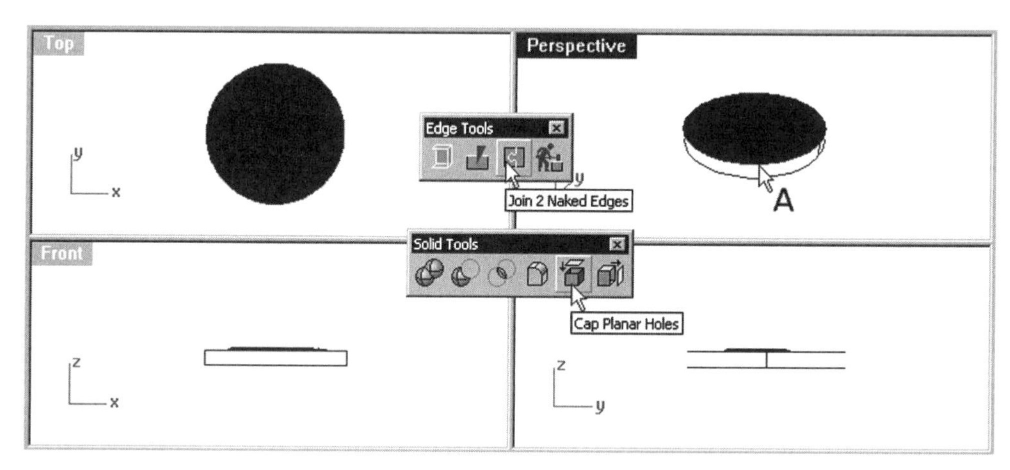

Fig. 9-69. Edges being joined and planar hole being capped.

Material and Rendering

To complete the model, add a ground plane and material properties, as follows.

1 Add a ground plane that is –3 units from the XY plane.

2 Assign material properties to the ground plane and to the model.

3 Render the model.

4 Save your file as *Medel.3dm*.

The model is complete.

Rapid Prototype

Assuming you have a rapid prototyping machine, you may output an STL file and produce a prototype, as follows.

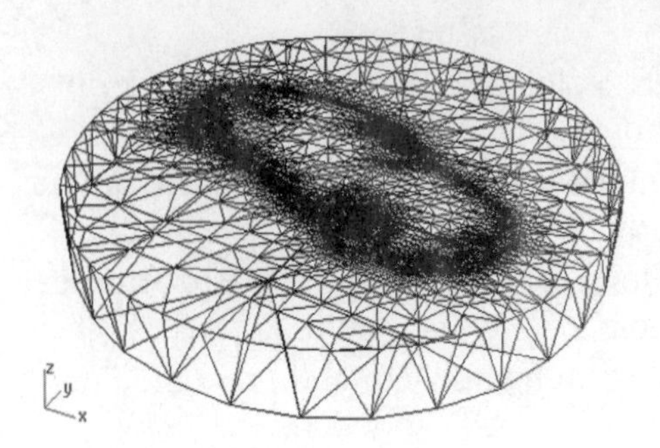

Fig. 9-70. STL file opened.

1. Select File > Save As.
2. In the Save dialog box, select Stereolithography (*.stl) from the *Save as type* pull-down list box.
3. Specify a file name and then click on the Save button.
4. In the STL Export Options dialog box, click on the OK button. In the Polygon Mesh Options dialog box, click on the OK button.

An STL file is saved. Figure 9-70 shows the STL file opened again in Rhino.

Now you can use the polygon mesh tools to make any necessary modifications. If there is no gap or opening, you can use the file to construct a rapid prototype. Figures 9-71 and 9-72 show, respectively, the STL file being processed and a rapid prototype being constructed.

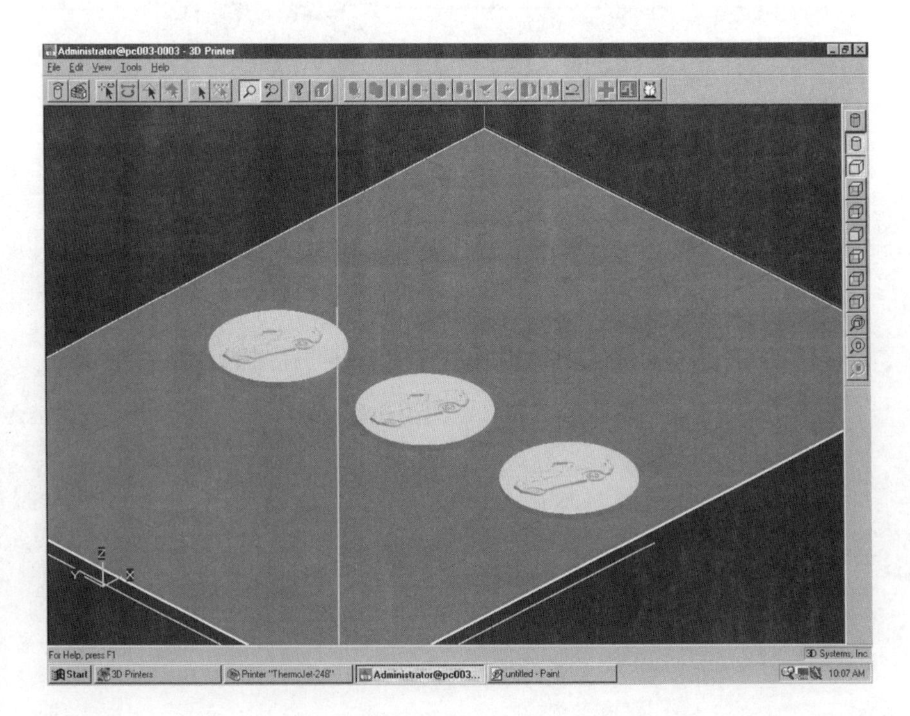

Fig. 9-71. STL file being processed.

Fig. 9-72. Rapid prototype being constructed.

Toy Car Body Project

Figures 9-73 and 9-74 show, respectively, the rendered image and surface model of a toy car's body.

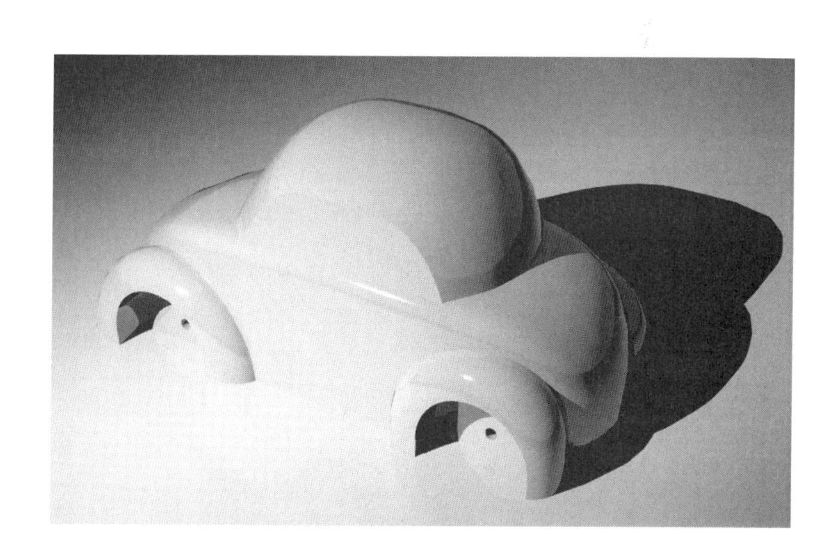

Fig. 9-73. Rendered image.

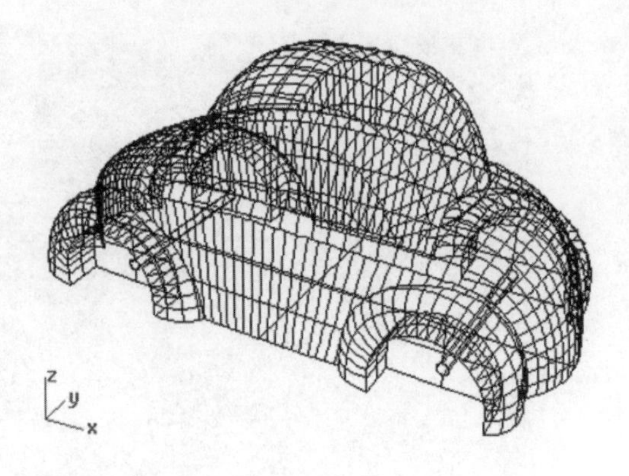

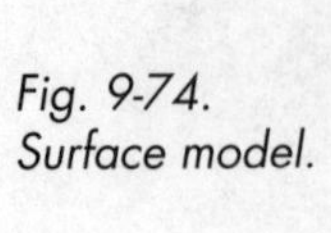
Fig. 9-74. Surface model.

Main Body

The main body consists of a sweep surface (top) and two lofted surfaces (sides). Because the model is symmetrical about its central plane, you need to construct one of the two lofted surfaces and then make a mirror copy from that surface.

Sweep Surface

The sweep surface that constitutes the top face of the main body consists of two symmetrical 3D rail curves and four planar profile curves. To construct the 3D rail curve, you will construct two curves in the Front viewport for easy reference, remap one of them to the Top viewport, and produce a 3D curve from the two curves. As for the profile curves, you will construct two profile curves in the Top viewport and a profile curve in the Right viewport, and then orient the curve in the Right viewport to its final location. Perform the following steps to construct a path curve for the sweep surface.

1. Start a new file. Use the metric (Millimeters) template file.
2. Set the grid extents to 120 mm.
3. Maximize the Front viewport.
4. With reference to figure 9-75, construct two free-form curves in the Front viewport.

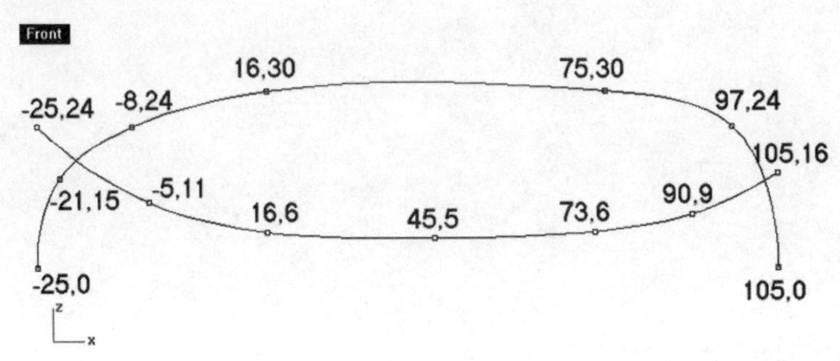

Fig. 9-75. Two curves constructed in the Front viewport.

5 Set the display to a four-viewport configuration.
6 Click on the Front viewport to set it as the current viewport.
7 Select Transform > Orient > Remap to CPlane, or click on the Remap to CPlane button on the Transform toolbar.
8 Select curve A (indicated in figure 9-76) and press the Enter key.
9 Click on the Top viewport.

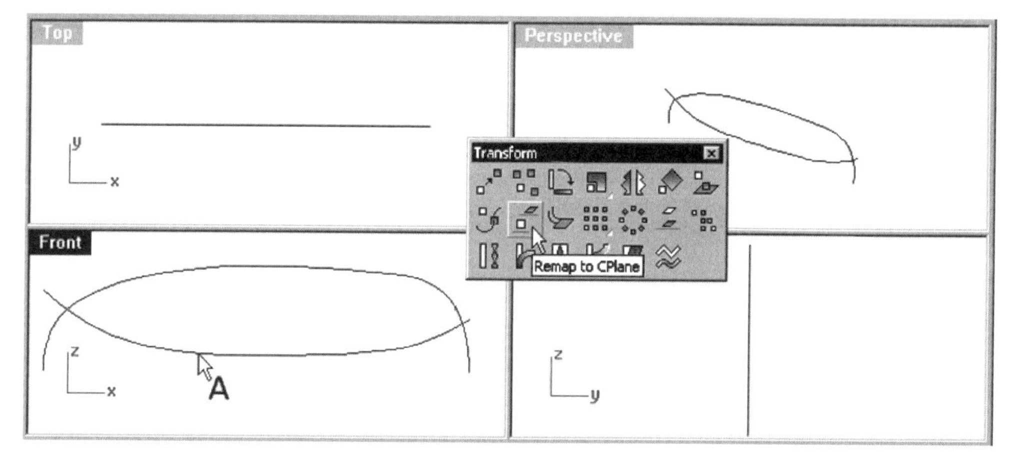

Fig. 9-76. Curve being remapped.

10 Select Curve > Curve From 2 Views, or click on the Curve from 2 Views button on the Curve Tools toolbar.
11 Select curves A and B (indicated in figure 9-77).

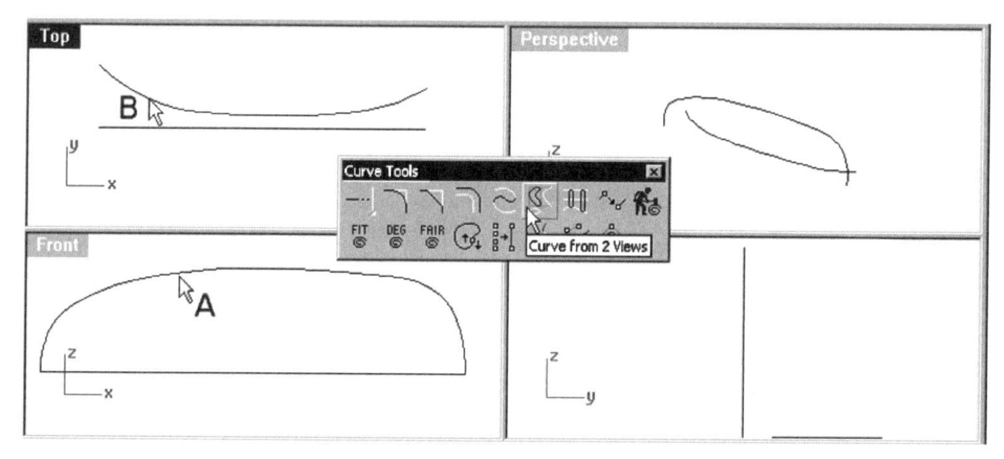

Fig. 9-77. Curve being constructed from two curves.

Depending on how you construct the source curves for making a curve from two curves, you might need to extend the end point of the derived curve. Continue with the following steps to extend the curve constructed from two curves.

12 Maximize the Perspective viewport.

13 Select Curve > Extend Curve > Extend Curve, or click on the Extend Curve button on the Extend toolbar.

14 Select curve A (indicated in figure 9-78) and press the Enter key.

15 If the option T is not equal to Arc, type *T* at the command line interface and then set it to Arc.

16 Select B and C (indicated in figure 9-78) and press the Enter key.

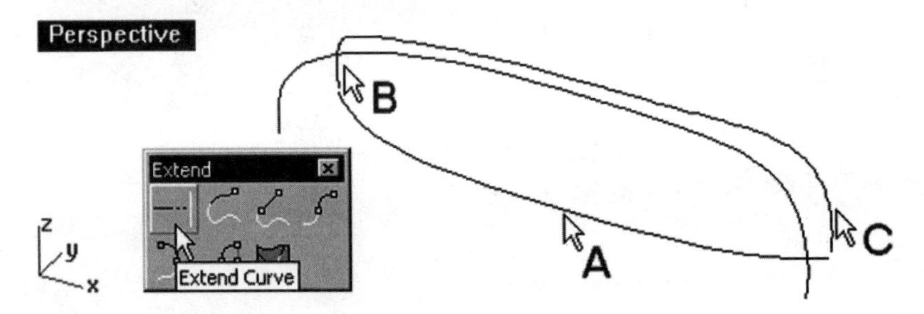

Fig. 9-78. Curve being extended.

Continue with the following steps to construct two profile curves.

17 Maximize the Top viewport.

18 With reference to figure 9-79, construct two free-form curves AB and CD.

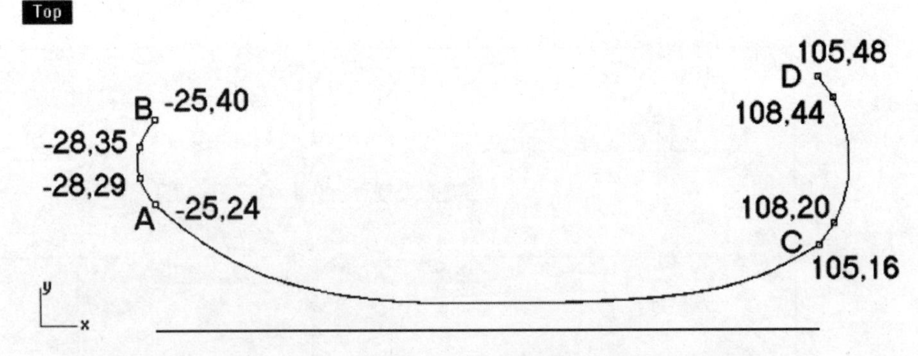

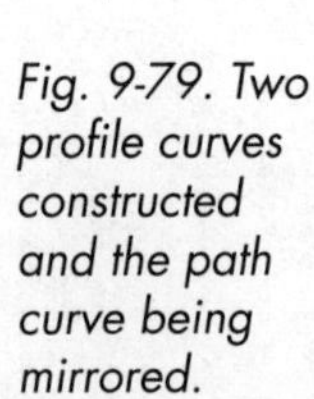

Fig. 9-79. Two profile curves constructed and the path curve being mirrored.

Continue with the following steps to mirror the 3D curve.

19 Maximize the Perspective viewport.

20 Select Transform > Mirror, or click on the Mirror button on the Transform toolbar.

21 Select curve A (indicated in figure 9-80) and press the Enter key.

22 Type *0,32* at the command line interface to specify the first end point of the mirror plane.

23 Type *r1 < 0* at the command line interface to specify the other end point of the mirror plane. Because the length of the plane is unimportant, the distance can be set to 1 (or any value) here.

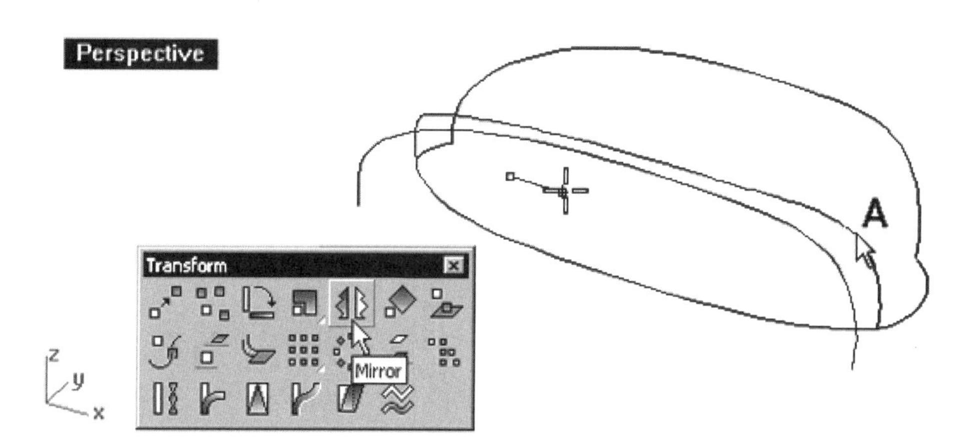

Fig. 9-80. Curve being mirrored.

Continue with the following steps to construct a profile curve in the Right viewport, make a copy of the curve, and orient the curves.

24 Maximize the Right viewport.

25 With reference to figure 9-81, construct a free-form curve.

26 Select Transform > Copy, or click on the Copy button on the Transform toolbar.

27 Select curve A (indicated in figure 9-81) and press the Enter key.

28 Type *i* at the command line interface to use the Inplace option.

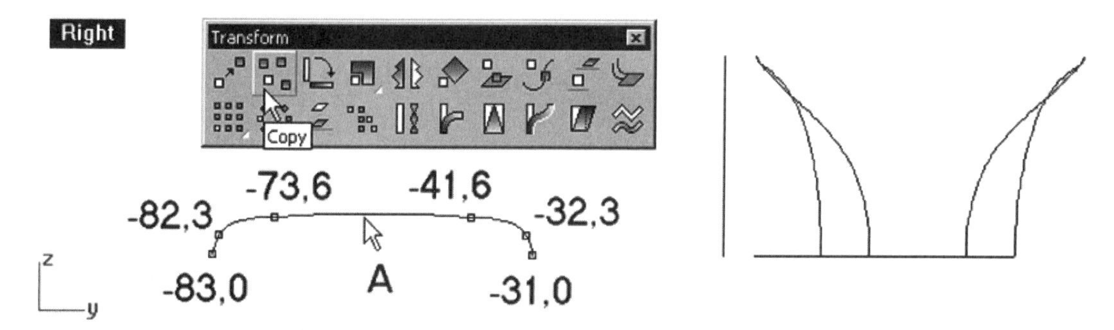

Fig. 9-81. Profile curve constructed and being copied.

29 Maximize the Perspective viewport.

30 Select Curve > Point Object > Divide Curve by > Number of Segments, or right-click on the *Divide Curve by Length/Divide Curve by Number of Segments* button on the Point toolbar.

31 Select curves A and B (indicated in figure 9-82) and press the Enter key.

32 Type *3* at the command line interface.

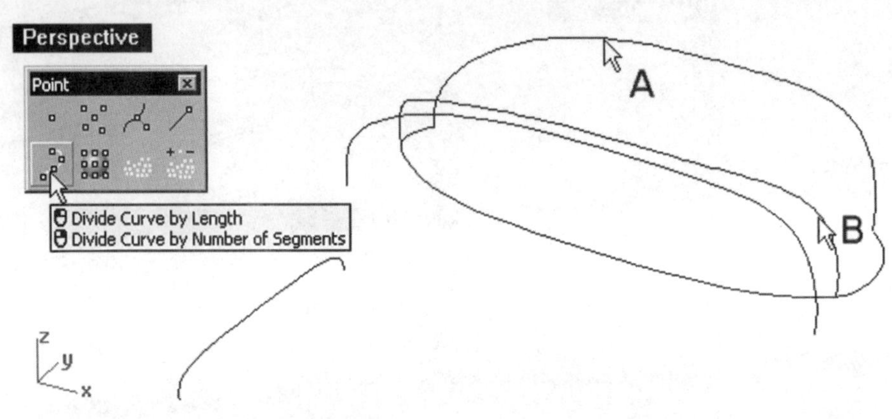

Fig. 9-82. Points being constructed along the path curves.

33 Check the End and Point boxes and clear all other boxes on the Osnap dialog box.

34 Select Transform > Orient > 2 Points, or click on the *Orient: 2 Points/Orient: 3 Points* button on the Transform toolbar.

35 Select curve A (indicated in figure 9-83). Because there are two identical curves at the same location, a pop-up dialog box will appear. Select one of the two curves.

36 Press the Enter key.

37 Select end points B and C and then points D and E (indicated in figure 9-83).

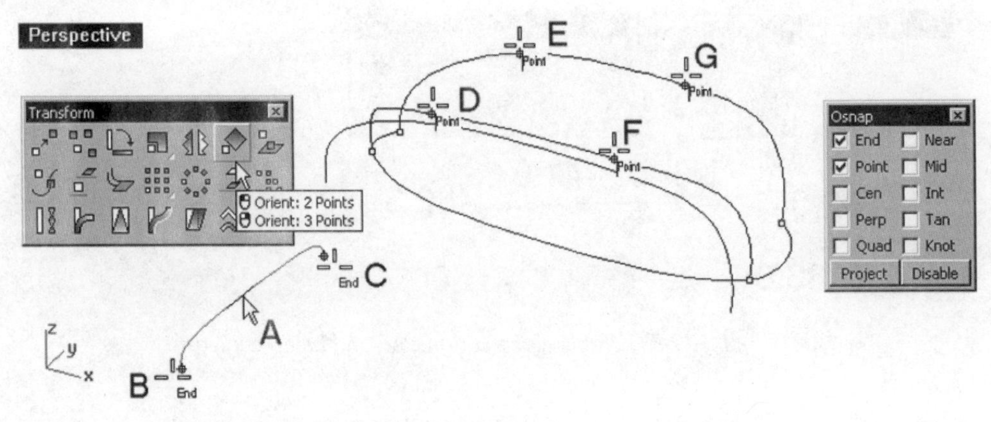

Fig. 9-83. Curves being oriented.

38 Repeat steps 34 through 37 to orient the other curve (located at A) to points F and G.

Continue with the following steps to construct a sweep surface.

39 Set the current layer to *Layer 01*.

40 Select Surface > Sweep 2 Rails, or click on the Sweep 2 Rails button on the Surface toolbar.

41 Select curves A, B, C, D, E, and F (indicated in figure 9-84) and press the Enter key.

42 In the Sweep 2 Rails Options dialog box, click on the OK button.

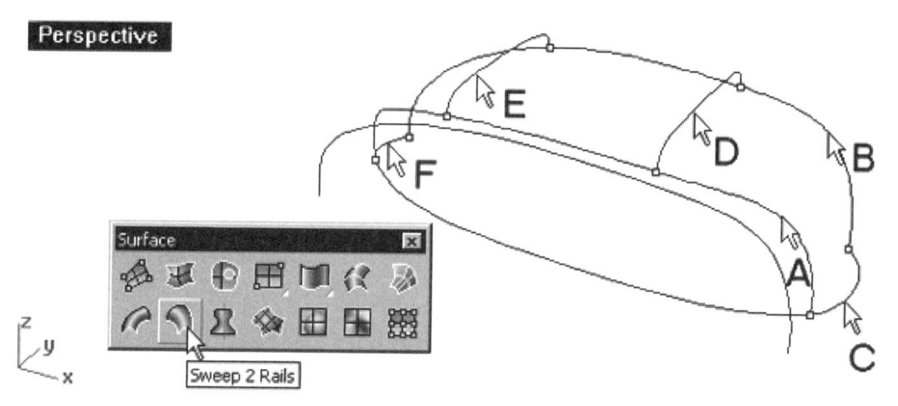

Fig. 9-84. Sweep surface being constructed.

Continue with the following steps to rebuild the surface.

43 Select Edit > Rebuild, or click on the Rebuild Surface button on the Surface Tools toolbar.

44 Select surface A (indicated in figure 9-85) and press the Enter key.

45 In the Rebuild Surface dialog box, set the point counts in the U and V directions to 50 and 16 (respectively), and click on the OK button.

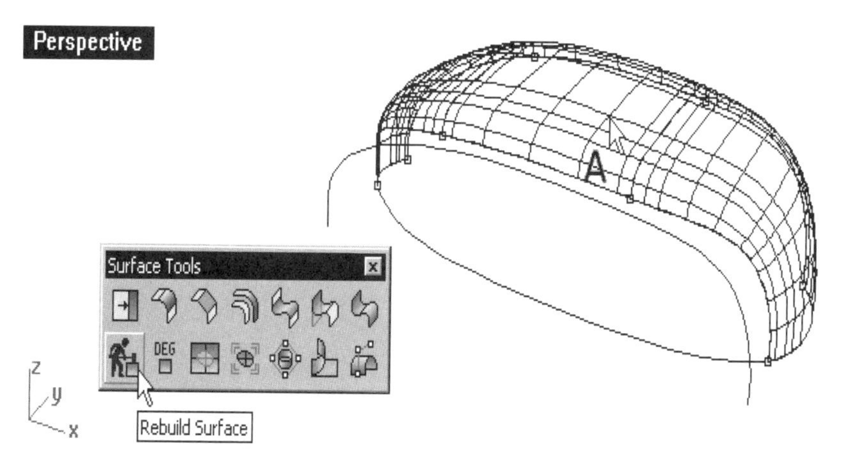

Fig. 9-85. Surface being rebuilt.

Lofted Surface

The curves for the lofted surface were present after you finished the sweep surface. Therefore, to create the lofted surface you need only use the appropriate command, as follows.

1. Select Surface > Loft, or click on the Loft button on the Surface toolbar.
2. Select curves A and B (indicated in figure 9-86) and press the Enter key.
3. In the Loft Options dialog box, click on the OK button.
4. Rebuild the surface with U and V point counts set to 6 and 60, respectively.

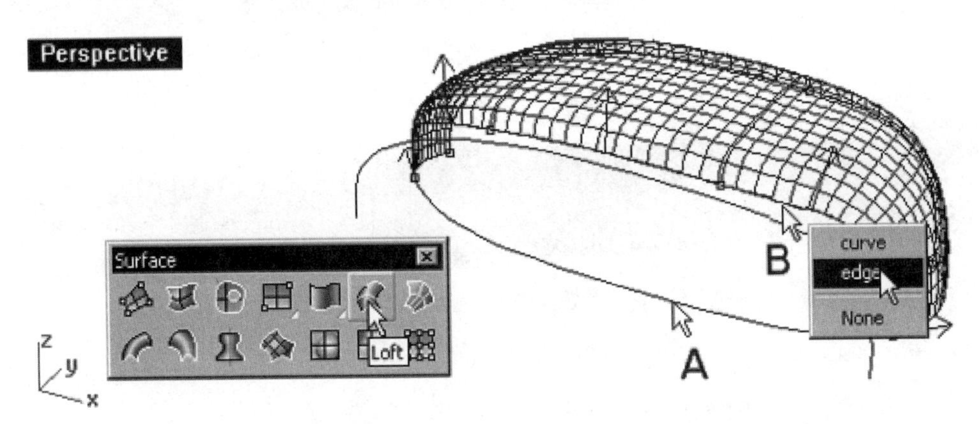

Fig. 9-86. Lofted surface being constructed.

Wheel Covers

The wheel covers for the model are sweep surfaces. Before making the surfaces, you will construct curves in the Front and Top viewports. Perform the following steps to construct the curves in the Front viewport.

1. Set the current layer to *Layer 02* and turn off *Layer 01*.
2. Select Curve > Arc > Center, Start, Angle, or click on the *Arc: Center, Start, Angle* button on the Arc toolbar.
3. Type *0,5* at the command line interface to specify the center of the arc.
4. Type *r14 < 0* at the command line interface to specify the start of the arc.
5. Type *180* at the command line interface to specify the arc angle.

6 Select Curve > Ellipse > Diameter, or click on the *Ellipse: Diameter* button on the Ellipse toolbar.

7 Type *–18,0* at the command line interface to specify the start of the first axis.

8 Type *23,0* at the command line interface to specify the end of the first axis.

9 Type *r18 < 90* at the command line interface to specify the end of the second axis.

An arc and an ellipse are constructed.

10 Select Curve > Extend Curve > Extend Curve, or click on the Extend Curve button on the Extend toolbar.

11 Press the Enter key to use the Dynamic Extend option.

12 If extension type is not Line, type *T* at the command line interface to set the extension type and then type *L* to specify Line.

13 Select end point A (indicated in figure 9-87) of the arc.

14 Type *r5 < 270* to specify the end point of the extension.

15 Select end point B (indicated in figure 9-87) of the arc.

16 Type *r5 < 270* to specify the end point of the extension.

17 Press the Enter key.

18 Select Edit > Trim, or click on the Trim button on the Main1 toolbar.

19 Type *U* at the command line interface if the UseApparentIntersections option is not set to Yes.

20 Select C (indicated in figure 9-87) and press the Enter key.

21 Select D (indicated in figure 9-87) and press the Enter key.

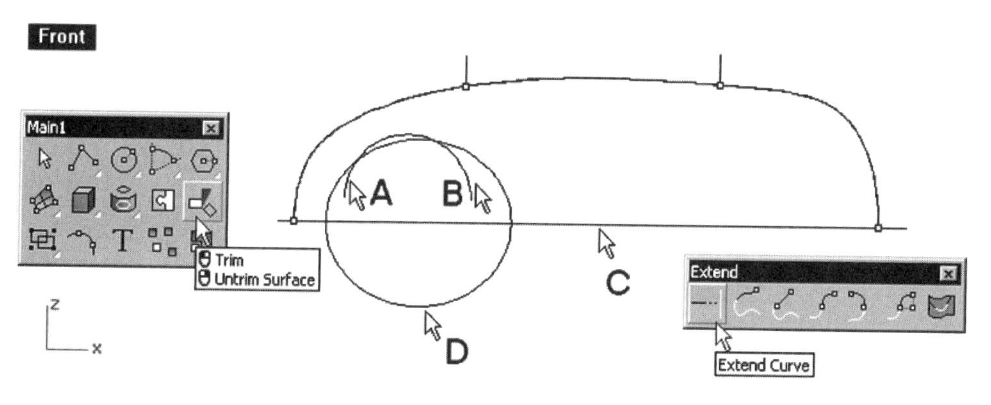

Fig. 9-87. Arc and ellipse constructed and arc being extended.

22 Select Edit > Rebuild, or click on the Rebuild button on the Curve Tools toolbar.
23 Select curves A and B (indicated in figure 9-88) and press the Enter key.
24 In the Rebuild Curve dialog box, set the point count to 12 and degree to 3, and then click on the OK button.
25 Select Transform > Copy, or click on the Copy button on the Transform toolbar.
26 Select curve A (indicated in figure 9-88) and press the Enter key.
27 Select any location on the screen.
28 Type *r87 < 0* at the command line interface.
29 Press the Enter key.

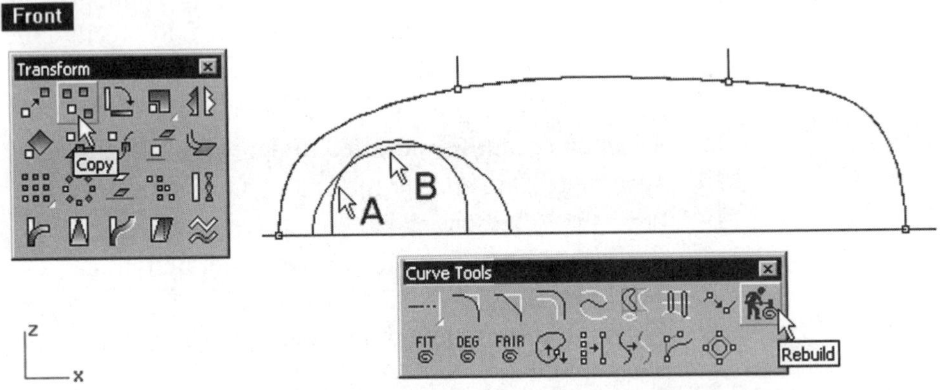

Fig. 9-88. Curves being rebuilt and a curve being copied.

Continue with the following steps to construct curves in the Top viewport and a curve from two views.

30 Turn off the default layer.
31 With reference to figure 9-89, construct five free-form curves: AB, BC, CD, EF, and GH.
32 Select Curve > Curve From 2 Views, or click on the Curve from 2 Views button on the Curve Tools toolbar.

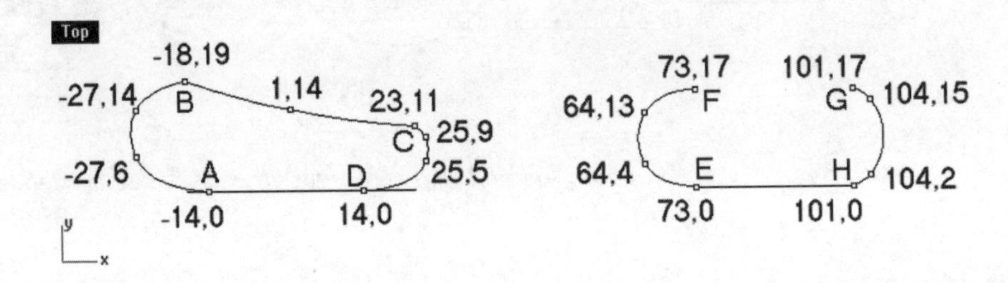

Fig. 9-89. Five free-form curves constructed in the Top viewport.

33 Select curves A and B (indicated in figure 9-90).

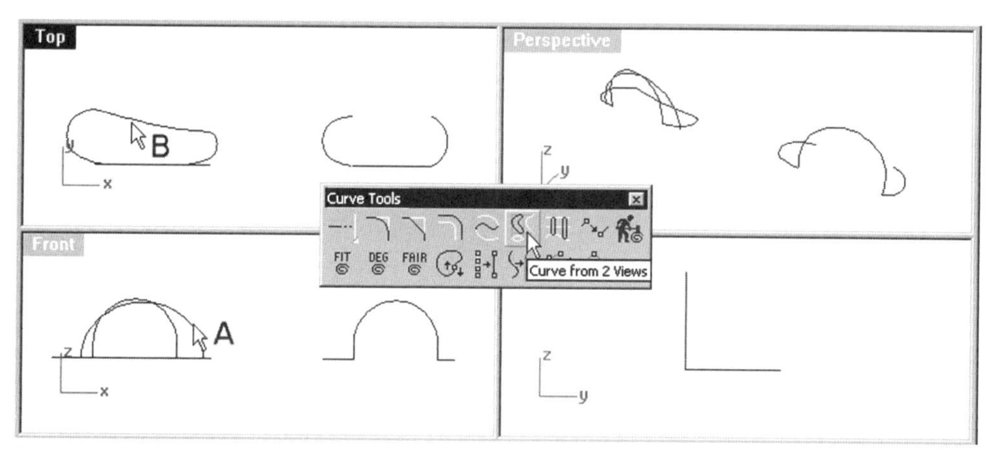

Fig. 9-90. Curve being derived from two curves in two views.

Continue with the following steps to construct two sweep surfaces.

34 Set the current layer to *Layer 03*.

35 Select Surface > Sweep 2 Rails, or click on the Sweep 2 Rails button on the Surface toolbar.

36 Select curves A, B, C, and D (indicated in figure 9-91) and press the Enter key.

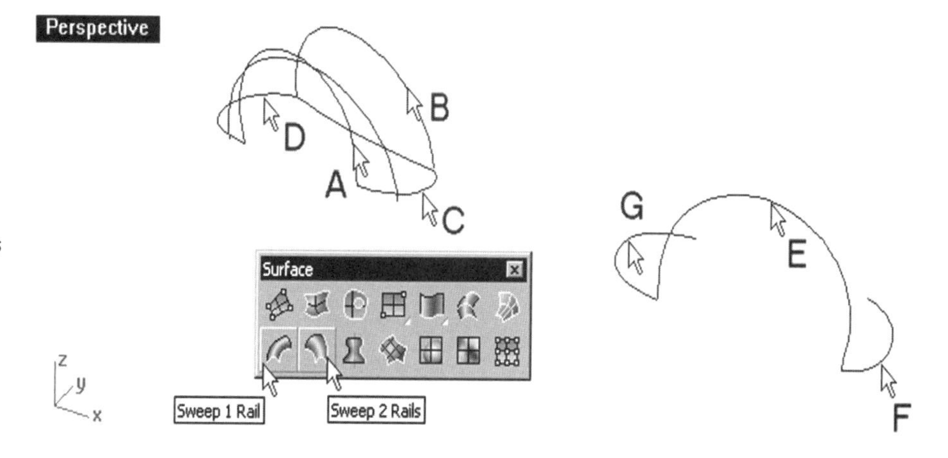

Fig. 9-91. Sweep surfaces being constructed.

37 Click on the OK button in the Sweep 2 Rails Options dialog box.

38 Select Surface > Sweep 1 Rail, or click on the Sweep 1 Rail button on the Surface toolbar.

39 Select curves E, F, and G (indicated in figure 9-91) and press the Enter key.

40 Click on the OK button in the Sweep 1 Rail Options dialog box.

41 Rebuild the surfaces with U and V counts set to 26 and 8, respectively.

42 Turn on *Layer 01* and turn off *Layer 02*, as shown in figure 9-92.

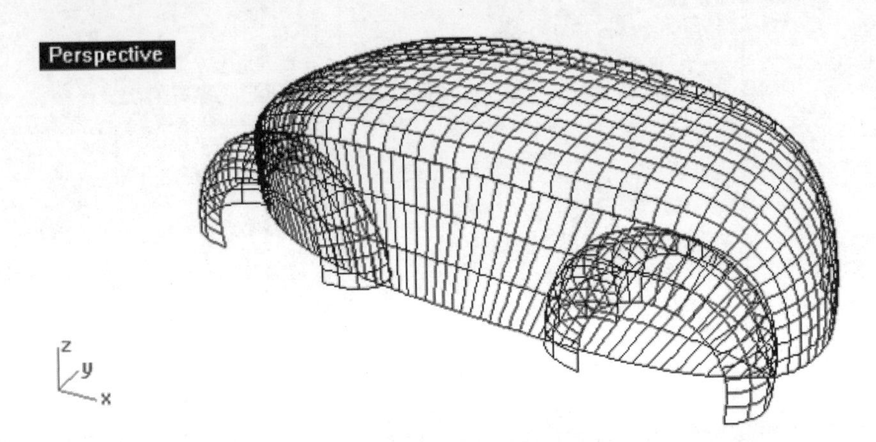

Fig. 9-92. Surfaces rebuilt and Layer01 turned on.

Continue with the following steps to trim and mirror the surfaces.

43 With reference to figure 9-93, trim surfaces A, B, and C.

44 Select Transform > Mirror, or click on the Mirror button on the Transform toolbar.

45 Select surfaces A, B, and C (indicated in figure 9-93) and press the Enter key.

46 Type *0,32* at the command line interface to specify the first end point of the mirror plane.

47 Type *r1 < 0* at the command line interface to specify the other end point of the mirror plane.

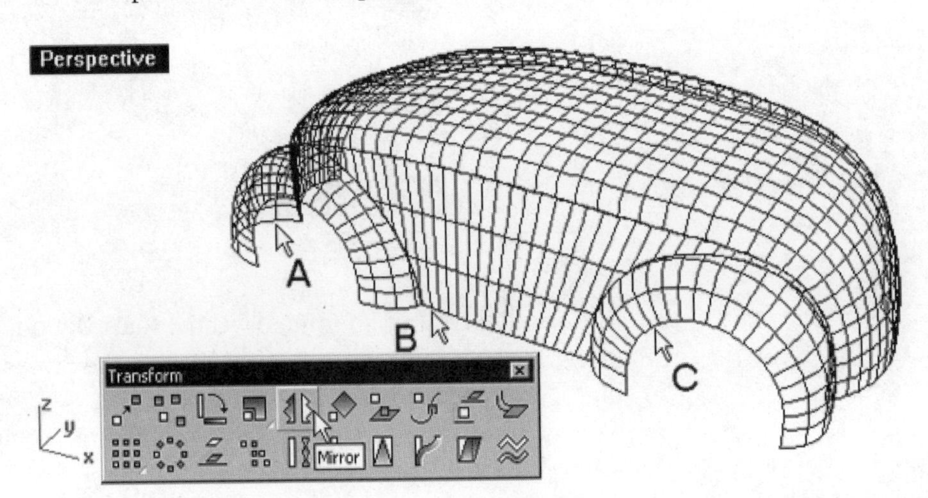

Fig. 9-93. Surface trimmed and being mirrored.

The main body is complete, as shown in figure 9-94.

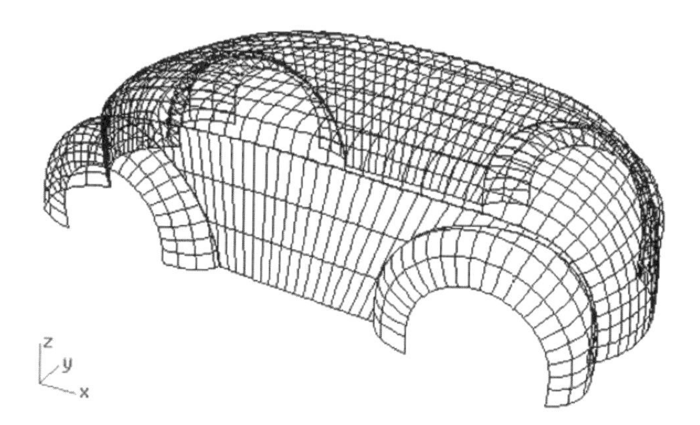

Fig. 9-94. Main body completed.

Upper Part of the Model

The upper part of the model is very similar to the main body. It has a sweep surface and two lofted surfaces. Perform the following steps to construct three curves on a construction plane at an elevation to the Top viewport.

1. Set the current layer to *Layer 04* and turn off *Layer 01*.
2. Maximize the Top viewport.
3. Select View > Set CPlane > Elevation, or click on the Set CPlane Elevation button on the Set CPlane toolbar.
4. Type *28* at the command line interface to specify an elevation of 28 units from the existing construction plane.
5. With reference to figure 9-95, construct three free-form curves.

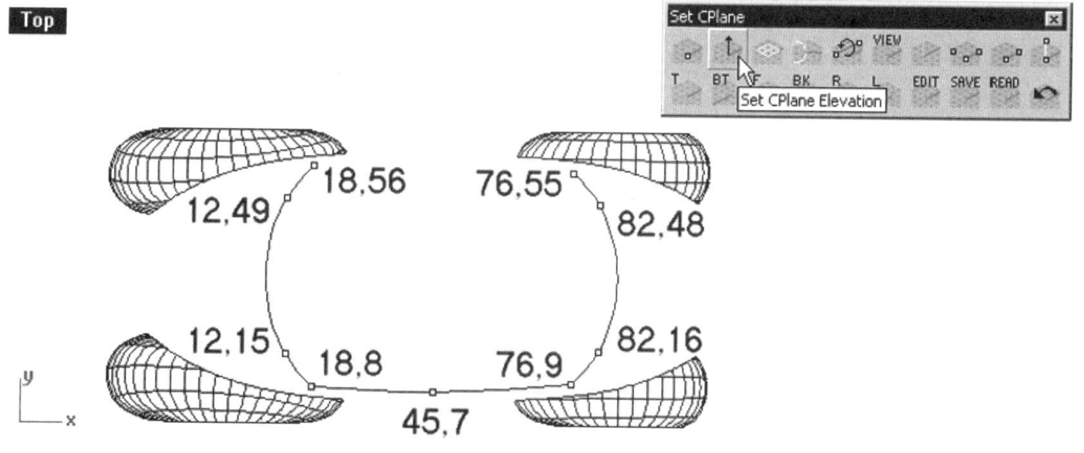

Fig. 9-95. Free-form curves constructed on a construction plane with an elevation.

Continue with the following steps to construct a curve from two views.

6 Maximize the Front viewport.

7 Construct a free-form curve in accordance with figure 9-96.

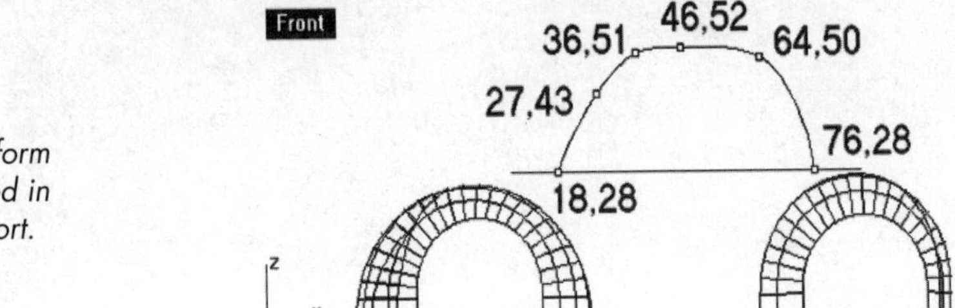

Fig. 9-96. Free-form curve constructed in the Front viewport.

8 Set the display to a four-viewport configuration.

9 Select Curve > Curve From 2 Views, or click on the Curve from 2 Views button on the Curve Tools toolbar.

10 Select curves A and B (indicated in figure 9-97).

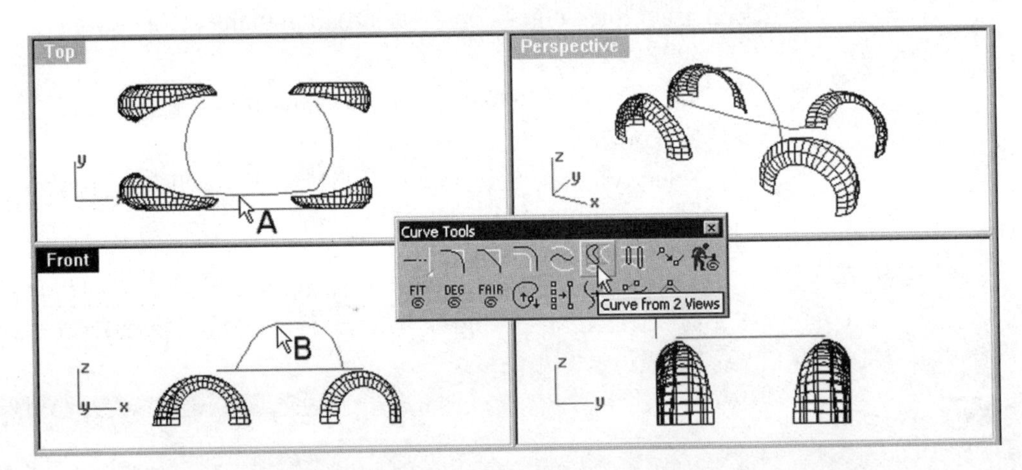

Fig. 9-97. Curve from two views being constructed.

11 Turn off *Layer 03*.

12 Maximize the Perspective viewport.

13 Select Transform > Mirror, or click on the Mirror button on the Transform toolbar.

14 Select curve A (indicated in figure 9-98) and press the Enter key.

15 Type *0,32* at the command line interface to specify the first end point of the mirror plane.

16 Type *r1 < 0* at the command line interface to specify the other end point of the mirror plane.

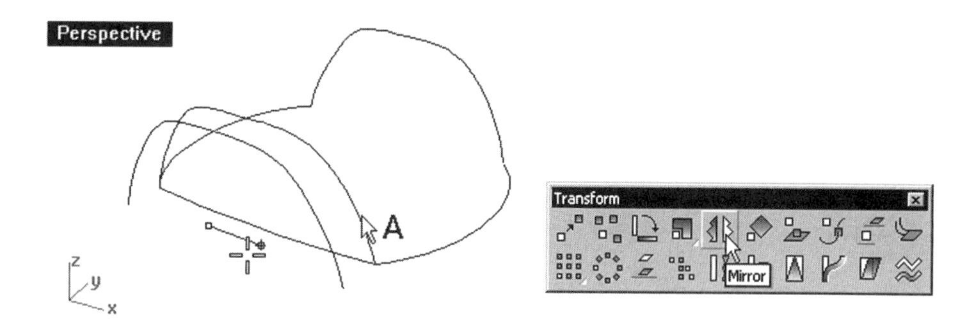

Fig. 9-98. Curve being mirrored.

Continue with the following steps to construct the surfaces.

17 Set the current layer to *Layer 05*.

18 Select Surface > Sweep 2 Rails, or click on the Sweep 2 Rails button on the Surface toolbar.

19 Select curves A, B, C, and D (indicated in figure 9-99) and press the Enter key.

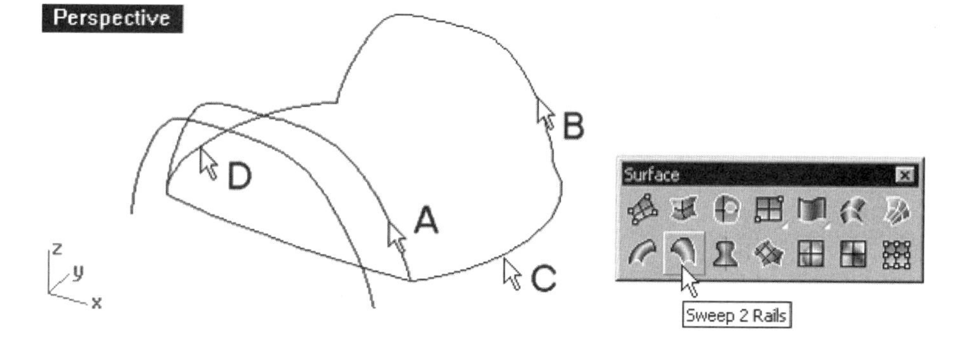

Fig. 9-99. Sweep surface being constructed.

20 In the Sweep 2 Rails Options dialog box, click on the OK button.

21 Rebuild the sweep surface with point counts of 32 and 8.

22 Select Surface > Loft, or click on the Loft button on the Surface toolbar.

23 Select edge A and curve B (indicated in figure 9-100) and press the Enter key.

24 In the Loft Options dialog box, click on the OK button.

25 Rebuild the loft surface with point counts of 8 and 34.

26 Turn off *Layer 04*.

27 Select Transform > Mirror, or click on the Mirror button on the Transform toolbar.

28 Select surface A (indicated in figure 9-101) and press the Enter key.

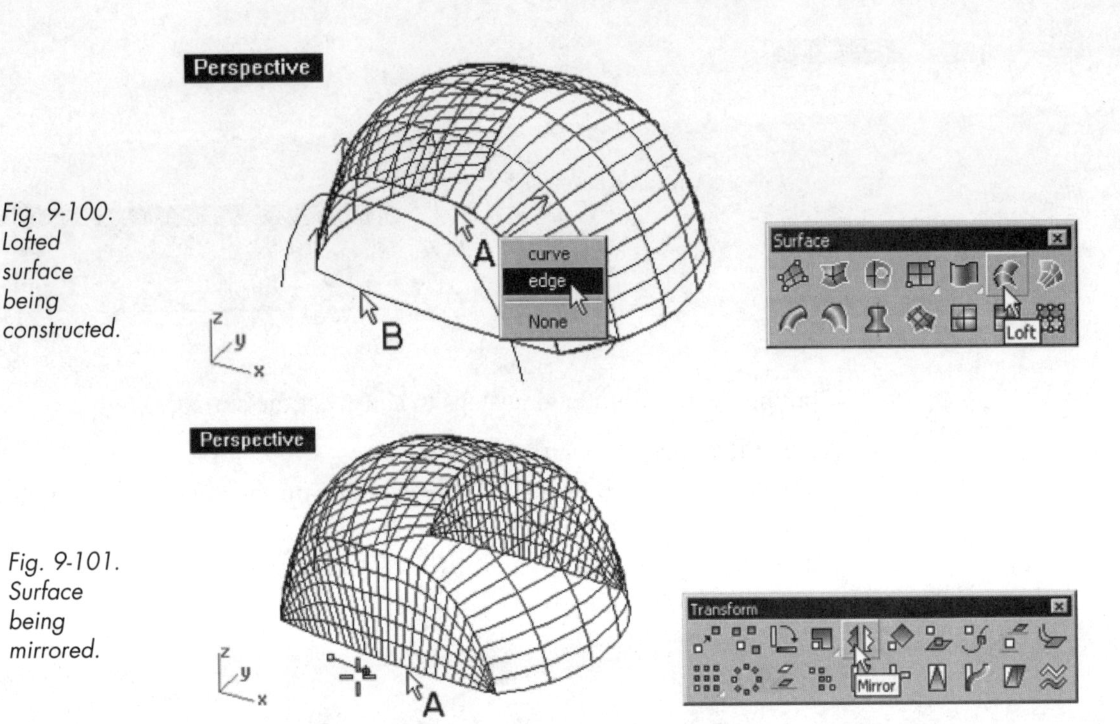

Fig. 9-100. Lofted surface being constructed.

Fig. 9-101. Surface being mirrored.

29 Type *0,32* at the command line interface to specify the first end point of the mirror plane.

30 Type *r1 < 0* at the command line interface to specify the other end point of the mirror plane.

31 Turn on *Layer 01*.

32 With reference to figure 9-102, hide the two surfaces.

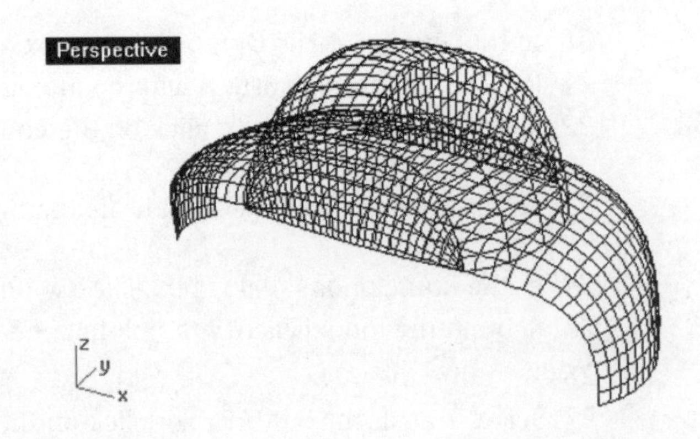

Fig. 9-102. Layer 01 *turned on and surfaces hidden.*

33 With reference to figure 9-103, trim the surfaces.

34 Unhide the hidden surfaces, as shown in figure 9-104.

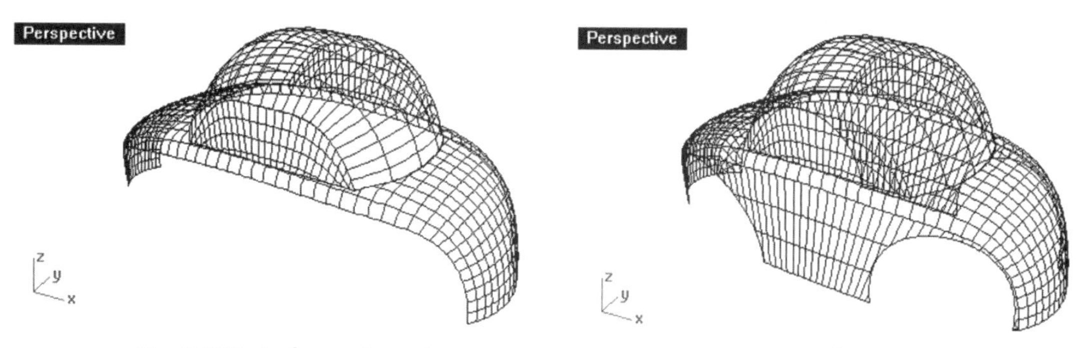

Fig. 9-103. Surfaces trimmed.

Fig. 9-104. Surfaces unhidden.

Completion

In the following you will add a few more surfaces to complete the surface model. Perform the following steps to construct two extruded surfaces.

1. Set the current layer to *Layer 03* and turn off *Layer 01* and *Layer 05*.
2. Select Extrude > Straight, or click on the Extrude Straight button on the Extrude toolbar.
3. Select edges A and B (indicated in figure 9-105) and press the Enter key.
4. Type *r10 < 270* at the command line interface to specify the extrusion distance.

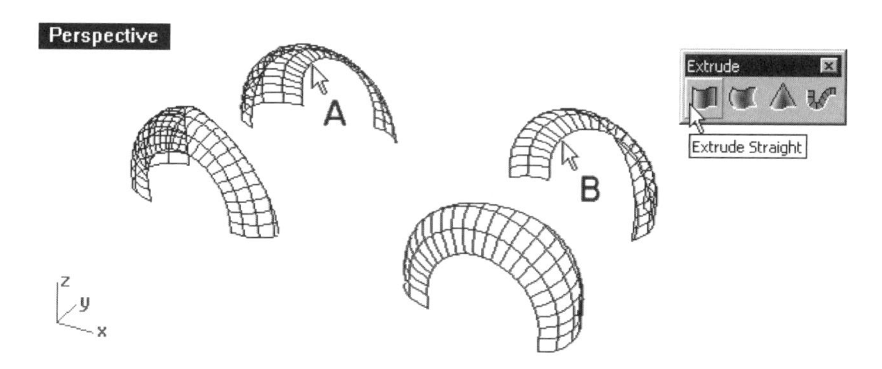

Fig. 9-105. Extruded surfaces being constructed.

5. Set the current layer to *Layer 06*.
6. Select Curve > Line > Single Line, or click on the Line button on the Line toolbar.
7. Select end points A and B (indicated in figure 9-106) and press the Enter key.

8 Repeat steps 6 and 7 to construct another line segment CD (indicated in figure 9-106).

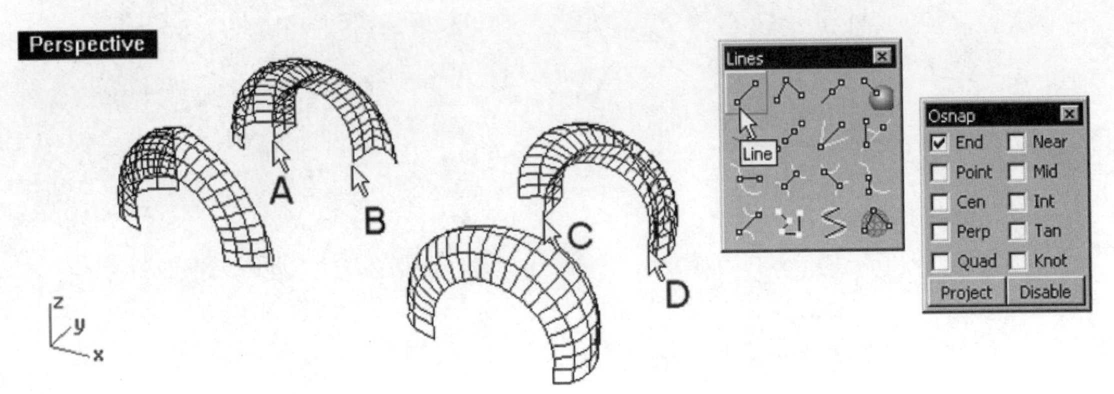

Fig. 9-106. Line segments constructed.

9 Set the current layer to *Layer 03*.

10 Select Surface > Planar Curves, or click on the Surface from Planar Curves button on the Surface toolbar.

11 Select edge A and curve B (indicated in figure 9-107) and press the Enter key.

12 Repeat steps 10 and 11 to construct another surface from edge C and curve D (indicated in figure 9-107).

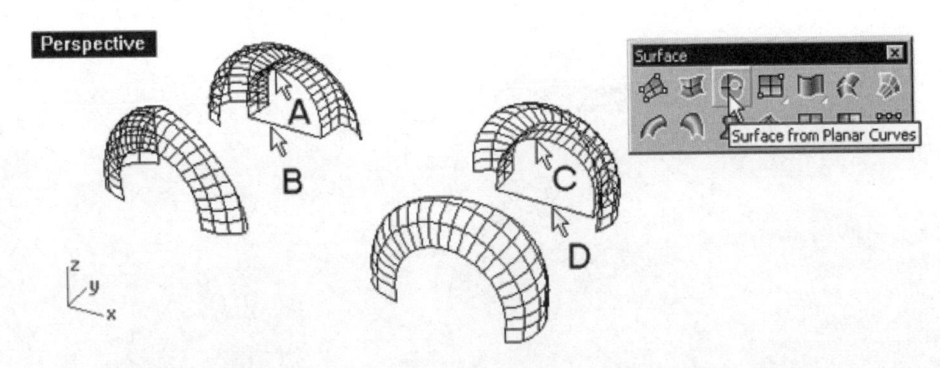

Fig. 9-107. Surfaces from planar curves being constructed.

13 Turn off *Layer 06*.

14 Select Transform > Mirror, or click on the Mirror button on the Transform toolbar.

15 Select surfaces A, B, C, and D (indicated in figure 9-108) and press the Enter key.

16 Type *0,32* at the command line interface to specify the first end point of the mirror plane.

17 Type *r1 < 0* at the command line interface to specify the other end point of the mirror plane.

18 Turn on *Layer 01* and *Layer 05*, as shown in figure 9-109.

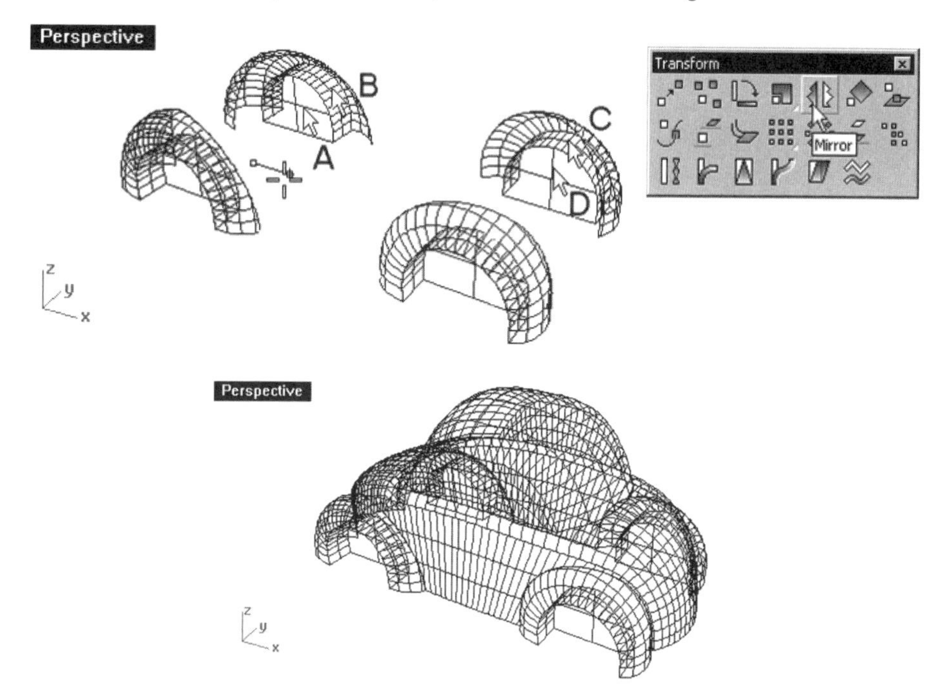

Fig. 9-108. Surfaces being mirrored.

Fig. 9-109. Surface model completed.

Solid Model

The surface model is complete. In the following you will cap the planar hole of the surface model to convert it into a solid model and add two cylindrical holes. Perform the following steps to join the surfaces into a single polysurface and cap the planar hole.

1 Slect Edit > Join, or click on the Join button on the Main1 toolbar.

2 Select contiguous surfaces one by one until all surfaces are selected.

3 Press the Enter key.

4 Select Solid > Cap Planar Holes, or click on the Cap Planar Holes button on the Solid Tools toolbar.

5 Select the surface model and press the Enter key.

A solid is constructed, as shown in figure 9-110. Continue with the following steps to add two holes.

6 Set the display to a four-viewport configuration.

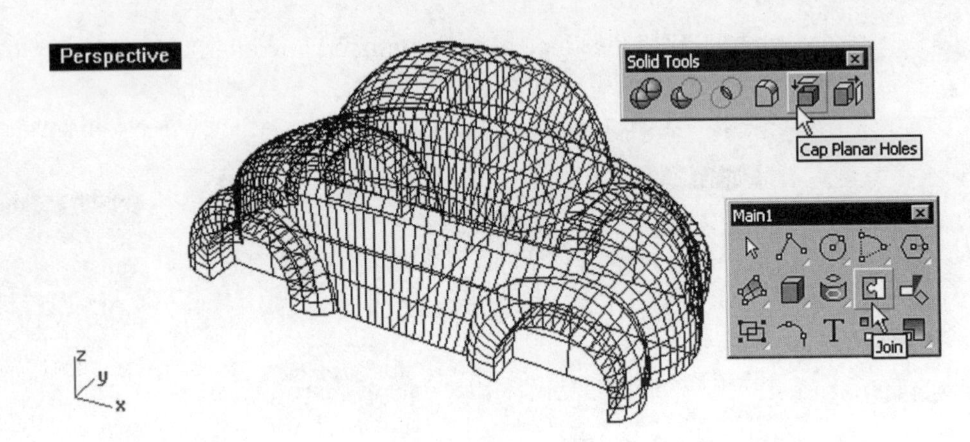

Fig. 9-110. Solid constructed.

7 Select Solid > Cylinder, or click on the Cylinder button on the Solid toolbar.
8 Click on the Front viewport.
9 Type *0,5* at the command line interface to specify the center location.
10 Type *1.5* at the command line interface to specify the radius of the cylinder.
11 Click on the Ortho button on the status bar.
12 Click on the Top viewport.
13 Type *r100 < 90* at the command line interface to specify the length of the cylinder, as shown in figure 9-111.

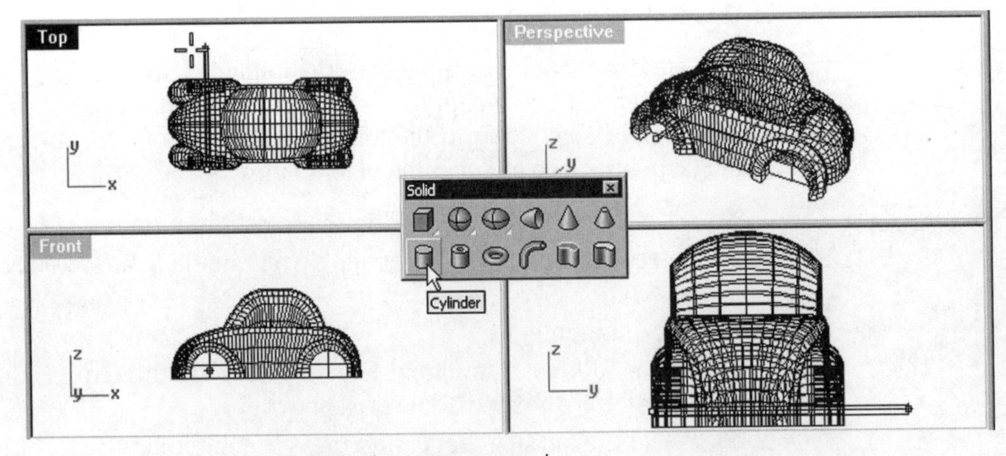

Fig. 9-111. Cylinder being constructed.

14 Maximize the Top viewport.
15 Select Transform > Copy, or click on the Copy button on the Transform toolbar.

16 Select cylinder A (indicated in figure 9-112) and press the Enter key.

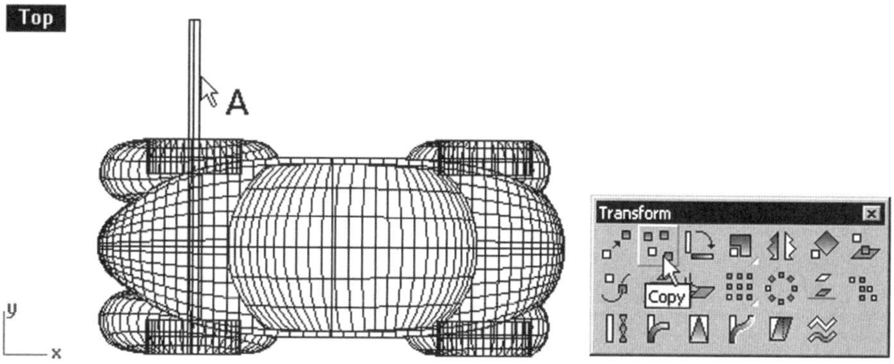

Fig. 9-112. Cylinder being copied.

17 Select any point on the screen.

18 Type *r87 < 0* at the command line interface to specify the copied distance.

19 Press the Enter key.

20 Select Solid > Difference, or click on the Boolean Difference button on the Solid Tools toolbar.

21 Select A (indicated in figure 9-113) and press the Enter key.

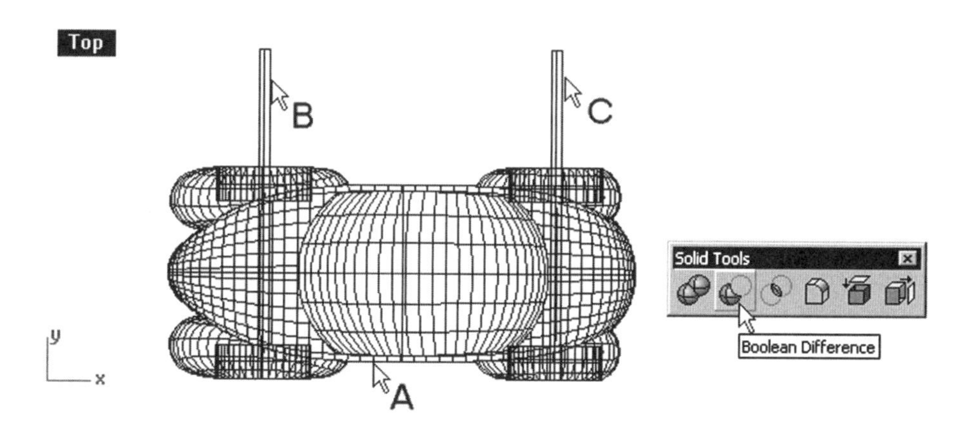

Fig. 9-113. Boolean difference operation being carried out.

22 Select B and C (indicated in figure 9-113) and press the Enter key.

23 Maximize the Perspective viewport.

Material Properties and Rendering

The solid model is complete. In the following you will add material properties to the solid and then produce a rendered image.

1. Apply material properties to the model.
2. Render the scene.
3. Select File > Export With Origin.
4. Select the solid model and press the Enter key.
5. Press the Enter key again to use the World origin.
6. In the Export dialog box, specify the file name *ToyCarBody* and then click on the Save button.
7. Save your file as *ToyCarBodyBak.3dm*.

The model is complete.

Toy Car Project

The toy car includes the component parts tire, wheel, and shaft. You will construct them one by one in individual files and then put them together with the body in a separate file. Figures 9-114 and 9-115 show, respectively, the rendered image and the surface model of the completed toy car.

Fig. 9-114. Rendered image of the completed toy car.

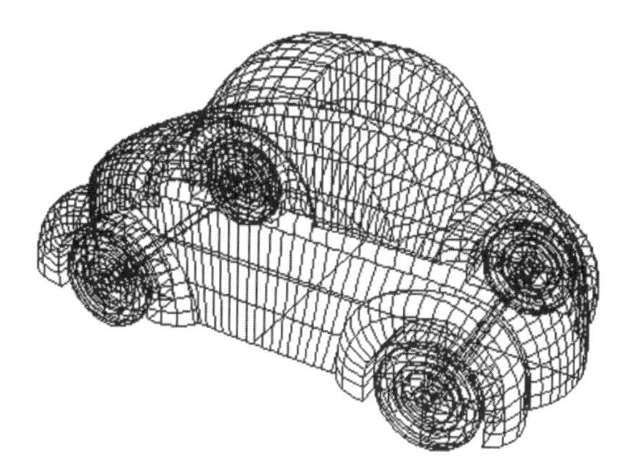

Fig. 9-115. Completed toy car model.

Shaft

Fig. 9-116. Shaft.

The shaft of the toy car is a simple cylinder. Figure 9-116 shows the rendered image of the completed shaft model.

In the following you will construct the solid model of the shaft of the car. It is a cylinder.

1. Start a new file. Use the metric (Millimeters) template file.
2. Select Solid > Cylinder, or click on the Cylinder button on the Solid toolbar.
3. Type *0,5* at the command line interface to specify the center location.
4. Type *1.5* at the command line interface to specify the radius of the cylinder.
5. Click on the Ortho button on the status bar.
6. Type *r64 < 90* at the command line interface to specify the length of the cylinder.
7. Apply material properties to the model and save your file as *Shaft.3dm*.

The model is complete, as shown in figure 9-117.

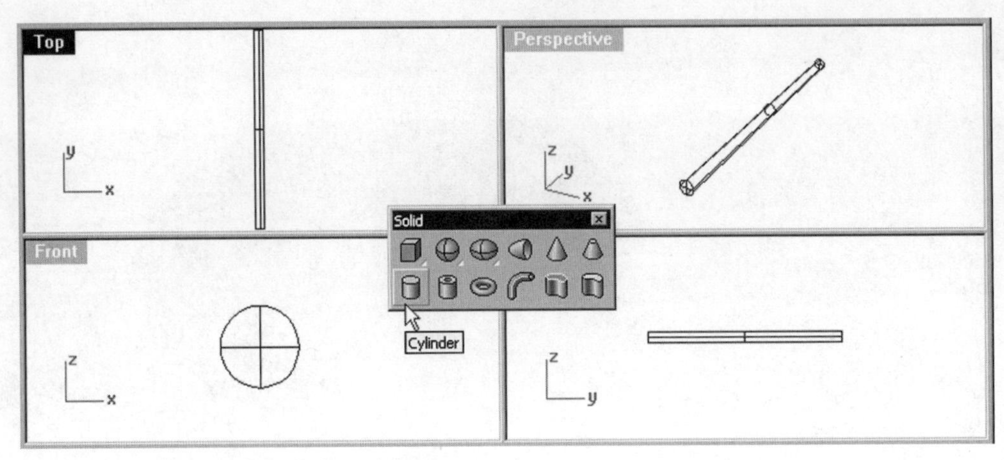

Fig. 9-117. Shaft model constructed.

Tire

Figure 9-118 shows the rendered image of a tire. The model can be constructed by making a tube and filleting the tube's edges. To construct the solid model of a tire, perform the following steps. You will construct a solid tube and fillet its edges.

Fig. 9-118. Rendered image of a tire.

1. Start a new file. Use the metric (Millimeters) template file.
2. Select Solid > Tube, or click on the Tube button on the Solid Tools toolbar.
3. Click on the Front viewport.
4. Type *v* at the command line interface to construct a vertical tube.
5. Type *0,5* at the command line interface to specify the center.
6. Type *12* at the command line interface to specify the first radius.
7. Type *8* at the command line interface to specify the second radius.
8. Click on the Top viewport.
9. Type *6* at the command line interface to specify the length of the tube, as shown in figure 9-119.
10. Maximize the Perspective viewport.
11. Select Solid > Fillet Edge, or click on the Fillet Edge button on the Solid Tools toolbar.
12. Type *r* at the command line interface to use the Radius option.

13 Type *2* at the command line interface to specify the radius.

14 Select edges A and B (indicated in figure 9-120) and press the Enter key.

15 Apply material properties and save your file as *Tire.3dm*.

The model is complete, as shown in figure 9-121.

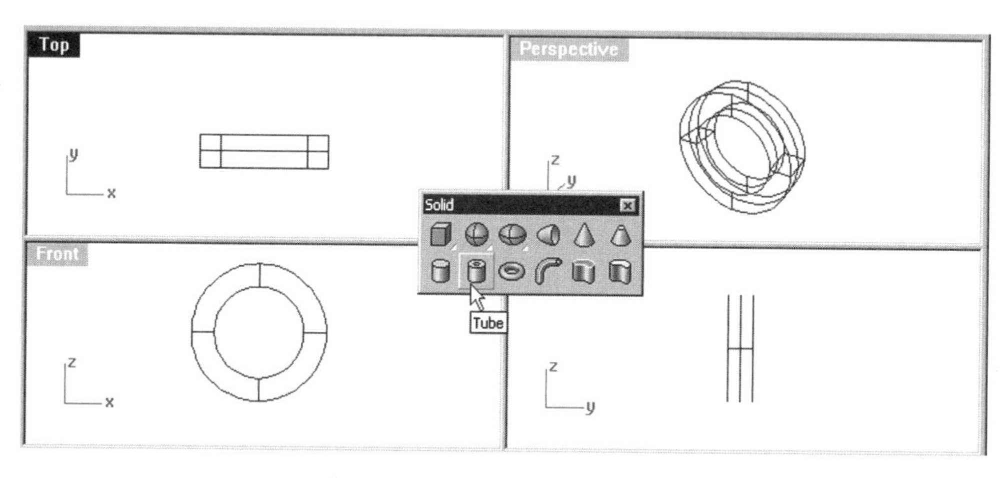

Fig. 9-119. Tube being constructed.

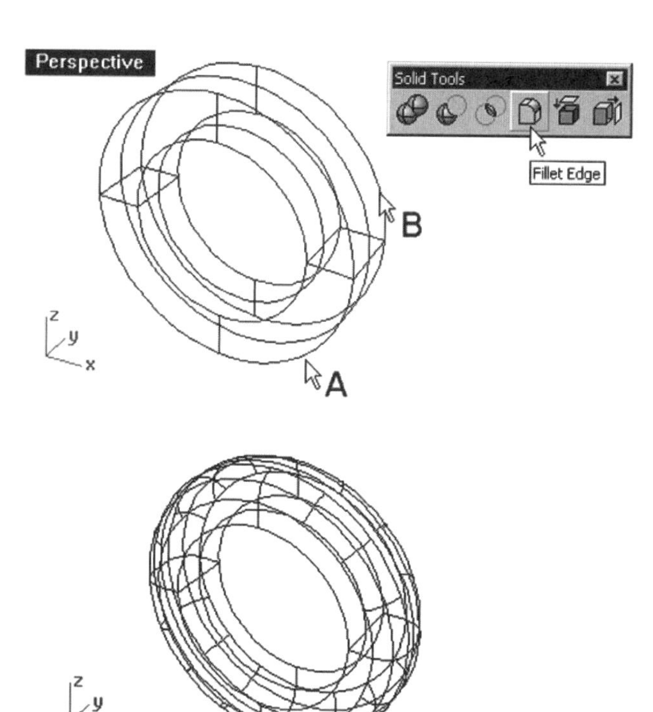

Fig. 9-120. Edge being filleted.

Fig. 9-121. Completed tire model.

Wheel

Fig. 9-122. Rendered image of a wheel.

Figure 9-122 shows the rendered image of a wheel. This model is the union of three tubes. Perform the following steps to construct two tubes.

1. Start a new file. Use the metric (Millimeters) template file.
2. Select Solid > Tube, or click on the Tube button on the Solid Tools toolbar.
3. Click on the Front viewport.
4. Type *v* at the command line interface to construct a vertical tube.
5. Type *0,5* at the command line interface to specify the center.
6. Type *1.5* at the command line interface to specify the first radius.
7. Type *3* at the command line interface to specify the second radius.
8. Click on the Top viewport.
9. Type *10* at the command line interface to specify the length of the tube, as shown in figure 9-123.
10. Repeat steps 2 through 9 to construct another tube of radii 6 units and 8 units, length 10 units, and with its center at 0,5 in the Front viewport.
11. Repeat steps 2 through 9 to construct one more tube of radii 1.5 units and 8 units, length 2 units, and with its center at 0,5. (See figure 9-123.)

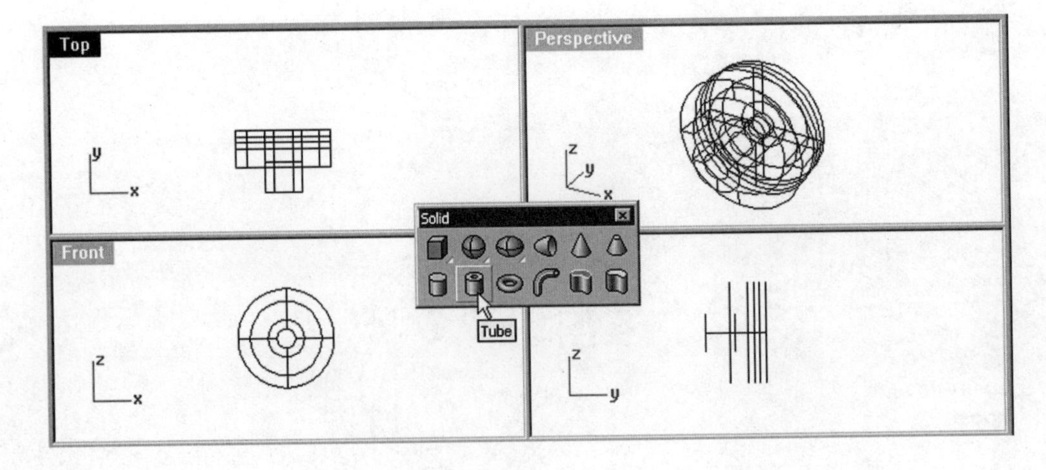

Fig. 9-123. Tubes constructed.

Continue with the following steps to unite the tubes into a single solid.

12 Select Solid > Union, or click on the Boolean Union button on the Solid Tools toolbar.

13 Select the tubes and press the Enter key. The result is shown in figure 9-124.

14 Apply material properties and save your file as *Wheel.3dm*.

The model is complete.

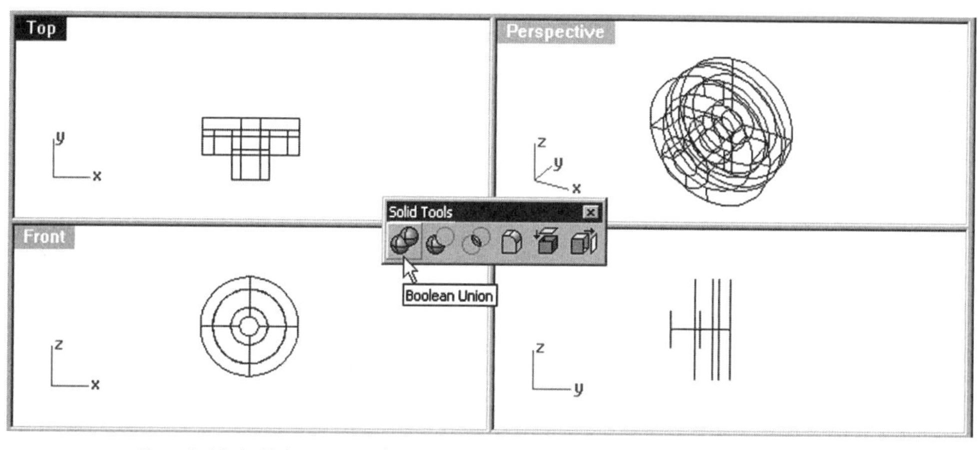

Fig. 9-124. Tubes united.

Toy Car Assembly

In the following you will construct an assembly of the toy car by merging the shaft, tire, wheel, and car body into a single file.

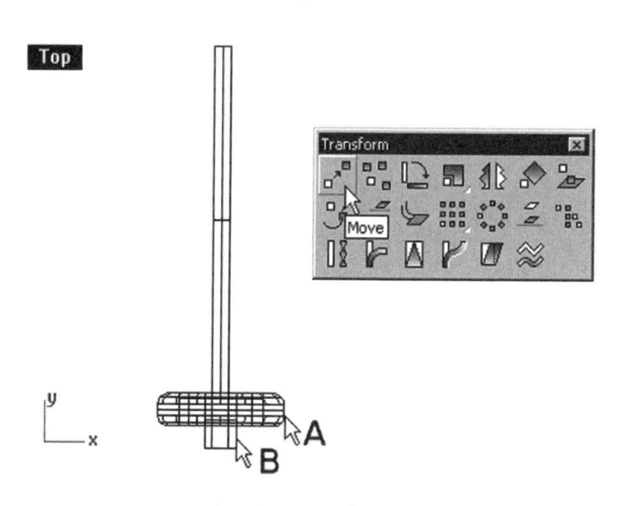

Fig. 9-125. Files imported.

1 Start a new file. Use the metric (Millimeters) template file.

2 Select File > Import.

3 In the Import dialog box, select the file *Shaft.3dm*.

4 Repeat steps 2 and 3 to import the files *Tire.3dm* and *Wheel.3dm*.

5 Maximize the Top viewport.

6 Select Transform > Move, or click on the Move button on the Transform toolbar.

7 Select A and B (indicated in figure 9-125) and press the Enter key.

8 Select any point on the viewport.

9 Type *r64 < 90* at the command line interface.

Continue with the following steps to mirror the tire and wheel and copy the shaft, tire, and wheel.

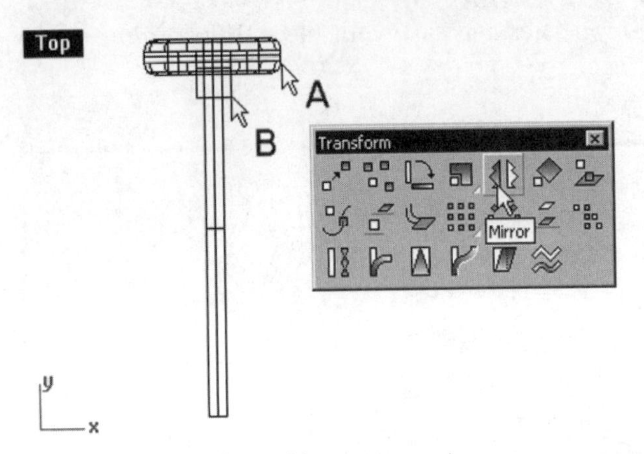

Fig. 9-126. Tire and wheel being mirrored.

10 Select Transform > Mirror, or click on the Mirror button on the Transform toolbar.

11 Select A and B (indicated in figure 9-126) and press the Enter key.

12 Type *0,32* at the command line interface to specify the first end point of the mirror plane.

13 Type *r1 < 0* at the command line interface to specify the other end point of the mirror plane.

14 Select Transform > Copy, or click on the Copy button on the Transform toolbar.

15 Select the tires, wheels, and shaft (shown in figure 9-127) and press the Enter key.

16 Select any point on the screen.

17 Type *r87 < 0* at the command line interface.

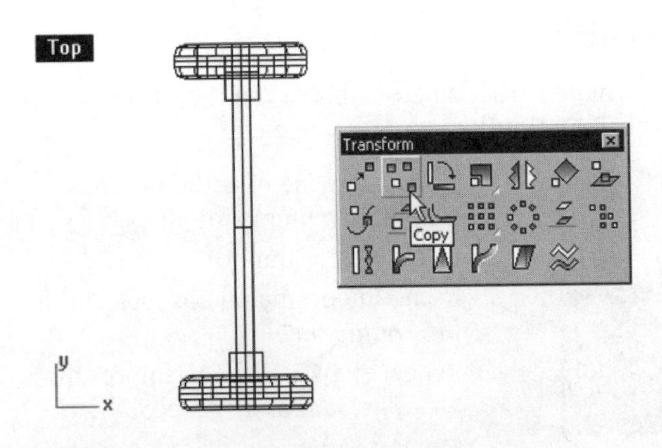

Fig. 9-127. Tires, wheels, and shaft being copied.

Continue with the following steps to complete the toy car by importing the car body.

18 Set the display to a four-viewport configuration.

19 Import the file *ToyCar-Body.3dm*.

The model is complete, as shown in figure 9-128.

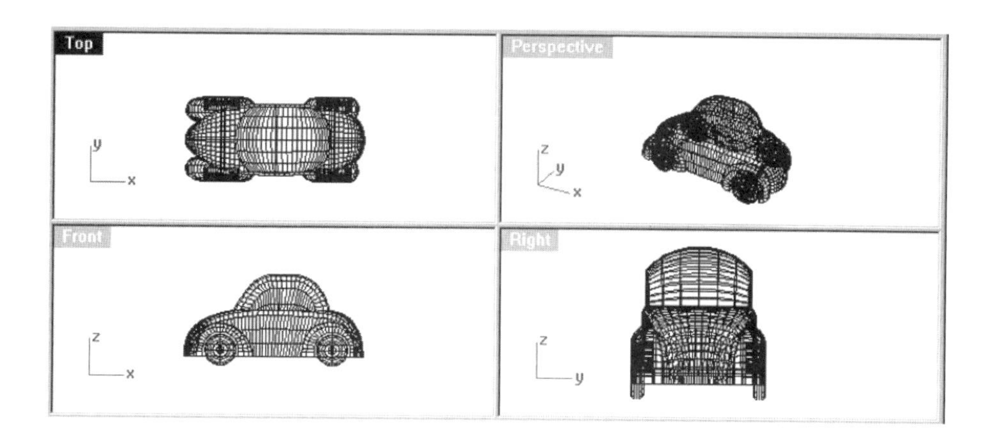

Fig. 9-128. Car body imported.

Shoe Project

Figures 9-129 and 9-130 show, respectively, the rendered image and the surface model of a shoe.

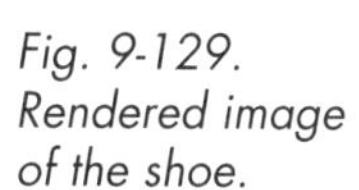

Fig. 9-129. Rendered image of the shoe.

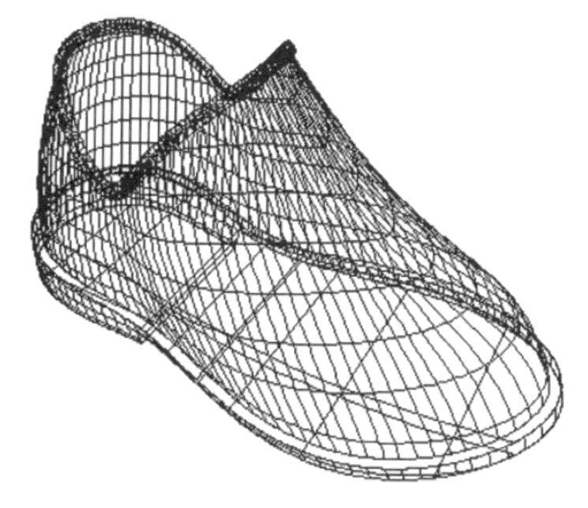

Fig. 9-130. Surface model of the shoe.

Main Body

The main body of the shoe is a sweep surface constructed from two path curves and a number of cross-section curves. To examine the relationships among the curves, you will construct the curves depicting the silhouettes of the top and front view of the shoe in the Front viewport and then remap some of the curves to the Top viewport. Perform the following steps to construct a number of curves in the Front viewport.

1. Start a new file. Use the metric (Millimeters) template file.
2. Set the grid extents to 200 mm.
3. Maximize the Front viewport.
4. With reference to figure 9-131, construct a closed free-form curve A and a line B.

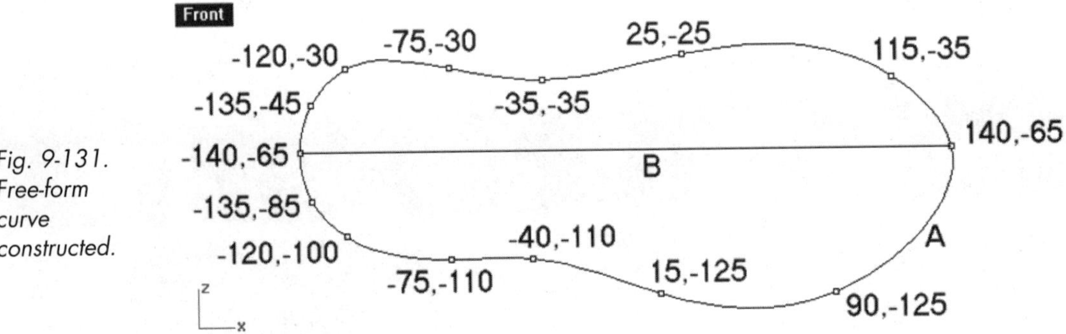

Fig. 9-131. Free-form curve constructed.

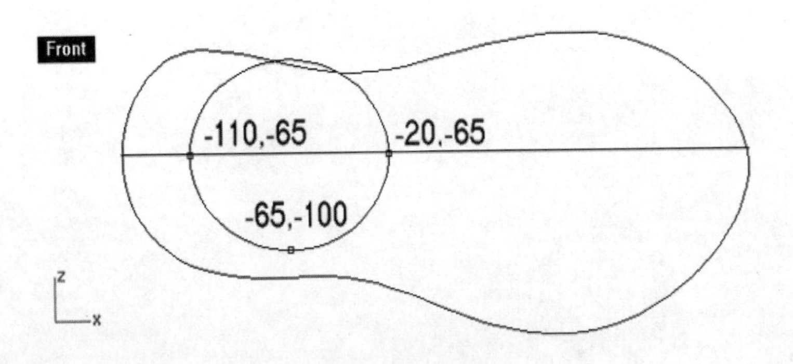

Fig. 9-132. Ellipse constructed.

5. Construct an ellipse in accordance with figure 9-132.
6. With reference to figure 9-133, in the Front viewport construct three free-form curves A, B, and C, and a line segment D.

Continue with the following steps to remap three curves and a line segment to the Top viewport.

7. Set the display to a four-viewport configuration.
8. Click on the Front viewport.

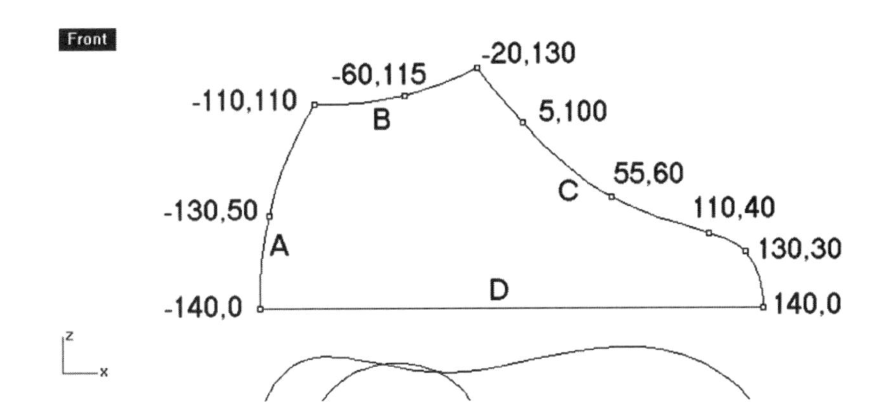

Fig. 9-133. Three free-form curves and a line segment.

9 Select Transform > Orient > Remap to CPlane, or click on the Remap to CPlane button on the Transform toolbar.

10 Select curves A, B, and C (indicated in figure 9-134) and press the Enter key.

11 Select the Top viewport.

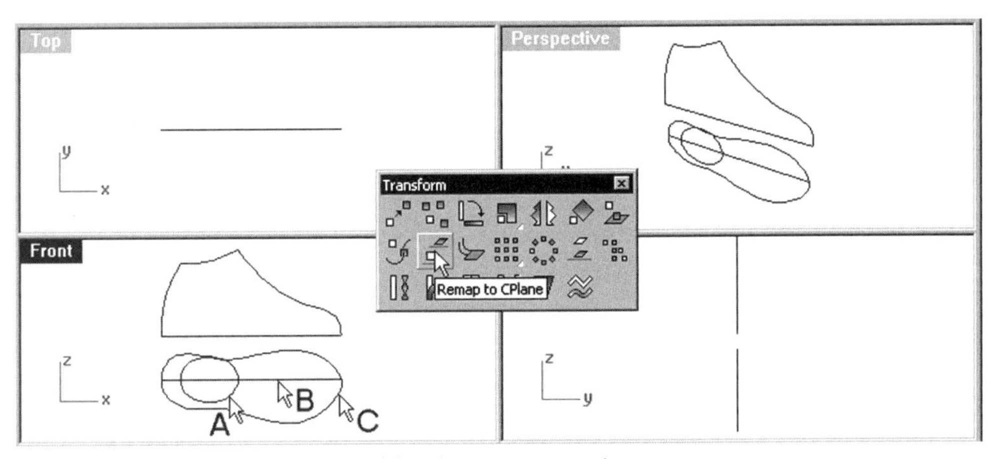

Fig. 9-134. Curves and line being remapped.

Continue with the following steps to construct a 3D curve from two planar curves residing in the Top and Front viewports.

12 Select Curve > From 2 Views, or click on the Curve from 2 Views button on the Curve Tools toolbar.

13 Select curves A and B (indicated in figure 9-135).

Continue with the following steps to move three curves and a line.

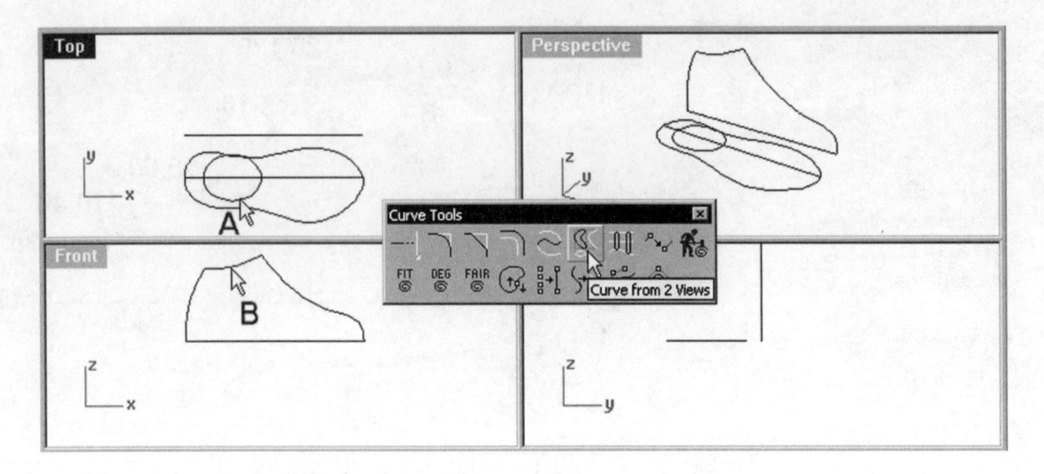

Fig. 9-135. 3D curve being constructed.

14 Maximize the Perspective viewport.

15 Select Transform > Move, or click on the Move button on the Transform toolbar.

16 Select curves A, B, and C (indicated in figure 9-136) and press the Enter key.

17 Select end point D (indicated in figure 9-136).

18 Select end point E (indicated in figure 9-136).

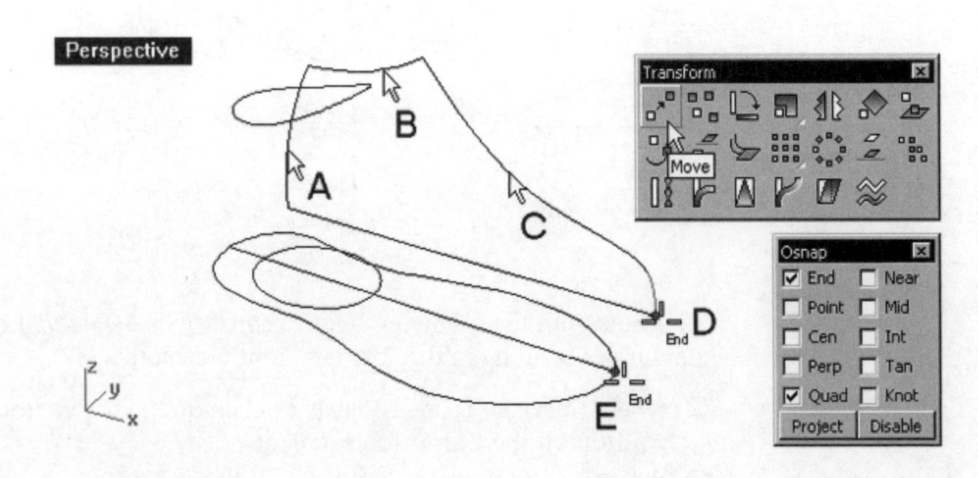

Fig. 9-136 Curves and line being moved.

Continue with the following steps to join two curves as necessary, and to hide three curves.

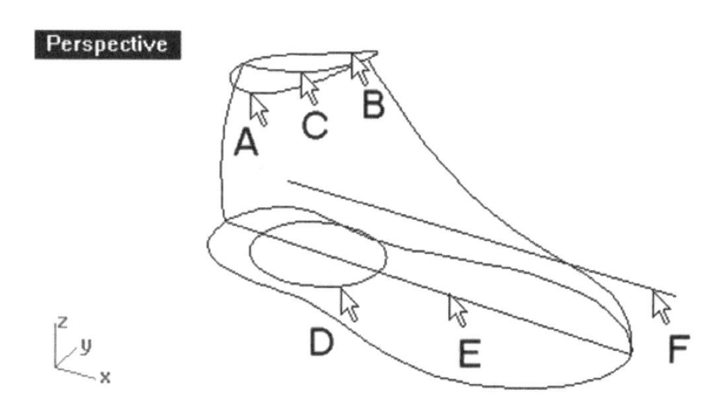

Fig. 9-137. Curves being hidden.

19 If curve A (indicated in figure 9-137) is broken into curves A and B, join them and rebuild the curve.

20 Hide curves C, D, E, and F (indicated in figure 9-137).

Continue with the following steps to construct a cutting plane rectangle, establish a construction plane on the cutting plane rectangle, and construct two curves on the construction plane.

21 Set the display to a four-viewport configuration.

22 Select Surface > Plane > Cutting Plane, or click on the Cutting Plane button on the Plane toolbar.

23 Select curves A and B (indicated in figure 9-138) and press the Enter key.

24 Click on the Front viewport.

25 Type *–75,190* at the command line area to specify the start of the cutting plane.

26 Type *–75,0* at the command line area to specify the end of the cutting plane.

27 Press the Enter key.

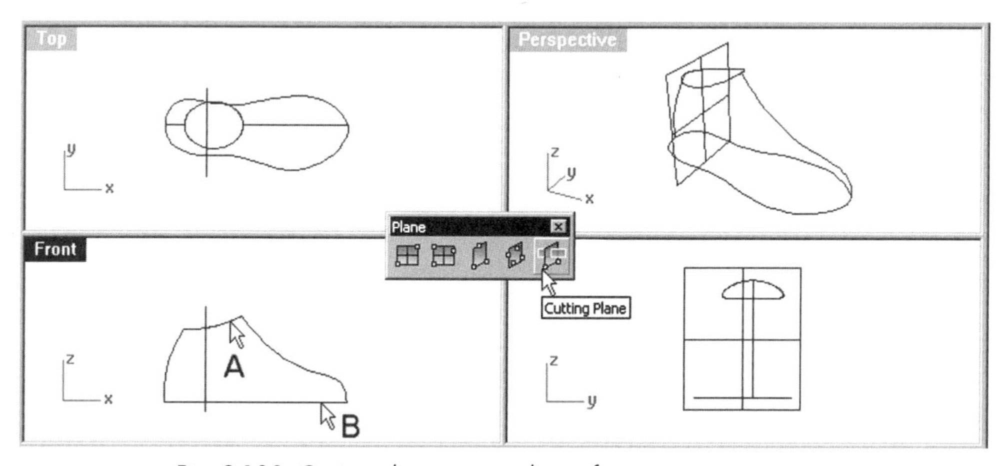

Fig. 9-138. Cutting plane rectangular surface.

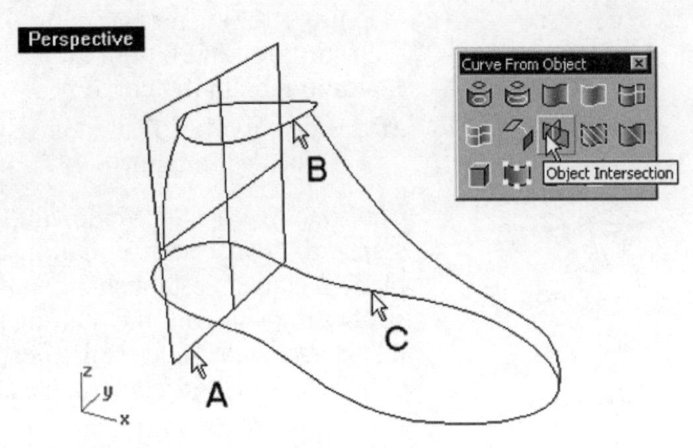

Fig. 9-139. Intersection objects being constructed.

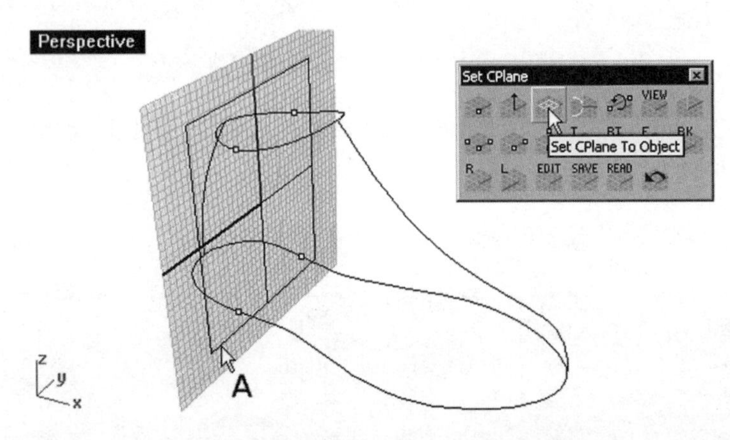

Fig. 9-140. Construction plane being set up and surface being hidden.

28 Maximize the Perspective viewport.

29 Select Curve > From Objects > Intersection, or click on the Intersection button on the Curve From Object toolbar.

30 Select surface A and curves B and C (indicated in figure 9-139) and press the Enter key.

31 Select View > Set CPlane > To Object, or click on the Set CPlane To Object button on the Set CPlane toolbar.

32 Select surface A (indicated in figure 9-140) to specify the CPlane origin.

33 Select Edit > Visibility > Hide, or click on the *Hide Objects/Show Objects* button on the Visibility toolbar.

34 Select the rectangular surface A (indicated in figure 9-140) and press the Enter key.

35 With reference to figure 9-141, construct free-form curve ABC passing through point A, location B, and point C, and free-form curve DEF passing through point D, location E, and point F.

36 Repeat steps 22 through 35 to construct two more curves in accordance with figure 9-142. The rectangular cutting surface for establishing the construction plane should pass through a point (–75,190) and another point (3,0) in the Front viewport.

Continue with the following steps to construct a sweep surface.

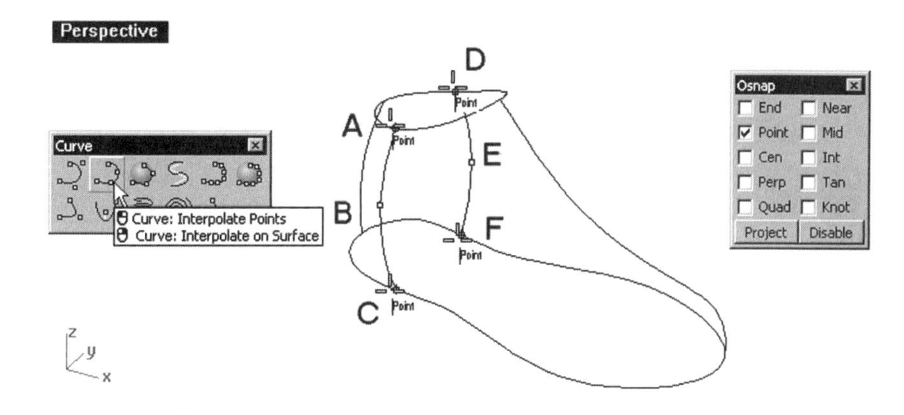

Fig. 9-141. Free-form curves constructed on the newly established construction plane.

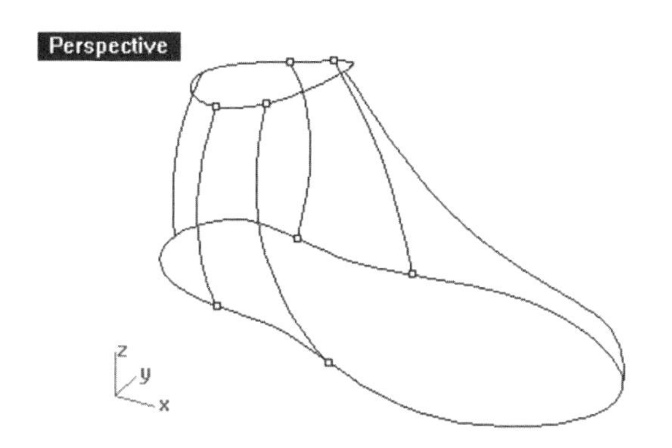

Fig. 9-142. Two more curves.

37 Set the current layer to *Layer 01*.

38 Select Surface > Sweep 2 Rails, or click on the Sweep along 2 Rails button on the Surface toolbar.

39 Select curves A, B, C, D, E, F, G, and H (indicated in figure 9-143) and press the Enter key.

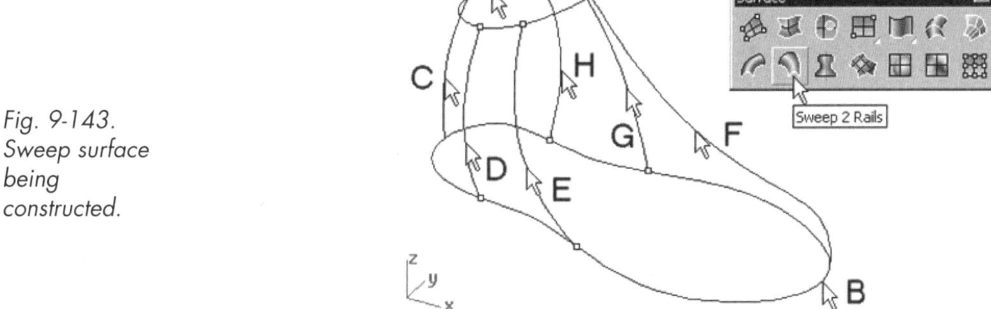

Fig. 9-143. Sweep surface being constructed.

40 In the Sweep 2 Rails Options dialog box, check the Closed Sweep option (if it is not already checked) and then click on the OK button.

41 Rebuild the surface, as shown in figure 9-144.

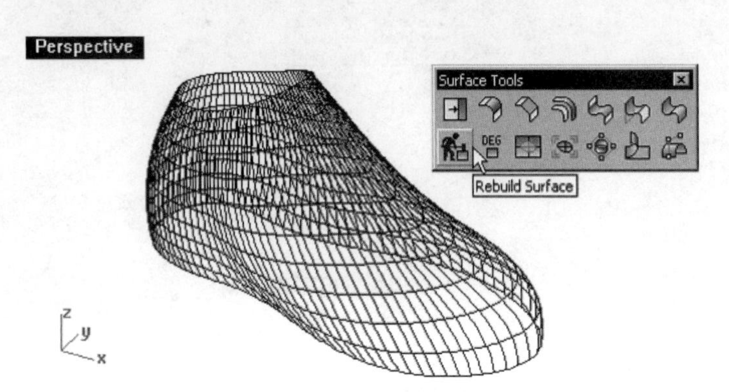

Fig. 9-144. Surface rebuilt.

Details

The main body of the shoe is complete. Using this main body, you may continue to design a shoe of your own. In the following you will add a base and a rim to the shoe, beginning with the construction of an offset surface.

1 Set the current layer to *Layer 02*.

2 Select Surface > Offset Surface, or click on the Offset Surface button on the Surface Tools toolbar.

3 Select surface A (indicated in figure 9-145).

4 If the normal direction of the surface is not the same as that shown in figure 9-145, type *F* at the command line interface to flip the normal direction.

5 Type *5* at the command line interface to specify the offset distance.

6 Press the Enter key.

Continue with the following steps to construct four extruded surfaces.

7 Turn off *Layer 01* and *Layer 02* and set the current layer to *Layer 03*.

8 Maximize the Front viewport.

9 With reference to figure 9-146, construct a free-form curve and two line segments.

10 Construct curve A, offset 5 units from curve B, as shown in figure 9-147.

11 Turn off *Layer 02*.

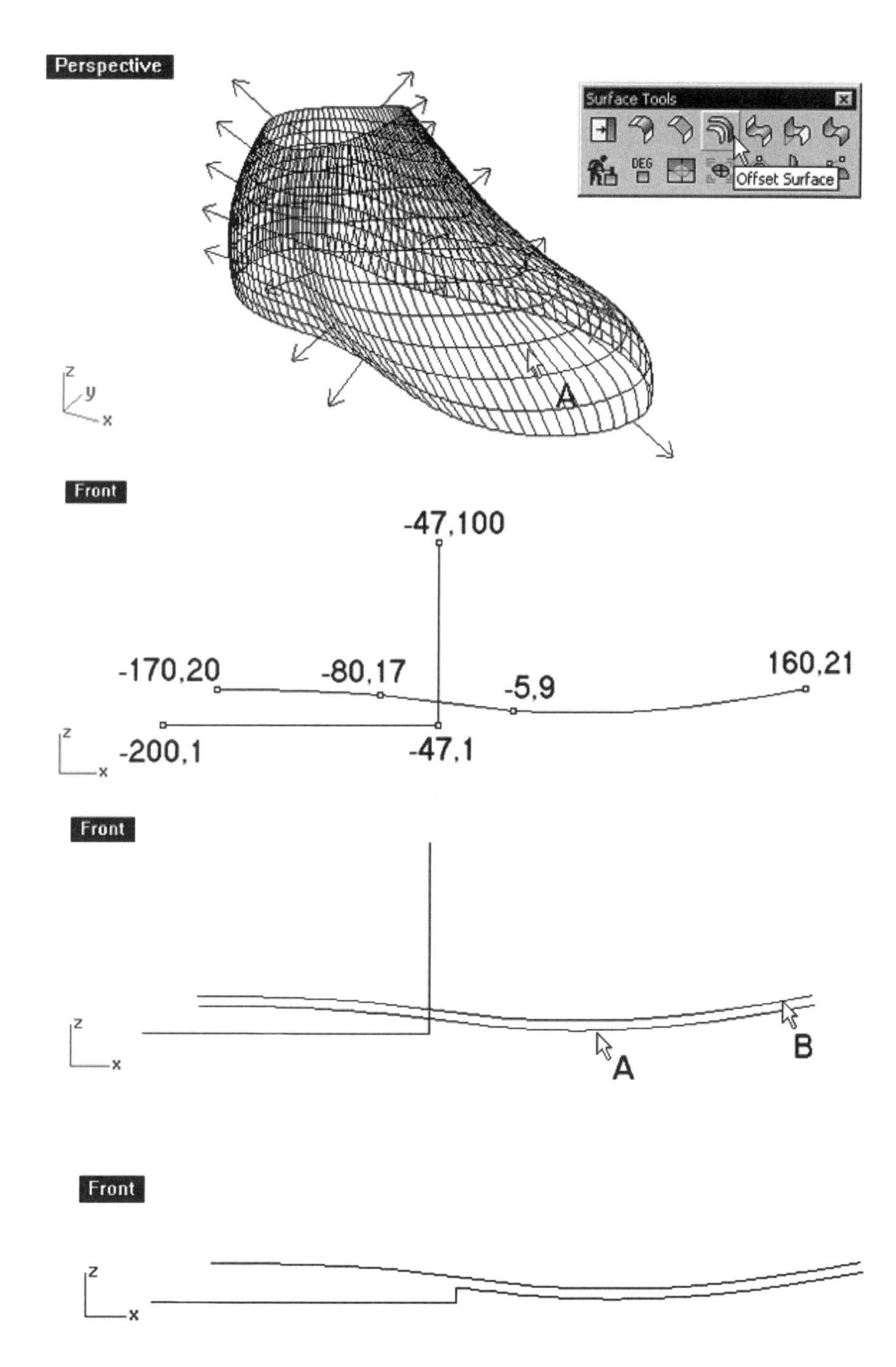

Fig. 9-145. Offset surface being constructed.

Fig. 9-146. Free-form curve and line segment constructed.

Fig. 9-147. Offset curve constructed.

Fig. 9-148. Curves and line trimmed.

12 Trim the curves in accordance with figure 9-148.

13 Set the display to a four-viewport configuration.

14 Set the current layer to *Layer 04*.

15 Construct four extruded surfaces with an extrusion length of 200 units, in accordance with figure 9-149.

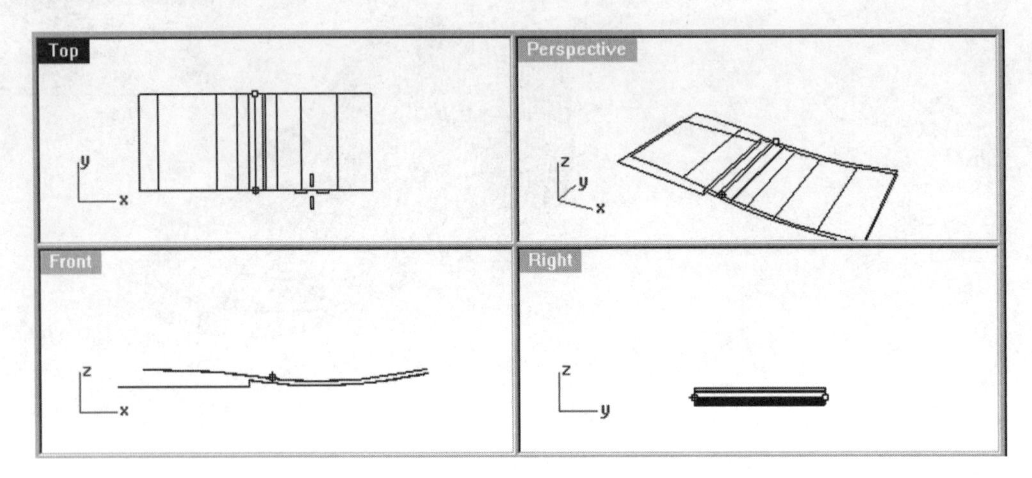

Fig. 9-149. Extruded surface.

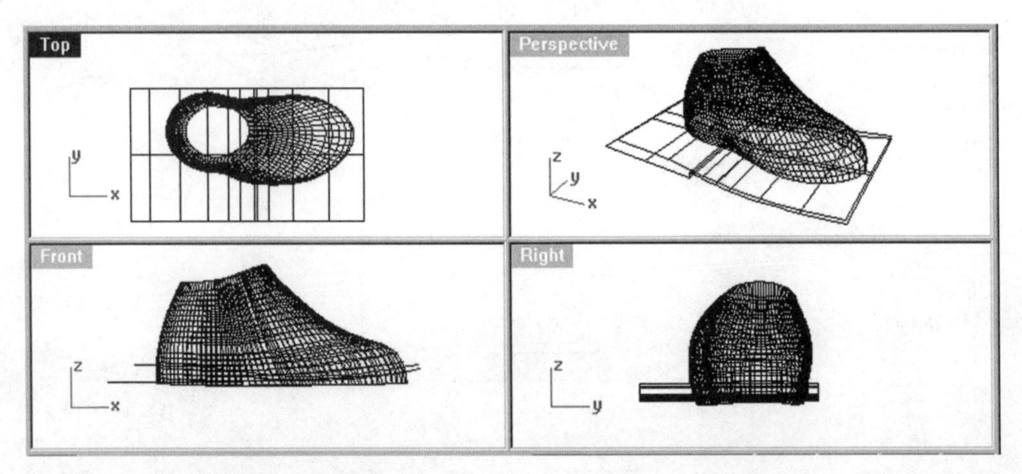

Fig. 9-150. Extruded surfaces.

16 Turn off Layer 03 and turn on *Layer 02*, as shown in figure 9-150.

17 Trim the surfaces in accordance with figure 9-151.

Continue with the following steps to complete the main body of the shoe.

18 Set the current layer to the default layer, turn on *Layer 03*, and turn off *Layer 02* and *Layer 04*.

19 Maximize the Front viewport.

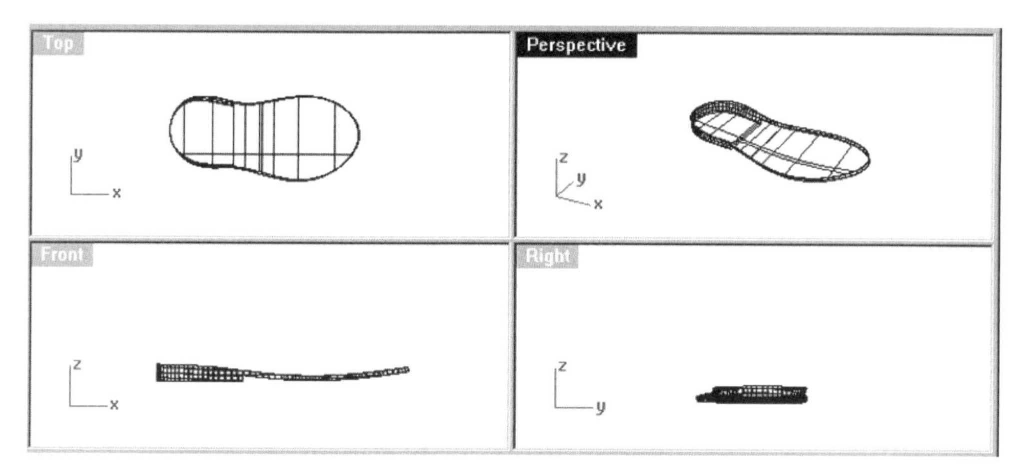

Fig. 9-151. Surfaces trimmed.

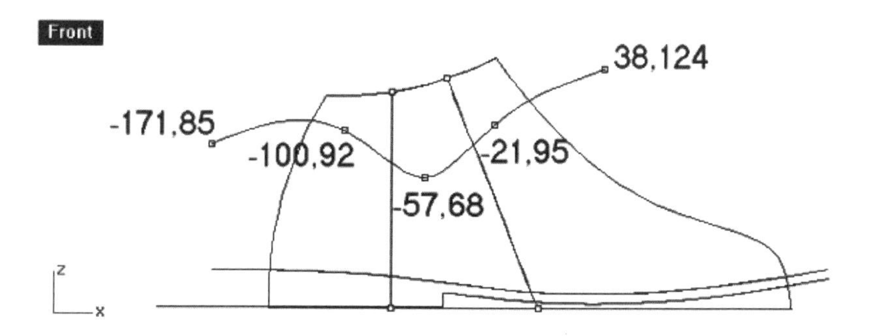

Fig. 9-152. Layers set and free-form curve constructed.

20 With reference to figure 9-152, construct a free-form curve.

21 Turn on *Layer 01*.

22 Select Edit > Trim, or click on the *Trim/Untrim Surface* button on the Main1 toolbar.

23 Select curves A and B (indicated in figure 9-153) and press the Enter key.

24 Select locations C and D (indicated in figure 9-153) and press the Enter key.

25 Set the current layer to *Layer 05* and turn off the default layer and *Layer 03*.

26 Maximize the Perspective viewport.

27 Construct a circle of 3 units radius, as shown in figure 9-154. The exact location is unimportant.

28 Select Transform > Orient > Perpendicular to Curve, or click on the Orient Perpendicular to Curve button on the Transform toolbar.

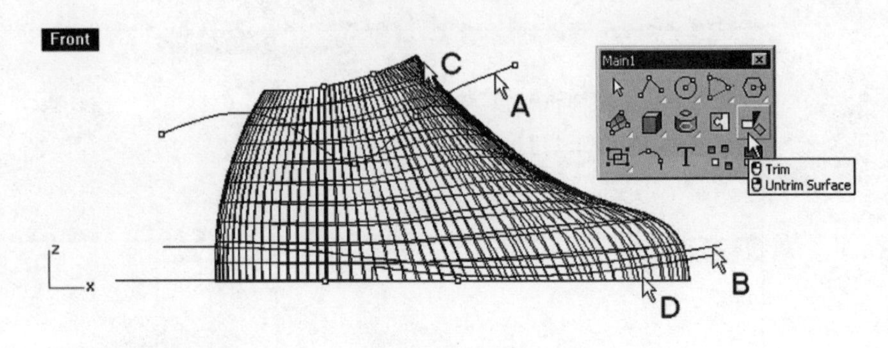

Fig. 9-153. Surface being trimmed.

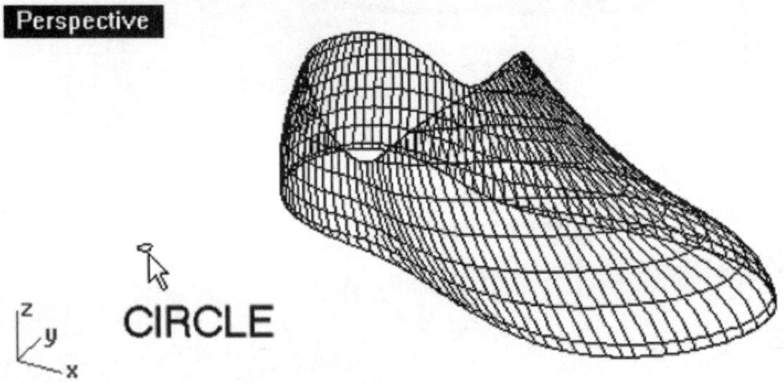

Fig. 9-154. Layers set and circle constructed.

29 Select the circle and press the Enter key.

30 Select center A (indicated in figure 9-155) of the circle.

31 Select point B (indicated in figure 9-155) on the edge of the surface.

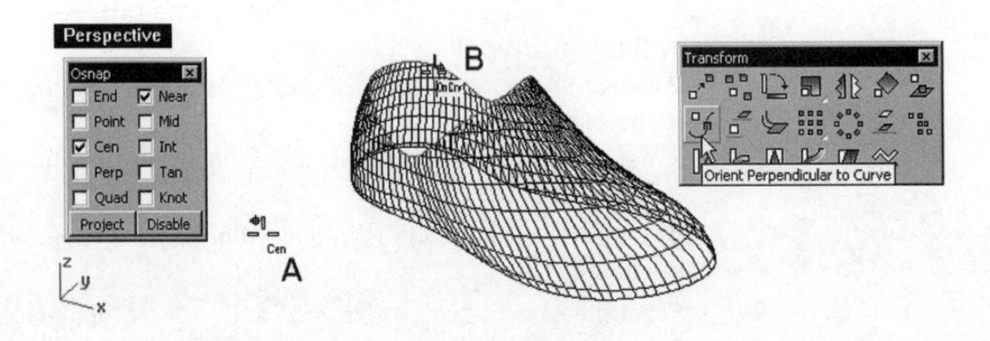

Fig. 9-155. Circle being oriented.

32 Set the current layer to *Layer 01*.

33 Select Surface > Sweep 1 Rail, or click on the Sweep 1 Rail button on the Surface toolbar.

34 Select edge A (indicated in figure 9-156).

35 Select circle B (indicated in figure 9-156) and press the Enter key.

36 Click on the OK button in the Sweep 1 Rail Options dialog box.

A sweep surface is constructed, as shown in figure 9-157.

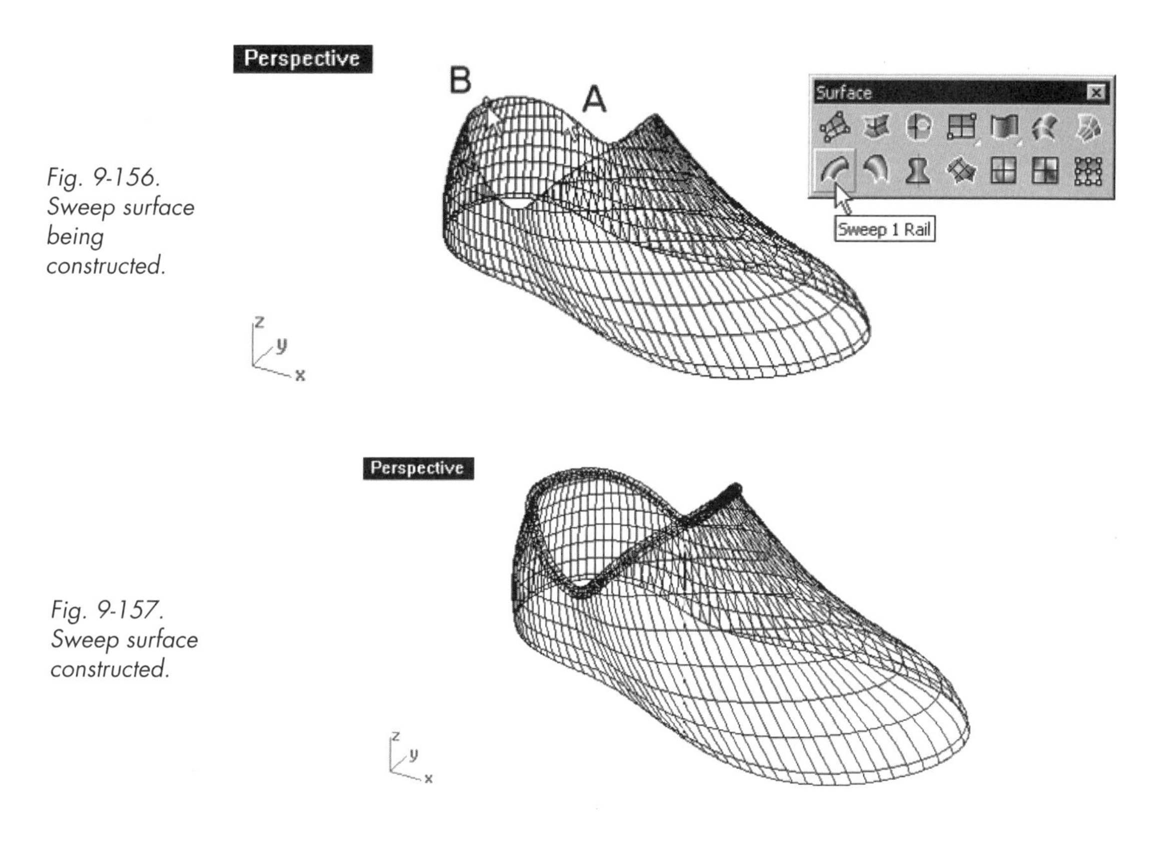

Fig. 9-156. Sweep surface being constructed.

Fig. 9-157. Sweep surface constructed.

37 Turn off *Layer 05* and turn on *Layer 02* and *Layer 04*.
38 Apply material properties to the model.
39 Save your file as *Shoe.3dm*.

The model is complete.

Summary

To construct a model of a 3D free-form object in the computer, you use a set of free-form surfaces. Before making the surfaces, you first study and analyze the 3D object to determine what types of surfaces are needed. You consider various types of primitive surfaces, basic free-form surfaces, and derived surfaces. Among the surfaces, free-form surfaces are those most commonly used. All free-form surfaces have one thing in common: they all need to be constructed from smooth curves and/or point objects.

It is natural to start thinking about the surfaces but not the points and curves while you are designing and making a surface model. However, because the computer constructs basic free-form surfaces from defined curves and/or point objects, the first task you need to tackle in making free-form surfaces is to think about what types of curves and/or points are needed and how they can be constructed. After making the curves and/or points, you then let the computer generate the required surfaces.

Review Questions

1. Write a brief account of the general surface modeling approach.
2. What types of curve construction tools do you use for constructing wireframes used in the subsequent construction of surfaces?
3. List the types of surfaces constructed on a framework of curves and use sketches to illustrate the form and shape of curves for making such surfaces.
4. What methods can you use to convert a set of surfaces into a solid?
5. What types of file formats are supported by Rhino?
6. What types of tools do you use to evaluate a surface model?

Index

F

G

N

O

T